Processor Design

Processor Design

System-on-Chip Computing for ASICs and FPGAs

Edited by

Jari Nurmi
Tampere University of Technology
Finland

Springer

A C.I.P. Catalogue record for this book is available from the Library of Congress.

ISBN 978-1-4020-5529-4 (HB)
ISBN 978-1-4020-5530-0 (e-book)

Published by Springer,
P.O. Box 17, 3300 AA Dordrecht, The Netherlands.

www.springer.com

Printed on acid-free paper

To Pirjo, Lauri, Eero, and Santeri

Preface

When I started my computing career by programming a PDP-11 computer as a freshman in the university in early 1980s, I could not have dreamed that one day I'd be able to design a processor. At that time, the freshmen were only allowed to use PDP. Next year I was given the permission to use the famous brand-new VAX-780 computer. Also, my new roommate at the dorm had got one of the first personal computers, a Commodore-64 which we started to explore together. Again, I could not have imagined that hundreds of times the processing power will be available in an everyday embedded device just a quarter of century later.

Little by little I delved into the design of digital circuits, and computer architecture. I finally learned my lessons in RISC philosophy when I was teaching computer architecture classes in early 1990s according to the famous groundbreaking book by Hennessy and Patterson. At that time, I had already started to design processors, first some simple configurable filters and then straightforward DSP cores. The story continued in a number of different kinds of design projects purely in academia, as academia-industry cooperation projects and as commercial developments in industry.

For me, this decade has meant the time to be back in academia, where I have taught processor-design courses since 1999. A characteristic feature to these courses has been the lack of a good course textbook. I have tried out a few books, and used a scattered set of my own material trying bridge the gaps that I perceived. Year after year I got more annoyed with the absence of a textbook, until, after gaining some editor experience in another book project, I decided that the book needed to be written.

I would like to thank my contact person at Springer, Mark de Jongh, who believed in me right from the start, and all the contributors of this book. A big part of the success of this project was that I knew some good people and asked for their contribution. I had worked with many of them previously in the annual International Symposium on System-on-Chip since 1999, without realizing what kind of assets they represented. Thanks also to all the people who used their valuable time to review the book chapters.

I hope that you will find this book to be beneficial to you whether you are a student, engineer, teacher or engineering manager. This book definitely fills the gap that I had recognized, so I hope that we shared the same gap.

In Tampere, April 2007

Jari Nurmi

Table of Contents

List of Contributing Authors

Anant Agarwal
Massachusetts Institute
of Technology
Computer Science and Artificial
Intelligence Laboratory
32 Vassar Street
Cambridge, MA 02139
USA

Tapani Ahonen
Tampere University
of Technology
Institute of Digital and Computer
Systems
P.O. Box 553
FI-33101 Tampere
Finland

Saman Amarasinghe
Massachusetts Institute
of Technology
Computer Science and Artificial
Intelligence Laboratory
32 Vassar Street
Cambridge, MA 02139
USA

James Ball
Altera, Inc.
110 Cooper St., Suite 201
Santa Cruz, CA 95060
USA

Ian Bratt
Massachusetts Institute
of Technology
Computer Science and Artificial
Intelligence Laboratory
32 Vassar Street
Cambridge, MA 02139
USA

Claudio Brunelli
Tampere University
of Technology
Institute of Digital and Computer
Systems
P.O. Box 553
FI-33101 Tampere
Finland

Fabio Campi
STMicroelectronics
FTM/CCDS/Configurable Logics
Viale C. Pepoli 3/2
40123 Bologna
Italy

Matt Frank
Massachusetts Institute
of Technology
Computer Science and Artificial
Intelligence Laboratory
32 Vassar Street
Cambridge, MA 02139
USA

Gene Frantz
Texas Instruments, Inc.
12203 Southwest Freeway,
MS 701
Stafford, TX 77477
USA

Steve Furber
School of Computer Science
The University of Manchester
Oxford Road
Manchester
M13 9PL
United Kingdom

Jim Garside
School of Computer Science
The University of Manchester
Oxford Road
Manchester
M13 9PL
United Kingdom

Dimitris Gizopoulos
University of Piraeus
Department of Informatics
80 Karaoli & Dimitriou Street
18534 Piraeus
Greece

Ben Greenwald
Massachusetts Institute
of Technology
Computer Science and Artificial
Intelligence Laboratory
32 Vassar Street
Cambridge, MA 02139
USA

Henry Hoffmann
Massachusetts Institute
of Technology
Computer Science and Artificial
Intelligence Laboratory
32 Vassar Street
Cambridge, MA 02139
USA

Jouni Isoaho
University of Turku
Department of Information
Technology
FI-20014 Turku
Finland

Paul Johnson
Massachusetts Institute
of Technology
Computer Science and Artificial
Intelligence Laboratory
32 Vassar Street
Cambridge, MA 02139
USA

Jason Kim
Massachusetts Institute
of Technology
Computer Science and Artificial
Intelligence Laboratory
32 Vassar Street
Cambridge, MA 02139
USA

Nektarios Kranitis
National and Kapodistrian
University of Athens
Department of Informatics
and Telecommunications
Panepistimiopolis, Ilissia
15784, Athens
Greece

Juha P. Kylliäinen
Tampere University
of Technology
Institute of Digital and Computer
Systems
P.O. Box 553
FI-33101 Tampere
Finland

Walter Lee
Massachusetts Institute
of Technology
Computer Science and Artificial
Intelligence Laboratory
32 Vassar Street
Cambridge, MA 02139
USA

Steve Leibson
Tensilica, Inc.
3255-6 Scott Blvd.
Santa Clara, CA 95054
USA

Johan Lilius
Åbo Akademi University
Department of Information
Technologies
Joukahaisenkatu 3-5
FI-20520 Turku
Finland

Grant Martin
Tensilica, Inc.
3255-6 Scott Blvd.
Santa Clara, CA 95054
USA

Jason Eric Miller
Massachusetts Institute
of Technology
Computer Science and Artificial
Intelligence Laboratory
32 Vassar Street
Cambridge, MA 02139
USA

Claudio Mucci
ARCES, University of Bologna
Viale C. Pepoli 3/2
40123 Bologna
Italy

Sanna Määttä
Tampere University
of Technology
Institute of Digital and Computer
Systems
P.O. Box 553
FI-33101 Tampere
Finland

Jari Nurmi
Tampere University
of Technology
Institute of Digital and Computer
Systems
P.O. Box 553
FI-33101 Tampere
Finland

Tero Nurmi
University of Turku
Department of Information
Technology
FI-20014 Turku
Finland

Christian Panis
Catena Radio Design, B.V.
Science Park Eindhoven
Ekkersrijt 5228
5692 EG Son en Breugel
The Netherlands

Antonis Paschalis
National and Kapodistrian
University of Athens
Department of Informatics and
Telecommunications
Panepistimiopolis, Ilissia
15784, Athens
Greece

Juha Plosila
University of Turku
Department of Information
Technology
FI-20014 Turku
Finland

James Psota
Massachusetts Institute
of Technology
Computer Science and Artificial
Intelligence Laboratory
32 Vassar Street
Cambridge, MA 02139
USA

Rodric Rabbah
Massachusetts Institute
of Technology
Computer Science and Artificial
Intelligence Laboratory
32 Vassar Street
Cambridge, MA 02139
USA

Stefan Rusu
Intel Corporation
2200 Mission College Blvd.,
M/S SC12-408
Santa Clara, CA 95052
USA

Arvind Saraf
Massachusetts Institute
of Technology
Computer Science and Artificial
Intelligence Laboratory
32 Vassar Street
Cambridge, MA 02139
USA

Nathan Shnidman
Massachusetts Institute
of Technology
Computer Science and Artificial
Intelligence Laboratory
32 Vassar Street
Cambridge, MA 02139
USA

Volker Strumpen
Massachusetts Institute
of Technology
Computer Science and Artificial
Intelligence Laboratory
32 Vassar Street
Cambridge, MA 02139
USA

Tero Säntti
University of Turku
Department of Information
Technology
FI-20014 Turku
Finland

Michael Bedford Taylor
Massachusetts Institute
of Technology
Computer Science and Artificial
Intelligence Laboratory
32 Vassar Street
Cambridge, MA 02139
USA

Dragos Truscan
Åbo Akademi University
Department of Information
Technologies
Joukahaisenkatu 3-5
FI-20520 Turku
Finland

Joonas Tyystjärvi
University of Turku
Department of Information
Technology
FI-20014 Turku
Finland

Seppo Virtanen
University of Turku
Department of Information
Technology
FI-20014 Turku
Finland

David Wentzlaff
Massachusetts Institute
of Technology
Computer Science and Artificial
Intelligence Laboratory
32 Vassar Street
Cambridge, MA 02139
USA

Tomi Westerlund
University of Turku
Department of Information
Technology
FI-20014 Turku
Finland

George Xenoulis
University of Piraeus
Department of Informatics
80 Karaoli & Dimitriou Street
18534 Piraeus
Greece

1 Introduction

Jari Nurmi

Tampere University of Technology

Embedded systems are "computers in disguise," systems carrying along a computer or processor-based system that are not programmable by the user and in most cases do not have any observable resemblance to a computer [36]. The software is typically developed by the designers of the system, downloaded or hard-coded to the system at manufacture-time, and inaccessible to the end-user. Such systems can be found at home (washing machine, DVD-player, game console), in the office (printer, WLAN base station, building automation), cars (engine control, ABS brakes, security system), aeroplanes (fly-by-wire, navigation, autopilot), industrial, forest, and harvesting machines (process monitoring, process control, robotics), in your pocket (mp3 player, mobile phone, PDA) and even in your wallet (electronic bus tickets, credit cards, keycards). The most high-end embedded systems, e.g., game consoles and cell phone base stations, may have processing capacity supreme to the desktop PC. In the progress towards a more nomadic lifestyle, various mobile gadgets and ubiquitous services in the everyday surroundings have emerged, further emphasizing the importance of small size, low cost and low power consumption of embedded computation.

Due to rapid technology advancement in integrated circuit era, the design and implementation of embedded systems has increasingly been merged with circuit design. As a visible symptom of this, the term System-on-Chip (SoC) [225] has been coined and increasingly used wherever highly integrated embedded systems emerge. This trend has been one of the enablers for processor design to come at reach for digital circuit and system designers.

The demands for low cost and low power consumption are characteristic to embedded processors. These requirements have in part been drivers of higher degree of integration, but they are also pushing designers more and

J. Nurmi (ed.), Processor Design: System-on-Chip Computing for ASICs and FPGAs, 1–6.

more towards the design of application-specific processor architectures and customization of existing processors. By providing application-specific performance, the designers are able to keep the overhead of programmable processors at minimum while maintaining the flexibility they provide. The overhead refers to area overhead of the general-purpose structures of the processor which may be poorly utilized by a specific application, timing overhead stemming from the time-multiplexed nature of processor resources, and power consumption overhead due to the generalized control and high clock rate requirements of the multiplexed computational blocks. The flexibility means here mainly the programmability which allows changing or incrementally improving the functionality of the design. However, there exists a trade-off between the flexibility and application-specific performance; the more the architecture is tailored to an application, the more difficult it is to reprogram it to perform something completely different.

Integrated circuit technologies have evolved during the past four decades following Moore's Law [293] which states that the achievable transistor count on a single integrated circuit doubles every 18 to 24 months. This exponential growth has brought the technology over the border where it can now accommodate complete embedded systems on a single chip. While a typical circuit in early 1990s was fabricated in 1 μm technology and could contain merely 100,000 transistor devices, the contemporary 90 nm (0.09 μm) technologies can easily deliver 100,000,000 transistors. This 1000-fold improvement enables to fabricate not only a complete processor on a single chip (as was the case in early 1990s) but to include a substantial amount of its surroundings there, too. Also multi-core and multi-processor systems are reality and thus increasingly a target of research and development efforts. According to the International Technology Roadmap for Semiconductors (ITRS) [208] the advancement of technology will continue at least for yet another decade.

One consequence of the technological advances is that Field-Programmable Gate Array (FPGA) [54] circuits have also evolved to carry enough resources to implement complex embedded systems on a single device. FPGAs consist of an array of configurable logic modules (rewritable look-up table memories), configurable interconnects between the modules, and increasingly also larger memory blocks and hardwired arithmetic blocks [459]. FPGAs are programmed or configured by feeding into the configuration memory (distributed over the FPGA circuit) a bit-stream containing the values controlling the logic functions and interconnect switching points. The major advantage of FPGAs over custom-made circuits is that the designer is relieved from addressing the increasingly complicated back-end design of integrated circuits. The back-end refers here to the physical placement of logic cells, routing between them, power network generation, clock network generation, and test structure generation; in all these operations the practical impairment caused by crosstalk, transmission line effects, voltage drops, leaking, etc. have to be taken into account [311]. FPGAs together with desktop computer processors are the forerunners in the use of advanced technology, e.g., FPGAs fabricated in 65 nm technologies are already shipping while most of the Application-Specific Integrated Circuit (ASIC) designers still work with the 2–3 previous technology nodes. Another advantage of FPGAs is their inherent reconfigurability; this is what FPGAs are all about! Recently also dynamic partial reconfiguration has become feasible, enabling to use reconfiguration also to increase the system complexity beyond what can be placed on a device at a time. The drawback of FPGAs is that they carry unnecessary overhead in cost, speed and power consumption compared to custom-made circuits. However, mainly due to increasing Non-Recurring Engineering (NRE) costs of integrated circuit fabrication, the FPGAs can be applied to an increasing number of applications not only as prototypes but as part of the final product. Only very high-volume and highly power-sensitive products remain out-of-reach for FPGAs. The border of having cost advantage from custom circuits is raising to higher production volumes continually.

One of the enablers for the recent trend to push processor design to the ASIC designers' toolbox is certainly the degree of maturity that the software tools for processor architecture exploration, instruction set design, and both software development tool chain and hardware generation have achieved. Design methodologies approaching the processor design from different angles have emerged, including an approach based on an existing base architecture to be customized, compiler-based exploration of architecture, and processor description language based design. Independently of the approach, the design flows can now produce the processor hardware instance and the development software tool chain in a structured and predictable way.

The rest of the book is organized as follows. Where the authors of the chapter are coming from outside my own research group, the inclusion criterion is also briefly pointed out.

In Chapter 2, the terminology and basic foundations of embedded computing architectures are revisited to pave way to the rest of this book. Different parts of a computer are introduced, the addressing modes found in various processors overviewed, different architecture styles illustrated, and available forms of parallelism in a processor architecture explored. Also the memory subsystem characteristics and I/O operations are revisited.

In Chapter 3, put together by electronics industry experts Steve Leibson and Grant Martin, a historical perspective to the fallacies, pitfalls, and flaws in processor design is provided. It is very enlightening to notice that the ideas that appear excellent at some point of time may turn out to be dead ends once implemented.

In Chapter 4, an overview of the design flow of processors is given, to that end illuminating the path through different architectural styles, design approaches, and various steps of the design process that await in the following chapters. In that chapter the instruction encoding phase is emphasized since it will not be addressed in detail elsewhere in this book.

In Chapter 5, the first design case is described. A general-purpose embedded processor core design carried out at Tampere University of Technology (TUT), Finland, for open-source distribution is presented. This case tries to underline the design choices made when designing COFFEE RISC Core and the reasoning behind them. The whole hardware design in VHDL description language and the related software tools are also available to the readers on our web site.

In Chapter 6, one of the specific application areas is handled. Digital Signal Processors (DSP) have a different design space which is addressed by a pioneer in this field, Gene Franz of Texas Instruments. Specific properties requiring special hardware, such as circular buffer addressing, hardware support for loops, or multiply–accumulate instructions, are described.

In Chapter 7, exploitation of parallelism in a DSP-oriented application area is carried out. A broad class of new kind DSPs has recently emerged, using the Very Long Instruction Word (VLIW) approach to increase Instruction Level Parallelism (ILP). The architectural choices, design space exploration and compilation techniques for VLIW DSPs are described by Christian Panis, representing Catena Radio Design, The Netherlands. His approach is quite special since it employs heavily compilers in the design space exploration for a parameterized VLIW architecture.

In Chapter 8, Steve Leibson from Tensilica describes the customizable processor approach. He is a perfect choice to do this since Tensilica is the leading provider of customizable cores. Customizable processors are based on a base architecture that is augmented and customized to form an application-enhanced processor instance. On top of them, also application-specific processor families are addressed in that chapter.

In Chapter 9, reconfigurable processors are addressed. While customizable processors can be modified at design time, reconfigurable processor architectures can be dynamically reconfigured at run-time. The underlying philosophy, architecture, design tools, and application results are presented in another chapter by one of the pioneers of reconfigurable processor architectures, Fabio Campi from ST Microelectronics in Bologna, Italy.

In Chapter 10, we take a co-processor approach. While the approaches introduced so far concentrate on getting the most out of a single processor running a single instruction stream, a co-processor approach to accelerating multimedia applications is using parallel reconfigurable hardware to extend the single-processor capabilities. The chapter concentrates especially on the BUTTER accelerator developed at Tampere University of Technology, boasting with floating-point processing capabilities to enable fast 3D graphics and video applications.

In Chapter 11, James Ball from one of the leading FPGA manufacturers, Altera, addresses designing processors for FPGAs which is an art of its own kind. As said, the FPGA is an increasingly interesting alternative to implement complex systems especially in lower production volumes. The chapter describes efficient ways to implement processor structures on (especially Altera) FPGAs.

In Chapter 12, we target protocol processors which are devoted to handling packet data in communication networks. The design issues and solutions to them using Transport Triggered Architecture (TTA) based processor architecture template are addressed by people of University of Turku (UTU) and Åbo Akademi University (ÅA), Finland. The group is one of the few using TTA approach to processor design, and possibly the only one applying it to protocol processing.

In Chapter 13, Java co-processor design is another specific application area addressed by UTU reseachers. While Java processing is often implemented as a virtual processor emulated in software, the researchers at UTU are taking another approach and accelerating the Java processing in a specific co-processor.

In Chapter 14, we take a look at streaming applications. Stream multicore processors are a solution for data-intensive processing, and one of the leading sites researching this topic is Massachusetts Institute of Technology (MIT) in the USA, thus the approach is described by Rodric Rabbah and Anant Agarwal from MIT.

There is an ongoing debate on what is the most liable solution: to run the SoC and its processors tied to a single clock, using multiple clock domains, or completely asynchronously. In Chapter 15, issues arising in especially the high-end processors, clock generation and distribution strategies, are addressed by Stefan Rusu from Intel, who has been in charge of several Itanium and Xeon processors clock design.

In Chapter 16, another angle will be presented by a team from Manchester University, introducing us to asynchronous and self-timed processor design. This particular team has implemented, e.g., the asynchronous version of ARM processor family.

In Chapter 17, early estimation models of processors are discussed by researchers from UTU and TUT. The work described is one of the few early estimation approaches in a SoC environment, and is also taking the models to the granularity of processor sub-blocks which gives important feedback also to processor designers in an early stage.

In Chapter 18, high-level simulation models enabling designers to evaluate systems efficiently on a high abstraction level are described by researchers from UTU, TUT and ÅA. The work is closely related to the architectures described in chapters 5 and 12.

In Chapter 19, programming tools for custom processors are discussed by researchers from ARCES laboratory in Bologna, Italy, and TUT. These teams are two of a very few who are providing support for custom architectures using open-source software.

In Chapter 20, software-based testing of embedded processors is addressed by specialists from the universities of Piraeus and Athens, Greece. These groups have been very active in this area. Testing and debugging of processors that are embedded not only in the application device but also in the middle of a System-on-Chip circuit is an increasingly important area that the designers have to be aware of.

In Chapter 21, the book is concluded by an outlook of the future directions in processor design, including increasingly parallel architectures with a mixture of Process Level Parallelism (PLP), Thread Level Parallelism (TLP), Instruction Level Parallelism (ILP), and Data Level Parallelism (DLP).

2 Embedded Computer Architecture Fundamentals

Jari Nurmi

Tampere University of Technology

This chapter provides a condensed view of essential computer/embedded system architecture concepts and terminology are clarified. An experienced reader may wish to skip this chapter or browse through it quickly and return later, in case some clarification is needed when reading the following chapters. The chapter is not based on any particular reference, but the author has learned a major part of the issues presented here from the "bible" of computer architecture by Hennessy and Patterson [187]. Other good references include textbooks by Stallings [388], Tanenbaum [409], Heuring and Jordan [192], and Patterson and Hennessy [336]. Regarding parallel architectures the books by Flynn [128], Sima, Fountain and Karsuk [377], Silc, Robic and Ungerer [376], and Corporaal [92] are good reading.

We first discuss the components of an embedded computer: datapath, control, memory, I/O, and interconnects, and how these are represented in the Instruction Set Architecture. Next, we look at different processor architectures from the organization point of view. Different ways of introducing parallelism in a processor are also addressed. The memory subsystem issues such as memory hierarchy and virtual memory are discussed, and a quick overview of I/O operations and peripherals ends the chapter.

Components of (an embedded) computer

In general, a computer or an embedded system can be considered as a combination of a *processor* that is responsible for executing programs, some *memory* for storing programs and data, and input/output (I/O) functionality that provides an extension to peripheral devices of the processor.

J. Nurmi (ed.), Processor Design: System-on-Chip Computing for ASICs and FPGAs, 7–26.

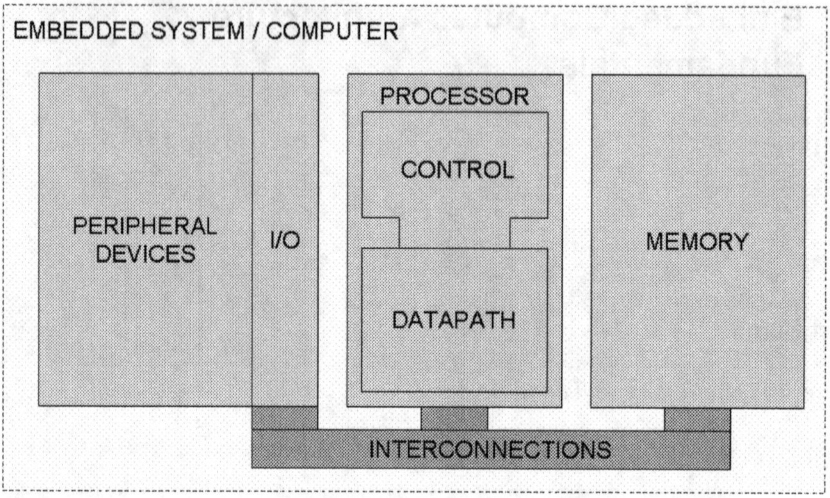

Fig. 2.1. Components of a computer or embedded system.

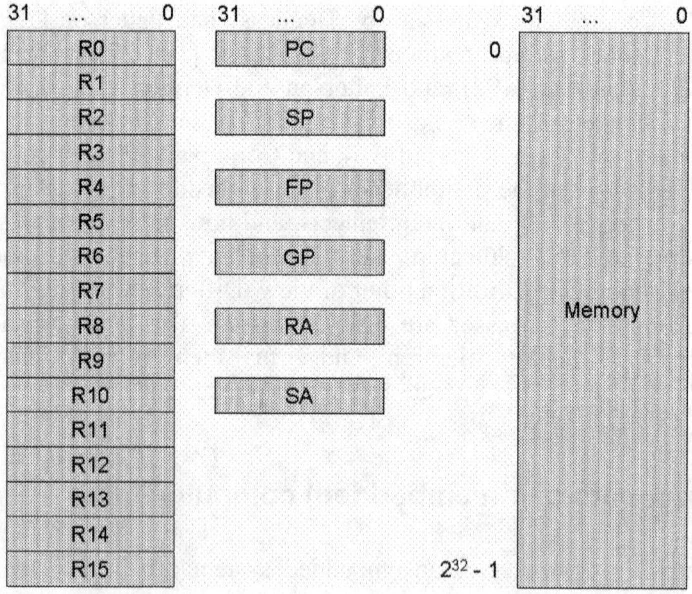

Fig. 2.2. Programmer view of a hypothetic processor.

The main components are interconnected in some way. Figure 2.1 illustrates this generic computing device. The processor is often called the Central Processing Unit (CPU). At this point, we do not care whether the memory is a single monolithic one or consists of a set of separate address spaces. Also, we do not give any thought to the peripherals and how they are accessed or addressed at this point.

The interface to the programmer is called Instruction Set Architecture (ISA). The ISA defines the *instruction set* of the processor, including available *addressing modes* and *data types*, and the set of *registers* and *memory spaces* visible to the programmer. A hypothetical programmer view of a processor is shown in Figure 2.2.

In addition to possible hardwired special functionality of some registers, there are often *conventions of use* for the registers; the compiler might, e.g., use certain registers as subroutine arguments, return values, stack pointer, frame pointer, and pointer to the global variable area (for a programmer these may be obvious; the others may as well ignore the details but catch the key point that there has to be a convention for allocating even "general-purpose" register use).

The processor can be further divided to datapath and control, as shown in Figure 2.3, again for a hypothetical processor. As can be imagined, the datapath assumes the role of data processing, while the control unit fetches instructions, decodes them, controls the operation of the datapath, and takes care of the control transfer operations (such as jumps, branches, subroutine calls, and returns). In very advanced processor architectures, the control part may consume a considerable part of the processor implementation and perform a variety of very complicated tasks such as instruction prefetching, cache control, branch prediction, issuing multiple instructions per cycle, and/or reordering results for register write-back. This is somewhat compensated by the increasing word lengths on the datapath side, but nevertheless a complex control occupies a significant amount of the silicon real estate. The datapath (or sometimes multiple datapaths) take care of the arithmetic, logic, string manipulation, etc. operations done on data. The heart of the datapath where the actual processing takes place is an Arithmetic Logic Unit (ALU). In case of multiple datapaths, the individual processing units are often called Execution Units (EXU) or Processing Units (PU).

The instruction set of a processor consists of all the different types of executable operations. They can be divided into arithmetic and logic instructions, load and store instructions, change of flow (control) instructions, and miscellaneous instructions like system calls, mode setting, etc.

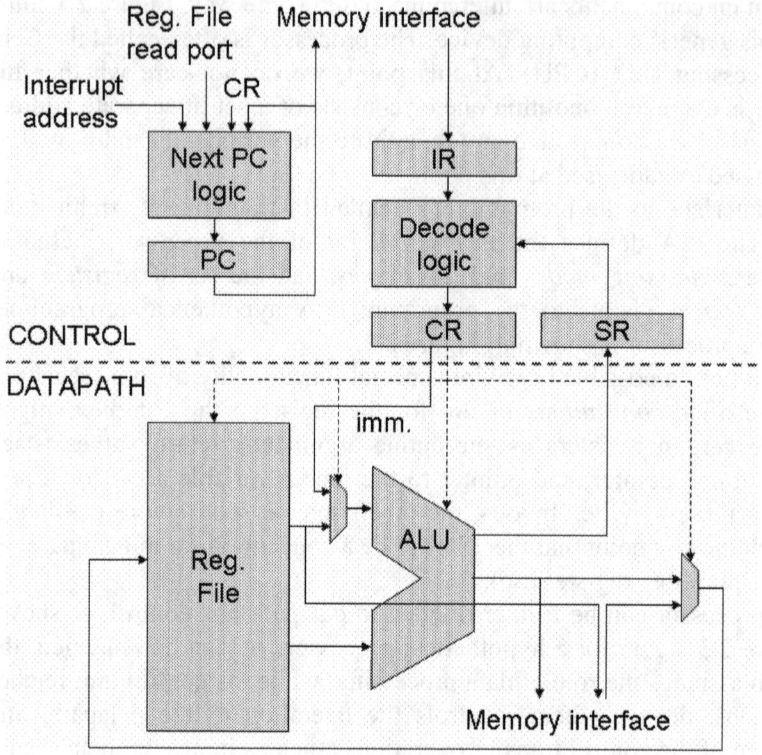

Fig. 2.3. Datapath and control.

In processors supporting vector data processing or specific multimedia instructions (see later in this chapter), these can be separated from the arithmetic instructions group. Depending on the architectural style, the instructions might also be compound instructions consisting of several distinct operations.

The addressing modes deserve a bit more attention at this point. Typically, there are different addressing mechanisms for program and data memories, while the latter are normally getting more attention. The program addressing, first of all, can be direct (or *absolute*), *relative*, register-based, or (as a kind of mixture) pseudo-direct. Direct addressing using a field from the instruction is seldom possible unless the address range is very limited. The full address range can always be accessed through register-based or *register indirect* addressing (assuming of course that the register length is at least equal to the address length). Register indirect addressing for program memory is used intensively in returns from subroutines and interrupts, though it also can be used for jumps to run-time computed addresses. Relative addressing is frequently used because it allows the use of small constants as offsets from the current program address, since the offset can then easily be incorporated to the instruction word. *Pseudo-direct* addressing uses a segment register or the upper bits of the program counter concatenated with a direct address from the instruction field.

A processors's data-addressing modes are usually far richer than its program-addressing modes, reflecting the need for a variety of ways to address various data items. The simplest form is *immediate* addressing which is no addressing at all; the data to be used in the operation is embedded in the instruction word itself. *Register addressing* denotes the register numbers to be used as operands. *Register indirect* addressing uses a register to hold a memory address, e.g., for a load or store operation. A slight but useful extension of it is *offset* (or base) addressing, which uses a register value plus an offset from the current instruction address to determine the branch target. This is especially useful when the register contains a pointer to the start of a data structure, e.g., an array, in memory and the offset is used to address the individual array elements.

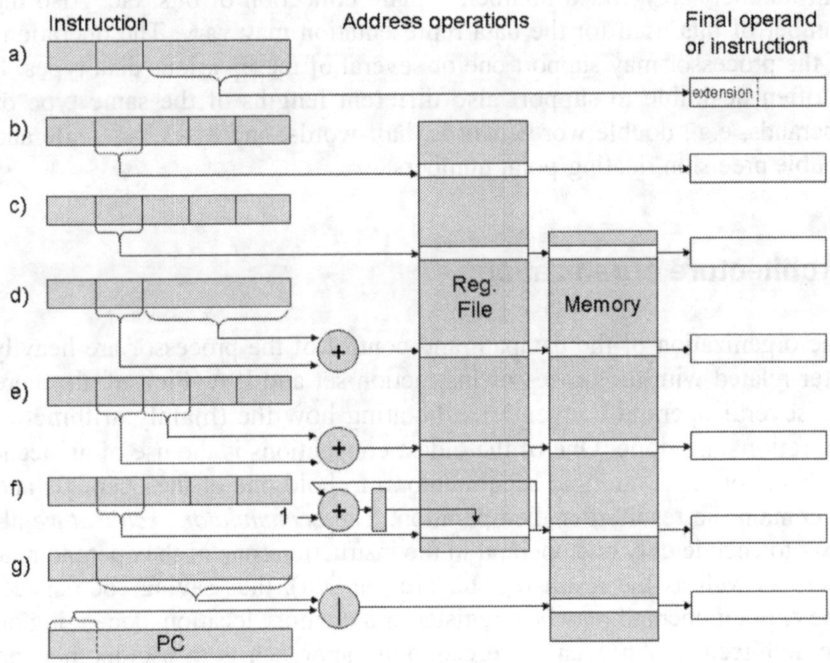

Fig. 2.4. Addressing modes: a) immediate, b) register direct, c) register indirect, d) offset (base), e) indexed, f) post-increment, and g) pseudo-direct addressing.

A further modification is to use two registers to compute the address. This is called *indexed* addressing. In some very complex addressing modes e.g., a scaling factor to multiply the contents of one register is used, another register is added to the result, and an immediate offset can be added on top of that. Especially in signal processing some special addressing modes, like modulo addressing and bit-reversal addressing are used frequently. In some architectures, a pre- or post-calculation can be combined with the register indirect addressing to also update the address register before or after the actual memory addressing, e.g., *pre-decrement* or *post-increment*. The data addressing modes are illustrated in Figure 2.4.

There are also equivalencies between data and program addressing, e.g., absolute addressing of programs and immediate data use similar encoding, and PC-relative addressing is a special case of offset addressing with the PC as an implicit base address register.

Data representations or data types create another issue that is also related to the selection of number format used in the processor in general. A binary number can be interpreted as an unsigned integer, signed integer, signed fraction, floating-point number, character string, binary-coded decimal number, grey coded number, a plain collection of bits, etc. Also the number of bits used for the data representation may vary. The operations of the processor may support one or several of these various data types. It is often desirable to support also different lengths of the same type of operands, e.g., double words, words, half-words, and bytes, or single and double precision floating-point numbers.

Architecture organization

The organization of the datapath and control of the processor are heavily inter-related with the processor instruction set and ISA. First of all, there are several operand conventions dictating how the (mainly arithmetic) operations are done. One of the oldest conventions is the use of an accumulator (or later, multiple accumulators) to hold one of the operands and to contain the result after the operation. The *accumulator architecture* allows to encode only one operand in the instruction (one of the operand registers as well as the result register are implicit), thus saving code space. The explicit operand may be a register or a memory location, depending on the architecture. An even more economic approach is to use another old convention, the *stack architecture*, where the operations are always performed on the top two items of a stack, and the result is saved on the top of the stack. However, special push and pop operation are needed to browse through the data of the stack if the topmost items are not the ones desired to be used for the next operation.

The (general-purpose) *register architecture* is commonly used in modern architectures. An instruction in a machine with a register architecture may contain two or three register numbers; one of the operands might be the same as the destination register. This architecture is also called the load–store architecture or Reduced Instruction Set Computer (RISC) paradigm; one of the key RISC features is that the arithmetic operations are performed on register values. This requires load operations to be performed prior to the arithmetic calculation, and a subsequent store to save the data to memory afterwards; thus the name load–store architecture. This architecture could be also called a register–register architecture in contrast to *register–memory* and *memory–memory* architectures where one or both of the operands may reside in memory. This implies that the instruction contains quite a lot of information for calculating the memory references in addition to the operation itself, making the instruction long and often also variable length, avoiding unnecessary sacrifice of precious memory space. The architecture styles are depicted in Figure 2.5.

It is worth coming back for a while to the RISC definition. In contrary to Complex Instruction Set Computer (CISC) philosophy, RISC is all about keeping it simple. The features that are associated with RISC are:

- Provides basic primitives, not “complete solutions” as instructions; this leads to the “reduced” instruction set
- Orthogonality and regularity in the instructions as much as possible (in real life compromises have to be made)
- Single-cycle execution of most instructions
- Easy to pipeline
- A lot of general purpose registers
- Arithmetic and logic operations (and address computation) are done for register operands or immediates (the load–store architecture principle)

The beauty of primitives is that you can build several different complex operations by combining the primitive operations in different ways, while complex instructions tend to be utterly inflexible. The CISC solution is to try and provide complete solutions presumably matching high-level programming constructs (often failing in this, too). One of the most successful and somehow prevailing way to apply CISC principles are signal processors which tend to pack a lot of small operations in one instruction to complete, e.g., a filter tap computation in one instruction. In restricted application areas the approach may be more feasible than in general-purpose processors where CISC concepts are, if not completely absent, at least merged with the RISC principles in the modern high-end CPUs.

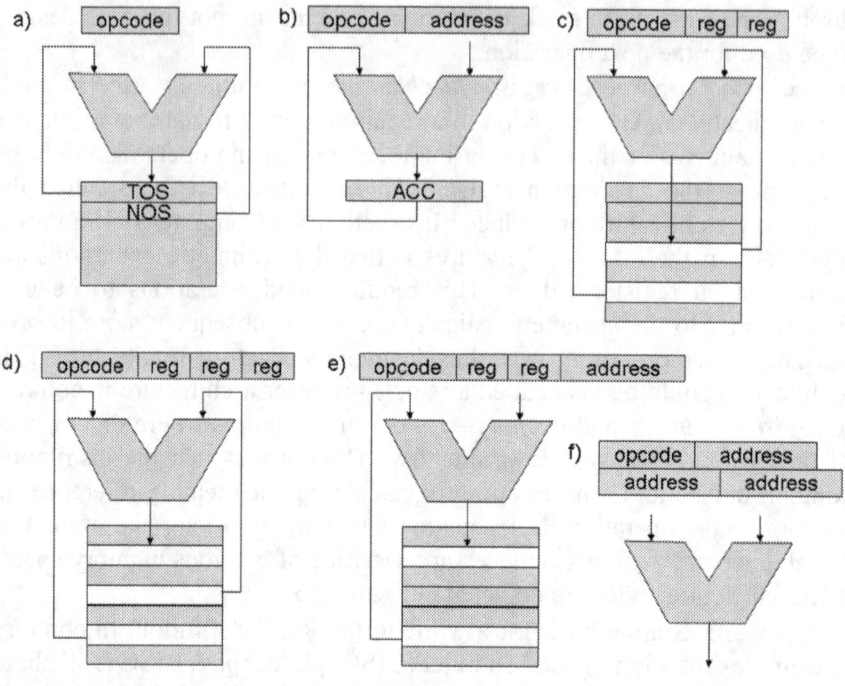

Fig. 2.5. Architecture styles: a) 0-address (stack), b) 1-address (accumulator), c) 2-address (register), d) 3-address (register) architectures, e) register-memory, and f) memory–memory architectures. In f) we are assuming the ultimate case that both operands and the result reside in the memory. The top part of each subfigure illustrates the instruction format corresponding to the architecture style.

Also customizable processors are stretching the architecture style towards the CISC philosophy.

Regularity means, for instance, that the instructions are of the same length, and that the operation code and operand fields are found at the same location in the instruction independently of the instruction at hand. Orthogonality means that the same addressing modes and data types can be used in several instructions, and the register set usable for an instruction is not restricted (by anything else but the use conventions) but is as similar as possible from one instruction to another. This also provides a large number of registers to be used by the many simple instructions in a RISC architecture. In CISC, the registers tend to be more specialized for limited purposes, one reason for that being reduction of the instruction lengths by allocating a special register to be used exclusively for a specific purpose. Orthogonality and a large number of general-purpose registers thus go hand in hand.

Ways of parallelism

The most straightforward way to increase the parallelism of a processor is to *pipeline* it, thus interleaving the execution of successive instructions. RISC architectures, in particular, use this type of parallelism, which is well in line with splitting the program functionality first into (temporally) short instructions and then just further splitting the execution into natural sequential stages. This can be further continued to *super-pipelining* which means pipelining beyond the functional block level, i.e., splitting "natural" execution stages such as instruction fetch into smaller sub-tasks each in their own pipeline stage.

Instruction Level Parallelism (ILP) is another widely used technique and it refers to parallel execution of complete instructions or operations. There are several approaches to ILP, including superscalar, Very Long Instruction Word (VLIW), Explicitly Parallel Instruction Computer (EPIC), Transport Triggered Architecture (TTA) and dataflow. These will be described briefly in the following paragraphs. The main difference of the approaches is in how much they rely on hardware (dynamic operation) in the implementation of the instruction issue, as shown in Figure 2.6.

Superscalar processors fetch several instructions into their instruction queue at a time and dynamically issue a pre-defined maximum number of them each clock cycle. The number and type of parallel instructions varies from architecture to architecture.

An example could be that the processor issues two integer and two floating-point operations each cycle. One major advantage of superscalar is binary compatibility between a single-issue processor and different multi-issue implementations of the same ISA. Dynamic instruction issue introduces one problem, the possibility for out-of-order execution. To properly execute programs, the processor must incorporate a reorder buffer or an extended register file (with a mapping table to map logical registers to physical ones) to write back the results to user-visible registers in the program order.

VLIW executes packets of operations packed into a very long compound instruction (as the name states). The instruction issue relies on static scheduling, determined by the compiler when packing the operations into a single instruction word. The operations are often executed in lockstep, meaning that the parallel instructions are not allowed to overtake each others but must wait if some of the operations in the VLIW packet will be stalled. This limitation and the large number of register file ports needed are the major drawbacks of VLIW. Clustering is used to fight the large number of ports, which makes the architecture less orthogonal.

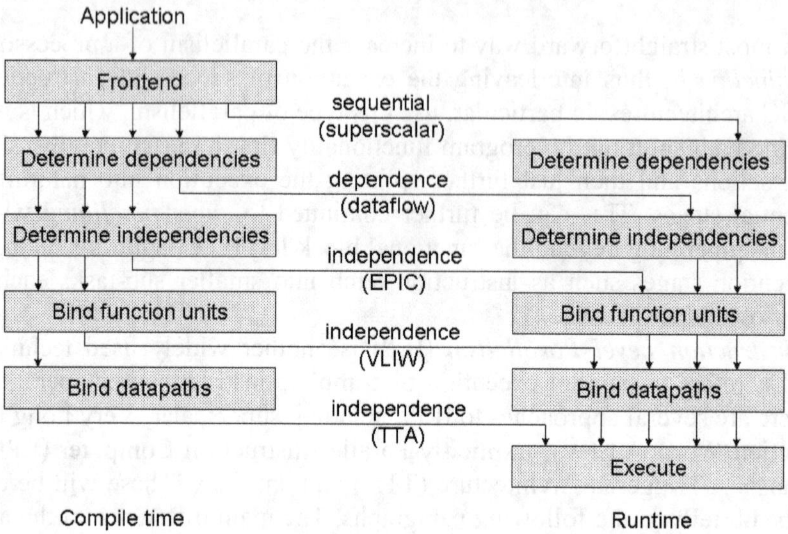

Fig. 2.6. ILP classes based on how much of the functionality are passed to be solved at run-time by the hardware.

Yet another drawback to older VLIW architectures was the wasting of instruction memory space. When the whole VLIW instruction could not be filled with some meaningful operations, the compiler added no-operation instructions to pad the VLIW packet. These NOPs occupied a lot of the instruction space, which was typically overcome by using different instruction templates with varying number of operations. To create a long-lasting solution that can also support different number of execution units in future implementations led to the EPIC approach, which might still be called (somewhat misleadingly) VLIW. In the EPIC architecture, the instructions are more clearly put together from a set of independent instructions, with some indication as to which of them can be executed in parallel. Thus, instead of a readily packed VLIW instruction, the actual issuing of the instructions may or may not happen in parallel, depending on the processor implementation. The actual binding to function units is done dynamically at run-time [407].

In the *dataflow* architecture, information on the next consumer of the processed data is encapsulated together with the data. These “tokens” (processing order information) and data are then moved around the inherently parallel architecture to complete the sequence of operations. The dependencies are thus implicit from the code itself.

TTA seems at first glance to be very different from the above mentioned ILP solutions. In TTA the key idea is to code the date movement between different functional units within the instructions instead of separately as functions and their operands. The moves to the local operand registers of the functions automatically trigger function execution. To make this architecture work properly, dependencies and independencies must be completely resolved at compilation time.

Another form of processor parallelism is *Data Level Parallelism* (DLP), which means that the operation is applied on several data items instead of one. Typically this is implemented with split datapaths, e.g., a 64-bit datapath could be used to carry out two 32-bit operations, four 16-bit, or eight 8-bit operations instead. Such a configuration is shown in Figure 2.7. This principle is also called *sub-word parallelism.* Such short-data parallel operations are especially useful in multimedia processing (like multi-channel audio or Red–Green–Blue (RGB) video signal). In this context the approach is also called *short vector processing*. In actual vector processors, the datapath is often time-multiplexed between the vector elements, thus only saving instruction memory rather than considerably speeding up the operation in case of vectors instead of scalars. In this sense the DLP in form of short vectors is completely different from the vector processor approach.

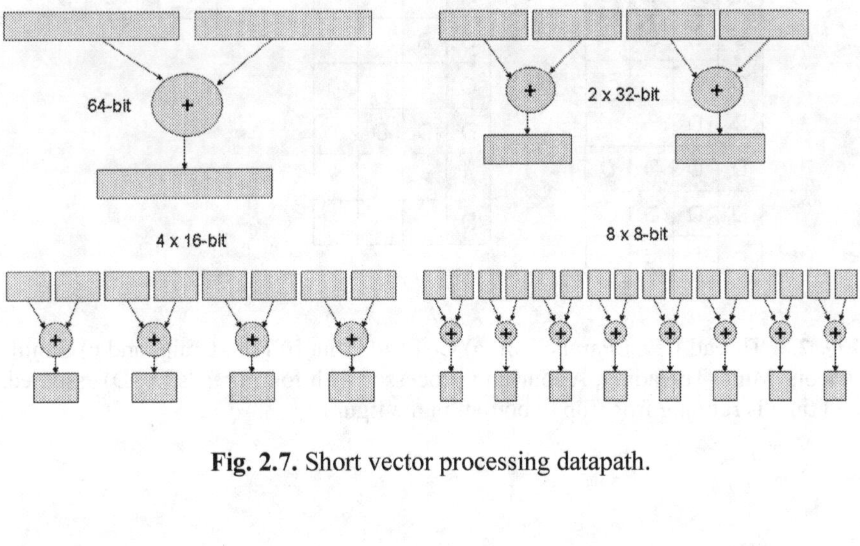

Fig. 2.7. Short vector processing datapath.

Thread Level Parallelism (TLP) and Process/Processor Level Parallelism approach larger parallelism granularity. In TLP, the execution units of a processor are shared between independent threads of the process (or even threads from different processes). TLP can be further divided into coarse-grain, fine-grain, and Simultaneous Multi-Threading (SMT). In coarse-grain multi-threading, the threads are context-switched infrequently, typically when the processor would otherwise stall to wait for a cache miss.

In fine-grain multi-threading, the thread in execution can be changed every clock cycle. SMT gets even further by allowing multiple threads to share each execution cycle, i.e., different threads are actually co-executed as opposed to time-multiplexed threads. Thus, in SMT there is physical

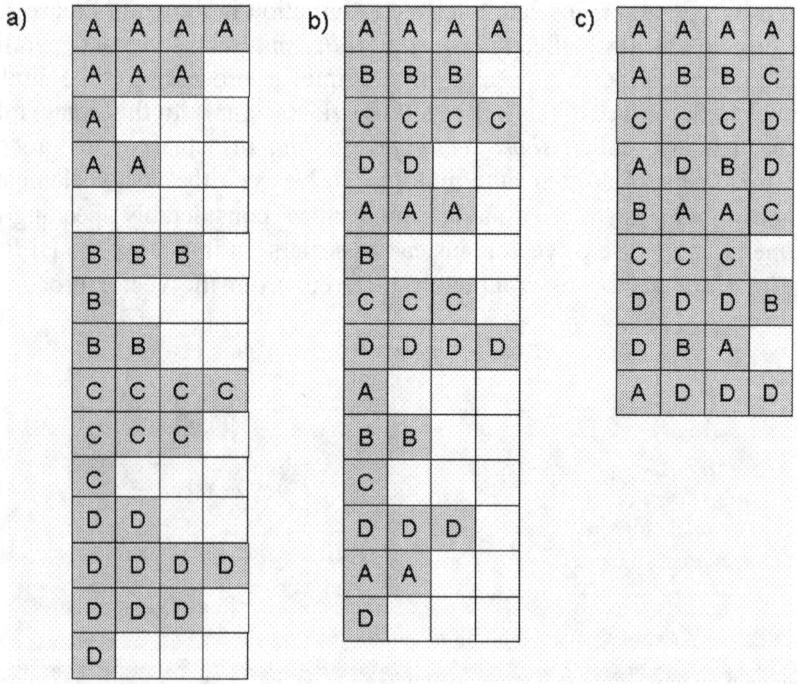

Fig. 2.8. Thread Level Parallelism: a) Coarse-Grain, b) Fine-Grain, and c) Simultaneous Multi-Threading. A four-unit processor with four threads (A–D) assumed, and time is running from top to bottom in the figure.

concurrency, not just virtual. The differences of the TLP approaches are illustrated in Figure 2.8.

In *Process Level Parallelism* (PLP), different processes are executed on parallel processors or processor cores. This brings us to real multi-core and multi-processor systems. A classification according to Flynn [127] divides (parallel) processing approaches to four classes, according to how many parallel instruction and data flows are handled in parallel. The Multiple-Instruction Multiple-Data (MIMD) class is reserved for the independent parallel multiple processor case. Single-Instruction Multiple-Data (SIMD) refers to data parallel or short vector processors. Single-Instruction Single-Data (SISD) describes a uni-processor case. Multiple-Instruction Single-Data (MISD) is often considered an anomaly of the Flynn classification. The closest candidates to that principle can be found in systolic arrays which process streams of data spatially in a pipeline.

Memory

First of all, there are physically different kinds of memories, each having their characteristic properties regarding speed, area, power consumption, and volatility. The fastest option is a *register*, which is constructed out of flip-flops or latches. Building memories out of registers will turn out to be too expensive, but they are definitely to be used within the processor. In Static Random Access Memory (SRAM) the memory cell is typically made out of six transistors, four of which are used as two cross-coupled inverters and the rest as transmission gates to read and write the cell. While registers can be accessed in a fraction of a clock cycle, a reasonably sized SRAM can be accessed in most cases in a single cycle. Read-Only Memory (ROM) can be used for constant tables or boot-up code, it is very dense with only a single transistor (or none, but the area is the same) per storage bit. While ROM is fixed at design-time, there are also some other similar type of *non-volatile* (meaning that the value is not lost in power-down) memories. Electronically Erasable and Programmable ROM (EEPROM) and FLASH memories can be also written and are also available in some digital integrated circuit technologies for on-chip implementation. They are slower and larger than the volatile SRAM.

Dynamic Random Access Memories (DRAM) are smaller since the storage element is typically a trench capacitor that is accessed through a single pass transistor. On the other hand, large DRAM arrays suffer from the access delay; the row decoding and column selection are done sequentially. To cope with this, Synchronous DRAM (SDRAM) utilizes the latched data from row access to deliver several subsequent data items in a burst. Double Data Rate (DDR) DRAMs use both rising and falling edge of the synchronizing clock to deliver data. *Rambus* technology uses a lower voltage swing to deliver data more promptly from the synchronous DRAM. Despite of all the various attempts to speed up dynamic memory, it requires several clock cycles to access data in them.

Due to physical characteristics of memory circuitry, large memories cannot be very fast. To create an illusion of a large and fast memory, a hierarchy of memories is often used. The illusion is based on the *principle of locality*. There exists both temporal and spatial locality in memory references. Temporal locality means that a referenced data item or instruction in the memory likely will be referenced soon again, as in the case of several subsequent executions of a loop body, frequent calls to the same subroutines, or loading of a variable for update and storing. Spatial locality means that if we read an instruction, we are also likely to read also the next one, and maybe even the one after that. In data access, various data structures like arrays, First-In-First-Out (FIFO) queues, and stacks can be easily seen to follow a similar pattern. Thus, it makes sense to have the frequently used items close-by in a fast but small memory, and keep the off-the-beaten-track ones available in a larger but slower storage. This principle is embodied in the *memory hierarchy*.

In computers, the memory hierarchy consists of the processor registers, level one data and instruction caches, level two (unified) cache, main memory, disk storage, and often even an off-line backup medium as the last resort. This is illustrated in Figure 2.9. In embedded systems the hierarchy may be lower, but also there the use of caches is increasing as the applications become more complicated. Again, for mere physical reasons it is reasonable to keep the closest memory fast enough to allow for single-cycle access. The increasing memory requirements (by the application) have been hampering this target for a while, thus driving the take-up of memory hierarchy in integrated embedded systems. Another reason is to allow the system to use more cost-effective memory technologies (such as different DRAM variants) for the off-chip (main) memory of the system.

Coming to cache memories, there are various parameters which determine the efficiency of the caches in terms of hardware area or operation speed. The common targets of cache design are high hit rate (in other words, low miss rate), low miss penalty (in number of cycles), and low access latency for hits. The misses are due to three main reasons: *compulsory* or cold-start misses when there are no items yet in the cache blocks that we are interested in, *capacity* misses when the cache is already full and does not yet contain the item we are looking for, and *conflict misses* when two or more blocks are competing over the same location in the cache.

The data is retrieved from the cache using a *tag* that is stored together with a *block* of data. Usually the block is more than one word long to exploit also the spatial locality of instructions and data, and to save tag memory space which is an overhead to the actual payload of the cache. A typical block could consist of four to eight words. In practice the tag section also contains some status bits, such as a *valid* bit. There may be more bits to be used, e.g., as guidance for the replacement algorithm, or for a cache coherence protocol in multi-processor systems.

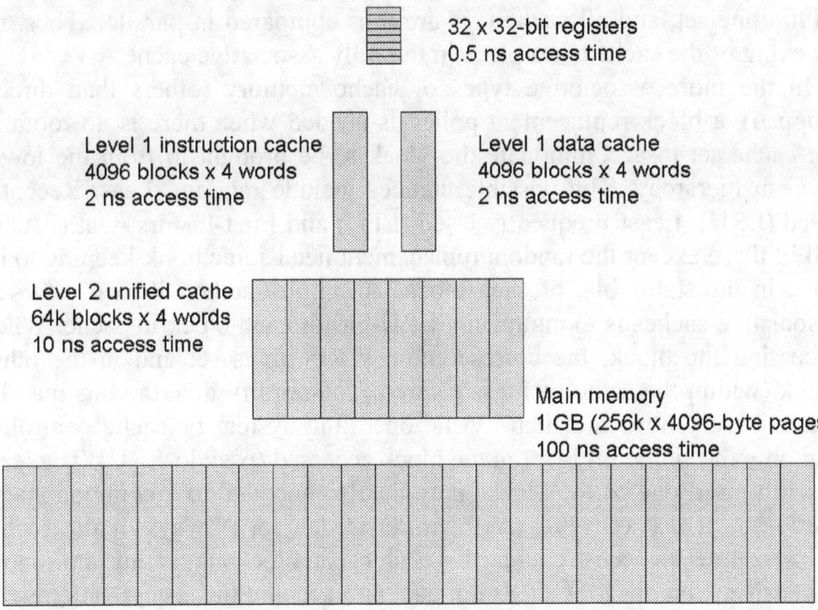

Fig. 2.9. Example memory hierarchy.

The design parameters include cache capacity (size), associativity, block replacement policy, write policy, and allocate-on-write-miss policy. The capacity is simply the amount of SRAM memory for the data storage in the cache. It can be utilized in different ways depending on the associativity of the cache. In a *direct-mapped* cache memory there is a single location to which the data can be mapped directly according to the least significant part of the address (address mod cache size). In a *set-associative* cache the memory has been divided into two or more ways. The data can be mapped into one of these ways according to a (re)placement algorithm. The set (cache area where a particular memory block can be placed) is a cross-section of the ways at the address formed by the modulus of the address and the way size. So, only one Nth of the cache indices are used in an N-way set-associative cache compared to a direct-mapped one with the same capacity. In a *fully associative cache* the data can be placed anywhere. This sounds like a good idea, but it also means that every location has to be searched when accessing the cache. Figure 2.10 illustrates the different associativity types. The required complexity is illustrated by showing the tag memory sections and comparison logic needed. The possible locations for the desired block are shown shaded. The fully associative cache extends from the N-way set associative one by including all cache blocks within one set, and all cache tags are thus compared in parallel. Thus, no indexing of the cache takes place in the fully associative cache.

In the more associative types of cache memory (others than direct-mapped), a block replacement policy is needed when there is no room in the cache set to accommodate the block to be brought in from the lower level in hierarchy. The possible choices include random, Least Recently Used (LRU), Least Frequently Used (LFU) and First-In-First-Out (FIFO). All of these except the random replacement need some book-keeping to be done in the status bits of each block. A simple method in two-way set-associative caches is to maintain a LRU-bit for each block in cache. When accessing the block, the corresponding LRU bit is set and in the other block (within the same set) it is cleared. To simplify this, the bits may be cleared every now and then by the operating system or cache controller and only the single bit is set in the block accessed (pseudo-LRU). The least recently used one of the blocks may be also encoded to a common status section such that only $\log_2(N)$ bits are needed for N ways of the cache. LFU requires an access counter for each cache block and is thus a bit more expensive solution. FIFO does not take into account the access patterns but replaces always the oldest block. Even random placement, despite its simplicity, yields relatively good performance.

The cache write policy can be write-back or write-through. In a write-back cache, the modified block is not written to the lower hierarchy level until it is replaced because of a cache miss. This will reduce the bus traffic to the memory. To know whether the block has been changed or not, a *dirty* bit is used to mark an updated block. There are also some drawbacks in the write-back method. It will increase the miss penalty in the case of a dirty block being replaced, since the old block has to be written to memory before the new block can be brought in. Also, in multi-processor systems

a)

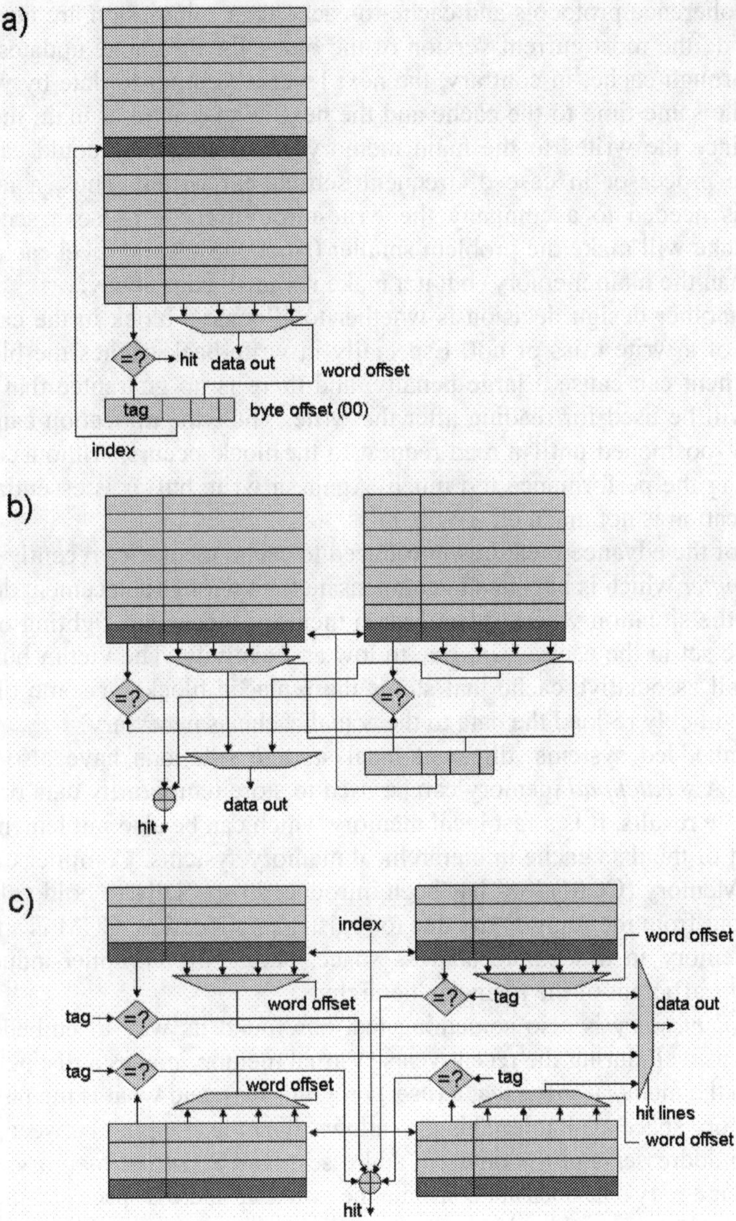

Fig. 2.10. Cache associativity: a) direct-mapped, b) two-way set-associative, and c) four-way set associative cache. The instruction word is divided into the fields of tag, index, word offset (within a block), and byte offset (within a word).

cache-coherence protocols and cache-to-cache data submission are needed to provide the most current version of the block for the cache updates. In write-through cache, in contrary, the next level is kept up-to-date by writing at the same time to the cache and the next level (cache or main memory). Since the writes to the main memory are slower they could easily stall the processor in case of frequent subsequent writes. Thus, a *write buffer* is needed to accompany the write-through cache. Also a second level cache will make the problem smaller (since the second level cache is faster than the main memory) but not make it vanish completely.

Yet another design decision is whether to allocate a block in the cache in case of a write miss or not. Especially in write-back caches the block replacement can cause a large penalty, and there is no guarantee that the block will be used for reading after the write. Thus, the allocation can be possibly postponed until a read request to the block occurs, without compromising the performance too much. Again, a write buffer is essential if the allocation is not made on a write miss.

One of the advanced features introduced to cache memories recently is a *victim buffer* which is intended to compensate for a wrong replacement decision or the situation where several main memory blocks are fighting over the same set in the cache, e.g., due to low associativity. The victim buffer is a small associative cache that stores the replaced block for some time, and can quickly re-load the data to the actual cache as necessary.

In embedded systems, different local storage solutions have also remained. A *scratchpad* memory can be used to store temporarily data or intermediate results. It is a fast local memory which can be also implemented as a part of the data cache in hierarchical memory systems. Compiler Controlled Memory (CCM) [88] has been introduced especially to hold values spilled out from the register file due to register re-allocation. CCM is a fast local memory in a separate address space, taking the compiler-induced memory traffic out of the memory hierarchy.

Virtual memory is also something that has found its way to embedded systems as well during the recent years. Virtual memory makes a distinction between the memory space that a (user) program sees, and what is the physical memory space available. This also allows dynamic mapping between the program addresses and the physical addresses. Due to virtual memory, the physical memory size does not constrain the available address space any more.

In virtual memory the address space is divided into *pages*, which is a lot larger than a cache block. *Segments* can be used instead of (or sometimes on top of) pages. Segments can be of any size, while the page size is a fixed system parameter. The mapping is conventionally done by using page tables in memory. A part of the virtual address is the address within the page table, pointing to the physical page number stored in the table. The lower end of the virtual address is typically the offset within the page directly. The virtual-to-physical conversion is illustrated in Figure 2.11.

The drawback of virtual memory becomes clear when we notice that the page table is stored in the main memory. Each memory access requires two accesses, one to the page table and another to the actual program or data memory (or cache, in the best case). A common way to speed up the address translation is to use a *Translation Lookaside Buffer* (TLB) which contains the virtual–physical address pairs. This is a small associative cache memory, where the virtual page number forms the tag.

Another issue arising with virtual memory is whether the caches should be addressed by virtual or physical addresses. A good solution is to use *virtually indexed, physically tagged* cache. It allows making the TLB

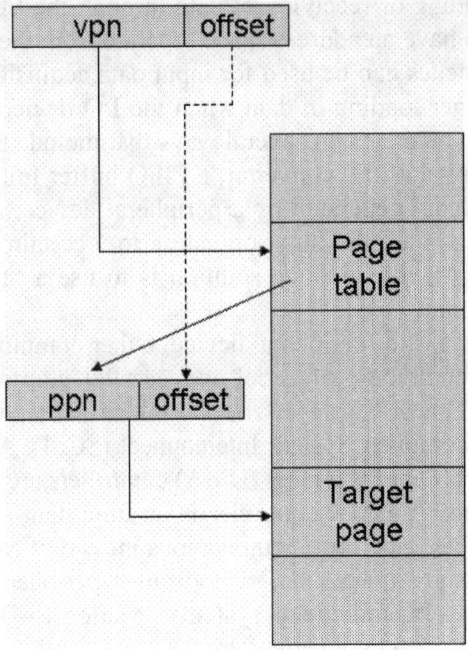

Fig. 2.11. Virtual address conversion to a physical address. vpn = virtual page number, and ppn = physical page number.

lookup in parallel with the (L1) cache, but on the other hand this requires that the index to cache is within the part of the address that is equal between the two memory spaces (virtual and physical). In direct mapped caches this means that the cache size cannot be any larger than the page size. In set-associative caches each of the ways must stay below this limit, allowing for larger capacity.

I/O operations and peripherals

In communicating with the outside world, there are a couple of different approaches on how the I/O will be seen by the programmer. One way is to use *memory-mapped* I/O, where the input and output registers of the peripheral devices are mapped to memory locations in (one of) the data memory space(s). Thus, the I/O operations become loads and stores. The opposite approach is to use *explicit* input and output commands which carry out the sending or receiving of data through the I/O device. The I/O devices may also have an address space separate from the memory space.

Several approaches can be used for input data acquisition. One is to use *interrupts* to trigger loading in data when the I/O device is ready. The interrupt signal can be driven in several ways that include the clock signal of an Analog-to-Digital (A/D) converter, a FIFO buffer full signal, or a dedicated interrupt signal generated by a peripheral device. Another approach is to *poll* the data or tie the input operation to a certain phase of the program execution. An intermediate solution is to use a timer to trigger the data input operation.

Timers are one sort of peripheral device. Other commonly used peripherals include different type of serial and parallel interfaces like RS-232, Universal Serial Bus (USB), Firewire, Peripheral Component Interconnect (PCI), or Small Computer System Interconnect (SCSI). Also interrupt controllers and Direct Memory Access (DMA) controllers are used frequently in embedded systems. A DMA controller is an important system component that allows the CPU to perform computations instead of carrying out routine data transfers from/to main memory to/from a peripheral device or local working memory. In embedded systems, Analog-to-Digital (A/D) and Digital-to-Analog (D/A) converters are also frequently used to communicate with the surroundings of the device. The signals to be converted may be sensor data, speech, audio, image, video, or Radio Frequency (RF) signals.

3 Beyond the Valley of the Lost Processors: Problems, Fallacies, and Pitfalls in Processor Design

Grant Martin and Steve Leibson

Tensilica, Inc.

During the entire 70 years of computer development, a huge variety of discrete and embedded processor species have emerged, evolved, and sometimes died out. Many strange and wonderful designs resulted from this evolution. Sometimes these strange and wonderful concepts lived on, some died out almost immediately, and some lived for only a short while – only to die off and then reappear as their gene lines re-emerged in later species.

This chapter surveys thirteen failed processor species (a baker's dozen) and explores the major design errors that caused their demise. Each major design mistake is also illuminated with examples. However, given the endless recycling of many old ideas as technology makes them shiny, bright, and new once again, who knows when the intrepid explorer/designer will next meet up with a descendent of one of these species?

To honor the fact that these species may be thought of as "dinosaurs," we solicited from our colleagues suggestions for good saurian-style names for each one. However, the analogy with dinosaurs also holds in another sense: many processors based on these concepts were the dominant species of their day, or had sufficiently bright coloring and loud roaring to attract a huge amount of attention. Just because the evolving world caused a corollary evolution and die-out of our processor species does not mean that a processor species was not a reasonable adaptation to the world as it existed during their heyday.

J. Nurmi (ed.), Processor Design: System-on-Chip Computing for ASICs and FPGAs, 27–67.

Problem 1: Designing a high-level computer instruction-set architecture (ISA) to support a specific language or language domain (Myopicsaur)

Since the earliest days of computing, there have been repeated efforts to drive the programming problem up in abstraction level – from patch-board programming, to toggle-switch entry, to machine-language entry, to assembly language, and then to a whole host of "high-level" programming languages (HLLs) starting with Fortran and COBOL in the 1950s and working through hundreds or thousands of programming languages over the next half century.

As soon as HLLs were developed, people began worrying about the semantic gap between HLL descriptions used to capture solutions to programming problems and the actual instructions generated by the HLL compiler to be executed on a target machine. Early compilers often produced poor results – sometimes very poor. Even today, despite more than five decades of compiler development, the quality of results that can be achieved for many algorithms by the most highly skilled human assembly-language coders can be an order of magnitude better (or more) than code produced by the best HLL programmers combined with the best available optimizing compilers.

Inevitably, computer researchers and commercial computer vendors began to investigate the feasibility of tuning a particular processor to specific HLLs or language classes in an attempt to more closely match the processor's instruction set with the language's requirements and narrow the semantic gap. The theory was that programs written in those targeted HLLs would execute much more efficiently on these tuned machines.

Many decades of experience and the attempts to implement this approach during every major processor era – mainframe, mini/midi, discrete microprocessor IC, and embedded processor core – have repeatedly established that this approach is a major architectural mistake. Indeed, it is one of the classical "fallacies and pitfalls" found in Hennessy and Patterson's seminal book on computer architecture ([188], p. 142).

The basic problems with this approach are many:

- Although tuned to one language, the processor may (and very likely will) be used to run programs written in other languages. The tuned processor will run programs written in these other HLLs relatively poorly due to its language-specific tuning.
- In earlier eras, scarce hardware resources were expended on efficient execution of rarely used instructions – a poor application of valuable architectural capital.

- A language-specific instruction may end up being implemented for some very specific use of an HLL construct and not be useful for the typical and most frequent uses. Thus the hardware for this instruction will essentially be wasted.
- Languages evolve. A computer architecture based on fixed, HLL-specific hardware tends to remain fixed for a much longer time than the language itself because software is much easier to change than hardware.
- The HLL-specific processor's popularity is tied inexorably to the popularity of the target HLL. Minority tastes in languages produce a processor with minimal market appeal.

Nevertheless, it took a long time for the pitfalls in this architectural approach to become apparent. From the 1960s through to the mid 1980s, before the rise of the Reduced Instruction-Set Computer (RISC) architectural approach, Complex Instruction-Set Computer (CISC)-based, HLL-specific computer architectures sparked tremendous interest. Researchers wrote hundreds or thousands of papers, dedicated conferences and symposia on this topic were quite popular, and companies introduced real machines into the marketplace based on this design philosophy.

The Burroughs "E-mode" machines were perhaps the most famous series of machines designed to support a specific language. This series included the B5000/6000/7000 and A-series machines from the early 1960s through the 1990s (some compatible processors are still available).[1] These machines were designed for direct execution of ALGOL 60. This computer family had many other significant features as well, including stack-based architectures, non-flat memory utilization, no assembly language, operating systems and specialized supervisory subsystems written directly in dialects of ALGOL 60, and a 48-bit memory word (plus tag bits). Indeed, during the 1960s and 1970s, Burroughs almost made a fetish of the HLL-specific computer-design approach.

During this period, the company produced medium-scale and small COBOL-specific mainframes (B2000/3000/4000) and an interesting microcoded architecture used in the B1700/1800 machines that included a set of interpreted instruction sets that could be swapped in and out to match different languages. As Earnest remarks about the B5000, Burroughs had a "dedication to the use of higher-level programming notation to the practical exclusion of machine or assembly languages" [110].

---

[1] A caveat lector: One of the authors (Grant Martin) worked for Burroughs. Repeated use of the E-mode machines as an example in this chapter can be thought of as a somewhat nostalgic and affectionate reminiscence of interesting times past.

Unfortunately, Burroughs' E-mode machines[2] suffered several of the disadvantages of HLL machines. Their performance on the standard scientific and business processing languages – Fortran and COBOL – was positively anemic. Later attempts to build C compilers for these machines and to port Unix onto their underlying architecture proved difficult, due in no small part to the architecture's hierarchical memory structure. Attempts to extend the HLL-specific instruction set to lower end machines (including a single chip implementation introduced by Burroughs' successor Unisys in 1989 called the Single-Chip A-series Mainframe Processor (SCAMP) [350]) required huge amounts of microcode. Unfortunately, ALGOL 60 never really took off as a popular programming language and this no doubt curtailed the popularity of the Burroughs machines.

As mentioned, Burroughs continued its HLL-specific design philosophy with the COBOL-oriented B2000/3000/4000 computers, which at least had the advantage of targeting a more popular, business-focused HLL.

The attractiveness of HLL-specific processor design also resulted in the development of machines that directly ran programs written in APL [182], Lisp [454], Prolog [124], and others that directly targeted BASIC, FORTRAN, PASCAL, PL/I, and SNOBOL [103]. Indeed, problems with the HLL-specific computer-architecture design approach led to a serious retrospective on them held in 1980 [103] just prior to the rise of the CISC workstation and the later rise of RISC processors and workstations in the mid 1980s.

Moving from the mainframe era to the midi/minicomputer era saw the aforementioned HLL-specific computer-architecture design approach repeated with the Burroughs B1700/1800, which provided microcoded instruction sets for several languages (COBOL, RPG, and Fortran among them) [319], and a number of specialized workstation-class machines. Machines designed to directly execute LISP are an especially notable example (LISP Machines, Symbolics).

The discrete microprocessor era also saw a few HLL-specific microprocessor architectures including the Inmos Transputer designed to run Occam, the CRISP processor designed at Bell Labs for directly executing C programs [105,106], and perhaps the most famous (or infamous) of all such microprocessors: Intel's 432, which was designed to run programs written in Ada [144]. The Transputer and its Occam language illustrate one of the "features" of an HLL-specific processor – the sometimes religious or quasi-religious devotion by its developers to a particular theory of computing that manifests itself in a slavish dedication to one programming

---

[2] Note: There are several references available on the E-mode machines – Organick [318], Carlson [69] and Doran [108] are just a few of them.

language and the exertion of substantial effort to develop a machine to support it. Although Transputer compilers emerged for more conventional HLLs later, the Transputer was introduced with Occam, a language based on Tony Hoare's communicating-sequential-processes concepts.

The Transputer was purpose-built for Occam. Iann Barron, head of INMOS, was Occam's high priest. The Transputer's history illustrates one of the problems with HLL-specific architectures listed earlier. Its success depended highly on finding a market interested enough in Occam to buy the processor designed for it – or interested enough in the Transputer to adopt its unusual language. This sounds a lot like a religious conversion.

Intel's 432 was designed to execute Ada and, in a more general sense, object-oriented languages. The Intel 432 perhaps represents the extreme for HLL-specific processors. It failed to achieve adequate performance for any language including Ada, the one it was designed for. In fact, Intel's 432 microprocessor suffered from a whole litany of design mistakes. Among those cited [144], we find:

- The Ada compiler generated spurious instructions.
- The Ada compiler did not perform common-subexpression elimination.
- The compiler passed parameters by value/result, even for large arrays (rather than by reference).
- The compiler used the much slower intra-module call all the time, even when unnecessary.
- Instructions were bit-aligned, thus slow to decode.
- No more than one instruction stream literal was allowed.
- The machine had extremely inefficient procedure calls – more than 1000 clock cycles including 282 wait states, compared to fewer than 100 clock cycles for other processors of the era.

As a result, Intel's 432 executed common benchmarks 10x to 26x more slowly than a Vax 11/780, and 2x to 23x more slowly than an 8-MHz 8086. Luckily for Intel, its success with x86 processors and all its successors used in the evolution of the IBM PC allowed the Intel 432 to simply disappear, mostly forgotten by today's computing practitioners.

The final foray for HLL-specific processors has been in today's embedded era, with specific hardware designed by Sun, ARM, and other vendors for executing Java (Sun's picoJava processor, ARM's Jazelle coprocessor, etc.). These Java-specific processors prompted some interest, but not a lot of enthusiasm. Interpreting Java on conventional high-performance processors and just-in-time (JIT) compilation have proven to be more interesting routes for designers delivering Java applications in the embedded world today. In addition, continued improvement in embedded processor performance has often proved quite adequate for many Java applications in embedded products, which are mainly control and user-interface oriented.

If the HLL-specific processor route has shown itself through four computing eras to be a mostly misguided approach, what are the other options open to those wishing to use hardware in ways that go beyond general-purpose processors to accelerate languages and applications written in those languages?

One of the first concepts to discard must be that "it's all about the language." Indeed, for data-processing intensive applications, it is much more "all about" the computational and communications kernels and algorithms embedded in the programs. If an application involves taking the dot-product of large vectors repeatedly, then a processor without an appropriately sized hardware multiplier, or better a multiply–accumulate (MAC) unit, will perform this application poorly whether the program is written in Fortran, Ada, C, Java, BASIC, or COBOL. If the processor has the right functional units and a good HLL compiler (or interpreter) for the language(s) being used, the algorithm expressed in just about any of these languages should execute roughly as fast no matter the language.

It is the algorithm's characteristics – rather than the language's characteristics – that should be used to design, modify, or choose the right processor. One could either search for a good processor with multipliers or MAC units (and perhaps zero overhead looping) for this application – a DSP might be a good choice – or perhaps even better, one could use instruction-set extensions to tailor a configurable processor core more precisely to the performance and communication requirements of the application. In this sense, the search for an HLL-specific computer architecture can be replaced today by a search for an application-specific instruction-set processor (ASIP).

Problem 2: Use of intermediate ISAs to allow a simple machine to emulate its betters (Rubeus Goldbergicus)

One of the major methods used over the years to implement HLL-specific processors, as discussed earlier, is to tailor an intermediate ISA to an HLL and then use or develop a simpler processor that emulates the defined ISA through microprogramming. Microcode, in which a defined sequence of basic processor instructions implements an intermediate ISA, is compiled from some simplified intermediate language or hand written, and is made available to the processor either through on-chip local memory or fast-access memory with a relatively low latency – often on-chip ROM/PROM or perhaps ROM/PROM designed into a multichip module in combination with the processor.

There are several advantages to this approach:

- It provides object-code compatibility with other processors within a family or with previous processor generations.
- It allows a family of processors with various price–performance characteristics to be built. High-end processors in the family implement the intermediate ISA more directly or even accelerate it through multiple function units, exploiting instruction-level parallelism (ILP). Low-end processors in the family map the intermediate ISA onto more restricted hardware that would execute programs more slowly but would also be much lower in cost.
- Compilers written to the intermediate ISA could be used on multiple processors in the family. The mapping from the intermediate ISA to the actual instruction set of the lower-end machines, embodied in microcode, would be written as a separate layer and would possibly even avoid the use of a compiler, or at least require a very simple one. Furthermore, such a mapping might only need to be done relatively infrequently because the intermediate ISA would not be exposed to users and would not necessarily evolve the way an HLL might evolve.
- A microcoded processor, through use of multiple ISAs and multiple microcode sets, could be dynamically adapted to different HLLs at runtime, and thus offer better performance to programs written in different languages.
- For languages that rely on interpreters, the formal development of an appropriate intermediate ISA and microcode mapping of that intermediate ISA to the target ISA might speed the availability of language interpreters by dividing their development into two simpler stages (the classic divide-and-conquer approach to engineering design).
- As discussed earlier, by separating the process of implementing a language compiler into two stages, it might be possible to provide language support for a new target machine more quickly than by writing a specific targeted compiler.
- Code size may be reduced using one intermediate ISA instruction instead of two or more target ISA instructions. In addition, performance may improve by reducing the number of instruction fetches from main memory.
- Supporting only part of an ISA rather than the whole thing may simplify compiler writing for infrequently used parts of a language. The hardware design for a new processor to support the ISA may also be reduced in complexity, design effort, and project risk. Consequently, execution

of well-tested microcoded implementations for certain functions might be a much better alternative to direct hardware implementations.

- This technique might better take advantage of more modern process technologies and much faster clock rates to offer backwards compatibility for older machines and instruction sets on newer processors – a kind of "virtualization" of the older ISA. The next step up would be to perform this conversion entirely in software, rather than involve any microcode at all.

There are also several disadvantages to this design approach:

- The performance of intermediate-ISA machines at the low end is often very poor compared to machines with simpler ISAs. The layering in intermediate-ISA machines often proves to be a less-than-optimal use of computing resources.
- Compilers that generate code in the intermediate ISA cannot optimize to the degree that a compiler targeted to the underlying simple ISA of the real target machine can. Optimization of the compilation can only be done independently in the two separate layers. HLL compilers aimed at high-end processors in a family that directly implement the intermediate ISA cannot optimize for low-end processors in the family unless specially modified for them, which negates some of the advantages.
- A machine that can be targeted to several different ISAs for several different languages might incorporate uneasy design compromises offering poor performance for all target languages.
- The microcode compiler, translator, or generator (translating the fixed intermediate ISA into the underlying target simple ISA) might be overly simple or hard to use because it is not intended to be run very often. In addition, the microcode might be difficult to change, especially if placed in ROM.

Some of the leading proponents of the intermediate-ISA concept embodied them in Burroughs processors, as mentioned in the previous section, but many other attempts can be found in the literature as well, supported by the availability of many different microprogrammable computers built over the years. Carlson [69] discusses a microprogrammed FORTRAN computer that represented a near-direct implementation of the FORTRAN language and required only a simple translator. Carlson also discusses an EULER processor (EULER being a variant of ALGOL 60) and APL machines [182] that all rely on microprogramming.

In 1980, Flynn [127] surveyed a number of architectural approaches – including the microcode concepts – and tried to define ideal language machines that would directly execute HLLs. Moulton [298] studied the general design of microprogrammed machines to support HLL compilation and execution. Among many other HLLs supported with microprogramming (see the previous section for a greater discussion) were LISP [454] and Prolog [124]. The Burroughs machine that perhaps was the greatest early embodiment of this concept was the B1700/1800 series, supporting intermediate ISAs for COBOL, FORTRAN, and RPG [319].

One might have supposed that the advent of VLSI in the 1980s would have tended to curtail the microprogramming approach and, indeed, the rise of industry-standard microprocessor ISAs, multiple generations of implementations of those ISAs, and the sheer transistor count available with modern IC-fabrication processes seem to have reduced the use of these microcoding methods. However, a few vestigial remnants of this technique have surfaced in recent years. For example, in the late 1980s, the Burroughs SCAMP processor, discussed earlier, combined a relatively low-end RISC processor of a similar kind to that used in the company's small, low-end A3 and A4 mainframes, with hundreds of Kbytes of microcode that implemented the Burroughs "E-model" mainframe instruction set – in use since the original B5000 appeared in the late 1950s. SCAMP was used in the "Micro-A" computer, where the SCAMP chip was assembled with a number of microcode ROM chips into a two-by-two-inch multichip module.

Another interesting vestige of this approach, and a counter-example for this problem, is to be found in current Pentium-class processors starting with the AMD K6 [172]. In these processors, the CISC instructions of previous x86 processor generations are implemented using an RISC instruction set. The processor's instruction-decode unit decomposes the CISC instructions into RISC operations and then assembles and issues groups of these simpler operations to the processor's parallel execution units. It's not exactly microcode, but something clearly akin to microcode.

This design approach also eases the creation of new CISC instructions for newer processors. It creates a hybrid CISC/RISC architecture. Clearly, microarchitected/microcoded machines still have a role and a place, one that may rise and fall as semiconductor technology and processor architectures continue to evolve. Perhaps a vestige of this saurian family tree will remain among the nimbler mammalian machines of the present day.

Problem 3: Stack machines (Stackadactyl)

Another computing dinosaur – the stack machine – evolved multiple subspecies. In fact, the stack machine actually may not be a dead end in computing evolution. To paraphrase a well-known US presidential quotation describing a somewhat questionable situation, “It all depends on what your definition of ‘stack’ (machine) is.”

A stack is simply defined as “a memory device in which data may be stored and from which they may be retrieved in, … , a “last in – first out” order” ([108], p. 64). Stacks have “been widely used by system programmers, especially for the implementation of compilers and interpreters” ([108], p. 63). The two general instructions used to access the stack are to PUSH a value onto the top of the stack to store it and to POP the value off the top of the stack to access it.

Koopman [236] has done an excellent job of defining and surveying stack-machine architectures, which fall into a total of 12 categories based on a three-axis taxonomy. The first axis represents the number of stacks: single or multiple (S or M). The second axis specifies the size of dedicated stack memory: small or large (S or L). The third axis represents the number of operands in the instruction format (0, 1, or 2). Thus SS0 represents a single stack machine with a small stack memory and a 0-operand instruction format – an example of which is the Burroughs B5000/6000/7000 family. SS2 machines include the Intel 80x86, which has a stack mode for floating-point computations.

Koopman describes stack machines in all 12 categories but concentrates on MS0 and ML0 machines. Representative SS1 machines include the HP300/HP3000 and the ICL 2900. SL2 machines include a number of early RISC processors such as the AM29000, CRISP, and RISC I. MS1 and MS2 machines include the PDP-11 and Motorola 680x0 respectively. ML0 machines include a number of different Forth machines, designed to directly execute the Forth language (which was the motivation for much of Koopman’s work). The ML1 category includes the Lilith (Modula workstation) and LISP machines.

Interest in stack machines seems to be based on the fact that:

- “Stacks are the most basic and natural tool that can be used in processing well structured code” ([236], pp. 18–19).
- “Machines with LIFO stacks are also required to compile computer languages” (Ibid).
- “Any computer with hardware support for stack structures will probably execute applications requiring stacks more efficiently than other machines” (Ibid).

Koopman distinguishes four uses of stacks: expression evaluation, return addresses, local variables for re-entrant or recursive code, and subroutine parameter stacks. Efficiency and simplification in stack operation motivate the use of multiple stacks rather than a single stack.

Hennessy and Patterson ([188], pp. 92–93) use a simple ISA taxonomy for stack, accumulator, and register type machines (register-memory and register-register/load-store). Despite the large number of stack machines and variants, "virtually every new architecture designed after 1980 uses a load-store register architecture" (Ibid, p. 93) – what they call a general-purpose register (GPR) computer. They ascribe this development to three factors:

1. Register accesses are faster than memory accesses (and with embedded processors, the discrepancy between internal-register and local-memory access, and system memory accessed over a bus has grown from a few clock cycles to potentially many tens or even hundreds of clock cycles as process technology has advanced).
2. With today's compilers, "registers are more efficient for a compiler to use than other forms of internal storage."
3. Registers can be used to hold variables rather than just operands and the results of operations, for reuse.

Ditzel and McLellan [104] give a good set of pros and cons for registers in their discussion of the C Machine "stack cache." Register-access time is at least an order of magnitude faster than main memory; register addressing requires fewer bits than a memory address, which may lead to more compact code; the processor can access registers in parallel with main memory, so they can be used to fetch multiple operands simultaneously. In general, registers reduce memory-bandwidth requirements.

The cons are:

- Context switching for procedure calls (the expense of which discourages the use of many small nested procedures in structured programming), which places the burden of register allocation on compilers.
- The fact that registers cannot be treated transparently as main memory – for example, in addressing modes.
- Contrary to the opinion above, at least for the C language, pointer aliasing for variables discourages the use of registers to hold variables (because memory would hold an old value if the register value had not been written back to it, and the compiler might not realize this due to use of pointers).

In their study, Ditzel and McLellan looked at using a stack cache for procedure calling, aiming to keep the top elements of the stack in high-speed registers.

As a particular and venerable example of a stack machine dating back to the early 1960s, many variants of the Burroughs E-mode machines held the two top elements of the stack in fast registers (the B7700 had a

32-register stack buffer). The rest of the stack spilled into memory. These machines used a single small-memory stack to hold operands for expression evaluation and stack frames for procedure calls (return addresses). This design approach was theoretically compelling but delivered relatively poor performance when trying to map other languages and large flat memory spaces to the architecture.

If we take our analogy of herds of various computer species at the metaphorical "dinosaur watering hole" and passively observe them over time, we would see a profusion of stack-based species and subspecies gradually dying out in the early 1980s to be replaced by the general-purpose-register (GPR) machines. It is interesting that this transition may have been driven more by a herd mentality in processor designers and a desire to emulate the more successful microprocessors and embedded processors than by clear technical superiority. Certainly, Koopman suggests the MS0 and ML0 stack machines represent a new breed or species that provide considerable advantages over the previous generations of SS0 and related machines. However, this new evolutionary branch of the stack-machine species seems to have died out.

Other technology factors evolved to make GPR machines good enough to survive and even flourish. Register windowing, for example, improved procedure-call efficiency and avoided much of the memory cost of procedure calls, thereby reducing part of the stack machine's comparative advantage. Process-technology evolution and the relentless application of Moore's law allow the number of registers at the processor's heart to grow sufficiently to accommodate normal programming needs. Registers are no longer a scarce and expensive resource.

Compilers improved in their ability to allocate registers in a more optimal fashion. Although GPR machines may lack some of the best features of their stack-based competitors, their environmental adaptations were "good enough" to allow them to survive, reproduce, and live on.

However, a large number of stack-machine variants have appeared over the years and the possible resurgence of Forth and related stack-oriented programming languages could reawaken interest in stack-based architectures. In fact, this very thing has already happened. The Java language and the idea of the Java virtual machine (JVM) reawakened some interest in stack architectures ([188], p. 149). The idea could take hold yet again with the rise in popularity of some future new programming language or the appearance of a revolutionary and unanticipated computing model.

Problem 4: Extreme CISC and extreme RISC (Microcodius Rex (CISC) and Reductius Rex (RISC))

In the beginning, there were no CISCs or RISCs – just computers. Computer processors and their instruction sets evolved in an increasingly Byzantine way over time as processor designers sought to improve performance through hardware extensions. The rapid and unchecked advance of Moore's law doubled the number of transistors with each integrated-circuit process generation and processor designers used this added capacity to create increasingly complex machines with increasingly more complex instruction sets. These new complex instructions had ever more complex addressing modes and included many variations of relatively simple basic instructions so that programmers could choose exactly the right instruction in each case. Extreme CISC ISAs could include many hundreds of instructions.

There were four intertwined motivations for creating increasingly more complex ISAs:

1. Early HLL compilers were not very good in generating optimized code. In theory, an ISA with more complex instructions would be an easier target for compiler writers. In practice, they weren't. For example, DEC's (Digital Equipment Corp's) extremely successful VAX ISA included hundreds of instructions, many of them quite complex. During the development of an 8-chip VLSI version of the VAX processor, DEC engineers discovered that 20% of the VAX instructions consumed 60% of the machine's microcode but represented 0.2% of the executed instructions [335].
2. Memory cost and memory bandwidth were major concerns for computer designers. Code size was also of concern because of memory costs and complex instructions, if effectively employed by compilers, greatly reduced code size. The big break for microcoded machines (really the start of the CISC movement) was the development and introduction of the IBM 360 mainframe family in the mid 1960s. This series of computers used magnetic-core memory for its main instruction and data storage. Thus the IBM 360's transistorized CPU was quite fast relative to its slow magnetic-core main memory. It therefore made a lot of sense to store small subroutines (really complex instructions) in a small, fast microcode memory (called Read-Only Storage or ROS by IBM) to minimize execution hardware in the less expensive members of the IBM 360 family. In these days before semiconductor memory, the low-end IBM 360 Model 30's microcode

memory was implemented with a circuit-board/punched-card capacitive sandwich called a CCROS (card capacitor ROS).

3. Performance could potentially be increased if efficient hardware could be produced to execute one complex instruction in one or a few clock cycles that replaced many simpler instructions executed over several clock cycles.
4. The quest for narrowing the abstraction gap between HLLs and computer architecture caused processor designers to focus on the creation of new complex instructions that encapsulated HLL operations into one machine instruction.

However, computer architects did not explicitly set out to create CISCs. Instead, CISC processors naturally evolved as the many evolutionary experiments in computer architecture and high-level language computers played out in the 1960s and 1970s. As we have seen, there were increasingly exotic computing species. The most extreme of them could definitely be seen as "extreme CISCs."

Interestingly, the concept of a CISC itself did not really appear until computer architects developed the RISC concept in the 1980s. CISCs required the invention of their antonym for people to recognize the CISC phenomenon. "In any science, taxonomy precedes causal analysis" ([123], p. 774).

The RISC concept started with John Cocke's group at IBM, which developed the IBM 801 to run a telephone-switching network back in 1974. The design team used a simple equation to determine the required processor speed. The network needed to handle 300 calls/second and it took approximately 20,000 instructions to handle a call. Combined with real-time response requirements, the team determined that they needed a processor that could execute 12 MIPS at a time when the IBM 370 Model 168 mainframe cranked out about 2 MIPS. Cocke's team developed the idea of a stripped-down ISA implemented in a pipelined machine coupled with separate, fast instruction and data caches and an optimizing compiler. It was the first implementation of an RISC machine [85].

However, IBM kept most of the IBM 801 project details undercover for years so the beginnings of the industry-wide movement towards RISC-based design can be traced to just two papers published in the early 1980s by Patterson and Ditzel [103, 334,]. These papers entirely document the emerging movement. The first paper arguably said goodbye to the past and the second hello to the future. A good summary of the RISC concept appears in Hennessy and Patterson [188], pp. 151–154.

The argument favoring RISC over CISC posits that simple computers with simple instruction sets are simpler to design and – although they require more instructions than CISC processors to execute more complex functions – RISC processors achieve much higher overall performance because they execute their instructions in many fewer clock cycles. In addition, because of their simple instructions, which simplify their hardware design, RISC processors can achieve much higher operating frequencies than can CISC processors in the same implementation technology, giving RISCs an extra speed advantage.

However, early RISC processors had a significant liability: they needed more instructions than CISCs – hence more instruction memory – to execute the same algorithms. Memory density is an advantage for CISCs with variable-length instructions, which is one of the reasons that the CISC microprocessor design style became so popular. Semiconductor memory was relatively expensive in the 1970s, during the first decade of the microprocessor. This factor, which favors CISC design, lessened throughout the 1980s as main-memory capacity grew rapidly and costs fell thanks to Moore's Law. At the same time, compiler writers improved the RISC compilers. These new compilers produced significantly better-optimized code with smaller memory footprints and, coincidentally, even better performance.

In addition, the HLL-specific computer architecture approach fell out of favor during the 1980s. By that time, some of the most extreme CISC instructions (like those of DEC's VAX, discussed earlier) consumed tens or hundreds of clock cycles. However, these specialized instructions would rarely be used by programmers writing in assembler or by compiler-generated code (possibly never). Consequently, expenditure of silicon area and power on these rarely used instructions in a high-performance ISA is clearly a bad idea because it wastes expensive resources on unused features. Ultimately, the notion that complex instructions made the compiler writer's work easier was disproved and abandoned. Complex instructions often proved insufficiently general to be truly useful and compiler writers elected to build code generators using a subset of the CISC processors' simpler, RISC-like instructions.

As discussed earlier, CISC remnants mostly exist as a vestigial part of the Intel x86 architecture – namely its original instruction set. Although modern x86 processors still accept these instructions for legacy-code compatibility, the instruction decoders in these processors immediately decompose each CISC instruction into one or more RISC operations, which are the modern x86 processor's true native instruction set. These RISC operations are then scheduled and issued to the processor's parallel execution hardware. Almost all other significant 32-bit microprocessors and embedded processor cores are now RISCs (IBM/Freescale PowerPC, Sun SPARC, MIPS, ARM, Tensilica Xtensa, etc.).

RISC processor designers are themselves just as guilty of extremism. One of the first avenues RISC designers explored to boost performance was through the use of parallel function units, which created superscalar RISC processors. With parallel resources, an optimizing compiler can look for and exploit additional ILP by assigning operations that can be executed in parallel to the additional execution units. Even with 4-way parallel execution engines, superscalar RISC implementations generally achieve no better than 1.5 instructions per clock (IPC) and 6-way superscalar processors achieve no more than 2.3 IPC [269]. That's a lot of additional execution hardware for relatively little gain (as wasteful as the infrequently used instructions in CISCs). The extra function units on superscalar RISC processors are idle a lot of the time because the optimizing compiler just can't find enough parallelism in the code to keep the extra function units busy.

Having exhausted all of the avenues for extracting parallelism from compiled code, RISC designers have started putting crystal balls in the form of speculative circuits into their designs. The earliest such circuits were branch predictors. If the processor can predict which way a program will branch, it can prefetch the first instruction at the branch target and avoid a branch bubble in the pipeline.

A static branch predictor "guesses" that the branch will always be taken and achieves about 65% accuracy, far short of the accuracy needed to make branch prediction worthwhile because of the time penalty for mispredicted branches. Dynamic branch-prediction circuits can achieve prediction accuracies of up to 95%, which can improve an RISC processor's performance in exchange for not much additional hardware.

However, designers of extreme RISC processors aren't content with just a little prognostication. After adding parallel execution resources to create superscalar RISCs, designers feel an overwhelming need to find something for these precious resources to do. One way to keep these function units busy is to let them speculatively execute code by letting the processor run code traces from both forks of a branch. The results from the fork not taken are then simply discarded (and the energy devoted to those calculations is therefore wasted).

Not content with just speculating on the code's execution path, some processor designers have developed designs for superspeculative RISCs, which speculate on the code's execution path and on the operand values the speculatively executed code will use. Superspeculation sidesteps data-dependency issues by assuming that operand values can be predicted before they are computed, pushing RISC processors even further into the realm of statistical operation.

Statistical program execution wastes execution cycles, which in turn wastes power. So, instead of wasting silicon resources on unused instructions the way extreme CISCs do, extreme RISCs waste execution cycles and energy all in the name of ultimately speeding program execution. Perhaps such speculation is warranted if there's a single task that requires more speed than the processor can supply and speculative execution can add the needed speed. Often, the extra speed is desired to give the processor enough headroom to execute multiple tasks, which is another flawed design philosophy held over from an earlier time that is discussed in the next section.

To badly paraphrase onetime USA presidential candidate Barry Goldwater: "Extremism in the pursuit of multitasking is a vice; moderation in the pursuit of processor efficiency is a virtue."

Problem 5: Very long instruction word (VLIW) (Medusius Horribilis)

Designers favoring the "one big processor" design approach noted that superscalar approaches – which rely on hardware to schedule parallel operations – quickly grow overly complex and cannot recognize opportunities for concurrency that compilers can statically recognize, optimize, and schedule prior to runtime. Thus was born VLIW: very long instruction word architectures. VLIW processor architectures represent a way to exploit ILP that fundamentally depends on software; that is, the amount of realized ILP depends on the quality of results that a modern compiler can produce and not on the number of transistors thrown at the problem. Given a plethora of execution resources, compilers can schedule multiple operations to be executed simultaneously and can place these operations into one long instruction word. The length of a VLIW instruction word is usually at least 64 bits but can be longer depending on the number of parallel execution units available in the VLIW processor's architecture and the particular layout of operations within an instruction.

Each VLIW operation usually requires many fewer bits than a standard 32-bit instruction so multiplying the number of operations in a VLIW word by 32 to determine the VLIW instruction length may be overkill, but a VLIW instruction may be sizeable nevertheless. For example, Silicon Hive's Avispa+ VLIW processor has a 768-bit instruction word [429], which controls 60 operation slots. This example is an outlier; more typical VLIW machines generally have three to five concurrent operation slots.

As discussed in Hennessy and Patterson [188] and Wikipedia [456], VLIW architectures date from the early 1980s, from work by Josh Fisher of Yale University. Fisher later co-founded Multiflow, which produced VLIW machines in the late 1980s. Multiflow's computer could issue 28 operations concurrently with each instruction, but unfortunately the machine was a commercial failure. Fisher then joined Hewlett-Packard Labs and continued his pioneering VLIW work.

Despite the initial Multiflow failure, VLIW lives on in various microprocessors. For example, NXP Semiconductors' (formerly Philips') TriMedia processor has a 64-bit instruction word containing five operations; Intel's Itanium 2 processor uses 256-bit double bundles containing as many as six operations; and TI's C64xx is a VLIW DSP with eight operation fields per 64-bit instruction word.

There are two problems with VLIW architectures of particular note:

- Depending on the application and compiler quality, the amount of ILP that can be extracted from an application may actually be quite low, making it nearly impossible to fully utilize the VLIW processor's multiple execution units. VLIW operation slots that can't be gainfully employed in a particular VLIW instruction are filled with no-ops (NOPs), which leads to code bloat. For example, if the average extractable ILP for an application program is three concurrent operations in a 5-issue VLIW machine, 40% of the potential concurrency is wasted.
- As a knock-on effect, a VLIW program containing many NOPs will be significantly larger than the same program compiled for a single-issue machine unless the VLIW processor designers have developed some very clever instruction coding techniques.

Clearly, problems of low ILP and the resulting VLIW code bloat depend a lot on the application, compiler quality, and the particular VLIW architecture. Data-crunching algorithms with a lot of regular loop nests are often amenable to SIMD processing and might map well into some kinds of VLIW architectures. Heavily control-dominated applications laced with irregular branching code may produce very sequential compiled code, forcing NOPs into most of the VLIW operation slots and resulting in bloated code.

Some approaches to VLIW design use clever instruction encoding or multiple fixed-instruction subsets to reduce code bloat, although many of the processor's execution units will still sit idle due to the NOPs. Clever instruction encoding coupled with VLIW design spends more gates both for the parallel execution units and for the special instruction-decoding hardware but does not necessarily consume much extra memory. The achieved ILP can still be somewhat disappointing, despite clever instruction encoding.

It has been suggested that VLIW architectures should have a maximum of three to five concurrent execution units (or some more irregular collection of execution resources – for example, some number of integer ALUs and perhaps a floating-point unit) to increase the probability of effective utilization, given the relatively low levels of ILP that are generally extractable from code. This view may be a natural lead-in to the concept of the ASIP, which matches the resources of the processor to an intended application or at least an application domain. This design approach makes the problem of creating a compiler more complex, but a combination of compiler smarts and manual invocation of task-specific instructions may be a viable solution for many well-understood application problems.

Another option open to modern SoC architects is to use the extra gates that might otherwise be consumed by a VLIW processor to instead create two or more simpler processors. A suitably partitioned application can exploit chip-level multiprocessing (CMP) without incurring some of the VLIW complexities mentioned above. A judicious combination of simple VLIW concepts in a multiprocessor SoC architecture might be the optimal combination of processors and processor complexity for the application. Automated creation of ASIPs coupled with the ability to create various Multi-Processor System-on-Chip (MPSoC) system architectures gives today's architects a wide range of choices and may allow them to avoid over-reliance on what may be a disappointing VLIW machine.

Problem 6: Overly aggressive pipelining (Canalisus Extremus)

The designers of the IBM 7030 (Stretch) computer were the first to use processor pipelining, back in 1961. They used this technique to raise the processor's clock rate by restricting the amount of logic operating within each pipeline stage during each clock period thus cutting the transit time through each stage. Because latches isolate each pipeline stage, each stage works on a different instruction during each clock period, which increases parallelism and therefore increases an important figure of merit for processors: IPC.

Pipelining became popular in microprocessor design with the advent of RISC processors, which standardized on 5-stage pipelines early on. A classical, 5-stage pipelined RISC processor might have the following pipeline stages:

- IF – Instruction fetch (usually from instruction cache or local memory)
- RD – Read the source operands from the register file
- ALU – Perform the operation specified by the instruction
- MEM – Read memory (for a load) or write to memory (for a store)
- WB – Write back, write operation result to the register file

Each of these stages performs a complete and well-defined task. However, it's possible to make pipelining work even faster by dividing each task into subtasks and each stage into sub stages, which further reduces the amount of logic within each stage and thus boosts maximum clock rates to ever more stratospheric heights. However, this approach also increases the number of clocks needed to complete the execution of each instruction. The net effect is one of diminishing returns with respect to the actual time needed to complete the average instruction.

Nevertheless, longer pipelines become quite attractive when the most important figure of merit for processor "goodness" is clock rate, as it was for IBM PC processors throughout the 1990s. This state of affairs explains the design of Intel's Pentium 4 and its NetBurst architecture.

The Pentium 4 processor is based on a microarchitecture that Intel calls P7. Looking at three succeeding Pentium microarchitectures – P5, P6, and P7 – clearly illustrates the measures Intel was willing to take to win the clock-rate war against AMD. The P5 microarchitecture, used in the original Pentium processor, had a 5-stage pipeline: instruction prefetch, two decode stages, an execute stage, and a write-back stage. This microarchitecture took Intel's Pentium to about 100 MHz.

Intel's PentiumPro (P6) microarchitecture represented a complete rethink of the x86 processor architecture. The P6 microarchitecture essentially transformed the Pentium into an RISC machine with a front-end chopper/shredder that transformed the original x86 CISC instructions into simpler, RISC-like operations. The processor fed these simpler operations into a reorder buffer and then distributed the RISC operations to a bank of parallel execution units, as is typical of superscalar architectures. The resulting P6 pipeline had 12 stages. The P6 microarchitecture served as the foundation for Intel's Pentium II and Pentium III processors. Over a three-year period (1999–2002), the P6 microarchitecture took Pentium clock rates from 133 MHz to 1.4 GHz – a 10:1 jump.

By then, Intel clearly had an edge in the clock-rate war against AMD, but the P7 microarchitecture – with an 8-stage CISC-to-RISC translation engine followed by a 20-stage, out-of-order execution pipeline for a total of 28 pipeline stages – was brought online to seal the deal [153]. The Intel Pentium 4 based on the P7 microarchitecture appeared in November, 2000. The processor's initial clock rate was 1.4 GHz, a speed not attained by the Pentium III until early 2002. Pentium 4 processors running at 3.8 GHz were available by November, 2005. The clock-rate climb ended there. Intel had planned a 4-GHz Pentium 4 but it disappeared from Intel's road map and was never introduced [243].

Two key factors ended Intel's hyperpipelined ascent. The first was a loss of processing efficiency [154]. The P7 microarchitecture's 20-stage out-of-order execution pipeline contains two stages where no tasks are performed. These stages are called "drive" stages and they exist simply to allow signals to travel from one part of the chip to another. Instead of allowing time in these pipeline stages for logic delay, the two drive stages accommodate wire delay. That means that the processor performs no computation whatsoever during 10% of each instruction's execution lifetime (two out of 20 stages inside the P7 execution pipeline).

Conditional branches cause big problems for long pipelines. Although all complex processors have branch-prediction logic, branches cannot be predicted with 100% certainty; every mispredicted branch causes pipeline bubbles. No useful work is performed inside a pipeline bubble; the longer the pipeline, the bigger the bubble. Overall, the P7 pipeline was far less efficient than the P6 pipeline, which made the Pentium 4 processor less efficient as well. Clock rate alone does not equal processor performance in the same way that an automobile's engine RPM alone does not equal vehicle speed. For a car, there's a transmission, differential, and tire diameter that translate RPM into MPH or KPH. In the case of a processor, the pipeline and the associated memory subsystems transform MHz (or GHz) into instructions/second.

The second factor that stopped the clock-rate climb was power dissipation, which rises superlinearly with clock rate due to the associated core operating voltage, which also depends on frequency. At 3.8 GHz, the Pentium 4 dissipated 115 W and drew 119 A from its motherboard. The 4-GHz processor was headed past 150 W before Intel killed it. At the same time, the amount of work done by the processor per clock had fallen precipitously. The Pentium 4's hyperpipelined design had gone well past the point of diminishing returns (in terms of real work done per second), and all was done for the sake of the fastest possible clock rate.

Consequently, the trend towards overstretched pipelines has hopefully been curtailed by Intel's experience with the P7 microarchitecture. It takes a lot of circuitry to keep a long pipeline running smoothly and all that logic can be put to better uses. One use that looks promising is to create additional on-chip processors. Intel is now devoting itself to multi-core Pentiums, AMD is concentrating on multiple-core Opterons, and the SoC industry is focused on MPSoC design.

Problem 7: Unbalanced processor design (Librius Tiltus)

Processor pipelines seemingly receive a disproportionate share of the limelight as the glamorous darlings of the processor-design community.

However, a processor's performance depends on more than just its execution pipeline. As with any engineering discipline, good processor performance depends on balanced design. Many factors contribute to a processor's (or a system's) overall performance and any of these factors can poison the operational efficiency of the "perfect" pipeline running real-world applications if it's not in balance with the others. Designers must employ an expanded range of design decisions and new technologies to produce balanced, cost-effective systems [60].

Advances made in processor designs over the past decade – both circuit advances, which have caused clock rates to rise at roughly 30% per year from 1985 to 2005 [129] and architectural improvements including wider instructions, VLIW architectures, and speculative execution – have increased microprocessor instruction-issue rate much faster than the rate of increase in main-memory bandwidth or the rate of decrease in main-memory access latency. Consequently, microprocessor accesses to bulk or main memory have become temporally expensive. This trend forces architectural and system-level design changes including:

- Wider connections to main memory (more pins)
- Larger and more efficient instruction and data caches
- Memory-centric system architectures

Each of these new approaches delivers benefits and incurs costs.

Burger et al. [60] divide a processor's execution time into three components, which help explain how a processor's design might be better balanced. The three components are:

- Processor time – when the processor is either fully utilized or partially utilized and partially stalled due to a lack of ILP
- Latency time – the time lost to contention-less memory latency (latency that cannot be reduced by increasing memory bandwidth between memory-hierarchy levels)
- Bandwidth time – the time lost to memory contention plus the time lost because of inadequate memory bandwidth between memory-hierarchy levels

Many "modern" processor-design techniques exacerbate problems for all three execution-time components. Speculative software and hardware prefetching techniques can improve a processor's performance by making sure that data is in fast cache memory when needed but these techniques increase traffic to main memory and waste bandwidth when they prefetch unneeded data, prefetch data that's evicted before it's used, or evict other data in the processor's caches before it can be used (forcing it to be refetched). Multithreading increases processor throughput by switching to a ready thread whenever a thread stalls due to a long-latency memory access or I/O operation but frequent thread switching thrashes the processor's cache and TLB. As a result, the gains achieved from multithreading can be partially or completely offset by reductions in cache efficiency. For a detailed explanation of caches used for processors, see [176].

The blind quest for high clock rate also drives a processor's design out of balance. As processors get faster, they consume instructions and operands at a faster rate, which puts additional pressure on main-memory latency and bandwidth requirements. A recent trend towards homogeneous, multi-core processors with coherent caches and common main memory also increases main-memory latency and bandwidth requirements. Experiments by Burger et al. [60] indicate that aggressive design techniques that place pressure on the processor-to-main-memory interface can result in processors that are stalled and waiting on memory as much as 50% of the time. Such systems are clearly out of balance.

Deep-submicron and nanometer circuit effects are also driving conventional processor design out of balance. Interconnect delay, previously insignificant, now dominates over gate delay because wire delay does not scale with feature size unless the aspect ratio of the interconnect wire's cross section changes. As the previous section discussed, designers of Intel's Pentium 4 microprocessor, which achieved a commercial clock rate of 3.8 GHz, were forced to devote two stages (10%) of the processor's 20-stage execution pipeline to accommodate on-chip wire delay.

As deep-submicron and nanometer design rules have allowed clock rates to rise, microprocessor designers have resorted to lowering core operating voltages in an attempt to limit power increases. However, doing so requires the use of transistors with lower threshold voltages to accommodate the lower core operating voltages. In turn, the lower threshold voltages increase leakage currents to the point that a processor's power dissipation due to leakage is roughly equal to its dynamic power dissipation at 90nm lithography levels and below. Such processors dissipate a lot of power even when doing nothing.

All of these choices and consequences force new directions in processor architecture. First among these is the need for more efficient caches. The fraction of data held in current caches that is "live" (will be referenced again before being evicted) is between 0.05% and 33% [60]. That means that most of the processor's caches are normally filled with dead, useless data and instructions. Improved cache-management logic can reduce the traffic between the cache and main memory by a factor of 2 to 100x and is therefore a wise expenditure of silicon [60]. One way to increase cache efficiency is to cache objects with a finer granularity than a cache line, which increases cache-tag overhead but reduces bandwidth requirements. Software-managed caches can be even more silicon-effective.

Moving a processor's main memory on chip with the processor can effectively make the entire on-chip main memory into a cache, which eliminates the memory hierarchy and substantially reduces memory overhead. Pushing design further in this direction produces memory-centric architectures that embed microprocessors into individual memory arrays. This is the domain of the MPSoC, which seeks to distribute the total processing load to a large number of small, inexpensive processors running at reduced clock rates. Consequently, MPSoCs may well be the high-performance processor architectures of the future.

Problem 8: Omitting pipeline interlocks (Recompilus Requirus)

Pipelined processors use hardware load-use interlocks to prevent instructions from executing until the required operands become available. Operands can be fetched from a register or register file, they can be fetched from memory, or they can be generated as results by earlier instructions. (For RISC processors, which have load/store architectures, memory operands are only associated with load and store instructions.) For operands located in the processor's register file, data forwarding and bypassing within the pipeline can avoid the hazards that might invoke a pipeline interlock. However, memory loads typically take too long to provide data to the instruction immediately following a load instruction because of the relatively long latency of memory-read cycles compared to register-read latency.

Data-hazard problems can be solved either by stalling the pipeline using a hardware load-use interlock or in software, using the compiler's instruction scheduler to schedule only instructions that do not need the result of the load operation to immediately follow that load instruction. If no such instruction is available in the existing compiled code, the scheduler inserts a NOP, which is guaranteed not to need any data at all.

Data dependencies inherent in all programs limit the amount of instruction reordering a code scheduler can perform. For single-instruction-issue processors, a scheduler inserts independent instructions after multi-cycle instructions (such as loads) to reduce pipeline interlocks. For multiple-instruction-issue processors, a scheduler must identify independent instructions that can be concurrently executed in addition to using instruction scheduling to reduce or eliminate interlocks. High-issue-rate processors provide performance improvements only when the scheduler is able to find sufficient instructions to concurrently execute multiple operations.

Jump and branch instructions also cause pipeline hazards. The address of the instruction to be executed after a jump or a taken branch must be computed and the instruction must then be fetched before execution can proceed. The processor cannot fetch the jump or branch instruction, decode it, compute the effective target address, and then fetch the instruction stored at the computed target address in time for the next instruction slot. Thus jump and branch instructions create control hazards, which can be solved either by stalling the pipeline with a hardware interlock or by creating a delay slot. The delay slot is the instruction slot following the branch instruction. Processors can make use of delay slots by always executing the instruction in the delay slot, regardless of the branch computation. It's the compiler's job to find something useful for the processor to execute in the delay slot.

Hardware load-use interlocks aren't strictly necessary. It is possible to simply organize executable code so that instructions and operands arrive at the right place in the processor's pipeline at exactly the right time. This is a job for the processor's compiler, which can rearrange instructions and insert NOPs where needed to prevent the data hazards that require interlocks.

It is possible to organize code to eliminate the need for interlocks, but in practice it's extremely difficult. In fact, optimal compile-time reorganization of code to eliminate the need for interlocks is an NP-complete problem although heuristic algorithms can do the job adequately [186]. In addition, even perfectly organized code is subject to interrupts and other exceptions, which lay waste to the compiler's carefully planned instruction sequences.

Nevertheless, two well-known commercial pipelined microprocessor architectures lacked interlocks, at least initially. These two processor architectures were the original 32-bit MIPS architecture and Intel's i860 processor. (The acronym MIPS originally meant "Microprocessor without Interlocked Pipeline Stages" [185].) The first MIPS processor, developed at Stanford University by Professor John Hennessy and his students, was designed in the early 1980s and was architected to deliver high performance within the constraints of the NMOS VLSI semiconductor processing available at the time.

This first MIPS architectural implementation used a 5-stage pipeline for speed and kept three instructions in the pipeline whenever possible. (Either the odd or even pipeline stages were active on each clock cycle.) The MIPS designers shifted as much of the processor's capabilities as they could from "expensive" hardware to the compiler to save transistors at a point in semiconductor development when transistors were relatively costly. Pipeline interlocks were one of the features that the MIPS designers moved to software. A piece of software called a "pipeline reorganizer" was responsible for examining instruction streams, identifying data hazards, and eliminating the hazards by reorganizing instruction sequences and adding NOPs where needed.

Normally, the MIPS processor's single-cycle instructions didn't create data hazards but load instructions had to be followed by a NOP if the next instruction used the value obtained by the load operation. Multi-cycle instructions would have created headaches using this scheme however and many more NOPs would have been needed. As a consequence, the MIPS designers culled multi-cycle instructions from the processor's repertoire. The original MIPS processor had no divide or floating-point instructions as a result. The original MIPS design used delay slots to avoid control hazards.

John Hennessy founded MIPS Computer Systems in 1984 to commercialize the MIPS RISC architecture. The first commercial MIPS processors (R2000, R3000) lacked most pipeline interlocks like the original MIPS design [217]. The R2000 and R3000 processors did have separate multi-cycle multiplication and division units that worked with special interlocked result registers named HI and LO. Subsequent MIPS processor generations added pipeline interlocks as semiconductor lithographic advances made the relatively paltry saving of interlock transistors superfluous.

Designers of Intel's i860 microprocessor, introduced in 1989, had similar goals: speed and design simplicity. However, the first i860 processor was a 1-million-transistor, 40-MHz, VLIW machine [30]. It wasn't all that simple. The i860 architecture, introduced in 1989, encompassed two separate pipelines: a 4-stage, 32-bit scalar pipeline (with one execution stage) and a 6-stage, 64-bit floating-point/graphics pipeline (with three execution stages). The processor's floating-point instructions did not follow the canonical RISC 3-operand instruction format (two source operands and one destination operand, all for *that* instruction). Instead, the i860 processor's floating-point instructions specified two source operands for current instruction *I*, but the instruction's destination operand specified the destination for the instruction that was executed *three cycles previously*. Intel named this approach "explicit pipelining."

The advantage of the i860 processor's explicit pipelining mode was that it allowed the processor to pipeline multi-cycle floating-point operations. It could start execution of a new floating-point instruction each cycle and the result of each instruction became available three cycles later. At that point, the floating-point instruction three instruction slots later would then store the result of the earlier instruction in the appropriate place.

Explicit pipelining placed the burden of scheduling and managing result latency squarely on the compiler and incidentally removed the need for hardware pipeline interlocks. If the compiler was already managing the process, no hardware assist was needed and the Intel i860 designers were happy to forego the interlock transistors. Unfortunately, interrupts and exceptions were the i860's downfall. The explicitly pipelined processor could not efficiently handle unpredictable instruction streams and its performance suffered in real-world applications. Intel subsequently discontinued the i860 processor family.

In the end, a lack of pipeline interlocks seems to cause more trouble than it's worth. Like many of the species of processor design discussed in this chapter, a liberal application of Moore's law has wiped out this particular branch in the tree of processor design.

Problem 9: Non-power-of-2 data-word widths for general-purpose computing (Datus Unusualus)

The world of general-purpose microprocessors and embedded processor cores has generally settled on power-of-two data-word widths: 8, 16, 32, 64, or 128 bits or the like. Extremely specialized processors or instruction-set extensions to configurable processors sometimes use a less regularly sized data word for specialized instructions, input and output word sizes, and internal registers (even quite irregular and odd sizes such as 13, 17, or 23 bits). However, these specialized, non-power-of-two word sizes are used only where required for very specific algorithms – often for data-intensive signal, audio, or video processing – where the required precision and range demands the use of a highly-optimized data size to meet system-level cost, power, and performance objectives.

For instance, 24-bit processing seems to be a standard data-word width for high-end audio applications. The 24-bit data word allows for extensive audio processing without audible degradation of the recorded sound and without the overkill of a 32-bit data word. Although the audio samples themselves tend to use only 16 bits (there are 20- and 24-bit audio ADCs and DACs, but they're not yet common), extra bits are needed to provide enough head room to maintain superior sound quality through the intermediate calculations executed by popular digital compression, decompression, and other audio-processing algorithms. However, for general-purpose computing, power-of-two data sizes seem to be the norm.

This was not always the case. The early and middle parts of the computer processor's evolutionary tree contain branches encompassing several different popular word sizes including 36- and 48-bit words. Early scientific computers often had a word length of 36 bits, which was long enough to allow positive and negative integers to be represented with an accuracy of ten decimal digits. In addition, 36 bits allowed six alphanumeric characters to be stored, if they used a 6-bit character encoding.

In the early days of computing, achieving desired accuracy while minimizing hardware costs was desirable because hardware was expensive. Computers using 36-bit word lengths included the IBM 701/704x/709x, UNIVAC 1100 and 2200 series, GE 600, Honeywell 6000, and DECsystem-10/20. Often these machines used 18-bit word addressing rather than byte addressing. A variety of character encodings were possible within a 36-bit word including six 6-bit characters, five 7-bit ASCII characters, four 8-bit characters (either 7-bit ASCII plus 1 unused bit per character or 8-bit EBCDIC (Extended Binary-Coded Decimal Interchange Code, a pre-ASCII character set derived from IBM punched cards)), or four 9-bit characters. Classical Burroughs mainframes, already discussed, used a 48-bit data word, which provided admirable scientific precision and could store six EBCDIC characters.

One of the main issues associated with using a non-power-of-two data word relates to character or byte indexing for a machine that accesses data memory using word-style, rather than byte-style, addressing. Finding out which word a particular character or byte lies in, when it's part of a structure, relies on dividing the byte or character index value by the number of bytes in a word. When the data word contains a power-of-two number of bytes, the index computation can be performed with a simple shift operation that can almost always be performed in one cycle. When the data word contains a non-power-of-two number of characters or bytes, this computation involves a far more complex division operation – for example, division by 5 or 6. Efficiently performing this more complex division operation requires either specialized hardware in the processor – divide by 5 or divide by 3 (a division by 6 is a divide by 3 followed by a shift to further divide by 2) – or the execution of a whole command sequence that accomplishes the same division using multiple arithmetic operations. Such a sequence might require several clocks for every character indexed. If programs use a lot of character or byte indexing, then the absence of specialized divide-by-*n* hardware, where *n*=3 or 5 or any other non-power-of-two number, results in slow and woefully inefficient implementations.

Low-end Burroughs E-mode-compatible machines definitely had this problem when used either for commercial processing (lots of COBOL code, with lots of character manipulation) or for general C execution, which uses a lot of byte addressing. The 36-bit machines tended to use 9-bit character encodings to simplify and speed up byte indexing when C became a popular programming language. Using 9-bit characters allowed byte indexing to be performed with a simple divide by 4 (2-bit right shift).

Such issues are not important for specialized algorithmic instruction extensions on configurable processors because data samples are manipulated directly via hardware compiled specifically for that purpose.

The general reduction in species diversity that the microprocessor and embedded processor core community has overseen since the 1980s has produced machines with power-of-two data-word widths. Consequently, the need to design specialized divide-by-*n* hardware to support general processing in languages such as C or COBOL has disappeared. This is one evolutionary branch that seems unlikely to return.

Problem 10: Too small an address space (Datus Minimus)

Microprocessors with abbreviated address spaces are very much like shoes that are too small – you can end up with blisters or abrasions when trying to shoehorn code and data into their small-memory spaces. If you're like one of Cinderella's sisters, you might even lose a toe or two. And like Cinderella's sister, the result is also not very pretty!

Bell and Strecker [46] express the problem of cramped address space this way: "There is only one mistake that can be made in computer design that is difficult to recover from – not having enough address bits for memory addressing and memory management."

Yet in spite of the difficult recovery, processor designers have made this same design mistake time and time again. This species of design error refuses to die. Hennessy and Patterson ([188], p. 501) provides a long but not exhaustive list of similarly afflicted processor designs: DEC's PDP-8, PDP-10, and PDP-11; Intel's 8080, 8086, and 80186; Motorola's 6800; MOS Technology's 6502; Zilog's Z80; and Cray's Cray-1 and Cray X-MP.

The reason that this particular design mistake is so critical is that in addition to determining the maximum addressable memory space for machines with fixed-length instructions, the bit width of a processor's address determines the minimum width of everything inside (and outside) of the processor having anything to do with the address including the size of the address word (and hence the instruction-word size), the size of branch offsets, the program counter, any other registers that must hold addresses, any arithmetic logic that computes effective addresses, the size of the MMU, and even the width of the address bus and the size of the IC package for packaged microprocessors. The repercussions associated with the processor's maximum addressable memory size permeate the processor's design. Once set, it's very hard to change such a bedrock architectural decision.

Processor designers limit address fields primarily to reduce instruction size, register size, and pin count (in packaged microprocessors). In making this decision, designers almost invariably defer to available memory costs and capacities of the day rather than considering long-term address-space requirements, in spite of the unfailing advance of Moore's law over more than four decades. They also tend to think only of the current kinds of programs, program sizes, and data-memory requirements rather than trying to think ahead to the future. Such short-term thinking always causes problems in the long term. We know from the history of Microsoft operating systems and applications that what seems adequate in performance and memory size for one generation of software proves to be totally inadequate for the next release.

For example, even though it was introduced way back in 1969 and was sold at the relatively low price of $5000, Hewlett-Packard's first desktop calculator, the HP 9100A, had a shockingly powerful 64-bit, floating-point, VLIW processor [260]. A major part of the HP 9100A development project was devoted to shoehorning all of the product's software into the machine's unbelievably small instruction-address space, which was a paltry 512 64-bit words (4 Kbytes). For its code storage, the HP 9100A used a specially designed inductive ROM, which was embedded in a 16-layer pc-board. (Note: The circuit board was the memory. Semiconductor memories had only just appeared and had nowhere near the required capacity.) The pc-board ROM's 4-Kbyte capacity represented the inductive technology's maximum storage limit, so that was the addressable-instruction capacity hard-wired into the machine's VLIW processor. No consideration was given to future expansion ability. Factory-cost considerations overrode all others.

The HP 9100A was so successful that HP decided to continue and expand its line of desktop calculators, which quickly evolved into desktop computers. However, when it came time to build the next generations of desktop computers, HP was forced to abandon the non-expandable HP 9100A's 64-bit, VLIW processor architecture. Instead, the company adapted the existing 16-bit HP 2116A minicomputer architecture, which had a 32-Kword (64-Kbyte) address space. In less than five years, HP's desktop computer designs hit the HP 2116A processor's address-space limits so HP expanded the architecture's address space with a series of block-switching circuits and MMUs. HP then adopted a third processor architecture – the Motorola 68000 with a 32-bit (4-Gbyte) address space, truncated to 24 bits (16 Mbytes) due to pin limitations on the original processor's package; then used a fourth, home-grown 32-bit CISC processor architecture for its desktop machines; and then a fifth architecture called PA-RISC for its workstations and servers.

A similar fate befell the IBM PC. The first IBM PC, named the model 5150, used an Intel 8088 microprocessor with a 20-bit (1-Mbyte) address space. Within that address space, the designers of the IBM PC devoted 640 Kbytes to application memory. Microsoft's Bill Gates is often quoted as asking, "Who needs more than 640K?" (After all, the IBM PC's 640-Kbyte application space was ten times larger than the 64-Kbyte space in the many successful personal computers that used 8-bit processors.) As it turns out, everyone eventually needed more than 640 Kbytes of memory in their personal computer. Subsequent models of the IBM PC used the Intel 80286 (16-Mbyte address space) and then the 80386 (4-Gbyte address space), but the damage had been done during the initial design of the first machine.

The IBM PC's 640-Kbyte application space determined:

- The structure and basic operation of the ascendant Microsoft DOS operating system
- The design of the original IBM PC expansion bus (eventually named ISA – industry standard architecture – not to be confused with the other ISA – instruction-set architecture)
- The architecture of memory cards designed for the ISA bus
- It even extended the longevity of the use of DIP memory chips

Eventually, with the introduction of Microsoft Windows, some hardware re-architecting, and an improved expansion bus, mainstream PCs started to use more than 640 Kbytes of application memory but it took a more than a decade to break all of the bounds set by the original model 5150's design.

In the embedded-design world, the 8-bit Intel 8051 microcontroller is perhaps the greatest example of the Frankenstein-like consequences of limited address space. The 8051 is the successor to Intel's first microcontroller, the 8048, which had a 12-bit (4-Kbyte) address space. The original 4-Kbyte address space was set just beyond the semiconductor-manufacturing capacities of the day, because the 8048 was designed to run code exclusively out of on-chip ROM, so there was no point in making its address space larger than the maximum on-chip ROM capacity.

The march of Moore's law put a quick death to that on-chip capacity limit and the 8048's architecture immediately fell short of system-design requirements. Intel revamped the 8048 architecture and produced the 8051, which has separate 64-Kbyte instruction- and data-memory spaces. The radically superior 8051 microcontroller rapidly became a popular choice for dot-matrix printer designs and its use then spread throughout the embedded-design world.

The 8051 architecture stayed popular for so long (currently more than three decades and counting) that its 128-Kbyte address space started to pinch. At first, embedded-systems designers added external block-switching circuits to extend the 8051's address space. Subsequently, several vendors of 8051-compatible microcontroller ICs incorporated block-switching into their devices, extending the address-space reach of their products into the multi-Mbyte region while making the embedded programmers' lives quite miserable as they tried to manage the many memory blocks without causing resource conflicts.

Even though saddling a processor with too small an address space is considered a design mistake, it can hardly be considered inevitably fatal. Note that each of the above examples of a shortchanged processor actually describes a commercially successful product. Consequently, processor designers will likely continue to make this mistake far into the future, thus needlessly complicating the lives of the people who design with and use these processors for many years to come.

Problem 11: Memory segmentation (Datus Minimus Rex)

Segmented-memory architectures often arise from patches to initial processor architectures that suffer from address spaces that were inadequate from the start or from address spaces that have become too small over the long useful life of a successful processor. By far, the most famous example of this design blunder is Intel's 8086. Developed in the late 1970s as an early 16-bit processor, the Intel 8086 designers sought a way of expanding beyond the 8-bit processors' 64-Kbyte (16-bit) address space without incurring the overhead of a large address field, which would fatten the processor's instruction-word width and increase the number of transistors needed for internal registers.

One approach that might have been used to solve this problem is block switching. The original 16-bit address space could have been designated block 0 and a 4-bit register could have been added to create a 16-block address space, giving an overall address space of 20 bits by combining the 4-bit block address (representing the most-significant bits of the full address) and the existing 16-bit address value for the least significant bits; crude but effective. This simple approach has since been used to expand the address space of Intel's 8051 microcontroller, but it's not the approach used by the designers of the 8086 microprocessor.

Instead, the 8086 designers decided to use segmented memory, which splits the 20-bit address into a segment address and an offset address. The offset address corresponds to the 16-bit address field universally used by 8-bit processors. The 8086 used a 16-bit segment address to modify the offset address, creating a 20-bit (1-Mbyte) address space by shifting the segment address left by four bits and then adding the 16-bit offset to produce a 20-bit address.

The 8086 contained four segment registers defining different active segments for code, data, stack, and "other" (the "extra" segment). This scheme allowed the 8086 processor to have four separate 64-Kbyte address spaces (a total of 256 Kbytes) active simultaneously and it allowed software to quickly switch from one memory space to another simply by changing one of the segment register's contents. Programmers quickly found the task of keeping track of four memory independent spaces to be more than four times harder than managing one large, flat memory space.

While memory segmentation did expand the 8086 processor's address space to 20 bits, it did not adequately solve the underlying problem. Two important reasons for giving a processor a larger address space are to allow the processor to run larger programs and to allow the processor to handle large data types such as image, video, and sound files. The 8086's segment registers allow the processor to handle larger programs and data files as long as the processor's operating system can reliably manage the jiggery-pokery needed to change the segment pointers as needed. However, this overhead saps the processor's bandwidth and invites programming bugs.

It's tempting to attribute the poor design of the 8086 processor's fixed-size, 64-Kbyte segments to the cold realities of late-1970s semiconductor processing. Zilog's contemporaneous 16-bit Z8000 processor also had a segmented-memory address space. However, Motorola managed to produce the 32-bit 68000 microprocessor with a flat 32-bit address space during the same period. The Motorola 68000 was simply a more aggressive design.

Inferior design or not, IBM selected the Intel 8086 (actually the 8088 variant with an 8-bit data bus) for its IBM PC, instantly making Intel's segmented-memory approach highly successful in spite of its difficult programming model. The 32-bit descendents of the 16-bit 8086 microprocessor still support segmented-memory addressing, but their offset registers now hold 32-bit offset addresses (rather than 16-bit offsets), which solves the problem of fitting large programs and data files into one segment, at least for most application spaces. However, some programs already need more than 32-bit data addressing. For example, some electronic design automation programs – design rule checking, as one type – have a hard time fitting into a 32-bit data space for very large chip designs fabricated in very aggressive IC-manufacturing processes.

There is a way of advantageously using memory segmentation in connection with a paged MMU, which provides virtual-memory management and memory-space protection as well as address translation. MMUs provide a much better, more comprehensive approach to memory management than simple segmented memory.

Problem 12: Multithreading (Clotho Replicatius)

Before discussing why multithreading is a problem in processors (or at least, not the right solution for many processing problems), we need to define what it is. From Wikipedia [455]:

> A thread in computer science is short for a thread of execution. Threads are a way for a program to split itself into two or more simultaneously (or pseudo-simultaneously) running tasks. The distinctions between threads and processes differ from one operating system to another, but in general, the way that a thread is created and shares its resources is different from the way a process does.

And then under Simultaneous Multithreading:

> Simultaneous multithreading, often abbreviated as SMT, is a technique for improving the overall efficiency of superscalar CPUs. SMT permits multiple independent threads of execution to better utilize the resources provided by modern processor architectures. .... Simultaneous multithreading allows multiple threads to execute different instructions in the same clock cycle, using the execution units that the first thread left spare.

Processor designers added multithreading to their designs as a remedy for diverging processor clock rates and memory-access times, caused by process-technology scaling. When a memory access took only one or just a few processor cycles, stalling the processor during loads and stores seemed a bit wasteful but did not constitute a crisis. Intelligent instruction scheduling and reordering by compilers operating on a single sequential program could often soak up the wasted cycles through out-of-order execution of instructions that weren't dependent on the results of the load or store. However, when system-memory accesses began to take tens or even hundreds of processor cycles, no amount of intelligent instruction scheduling by a compiler or by a hardware-based, out-of-order instruction scheduler in a superscalar architecture could fill all of the wasted cycles with useful work. The result: potential performance loss and energy waste.

To resolve this crisis and waste, processor architects developed multithreading strategies to maintain high processor utilization even when an execution thread stalled somewhere in the processor's pipeline. Moore's law permitted new generations of processors to run more tasks simultaneously by adding hardware state that allowed pipeline stalls to be sidestepped simply by switching to other program threads during the stall. Hardware support for rapid thread context switching allows multiple active and stalled threads to run on a single set of processor resources.

Hardware support for thread scheduling can do more than implement a simple round-robin scheme – it can take process or task deadlines and task-completion estimates into consideration to better meet real-time or near-real-time processing goals. Simultaneous multithreading can take instruction scheduling even further in a superscalar processor with multiple execution units and other processing resources that permit execution of two or more threads at once. SMT processors need not wait for threads to stall before beginning a second or third thread, which can take advantage of idle function units and other resources.

Various multi-core processors that support multiple thread execution recently announced include Sun's Niagara and its follow-on, IBM's Power5, Intel's Core 2 Duo Pentiums, AMD's Multi-Core Opterons, MIPS' 34K, and others. These processor vendors all seem to see multithreading – whether SMT or fast thread context switching, potentially every cycle, or even "chip multithreading" (where each core is running a different thread, but only one at a time) – as a key part of their multi-core evolution strategy. Clearly, if it was easy to partition applications into multiple concurrent threads, or easy to switch between threads on stalls without putting pressure on other resources, then multithreading would be a viable strategy to both exploit Moore's law and to prevent resource and energy waste.

However, there is another kind of multithreading that employs chip multiprocessing (CMP, another form of multi-core technology) without actually splitting an application into concurrent execution threads. That is, if each task has a single execution thread and processor cores can quickly switch from one thread to another, and all of the cores or processors share a common view of memory (otherwise known as Symmetric Multi-Processing or SMP), then idle threads can sit in a memory pool and be dispatched to any processor that stalls while running a thread. Because all processors have a shared and coherent memory view, any waiting thread can quickly resume execution on a new processor until that thread itself stalls. One thread runs per processor at any given moment, which results in a kind of chip multithreading (CMT) as described above.

As a concept, CMT running on an SMP multi-core machine may make a certain amount of sense for applications with the following characteristics:

- General-purpose applications that may not be known at the time the processor is chosen for the product or design.
- Applications with large, complex control structures rather than data-intensive computation.
- Applications without hard real-time task deadlines or with very soft task deadlines or soft quality-of-service requirements.
- Applications intended to be run on large, general-purpose servers, powered by line plug rather than battery.

Applications such as Web queries and other Web-based services are good candidates for this kind of machine. These applications do not have particularly hard quality-of-service characteristics as long as a certain responsiveness level is maintained; the applications may not even exist when the servers are designed. In such circumstances, the amount of processing per query is relatively low; applications may stall on database lookups and other memory intensive queries; when one user thread stalls, there may be hundreds of others waiting to be serviced. Large server farms are now built with hundreds, thousands, or even tens of thousands of servers, each of which may have several dual- or quad-core processors. Such servers may be more effectively built with a CMT approach rather than SMT approaches, because it may prove more efficient to have cheaper cores and more of them than more expensive SMT cores.

Although such applications may run reasonably well on SMP multi-core machines, such processor designs do not exhaust the application space of interest. In particular, deeply embedded applications in consumer products – for example, audio/video/multimedia and image processing – have a rather different set of characteristics from those discussed above:

- Data-intensive processing, often incorporating algorithms for known standards (for example, audio–video codecs), possibly written ahead of time or falling within reasonably well-known application domains.
- Applications with hard real-time constraints or with tight quality of-service demands.
- Applications that must run on portable consumer products, usually battery powered, where energy consumption is a very big concern.
- Applications that can be partitioned into pipelined dataflow tasks.

For such applications, multiprocessing rather than multithreading is the better architectural approach. And not just general-purpose multiprocessing (as in the SMP model), but application-specific, asymmetric multiprocessing (AMP) may well be the best solution and the best utilization of Moore's law. Rather than loading all of the application's tasks onto a single multithreaded core – for which a high clock rate (and very high energy consumption) will be required to meet processing deadlines – a set of application-specific, configured processors (ASIPs), running at much lower clock rates may be a superior solution.

If the processors are extended with specialized, application-specific instructions, they may run at clock frequencies far lower than would be required for one or two general-purpose cores, especially when the general-purpose processors are loaded with multiple tasks and threads. The reduced clock rate leads to lower required processor-core operating voltages and, consequently, much lower energy consumption.

Processor subsystems in AMP designs can be tailored for and dedicated to specific application tasks. Instruction tailoring reduces power dissipation by allowing a processor to execute a task in fewer clock cycles. When these tasks aren't running, the associated processor cores can be shut down completely, which saves additional power. This strategy is more difficult to achieve with a general-purpose, multithreaded core because such cores must run all relevant tasks and must therefore operate as "always-on" processors, as long as there are tasks to run. ASIPs designed for a specific application domain, if well-designed, can efficiently run both current and future versions of task-specific application code – such as new audio or video codecs.

The gates required to support multithreading on a general-purpose processor may be better used if dedicated to an ASIP. Thus a multiprocessor network of ASIPs can be a much better way to exploit Moore's law and semiconductor technology evolution than a general-purpose, multithreaded processor for many application domains of interest.

Problem 13: Symmetric multiprocessing (Multisaur Symmetrius)

Symmetric multiprocessing computer systems have been around for a long time. Many high-end supercomputing systems – built from multiple homogeneous processors with a high-speed interconnect network – use SMP as a fundamental architectural concept. More recently, as discussed earlier, on-chip multi-core designs such as Sun's Niagara, IBM's Power, AMD's Multi-Core Opteron, ARM's ARM11 MPCore, and Intel's dual and quad core Pentium designs are based on SMP architectures, often in conjunction with some form of CMT or other multithreading approaches.

SMP systems employ multiprocessing architectures with a globally coherent view of system memory. That is, every processor in the SMP system sees the same global-memory range. By using this arrangement, SMP systems offer a uniform set of resources to any task or application program; any task can execute on any processor and, if a task is suspended or interrupted, it can be rescheduled on any processor. When it resumes, the task finds its context (stored in global memory), reloads it, and restarts execution. This scheme greatly simplifies the creation of multiprocessor task schedulers and operating systems and it reinforces the system designer's desire to use identical processors in the SMP system's design for simplicity's sake.

The SMP system's multiple processors access global memory over some kind of system bus. A simple SMP system could be built from processors (including multiple processors on a single die) that directly access global memory via a system bus without any intervening memory hierarchy. When off-chip memory could be accessed in one or a few cycles of processor execution and the system bus added a relatively small delay, this sort of system architecture could be tolerated. However, differential process-technology scaling for processors and DRAM has produced an imbalance, where processors fetch and execute instructions from single-cycle caches and local memories at speeds ranging from hundreds of MHz to a few GHz but accessing off-chip global memory directly consumes tens to hundreds of processor clock cycles. An SMP system in which every processor is likely to be stalled waiting on memory would of course defeat the whole purpose of SMP.

Modern processors deal with the problem of slow global-memory access through the use of on-chip cache memories that can be accessed extremely rapidly, often in one cycle. The use of multiple cache levels in a hierarchy is also very common, as is the separation of caches into instruction and data banks. These approaches improve performance because the memory-access patterns may be very different for instruction fetching versus data-operand fetching and storage.

Caching strategies work well for individual processors because they effectively mask off-chip memory latency for large numbers of applications running on such systems. On-chip caches usefully exploit the transistor density offered by the application of Moore's law. However, for SMP systems, caches introduce a further wrinkle: cache coherency problems.

Within a cache, the valid value for a particular target memory location might be located in one of several places: the register where the value was stored after an instruction has fully executed; the level-one, -two, or -three cache; or in global memory. If another processor wishes to access the valid value for that memory location, it may be quite inaccessible until it is written back into global memory. Unless that value has been marked in some way, a processor could easily end up with an incorrect copy of the target value. Marking where the valid value for a memory location resides is required for even single-processor systems with caches. To build an effective SMP system, the concepts of cache coherency, marking, and updating must be extended to all of the SMP processors in the system.

Cache coherency of course incurs a cost and may degrade access latency depending on how it is implemented. If coherency is maintained by software, via a multiprocessor operating system and its services, access latency may increase substantially for many memory accesses, if not all of them. Most SMP systems – certainly most on-chip multi-core systems with two, four, or eight cores – implement cache coherency using specialized hardware that accelerates global access to memory items whose correct value may lie in one processor's cache hierarchy. This feature is offered in ARM's MPCore, for example. Such hardware cache coherency support comes at some considerable hardware cost, depending on the particular SMP system architecture and processors.

SMP can be regarded both as a problem and a solution. SMP together with a multithreading strategy may be an appropriate design solution for high-end server farms running general-purpose computing applications that are heavily control dominated or with considerable amounts of database searching, for which reasonable but not hard real-time response is required, and for highly variable computing loads. Web searches are the now-classic application in this space. For example, Google compute-server farms contain thousands of rackable SMP computer systems. Many other large-scale problems have similar characteristics.

To some extent, SMP systems offer the opportunity to dynamically scale processing power to load and to move applications and stalled threads from processor to processor in an ad hoc manner without worrying about moving the threads' memory context with them. The "system" automatically takes care of all this bookkeeping. Here, flexibility of system architecture, an ability to incrementally upgrade the system and to keep it running, and an ability to guarantee various qualities of service to different application classes may trump other considerations including compute efficiency and power consumption. Such general-purpose applications are usually not data-computation intensive. For example, they may not be used in deeply embedded deeply embedded applications such as in consumer products and the like.

SMP can be regarded as a problem if it is always regarded as the only solution to the need for multiprocessing or if it is applied to the wrong problem, the wrong product, and at the wrong time. With so many large processor companies and real-time operating system (RTOS) vendors advertising their multi-core SMP solutions, system designer's might well assume that SMP is the only solution for higher performance SoC designs.

There is an alternative system-design approach that makes sense for many data-intensive, deeply embedded applications, as found in many multimedia (audio and video), image-processing, and signal-processing applications. The alternative approach is AMP, where heterogeneous processors, configured for different parts of an application, are linked together in an architecture designed for effective performance and power-efficient computation. The classical AMP SoC example is a modern cell phone: one or more control processors handle user-interface code and wireless protocol stacks and one or more DSPs plus dedicated hardware blocks handle voice and image processing – encoding and decoding, as well as other kinds of processing.

By exploiting precisely the right kind of processor for each task, and by carefully designing a system architecture that allows these different processors to communicate via a variety of mechanisms (not just SMP's shared global memory), AMP solutions may offer the right application performance level at a small fraction of the energy consumption of SMP systems and without the need for cache-coherency hardware. Instructions matched to the application allow a configurable processor to consume far less energy than general-purpose processors running the same specialized tasks (on the order of one to two orders of magnitude less energy) and there may be many fewer processors needed as a result of the specialization.

AMP systems can be built from standard embedded processors of various kinds, as well as ASIPs that could be standard Intellectual Property (IP) blocks, custom designed, or based on configurable and extensible processor IP. In all cases, effective use of heterogeneous processors requires that design teams have some knowledge of the application tasks that will run on the processors – if not the actual code, then at least an awareness of the code's general characteristics (audio or video, for example) – and the kinds of ISAs most suited to each application domain.

Of course, it may well be that the right architecture for a product could be a combination of SMP and AMP. Products may have both application-oriented, data-intensive subsystems – such as for audio, video, and network processing – and a need to run general-purpose applications such as various user-interface functions, data searches, etc. These general-purpose applications may not all be known in advance. The evolution of the cell phone and PDA into a general portable digital appliance may well point in such a direction. In this case, an SMP multi-core subsystem combined with several AMP multiprocessor subsystems – the former to run general-purpose, non-real-time, non-data-intensive tasks; the latter to run all the highly data-intensive and real-time tasks organized into a constellation of processing subsystems – may be a good architecture. Clearly, SMP can be both a problem and (part of) a solution in such cases.

4 Processor Design Flow

Jari Nurmi

Tampere University of Technology

In this chapter, an overview of a generic processor design flow will be presented. The chapter is based on the author's experiences and also somewhat on existing literature. The design flow is illustrated in Figure 4.1 and explained in more details in the following sections.

Capturing requirements

The design process starts by capturing the requirements for the processor. There are both functional and non-functional requirements. The functional requirements stem from the foreseen applications, and also from the operating environment of the processor.

In the case of an extremely application-specific processor, the functional requirements of the applications can be found quite easily. In the best case, all the algorithms to be run on the processor are known *a priori*. At least, one can easily identify a representative set of algorithms within the application area.

The next step is to find operations or *prototype instructions* that support efficient execution of the known algorithms. We also have to figure out how we can address the operands, and what type of data we will be processing by the instructions. In a simple case the operations can be sketched by looking at the algorithm descriptions, in a more complicated case some *profiling* is needed to find out how frequently some operations, operation patterns or common subroutines are executed. The profiling can be done for full algorithms or for the algorithm kernels or inner loops. Examples of the full algorithms in embedded systems could be GSM speech codec,

J. Nurmi (ed.), Processor Design: System-on-Chip Computing for ASICs and FPGAs, 69–82.

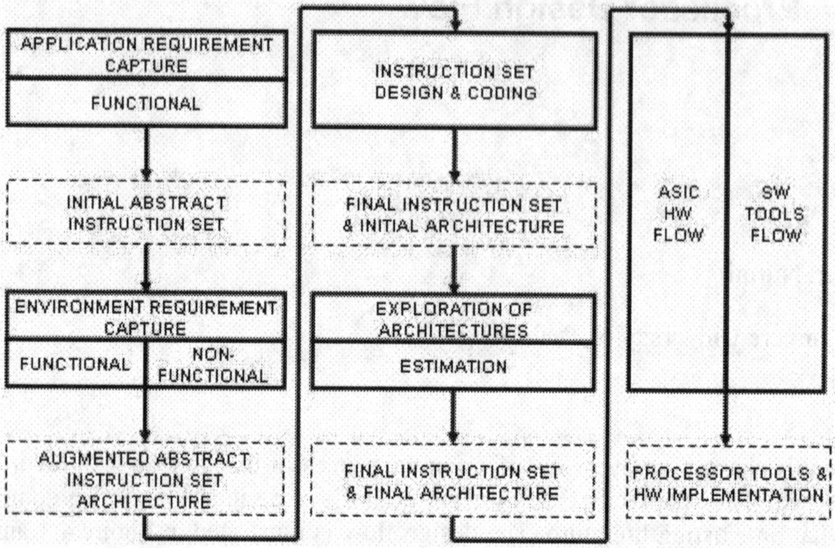

Fig. 4.1. Generic processor design flow.

MPEG-4 video compression and decompression, or Viterbi convolution decoder. The kernels from the algorithms could in turn be, e.g., codebook search, motion estimation, and path metrics computation.

There is one thing to be kept in mind if profiling a real source code with an existing processor when designing a new one. The results will be largely dependent on the instruction set of the old processor architecture, and also on the compiler ability to optimize for the target. This means that, in the worst case, the profiling results may be leading the designer astray. Here is an example from real life. We were profiling ADPCM encoder/decoder with an old Sun workstation whose processor did not support two important operations that were needed in the application: extraction of exponent (or finding the number of shifts to normalize a number) and multi-bit shifting based on a register value (the normalization once the right amount of shift has been found). This led to a huge number of cycles that were actually used for scaling the results with very primitive operations. Once this was finally noticed, we profiled the algorithm again using Analog Devices 2151 DSP with hardware resources for exponent extraction and a barrel shifter. This yielded completely different results and indicated better the requirements of the algorithm when implemented efficiently.

Another case illustrating the profiling process follows, originally published in [310]. In the development of a speech encoding/decoding processor for a mobile phone, the algorithm was described as a pseudo-code. We constructed a simulator to execute these pseudo-code instructions, and profiled the operation frequencies by running the given compliance test vectors in the simulator. The profile is shown in Figure 4.2. There are separate profiles for operations requiring multiplication and operations involving the ALU, and for the word length of the operations (32 or 16 bits). The third set of bars shows the control operations profile.

The profiles show that there is a significant number of both 32-bit and 16-bit operations, and many of the multiplication instructions (L_mac, mult_r) require both the multiplier and ALU to complete the result. Absolute value and normalization are required quite infrequently, but they are simple to implement in hardware. The division operation is not used very much and would require a complex hardware component, so it is definitely better to implement the division in software. A subroutine consisting of 14 instructions was constructed to execute the division. The ALU and multiplier were made to support both 16-bit and 32-bit data types. The multiplier and ALU are put in series, and some implicit shifts (not counted in the pseudo-code simulations) are implemented by pre- and post-shifters in the ALU.

There are a large number of assignments required, in practice this means loading data to the multiplier and ALU. Since the MAC (Multiply-ACcumulate) operations dominate, two buses are needed to efficiently feed the multiplier with two new values each clock cycle. There was also a number of various logic operations and bit-field masking required, which determined the need for a logic unit, capable to carry out the basic logic operations. The pre-shifter following the logic unit completes the bit-field masking operations. Thus, the ALU was split into a logic unit and arithmetic unit, separated by the pre-shifter.

The number of FOR loops may seem low at first glance, but the number refers to for loop instantiations rather than loop body executions, so there was a need for a hardware loop counter to efficiently take care of the FOR loops. There were also runs of maximum or minimum detection in the pseudo-code, which was easily added to the hardware (comparison and conditional loading of the new maximum/minimum accumulator). Following the common principle in DSP processors (as shown in a later chapter) the address registers are separate and have their own effective address calculation and register update logic.

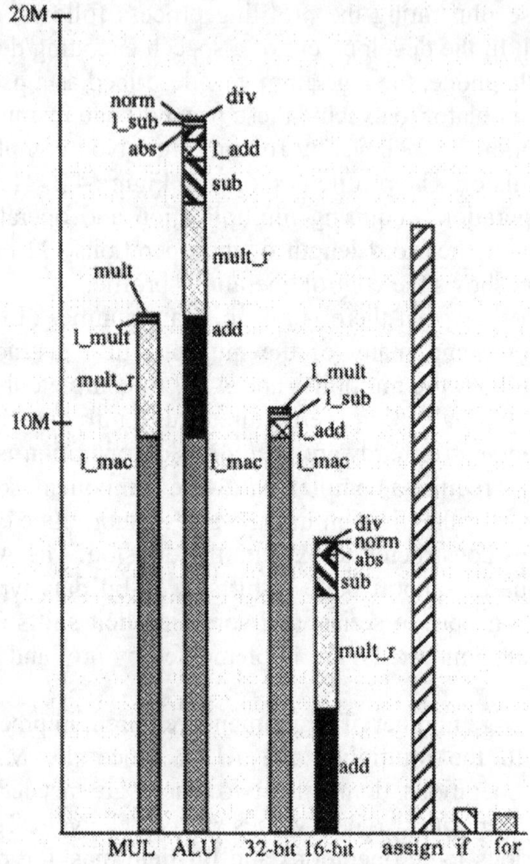

Fig. 4.2. GSM half-rate speech codec profiling data. © IEEE, 1994 [310].

Table 4.1 shows the instruction set of the resulting speech codec processor. L in the start of a mnemonic denotes a long (32-bit) operation. SLEEP, LDAS and STS are implementation specific additional instructions for power efficiency and saving and restoring an intermediate register in the multiply-accumulate pipeline. ENDL is a marker for the end-of-loop instruction, of which location is actually coded into the LOOP instructions. In DSP fashion, several operations can be executed concurrently in a single clock cycle, such as MUL, LADD, two MOVES and two INC/DEC post-operations on the address registers used at the same time in the moves.

Table 4.1. Instructions of the half-rate speech codec DSP.

Mnemonic	Meaning
ABS/LABS	Absolute value
ADD/LADD	Add 16-bit or 32-bit numbers
AND	Logical AND of 32-bit numbers
ASH/LASH	Arithmetic shift 16-bit or 32-bit numbers
CALL	Call a subroutine
DEC	Decrement address register
ENDL	End loop
INC	Increment address register
INC3	Increment address register by 3
JMP	Unconditional jump
JMPIF	Conditional jump
LDAS	Load accumulator with S-register
LOOP	Start a hardware-supported loop
MAX	Find maximum of two numbers
MIN	Find minimum of two numbers
MOVE	Load or store
MUL	Multiply 16 x 16 = 32-bit
NAND	Logical NAND of 32-bit numbers
NEG/LNEG	Negate (2's complement) 16-bit or 32-bit number
NOP	No operation
NOR	Logical NOR of 32-bit numbers
NORM	Find number of shifts to normalize accumulator
NOT	Logical NOT of a 32-bit number
OR	Logical OR of 32-bit numbers
RET	Return from subroutine
RETI	Return from interrupt
RND	Round from 32-bit to 16-bit number
SLEEP	Go to sleep
STS	Store accumulator to S-register
SUB/LSUB	Subtract 16-bit or 32-bit numbers
XNOR	Logical XNOR of 32-bit numbers
XOR	Logical XOR of 32-bit numbers

Some of the functional requirements are set by the operating environment rather than the applications themselves. For instance, if the processor will be running a (real-time) operating system, it will affect the design decisions, including the instruction set. Also, the memory subsystem connected to the processor may affect the requirements, and so will the I/O devices. In some cases even the software development tools may impose some functional requirements to the processor.

At the edge of functional and non-functional issues there are requirements for the clock cycle time, throughput and latency requirements set by the surrounding parts of the system.

Non-functional requirements include costs such as silicon area, pin count or actual manufacturing cost or retail price, power and energy consumption. Compatibility with the design tools may restrict the architecture, and compatibility with software tools or even binary compatibility of the instruction set to an existing family of processors may set additional requirements to be fulfilled by the processor designer. There might be also Electromagnetic Compatibility (EMC) issues arising, and manufacturability, testability, and maintainability are affecting the design decisions.

Yet another issue is how to prepare for the future. One approach is to maintain a certain degree of general-purpose characteristic even in the application-specific processors. This means to provide elementary support also for rarely operations (not judging them impossible by design), and elementary support for rare addressing modes, too. This follows the well-known postulate of Hennessy and Patterson: make the common case fast and the rare case correct [188]. However, there is some limit to the preparation for the future. Every addition that we make to the architecture to support hypothetic future needs will affect the cost some way. On the other hand, no matter how well we prepare, there will be more applications than what we can think of today. Processor designers are no wizards. Sometimes it is better to just leave room for future extensions – or use dynamic reconfigurability to respond to the future needs when they will arise. These alternatives will be further explored later in their own chapters.

Instruction coding

By now we have found out the required operations, addressing modes to access the operands, and data types that can be used to represent the data. Now we need to design the actual instruction set and its encoding to form binary instructions. This is an iterative task which is restricted by the cost and performance requirements.

The instruction coding is a compromise between a few different principles. Maximal utilization of parallelism would require to basically represent all the processor control signals in the instruction word. This would enable any combination of resources to be used in any one instruction, but on the other hand is in most cases intolerable because of the resulting long instructions. The other extreme is to encode the operation and its operands in as few bits as possible. This would, on the other hand, restrict the useful combinations of resources. Thus, the encoding is typically something between these *horizontal* and *vertical* code approaches.

Another issue is that different operations may require different number and different size of operands to be presented in the instruction. This leads to an attempt to use *variable-length* instruction coding. On the other hand, simplicity favors regularity which suggests having just a single instruction length with as fixed format as possible across the whole instruction set. Also in this aspect compromises will be made in the form of a few different instruction formats, possibly also in having a couple of different lengths of instructions. In the case of variable-length instructions, the lengths are typically multiples of the basic instruction length.

In addition, there are often requirements such as low cost, which can be affected by a small footprint of the programs. Low power consumption is one goal, which can be achieved by avoiding excess use of memory but also by choosing the binary codes to avoid as much as possible changes in the bit values in adjacent program memory accesses.

One goal is the low hardware complexity in instruction decoder. On the other hand, the assembling of programs should be easy, and the instruction set should be a good target for compilers. The instructions should perform as much work as possible, and have a good command over the processor resources. As can be seen quite easily, many of the goals or requirements are contradictory. It is all about making good compromises.

Let's take an example to illustrate the instruction encoding task. We have been told to save memory area, so the operations listed in Table 4.2 should be encoded in 16-bit instructions. The data word length is 16 bits, and the instructions should be all of the same length and enable single-cycle execution of the specified operations.

We have already decided that the processor is using a general-purpose register file with as many registers as can be practically addressed by the instruction word. We have also decided that the source and destination registers may be different (yielding at most three register operands in a single instruction). We can assume that at most 16 data registers can be used, since 3×5 bits to encode 32 registers would leave only 1 bit for the operation code. With 16 registers we have 4 bits for each register operand and four remains for the operation code. We have also decided that branches are PC-relative and have as long an immediate offset as possible in a single instruction word, and they have as many conditions as can be easily fit into the instruction. The shifts shall be full ± 15 bits to enable single-cycle shifting. The immediate constants to be loaded shall be as long as possible. The multiplication produces a double-length product, so we have decided to work with a pair of registers to store the result (meaning a split register file

Table 4.2. Operations specification.

Instruction	Addressing modes, limitations, etc.
Add	2 source and 1 destination reg
Add with carry	2 source and 1 destination reg
Subract	2 source and 1 destination reg
Multiply	Mixture of signed and unsigned operands. 2 source and 1 double destination reg
Logical shift	1 source and 1 destination reg. ±15-bit shifts from instruction
Logical shift on register	1 source, 1 destination, 1 shift amount reg
AND	2 source and 1 destination reg
OR	2 source and 1 destination reg
XOR	2 source and 1 destination reg
NOT	1 source and 1 destination reg
Unconditional branch	PC-relative branch in ±255 range
Conditional branch	As above but with 7 conditions
Unconditional branch on register	Branch address in register
Conditional branch on register	As above but with 7 conditions
Unconditional branch and link on register	1 branch address and 1 link reg
Conditional branch and link on register	As above but with 7 conditions
Load	Register indirect with possible post-inc/dec. 1 dest reg
Store	Register indirect with possible post-inc/dec. 1 src reg
Load immediate	Signed ±127 immediate, 1 dest reg
Register to register move	1 source and 1 destination reg
No operation	

register file implementation). These additional conditions have been taken into account in the table already.

We start the encoding from the regular operations. Operations ADD, ADDC, SUB, LSHR, AND, OR, XOR need three registers each. Three register fields are needed (4 bits to encode 16 registers), so there are 4 bits left for the regular operation codes. There are seven operations of this kind, so they need at most 3 bits of the opcode field to differentiate them from each other.

0xxx	REG	REG	REG

Then we encode two-register operations NOT, LD, ST, MOVE. Two register fields are used, plus space for additional information (LD/ST direction, post-modification, register move, not) in the third register space. We can use the 8th 3-bit operation code (0111) for these operations.

0111	4 bit specifier	REG	REG

The first 2 bits of the specifier can be used to encode the type of operation: LD, ST, MOVE, NOT (01, 10, 11, 00). So, the first 2 bits actually extend the operation code. The last 2 bits of the specifier field can denote the post-operation used in load/store instructions: increment, decrement, no change (10, 01, 00).

Multiplication needs to have the signed/unsigned information + double registers as the destination. There are two source registers, so two fields needs to be used for them. Signed/unsigned indication of both operands takes 2 bits. For double registers we need to restrict the addressing to register pairs 1:0, 3:2, etc. so we need 3 bits to encode the eight pairs.

100	s/u	s/u	REG PAIR	REG	REG

To incorporate the two signed/unsigned bits we had to "borrow" from the opcode field (we are kind of using two operation codes for the multiply instruction). We can note that in fact combinations US and SU are redundant (because the operands could be re-ordered), but there is no harm of the redundancy either.

For the shift instruction we need a 5-bit signed immediate for the shift value. Two fields are used for source and destination registers. Again, we need to borrow from the opcode to include the full-length shift value. The shift can be directly a 2's complement number denoting the power of two that is used to "multiply" the input value (which is also directly the shift amount, where negative values shift right and positive values shift left).

101		SHIFT	REG	REG

The branches are still remaining. If we allocate a 9-bit immediate offset and use seven conditions plus an unconditional branch (fitting in the remaining 3 bits), we end up with the following encoding.

1100	cond		OFFSET	

Another branch is on register also using conditions, and there is the subroutine call variation (branch on register and link), similarly with conditions. We use one register field for the branch target register and another one to denote the link register that will store the return address. We can use the one spare bit from the condition field to denote whether to link or not. In the case of a normal branch, the link register field in the instruction is unused. Notice that this encoding fits perfectly with the previous branch offset field length selection, even a 1 bit longer offset field (and thus four less different conditions!) would waste 1 bit in the register-based jump-and-link instruction.

1101	cond	link	REG	REG

The next instruction is load immediate, with an 8-bit immediate value. We need one register field for the target, and allocate yet another opcode.

1110			REG

Finally, NOP. There are several options to implement the NOP as a macro instruction, e.g.:

- MOVE register to itself
- OR R0, R0, R0
- AND R0, R0, R0
- JUMP to the next instruction

The drawback of such assembler macros is the power consumed in executing NOP operations. We could also use an actual binary code distinguished by the instruction decoder. Even there we have several additional options:

- LSH with shift –16 (0b10000) decoded from the 8 MSB bits
- any two-register instruction with both inc and dec bit set (decoded from 6 bits)
- the remaining opcode 0b1111 (decoded from the 4 MSB bits)

We can choose the latter to keep the decoding simple. The instruction coding is now completed.

1111			

The final instruction coding is shown in Table 4.3.

We can notice that there are very few unused bits (x) in the coding. Only in the NOP instruction there is a significant amount of them, in register-based branch (BR) there is one unused register field, and in MOVE instruction the two modifier bits are not in use.

This instruction coding is using the 16-bit instruction words very efficiently, especially if we think of how infrequently the instructions with some redundancy would be used in actual code. Practically BR is used in subroutine returns mostly; NOP is tried to be avoided and MOVE between registers is quite rare compared to loads and stores that dominate in load/store architectures. Loads and stores can easily comprise 30–50% of all instructions. This instruction set is very simplified and does not take

Table 4.3. Final instruction coding.

[15:12]	[11:8]	[7:4]	[3:0]	Interpretation
0000	rrrr	rrrr	rrrr	ADD
0001	rrrr	rrrr	rrrr	ADDC
0010	rrrr	rrrr	rrrr	LSHR
0011	rrrr	rrrr	rrrr	SUB
0100	rrrr	rrrr	rrrr	AND
0101	rrrr	rrrr	rrrr	OR
0110	rrrr	rrrr	rrrr	XOR
0111	00xx	rrrr	rrrr	MOVE
	01+−	rrrr	rrrr	LD
	10+−	rrrr	rrrr	ST
	11xx	rrrr	rrrr	NOT
1000	0RRR	rrrr	rrrr	MULUU
	1RRR	rrrr	rrrr	MULUS
1001	0RRR	rrrr	rrrr	MULSU
	1RRR	rrrr	rrrr	MULSS
101s	ssss	rrrr	rrrr	LSH
1100	ccci	iiii	iiii	Bccc
	111i	iiii	iiii	B
1101	ccc0	xxxx	rrrr	BRccc
	1110	xxxx	rrrr	BR
	ccc1	rrrr	rrrr	BRALccc
	1111	rrrr	rrrr	BRAL
1110	iiii	iiii	rrrr	LDC
1111	xxxx	xxxx	xxxx	NOP

into account the operations that are needed in interrupt processing, e.g., disabling and enabling interrupts, or in running even the simplest operating systems, e.g., system calls.

This step has a major effect on the organizational architecture, too. In fact, we need to have quite a good understanding of the organization to complete this phase of the design. The operations and operand access will have certain implications on the arithmetic and address calculation resources needed, type and number of data registers, how data transfers are carried out, and what kind of control structures and control registers might be needed. The final organization will require some further exploration and analysis.

Exploration of architecture organizations

Based on the basic incredients, we must start exploring the organizational architecture alternatives. That can be accomplished by pen-and-paper methods, with spread-sheet calculations of cycle counts, etc. within each alternative while executing the kernel operations, or by using processor specification languages with *automatic performance estimator* or *simulator generation*. If such tools are available, they can be strongly recommended especially in the case of more complex processor designs.

Independently of the method used, the estimation of the foreseen implementation(s) based on the architectures explored is of key importance. We are targeting at a certain performance level, at the same time trying to fit within the constraints of the non-functional specification (e.g., cost, power). Critical questions at this phase include: Are we achieving the target cycle time? What will be the critical paths? Are there any severe bottlenecks restricting the performance? There is another chapter on early estimation methodology and models to further clarify this stage.

At this point we can sketch the final architecture for the previous instruction coding example, as shown in Figure 4.3. This is in fact a refinement of the architecture of Figure 2.3 of Chapter 2. The architectural choices were pretty much done already interleaved with the instruction coding phase. The register file has now another write port D for extended write-back of double-length multiplication products. The multiplier and shifter are shown as separate blocks from the basic ALU which only contains add, subtract and logic functions. The shown setup implements a three-stage pipelined processor (fetch–decode–execute) which could be pipelined further as necessary by adding pipeline registers to the datapath and corresponding pipeline control, possibly with forwarding logic control. In the figure, the control is not shown in detail, and for instance operand latching could be added to the functional units to improve their power efficiency.

Hardware and software development

When the architecture has been finally chosen, it is time to start the implementation. The hardware development follows typically the normal ASIC or FPGA design flow, including high-level modeling, refinement of functional blocks in a hardware description language such as VHDL or Verilog, logic synthesis, floorplanning, back-end optimizations and verifications using simulators and static analysis tools.

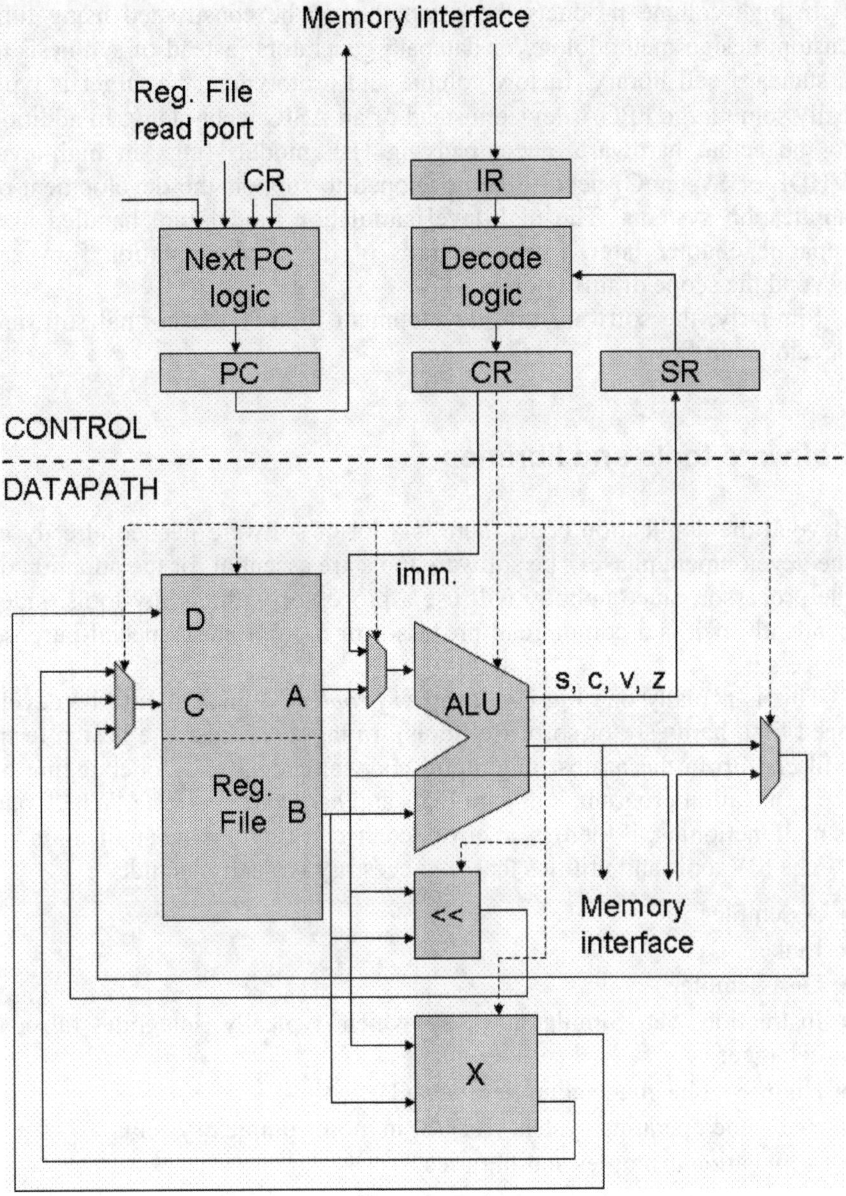

Fig. 4.3. An architecture to implement the example instruction set.

In high-volume products the datapath may be constructed using full-custom design methodology or datapath generators instead of synthesis to a standard cell library. In low-volume and prototyping the target is typically some large FPGA device instead of an ASIC technology. In addition to the actual hardware, accompanying HW models (e.g., in high-level VHDL or SystemC) need to be developed to support the development of application systems. The high-level simulation models are handled in a separate chapter later. Further details of the implementation flow are beyond the scope of this book.

Similarly, the software tool development follows the normal software development flow.

Software tools and libraries

Beyond the application code, there is a lot of software needed already in the development phase. The software tools are essential for the adoption of the processor, since nobody will use a processor without any tools, independently if it is a commercial processor or a proprietary special-purpose processor.

This is actually one big *egg–chicken problem*: Which should be developed first, hardware or software (tools). In the worst case the design team will end up in the utmost ungrateful codevelopment and codebugging of the application program, software tools and hardware platform. Which one is malfunctioning, if there is an error detected in running the application?!

The SW tools and utilities that are necessary typically include

- Assembler
- Linker
- Disassembler
- Instruction Set Simulator (ISS) which typically integrates also a debugger
- High-level language compiler (e.g., C/C++)
- Real-time operating system (RTOS) in more complex systems
- Application examples and libraries

Actually a processor development project can mainly consist of software development, although the processor is primarily associated with hardware in the engineering mind-set. Do not forget the importance of software! There is another chapter on software development tools for reconfigurable processors in particular to underline the importance of the tools.

5 General-Purpose Embedded Processor Cores – The COFFEE RISC Example

Juha P. Kylliäinen, Tapani Ahonen, and Jari Nurmi

Tampere University of Technology

As a starting point for the exploration of embedded processor architectures, we describe a general-purpose embedded processor core called COFFEE RISC. The reasoning behind its development and the design decisions made are discussed throughout the chapter. Application and application domain-specific issues will be treated in the following chapters.

Introduction

The complexity of processor architectures varies from multi-million gate designs to designs with a few tens of thousands of gates. Power consumption varies from milliwatt range to hundreds of watts. In order to set the scope for comparison, we need to classify processors and define what we mean by a processor core in this context. Processors targeted to personal computers and mainframe computers form a class with similar requirements. Computing performance and hardware support for operating systems are key requirements in that class whereas power consumption has not been an issue in the past but because of increasing power densities has become a major performance limiting concern.

Processors targeted to embedded systems belong to another class. Embedded systems are products other than general-purpose computing machines. Processors in such systems are used to implement certain functionality of a product, the capabilities of the processor are not important to the user as long as the product fulfils certain requirements. Some level of performance is needed and usually a processor which has just enough performance, but no more, is selected. Especially in battery operated mobile

J. Nurmi (ed.), Processor Design: System-on-Chip Computing for ASICs and FPGAs, 83–100.

devices, it is essential to select a processor which is 'not too good' in order to avoid excessive power consumption. We were targeting such an embedded processor for FPGA and ASIC implementations.

Independent of the target application domain, every processor has an execution core. Here we define the core of a processor as follows. The core is the unit responsible for interpretation and execution of the instruction set of the processor in question. If we rip off all peripheral components, buses and cache memories, only the core is left. In many contexts, processor cores contain also cache memories but we prefer to leave them out. This is because memory architecture affects drastically to performance as well as power consumption and chip area. Also, there is not one optimal solution for memory hierarchy, but the design of the memory architecture is guided by the application.

We can roughly divide instruction sets to reduced instruction set computer (RISC), complex instruction set computer (CISC) and digital signal processing (DSP)-like instruction sets. DSP processing cores belong to the application-specific group. RISC and CISC cores are suitable for general-purpose processing. RISC and CISC instruction sets differ mainly because of different design approach. We can argue that a CISC-like instruction set is designed for human and an RISC like instruction set is designed for a compiler.

In the early stages of minicomputers, programming was carried out using symbolic machine code, assembly code. It was advantageous to have instructions understandable by humans and instructions which performed 'more'. Nowadays the compilers produce RISC-like instructions even for CISC processors. Instructions not used by a compiler are mostly just a waste of resources reducing performance and increasing chip area as well as power consumption. This is one of the reasons why current trend is towards RISC type processors. In our COFFEE RISC Core[1] we have adopted the RISC philosophy to derive a good processing engine, amenable to compiler-based software development for embedded computing [248].

Implications of RISC design philosophy

As described in the chapter about embedded computer architectures, we can point out a few rules which are followed by RISC designers. The abbreviation RISC focuses on reducing the number and complexity of instructions but modern RISCs may have quite complex instructions included

[1] COFFEE RISC Core is a trademark of Tampere University of Technology, Tampere, Finland.

in their instruction set. The following ones were the principles that we followed in specifying our COFFEE RISC Core architecture and instruction set.

One instruction per cycle. This requirement does not make sense unless we refer to a pipelined design. If multiple parallel execution pipelines are used, more than one instruction can complete each cycle. The execution time of a program depends on the throughput of the pipeline, not the latency of individual instructions. Increasing the number of pipeline stages reduces the clock cycle time but at the same time the number of stall and flush cycles (wasted cycles) is increased. The number of wasted cycles can be reduced by careful scheduling of instructions but cannot be fully eliminated. This is a consequence of the fact that software execution is sequential and operations tend to have strong dependency on results of previously executed operations.

Also the penalty of branching becomes significant with very deep pipelining. During the evaluation of branch condition and branch target address, several instructions may enter the pipeline. If the branch is taken, those instructions have to be flushed. There are several ways to alleviate this problem. Delayed branching is quite efficient because a good compiler almost always finds instructions to be placed in delay slot(s) after the branch instruction. Delayed branching together with efficient hardware for address calculation and condition evaluation are enough in designs with less than ten pipeline stages. With deep pipelines, pre-fetching and speculative execution are most often used.

Fixed instruction length. This requirement aims at simplicity of decoding an instruction. In addition the requirement of issuing one instruction per cycle cannot be usually achieved if multiple memory accesses are needed to fetch a complete instruction word. An RISC instruction word usually contains all the information needed for its execution. The width of the instruction is most often 32 or 64 bits.

Only load and store instructions access memory. This requirement aims at utilizing the pipeline in an optimal way as well as minimizing memory traffic. Modern RISC processors usually exploit a pre-fetch mechanism: The address of the needed data is passed to cache memory well before that data is actually needed. A compiler can schedule pre-fetch commands in order to minimize cache misses which cause stalls. RISC processors usually have many general-purpose registers (from 16 to 128 typically), which make it possible to handle most of the processing inside the core and use load and store instructions only to move ready results to memory and new data in. However, this is a simplified view since the

Actual amount of memory traffic depends heavily on the application (and compiler).

Simplified addressing modes. Compilers hardly ever use complex address arithmetic supported by CISC hardware. Complex address calculations in hardware only extend the clock cycle, so why should they be used. If very complex address arithmetic is needed, it can be synthesized using a few simple RISC instructions.

Fewer, simpler instructions. Simpler operations imply shorter clock cycles. Simple instructions also fit better to RISC-like pipelines. Many simple RISC designs perform most of their operations in one execution stage, that is, once the data has been fetched from the register file, it takes one clock cycle to evaluate the result. Demanding instructions, such as multiplication, use several cycles in the execution stage and in effect stall the rest of the pipeline. This approach is adopted by for example Advanced Risc Machines (ARM) [133]. In the COFFEE RISC Core, multiplication is also pipelined in order to increase the throughput.

The COFFEE RISC Core instruction set architecture

One cannot prove that a certain set of instructions is better than another. We can easily measure or compute the number of instructions executed per time unit but we can only compare cores which execute the same instruction set. Sometimes it is not even possible to do this because deeply pipelined architectures usually expect the compiler to schedule instructions in an optimal way. Measurements depend on the compiler and application. Several benchmarks have been developed in order to facilitate performance measurements. These benchmarks are usually a set of programs which are executed on the target processor and the execution time is measured. Even though it might be justified to compare benchmark results between different processors, it is clear that they are a measure of a complete system composed of compiler, processor core and the memory architecture.

The scenario for the use of the COFFEE RISC Core was that the core will be used to set up embedded systems for mainly telecommunications and multimedia applications, and the most computationally intensive tasks of the application will be accelerated by coprocessors if needed. The workload characterization of embedded RISC processors has been conducted by several processor vendors with similar conclusions. Thus, the instruction set of the COFFEE RISC Core was designed based on instruction sets of RISC processors currently on market. Instructions which were available in most of processors were included and the rare ones were excluded after careful consideration. Instructions which enable coprocessor support (dedicated instructions to move data and coprocessor instructions over a coprocessor bus) were also added as a way to extend the instruction set if needed. This approach provided a good starting point for the development regardless of not being a highly analytical. The most important inclusion was the hardware support for multiplication to speed up algorithms with DSP flavor.

The penalty of implementing a particular instruction is not known before modeling the execution of that instruction with hardware timing. This makes it extremely difficult to decide whether an instruction should be included or not prior to the implementation phase of the design. Some instructions present in some RISC architectures were easy to exclude, for example division. Division is not a deterministic process, that is, its execution time cannot be predicted. This implies iterative execution which means in practice stalling the pipeline until the result is ready. If needed, division shall be done in software, in a coprocessor, and is best to be avoided in time critical algorithms.

The regular COFFEE RISC Core implementation has 66 instructions. A special instruction was included for future extensions: switch mode (**swm**). It is used to switch to a different instruction decoding hardware. It can be used to implement application-specific instruction sets and develop 'better' instruction sets in the future without giving away compatibility with old software. The execution pipeline of the COFFEE RISC Core (explained later in this chapter) together with **swm** instruction makes it possible to integrate for example multiply and accumulate (MAC) instruction in the pipeline without deteriorating performance. Currently **swm** instruction is used to switch to compressed instruction mode where each half of the 32-bit instruction word is interpreted as an individual 16-bit instruction.

The data processing instructions operate on two register operands, or alternatively, one register operand and one immediate operand. Instructions which produce data can write their result to any general-purpose register. Three register indexes can thus be specified in one instruction word.

There are fourteen arithmetic instructions, ten-bit field manipulation instructions (bytes, halfwords, and arbitrary bit fields), six Boolean operations, eight conditional branches, four other jumps (linking jumps, absolute jumps etc.) and six shift instructions.

Conditional branching is implemented using two instructions: compare and branch. Compare instructions of the COFFEE RISC Core produce condition flags which can be saved to one of eight possible condition flag registers for later use. Branch instructions evaluate branch condition based on those flags. A delay slot of one instruction is present after any jump or branch. This delay slot can be used to execute a meaningful instruction instead causing a pipeline bubble (a stall cycle) for every branch.

Most of the instructions can be executed conditionally making it more efficient to implement short conditional statements. With conditional execution there is no need to jump over code in order to avoid its execution.

It was decided that COFFEE RISC Core shall support real-time operating systems (RTOS). To this end, separate user and supervisor modes were introduced. System call and trap (programmed interrupt) instructions are provided to transfer the control to the supervisor which in most cases means an operating system. The COFFEE RISC Core instructions are listed and shortly explained in Table 5.1.

Software view of the COFFEE RISC Core

The COFFEE RISC Core is a so-called load-store machine: Memory operands have to be loaded into registers before performing any operation on them. Similarly, a result of an operation is stored into a register from where it can be written to memory using data transfer instructions. As in most of RISC architectures, a large general-purpose register bank is provided to avoid excessive memory traffic.

Figure 5.1 depicts the programmer's view of the core's registers. The general-purpose register bank is divided into two sets. This division enables fast mode and context switching as well as facilitates the implementation of separate user and supervisor modes for RTOS support. Each register set contains 32 registers. Both sets are available in the privileged supervisor mode, but only one set is accessible in the user mode.

Software configurability of such features as protected memory areas and peripheral address space was foreseen to facilitate system development. To enable the desired software configurability, an internal memory mapped configuration register bank called core configuration block (CCB) was provided. The CCB itself can be freely relocated within a 32-bit address space.

The peripheral devices around COFFEE RISC Core can be configured and communicated with through an external module called peripheral control block (PCB). Just like the CCB, the PCB provides software configurability through a register set interface. Control and data registers of peripherals can be placed into one register bank having a single decoding logic or they can reside inside each peripheral device just sharing the bus.

Table 5.1. COFFEE RISC Core instruction set.

Mnemonic	Meaning	Mnemonic	Meaning
Integer arithmetic			
Mnemonic	**Meaning**	**Mnemonic**	**Meaning**
add	Add 32-bit integers (in registers)	muli	Multiply 32-bit register with 16-bit immediate
addi	Add 32-bit integer with 16-bit immediate	muls	Multiply signed 32-bit integers (in registers)
addiu	Unsigned version of Addi	mulu	Multiply unsigned 32-bit integers (in registers)
addu	Unsigned version of Add	mulus	Mixed unisigned-signed multiplication
sub	Subtract 32-bit integers (in registers)	muls_16	Multiply signed 16-bit integers (in registers)
subu	Unsigned version of Sub	mulu_16	Multiply unsigned 16-bit integers (in registers)
mulhi	Get upper 32 bits of 64-bit multiply result	mulus_16	Mixed unisigned-signed 16-bit multiplication
Byte and bitfield manipulation			
Mnemonic	**Meaning**	**Mnemonic**	**Meaning**
exb	Extract (immediate-specified) byte from word	lui	Load upper halfword from immediate
exbf	Extract bitfield from word (register mask)	sext	Sign-extend an integer (register mask)
exbfi	Extract bitfield from word (immediate mask)	sexti	Sign-extend an integer (immediate mask)
exh	Extract halfword from word	conb	Concatenate bytes
lli	Load lower halfword from immediate	conh	Concatenate halfwords
Boolean bitwise operations			
Mnemonic	**Meaning**	**Mnemonic**	**Meaning**
and	AND two 32-bit register values	or	OR two 32-bit register values
andi	AND 32-bit register with 16-bit immediate	ori	OR 32-bit register with 16-bit immediate
not	Bitwise NOT for a 32-bit register value	xor	Bitwise XOR two 32-bit register values
Conditional jumps (branches)			
Mnemonic	**Meaning**	**Mnemonic**	**Meaning**
bc	Branch if carry	bgt	Branch if greater than
begt	Branch if equal or greater than	blt	Branch if less than
belt	Branch if equal or less than	bnc	Branch if not carry
beq	Branch if equal	bne	Branch if not equal
Other jumps			
Mnemonic	**Meaning**	**Mnemonic**	**Meaning**
jal	Jump (based on immediate offset) and link	jmp	Jump (based on immediate offset)
jalr	Jump (on register) and link	jmpr	Jump (on register)
Integer comparison			
Mnemonic	**Meaning**	**Mnemonic**	**Meaning**
cmp	Compare registers (set flags)	cmpi	Compare register to immediate (set flags)
Shifts			
Mnemonic	**Meaning**	**Mnemonic**	**Meaning**
sll	Logical shift left	srai	Arithmetic shift right based on immediate
slli	Logical shift left based on immediate	srl	Logical shift right
sra	Arithmetic shift right	srli	Logical shift right based on immediate
Memory load, store, move			
Mnemonic	**Meaning**	**Mnemonic**	**Meaning**
ld	Load word	mov	Register-to-register move
st	Store word		
Coprocessor instructions			
Mnemonic	**Meaning**	**Mnemonic**	**Meaning**
cop	Coprocessor instruction	movtc	Move data to coprocessor
movfc	Move data from coprocessor		
Mode changing instructions			
Mnemonic	**Meaning**	**Mnemonic**	**Meaning**
chrs	Change register set	reti	Return from interrupt
di	Disable interrupts	retu	Return to user mode
ei	Enable interrupts	scall	System call
swm	Switch decoding mode		
Miscellaneous			
Mnemonic	**Meaning**	**Mnemonic**	**Meaning**
rcon	Restore condition from GP-register	trap	Software exception (programmed interrupt)
scon	Save condition in GP-register	nop	No operation, idle

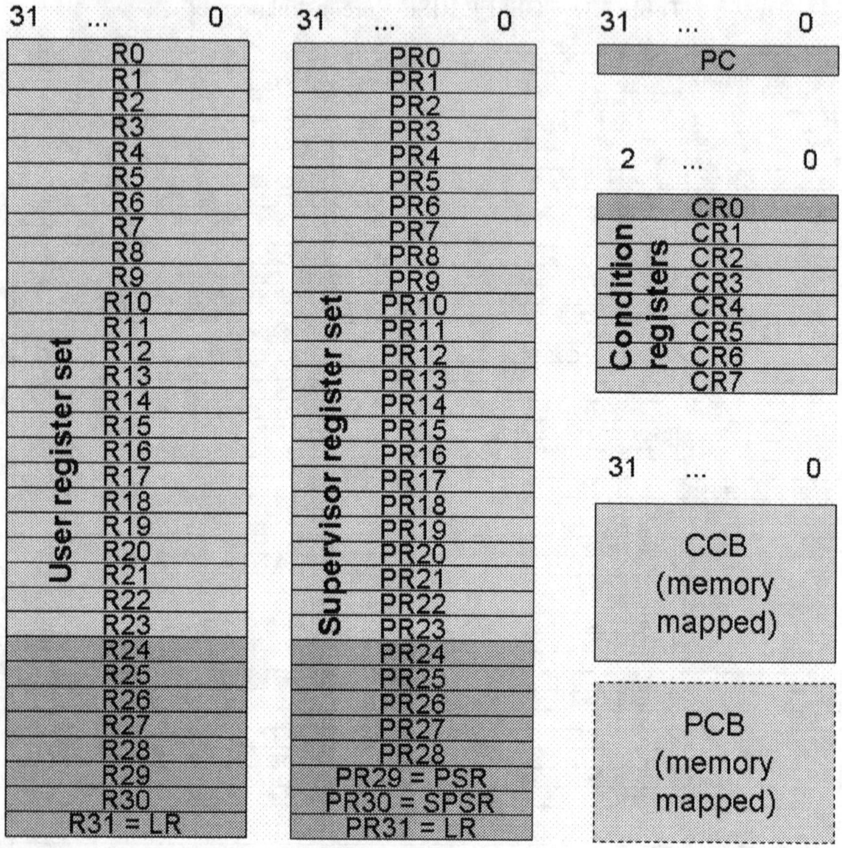

Fig. 5.1. COFFEE RISC Core programmer's view of the register set. The shaded registers are available also in the 16-bit mode.

Hardware view of the COFFEE RISC Core

The COFFEE RISC Core is a 32-bit architecture, that is, data is manipulated in 32-bit words. The memory interface is of Harvard type, having separate interfaces for data and instruction memory. This supports simultaneous access from the instruction fetch and memory access stages of the processor.

Figure 5.2. shows an example of interfacing the COFFEE RISC Core. Memories are interfaced with generic signals to allow integration with various types of memory parts. Multi-cycle accesses are supported enabling direct connection to large and slow memories. The access times are software configurable (in cycles) separately for both interfaces via the

CCB. The COFFEE RISC Core also incorporates support for sharing the data bus.

As shown in Figure 5.2, the PCB is attached to the COFFEE RISC Core's data bus interface. All accesses to the memory space reserved for peripherals will assert peripheral control block write (*pcb_wr*) and peripheral control block read (*pcb_rd*) signals directing the access to the PCB instead of data memory. Data memory accesses assert the general write (*wr*) and read (*rd*) signals.

The coprocessor interface is much like a memory interface. Addressing is limited to 7 bits, including two bits for coprocessor identification and 5 bits for coprocessor register indexing. The COFFEE RISC Core thus supports up to four coprocessors with the maximum register bank size of 32. In addition, a coprocessor can interrupt the core by asserting an exception signal included in the interface. An important feature of the coprocessor

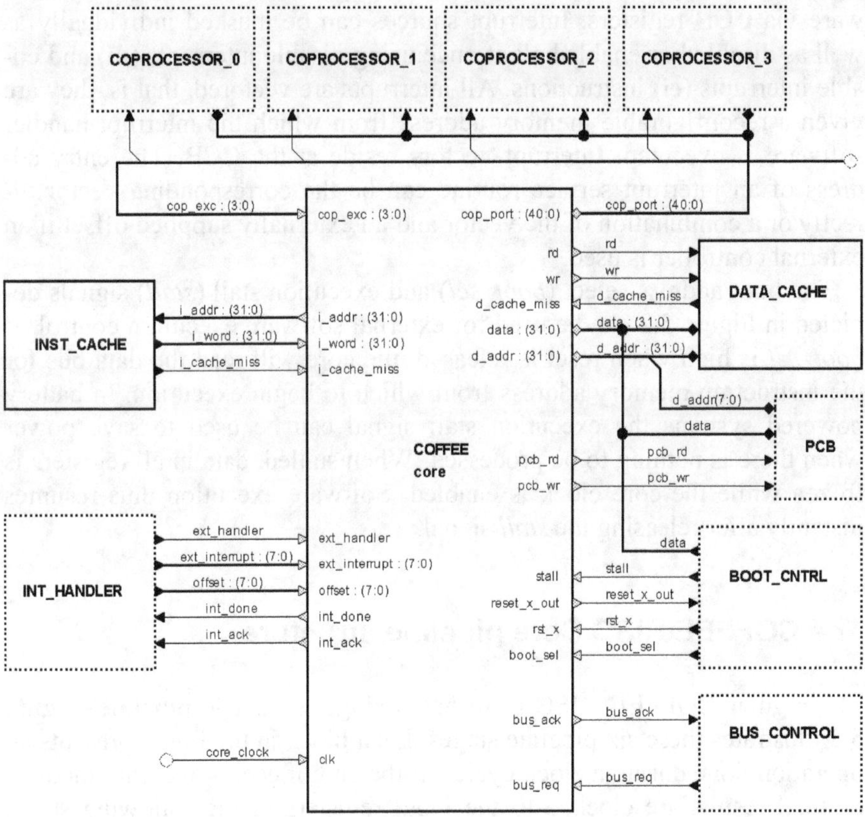

Fig. 5.2. Interfacing the COFFEE RISC Core. © IEEE, 2003 [248].

interface is its ability to connect to different clock domains. This is achieved by synchronizing also exception signals on core side and allowing data transfer times of up to 16 clock cycles. As with memories, the data transfer time is software configurable through the CCB. Synchronizing circuitry is required on the coprocessor side unless the coprocessor's operating frequency is an even multiple of the core clock frequency.

The COFFEE RISC Core provides an internal interrupt controller which is adequate for many designs but also a possibility to extend is provided. Eight external interrupt sources are supported by the internal interrupt handler. If coprocessors are not connected, the four inputs reserved for coprocessor exceptions can be used as interrupt request lines, giving the possibility to connect 12 sources. The internal interrupt controller has synchronization circuitry allowing asynchronous signals to be connected. If an external controller is used, synchronization is bypassed in order to reduce signaling latency. Priorities between interrupt sources can be set by software via CCB registers. Interrupt sources can be masked individually as well as disabled or enabled all at once using disable interrupts (**di**) and enable interrupts (**ei**) instructions. All interrupts are vectored, that is, they are given a reconfigurable memory address from which the interrupt handler software is executed. Interrupt vectors reside at the CCB. The entry address of an interrupt service routine can be the corresponding vector directly or a combination of the vector and an externally supplied offset if an external controller is used.

The boot address select (*boot_sel*) and execution stall (*stall*) signals depicted in Figure 5.2 can be used for external software execution control. If *boot_sel* is high when reset is released, the core will read the data bus for the instruction memory address from which to begin execution. In battery powered systems the execution stall signal can be used to save power when there is nothing to be processed. When stalled, data in all registers is frozen while the core clock is enabled. Software execution thus resumes instantly after releasing the *stall* signal.

The COFFEE RISC Core pipeline structure

The regular COFFEE RISC Core has a single six stage pipeline. Figure 5.3 illustrates these six pipeline stages. Each block in the figure presents an operation done during a clock cycle. At the end of each stage, intermediate or final results are clocked to the input registers of the following stage. Execution proceeds from left to right and instructions complete in order. In ideal conditions a new instruction enters the fetch stage and one instruction completes at the write back stage every clock cycle. The maximum throughput is thus one instruction per cycle. Each pipeline stage is described briefly in the following.

In the first pipeline stage, marked as *Fectch* in the figure, three operations are performed. A new 32-bit instruction is fetched from the location pointed by the program counter, *PC*. In the 16-bit mode, a 32-bit double instruction is fetched if the address is even. The address in PC is checked and an exception risen in case of a violation. Finally the program counter is incremented by two or four depending on the mode.

The second pipeline stage, marked as *Decode* in the figure, is the most important from the control point of view. This is where an instruction is identified and most of the decisions about its behavior in the next stages are made. If the core is in the 16-bit decoding mode, a 16-bit halfword is extended to an equivalent 32-bit instruction before passing it to the decode logic. Special fields inside the instruction word defining the condition for execution are evaluated against pre-evaluated condition flags. If the result of this evaluation is false, the instruction will simply be flushed on next rising edge of the clock. In parallel with the conditional execution check, signals needed during the current and following stages are decoded from the instruction word. Based on the signals evaluated in the decode stage

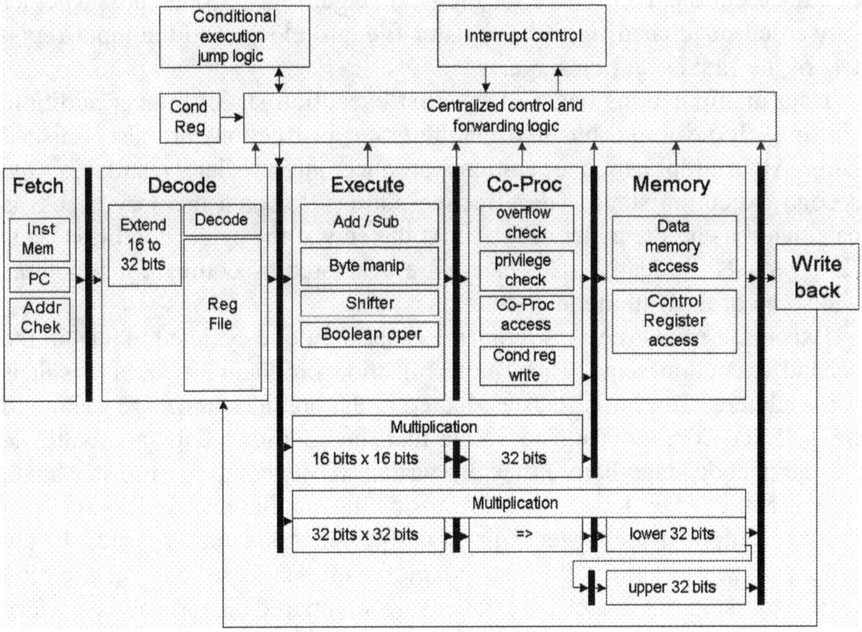

Fig. 5.3. The COFFEE RISC Core pipeline. © IEEE, 2003 [248].

and signals decoded from previous instructions currently on the pipeline, the control checks for data dependencies.

The COFFEE RISC Core resolves all data dependencies by forwarding the required data as soon as it becomes available. Fetch and decode stages are stalled only if source operand data is not available at the register file and it cannot be forwarded. Hardware support for resolving dependencies makes programming as well as compiler construction easier. As can be seen from Figure 5.3, there are several locations from which data is forwarded. In this simple six-stage pipeline, the forwarding logic has a delay of approximately one third of the clock cycle. Thus the forwarding logic itself does not reduce clock frequency, but only improves performance by avoiding unnecessary stalls. However, the wide multiplexers required to select input operands from several possible sources may form a performance bottleneck in some, especially FPGA-based, instantiations.

Other operations in the decode stage include extending an immediate operand, calculating the PC relative jump address and evaluating new status flags if needed. All jump instructions and conditional branches (PC relative and absolute) are executed in the decode stage, that is, at the end of the stage the target address is clocked into the PC register. Conditional branching is based on the same pre-evaluated condition flags as for conditional execution. To prepare for the next stage, register operands, whether forwarded or fetched from the register file, are clocked to the input registers of the first execution stage.

Data manipulations begin in the first execution stage. Integer addition, shifting, Boolean and bit-field manipulating instructions are also finished here. All multiplication operations produce intermediate results for the second execution stage. The effective address for data memory access is calculated using the adder of ALU. At the end of the cycle, condition flags (Z = zero, N = negative, C = carry) are evaluated by compare instructions and some of the arithmetic instructions.

Execution of instructions requiring more than one cycle continues in the second execution stage. 16-bit multiplication producing a 32-bit result is finished here. Condition flags evaluated in the previous stage are written to the selected condition register. Note that the condition flags are available for the decode stage before they are written to the condition register bank. This is achieved by forwarding data inside the condition register bank from input to output if the target register is the same as source register. In the second execution stage the data memory address calculated in the first execution stage is checked. Address is compared against the memory boundaries set for the user. It is also checked if the address points to the CCB in which case data bus access is not initiated. Address calculation overflow is detected and all coprocessor accesses are performed in the second execution stage. If the CCB configuration for coprocessor bus access time indicates multiple cycles the pipeline will be stalled during the required wait states. This implies that if a slow coprocessor is used, performance will be deteriorated unless an efficient interface wrapper is used.

Thirty two-bit multiplications are finished in the third execution stage. Data memory is accessed thereby finishing load (**ld**) and store (**st**) instructions. If the CCB configuration for data bus access time indicates multicycle latency, the rest of the pipeline is stalled during the required wait states to ensure that instructions are completed in order. This underlines the importance of fast data memory or data cache and pre-fetch capability.

In the last stage (*Write back*), data is written to the selected destination register thereby completing the execution of all data producing instructions. The register file has internal forwarding which makes data in this stage visible to the decode stage.

The COFFEE RISC Core implementation

The COFFEE RISC Core is a register transfer level (RTL) VHDL description, that is, a soft core. It is hence portable between technologies. The VHDL description is written in a way minimizing variation between different technology libraries. Arithmetic operations are coded at Boolean level which produces predictable results since a synthesis tool does not try to map operations to fixed hard implementations. The pipeline is balanced based on relative measures of the depth of the logic in each stage. This should ensure equal results between different synthesis tools. Also mapping directly to technology without optimization should produce acceptable results.

The type of architectural realization can be forced by hardware description style. The COFFEE multiplier is a good example of that. For ASIC realization, the description needs to be detailed enough to force the desired usage of a fast Wallace tree architecture. The synthesis tool may instantiate a less efficient architecture such as a Booth encoded multiplier, if the hardware description lacks detail. On FPGA, however, the same description cannot be mapped on the DSP block resources which include fast multiplication hardware. This expands the core area unnecessarily. For this reason a more abstract multiplication hardware description is used with FPGAs.

The testing strategy of COFFEE is based on comparisons against a cycle-accurate 'golden design'. Input stimuli can be synchronized to clock or instruction stream. Outputs from the design under test and the golden design are sampled and the samples from the same clock edges are compared against each other. Different test cases are defined as a set of input and output files in order to avoid any modifications to the test setup itself. Initialization of register resources is possible without additional test structures. Register and memory dumps are allowed at any instant of time. A VHDL disassembler is used to visually track down the cause of a bug.

The core was designed to be a general-purpose processing element suitable for most applications in either System-on-Chip (SoC) environment or in more conventional embedded systems. One might think that a general-purpose machine is too much of a compromise, that is, no good for anything. While this might be true in some cases, the COFFEE RISC Core makes an exception. It was designed to provide a platform which can be tailored to suit the application. In practice this means that the COFFEE RISC Core is not a fixed design and moreover, it is many designs. The regular version provides adequate resources and processing power for many applications but it can be enhanced in various ways. Designer of a system can choose the combination of modules to get the best trade-offs. Usually this means getting just enough performance while minimizing power consumption and silicon area. If none of the ready-made modules results in a satisfactory design, custom modifications can be made. In addition to tailoring the core, external modules can be connected to construct a suitable platform for an application. The core provides simple interfaces for expansion and communication.

Customizing the core is straightforward thanks to its design for modifiability. One example of such a modification is the recent integration of floating-point operations into the core itself from an existing coprocessor implementation. The integration work was carried out in a couple of weeks including verification. The original floating-point coprocessor design, described in another chapter of this book, is called MILK (which goes well with COFFEE). The COFFEE RISC Core version with integrated MILK floating-point capabilities (CAPPUCCINO) was developed to solve the processor–coprocessor bottleneck in floating-point intensive applications. Despite of that, the separate MILK serves well in situations where the coprocessor is just moderately used. Other coprocessor developments around the COFFEE RISC Core include the digital communications coprocessor set ESPRESSO [178], the reconfigurable floating-point capable accelerator array BUTTER [57,58], and the Reconfigurable Algorithm Accelerator RAA [351,352].

The COFFEE RISC Core was designed according to the guidelines for producing reusable IP (Intellectual Property) components [223]. A good IP is more than a good design; there are several things to consider. The importance of documentation cannot be stressed enough. An IP block without proper documentation is useless no matter how flexible or configurable it might be. Reusability and configurability were the main postulates for the COFFEE RISC Core design. Any design which does not constraint implementation technology has comprehensive documentation and is moderately easy to modify is reusable. If we add scalability and extendibility to the list, we have an IP block. In fact, our core is more than IP. Since it is published as an open source component, it can be referred to as Intellectual Commons (IC) which enables innovation to be incrementally built on top of what we provide. It is the Linux of computation hardware.

Modularity gives user the freedom to select the optimum modules or blocks for the design from a set of ready-made blocks. In addition, modular structure with well documented component interfaces allows custom blocks to be used. Module-wise synthesis allows each module to be optimized either for speed or area resulting in overall optimal design.

Thanks to its relatively simple interface, the COFFEE RISC Core is easy to instantiate. It was designed to work as a stand-alone unit without any additional circuitry. It can however easily be equipped with cache memories and unlimited amount of peripheral devices. Peripherals can be connected via direct register interface or a bus, such as AMBA. Memory interfaces make no assumptions about the type of memories. The user can map the address space freely since there are no fixed addresses for peripherals or configuration registers. Even the boot address can be defined externally. A series of VCI [446] interface wrappers are provided which allow easy connectivity to other VCI components.

The regular COFFEE RISC Core provides a starting point for developing suitable platforms for applications, such as the one described in [4–6]. It provides the common resources needed by every embedded system: built-in interrupt controller which supports up to 12 sources, simple memory potection mechanism and two timers. System designer selects memories and I/O peripherals as required for the application. Up to four coprocessors can be connected to boost for example floating-point operations [55] or DSP processing.

The COFFEE RISC Core characteristics

Results from automated implementations of the COFFEE RISC Core are discussed here. Figures are given for a 90 nm low-power ASIC technology as well as Altera and Xilinx FPGAs. The register file was implemented as a register bank for all these realizations. Area could be significantly reduced through SRAM-mapping of the register file.

Synthesis results for a low-power ASIC technology are shown in Table 5.2. The synthesis runs are trying to capture the worst possible case behavior by assuming a low supply voltage (0.95 V), high temperature (125°C), and worst case process variation to ensure very high manufacturing yield. The highest secure operating frequency is the inverse of total propagation delay plus the setup and delay times of registers, that is, about 150 MHz for the speed-optimized version. For example at 25°C temperature, 1.4 V supply voltage, and typical process variation for average yield, the operating speed would be more than three times faster (about 500 MHz). In practice clock rates of 300–400 MHz can be used in applications where the few slow chips can be binned out. The other synthesis targets at the minimum resource utilization, while the absolute worst case speed remains at around 35 MHz. Again, the practical frequency achievable in real applications is in the range of 100 MHz which matches the most optimized FPGA implementations.

Table 5.2. COFFEE RISC Core synthesis results for a 90 nm low-power ASIC technology.

	Coffee (small)	Coffee (fast)
Equivalent gates	60,665	88,074
Actual std cells	20,511	28,901
Leakage power	2.1364131 uW	3.9165367 uW
Nets	22,109	29,915
Average fanout	2.46524	2.217516
Worst path information		
Register setup	0.513 ns	0.251 ns
Propagation time	27.541 ns	6.430 ns
Clock frequency	35 MHz	150 MHz
Logic depth	46	33

Table 5.3. Post-fit (place-and-route) results on Altera Stratix-II.

Constraint	30 ns	15 ns	10 ns	9.09 ns
Registers	5,158	5,760	5,571	5,472
ALMs	5,462	6,808	6,534	6,267
DSP blocks	2	2	2	2
Memory bits	4,608	9,216	9,216	9,216
Cycle time (ns)	14.13	10.168	9.632	9.412
Frequency (MHz)	70.77	98.35	103.82	106.25
Logic depth	19	13	12	14
Highest fanout	31	32	82	82
Interconnect delay (% of cycle)	73.33 %	65.62 %	78.39 %	71.97 %

Table 5.4. Post-place-and-route results on Xilinx Virtex-4.

Constraint	30 ns	15 ns	10 ns	9.09 ns
Registers	5,207	5,273	5,272	N/A
Logic slices	6,503	7,030	7,022	N/A
DSP blocks	16	16	16	N/A
Memory bits	0	0	0	N/A
Cycle time (ns)	26.207	14.442	14.755	N/A
Frequency (MHz)	38.158	69.24	67.77	N/A
Logic depth	23	13	17	N/A
Highest fanout	119	353	19	N/A
Interconnect delay (% of cycle)	75.55 %	79.34 %	71.60 %	N/A

FPGA synthesis and place-and-route were carried out for both Altera EP2S130F1020C4 and Xilinx XC4VLX160FF1148-11 devices. The synthesis runs were constrained by different timing constraints. The tools used were Quartus II 5.0 SP1 and ISE 7.1i SP3, respectively, for Altera and Xilinx. Table 5.3 shows the results for Altera and Table 5.4 for Xilinx. The results reflect the different policies of the FPGA vendors. While Altera clearly targets at higher density and higher operating frequencies [12], Xilinx is more conservative and cares more for the power consumption of its devices [232]. It would be interesting to see the power consumption figures for the designs running at the same frequency. As a consequence of the policy, the speed of the COFFEE RISC Core implementation on the Xilinx device saturates at about 70 MHz while on Altera it achieves up to 106 MHz. Both devices are fabricated in a 90 nm process, which provides an interesting comparison to ASIC technology.

Software development tools for the COFFEE RISC Core have also been developed at Tampere University of Technology. GNU compiler collection [157] has been ported to the COFFEE RISC Core. The tool set includes also the GNU 'bin-utils', that is, assembler and linker. In-house assembler, linker and instruction set simulator have also been developed early in the project. Software tools for custom processors in general are addressed in another chapter of this book.

To support COFFEE RISC Core-centric system design, several simulation models of different degree of precision are provided. The highest level is a Transaction Level Model (TLM) written in SystemC [400]. This is a functional model of the COFFEE RISC Core ISA. The ISS, on the other hand, models the core in cycle-accurate level, except some pipeline effects. There also exists a C++ simulation model with cycle-accurate behavior. The lowest level is the RTL model used also in implementation.

Conclusions

It is quite straightforward to design and implement a general-purpose processing core by following RISC design guidelines. Here, we have presented the open source COFFEE RISC Core which can be used in SoC design or in conventional embedded systems. The core forms a good starting point to develop application-specific platforms.

If the performance is not at premium the circuit implementation work could be even automated to certain extent. Much of the work can be left to the synthesis tool if the implementation is done using hardware description languages. The RTL VHDL description of the COFFEE RISC Core enables to do this. Already as such, the COFFEE RISC Core is relatively powerful with its 100+ MHz FPGA and 300–500 MHz low-power ASIC operating frequency.

Development and research work for more automated processor generators is currently going on. The problem is that if we want to achieve a short time to market, we also have to be able to generate software tools for a new architecture version quickly. There, processor design tools such as the Processor Designer by CoWare would provide excellent support. We are also working on Linux support, including the integration of a Memory Management Unit (MMU) to the core.

The tools and models provided are currently sufficient to carry out application software development for the COFFEE RISC Core. The reader is referred to the later chapters of this book to probe further the tools and models.

The COFFEE RISC Core is, in fact, several designs with different characteristics, ranging from a stripped-off version without the hardware multiplier to the full version with integrated floating-point capabilities. In the future the development continues towards multi-issue, multi-threaded, multi-core, multi-processor direction to deliver more performance. Another way to add application-specific processing power to a COFFEE RISC Core-centric system is to include a loosely coupled streaming-oriented corprocessor, such as the BUTTER accelerator array described in another chapter of this book.

Readers interested to explore COFFEE RISC Core and its tool set in practice are referred to the COFFEE web site, coffee.tut.fi.

6 The DSP and Its Impact on Technology

Gene Frantz

Texas Instruments, Inc.

In this chapter, an overview of digital signal processing (DSP) is presented, starting from its history, stating the important problems in the field, and showing what kind of architectures can be used for processing signals.

Introduction

The early beginning of DSP

DSP is a relatively new science. It has its roots in a group of universities following the discovery (or rediscovery) of the Fast Fourier Transform (FFT) in the mid-1960s by Cooley and Tukey, see Figure 6.1. At the time, the only computing resources available were mainframe computers, Figure 6.2 shows as an example the computer used at MIT. Because signal processing requires a significant number of multiplications and additions, it was impossible to do any of the research in real time. Even though I'll define "real time" later in this chapter, let me attempt to give it some significance to the term at this point. The notion of real time is that there is no perceptive time delay from the signal's input to a processing system to the output of the processed signal.

J. Nurmi (ed.), Processor Design: System-on-Chip Computing for ASICs and FPGAs, 101–119.

An Algorithm for the Machine Calculation of Complex Fourier Series

By James W. Cooley and John W. Tukey

An efficient method for the calculation of the interactions of a 2^m factorial experiment was introduced by Yates and is widely known by his name. The generalization to 3^m was given by Box et al. [1]. Good [2] generalized these methods and gave elegant algorithms for which one class of applications is the calculation of Fourier series. In their full generality, Good's methods are applicable to certain problems in which one must multiply an N-vector by an $N \times N$ matrix which can be factored into m sparse matrices, where m is proportional to $\log N$. This results in a procedure requiring a number of operations proportional to $N \log N$ rather than N^2. These methods are applied here to the calculation of complex Fourier series. They are useful in situations where the number of data points is, or can be chosen to be, a highly composite number. The algorithm is here derived and presented in a rather different form. Attention is given to the choice of N. It is also shown how special advantage can be obtained in the use of a binary computer with $N = 2^m$ and how the entire calculation can be performed within the array of N data storage locations used for the given Fourier coefficients.

Fig. 6.1. The abstract from the paper by Cooley and Tukey on the FFT [87].

Fig. 6.2. An example of the early computer system used to do research in digital signal processing.

Advancements in computer architectures and the invention of the microprocessor began to allow some very simple signals to be processed in real time. But during this early phase of DSP, the array processor and minicomputer were the state of the art solution for processing signals and became the tool of choice in the DSP research community. To indicate the state of the art at that time, Dr. Tom Barnwell, now a professor at Georgia Tech, was credited with the following quote about DSP:

"That discipline which has allowed us to replace a circuit previously composed of a capacitor and a resistor with two anti-aliasing filters, an A-to-D and a D-to-A converter, and a general purpose computer (or array processor) so long as the signal we are interested in does not vary too quickly."

But, while the research community was busily discovering new algorithms and concepts in digital signal processing, there was another community of technologists who were on the path to the creation of the DSP. Simply described, the breakthrough that made the DSP possible was the combination of a microprocessor and a hardware multiplier. The high points of this path towards the DSP were:

- The invention of the transistor in 1948 at Bell Labs [330]
- The invention of the Integrated Circuit in 1958 at Texas Instruments [331]
- The invention of the microprocessor/microcomputer in 1970 [332]
- The invention of the Speak N Spell learning aid in 1978 at Texas Instruments [333]

As detailed above, when the research and circuits communities shared their individual successes/research/IP, etc. in the late 1970s and early 1980s, the DSP revolution truly began. Within approximately 30 years, the industry progressed from the invention of the transistor to the creation of the first commercially available DSP, thereby dramatically changing the future of the world.

The DSP revolution

If we look back at the approximately 40 years of the DSP revolution, we can summarize them in decades of progress. Simply, DSPs were primarily a university curiosity in the 1960s – a toy for university professors to play with. In the 1970s, DSP became a military advantage, as the available technology was prohibitively expensive to create and required costly hardware implementations. Only a few nations had the ability to take advantage of this new, expensive technology. The 1980s experienced the introduction of the cost-effective DSP device, resulting in its characterization as the decade of commercial success. Products such as voice band modems, Hard Disk Drives and 3D Graphics became successful by applying this new DSP technology to what had previously been a predominantly analog. That was followed by the 1990s, which can best be described as the decade of consumer expansion. The best example of this consumer expansion was the creation of the Digital Cellular Phone. Finally, that leads us to the 21st century. This ongoing period can best be described as the time when "DSP is everywhere."

If we examine the advent of the DSP according to its impact on society, we can characterize it as waves of the revolution. There seems to have been two distinct waves to date, with a third impending wave that has yet to fully materialize.

The first wave of the revolution was, and continues to be, communications. With the DSP, the analog wire-lined world of communications became digital. With voice communications digital, a secondary result was the rapid increase of the data rate of voice band modems. Within a short span of time the industry moved from 2400 bps modems to 56K bps modems. But, it didn't stop with wire-lined communications. It extended on to wireless communications. We now have digital cell phones everywhere. One of my favorite examples demonstrating this is when I was in India on a business trip. We were driving through the streets of Bangalore, and I noticed an ox drawn cart with a young man in it, leaning up against a bail of hay while chatting on his cell phone.

We have seen 802.11, DSL, cable modems, and Bluetooth become part of our everyday lives. DSP has truly revolutionized our lives regarding communications. And there is much more to come.

The second wave of the revolution is entertainment. We have seen our music, photos, TV, radios, and games go digital. There is a second phase to this story. Industry leaders have figured out that if we combine digital communications with digital entertainment, we can have streaming media. So, consumers now live in a world in which we can have our entertainment when and where we want it. And, it can all fit into our pockets.

This begs the question of whether there will be additional waves of this revolution. The simple answer is yes, of course. There are four obvious candidates for this next revolution:

Transportation. Our cars will become autonomous. This has already begun with the introduction of sensors on our cars to warn us of things happening behind us or to the side of us in our blind spot. It will only be a matter of time before our cars will drive from point A to point B without the need of a human.

Medicine. We will have longer lives with higher quality of life. We have already seen Cochlear implants created for profoundly deaf people to hear.

We are presently seeing similar things being done for the blind. Only our imaginations can predict the next capability that will impact our lives.

Security. We and our loved ones will be safe while maintaining their privacy. Products are already impacting our lives at the airport, on our streets and in our homes. We see the results on the TV news every night in the fight against crime.

Education. Learning will become a way of life. This is actually the sleeper among the four candidates. The first use of technology in the classroom seems to be students using cell phones to cheat. Although a creative use of technology, not an appropriate application. There will certainly be better uses of technology in the classroom in the future.

With this as a background, let's turn to the fundamental concepts of a DSP and how it differs from other processors.

Why a DSP is different

As we have evolved from the first DSP to the current embodiment of a DSP, the architecture of this processor has changed significantly. Later in this chapter I will spend some time describing this evolution. For now, it is important to note that even though the architecture has changed, the basic characteristics of a DSP are the same. These basic characteristics are as follows:

- Sampled data system
- Intensive mathematics
- Real time
- Deterministic
- Interrupt handling
- Accuracy
- Special hardware

As you read about each of the basic characteristics in the following discussion, you will be able to find more details, and perhaps better explanations, in references [246, 278, 378].

Sampled data system. A fundamental concept of DSP is that it is a Sampled Data System. The important fact to understand is that for a sampled data system to work, one must sample the signal at a rate that is at least twice the highest frequency of interest. This is known as the Nyquist rate. Further information about the Nyquist rate will be under the heading of "Sampling," "Sampled Data," or "Nyquist" in references [246, 278, 378].

Much of the remainder of this paper will depend on the reader knowing this much about a sampled data system.

Mathematically intense. The work horses of digital signal processing are the filter and transform. Filters take on two basic forms: Finite Impulse Response (FIR) and Infinite Impulse Response (IIR). Generally, the FFT is considered the basic transform in DSP, but there are many others. Simply put, the filter's purpose is to eliminate unwanted information (or to focus on specific information) and the transform's purpose is to look at the signal from a different perspective.

The basic concept of digital signal processing is the filter, specifically the FIR filter. When considering the basic concepts of DSP, note that you were likely introduced to it in the third grade. Then, it was called "averaging." When coupled with a dependency it is known as the "moving average" in other fields of mathematics. An example of an FIR filter used in following stock prices is shown in Figure 6.3. I will talk more about the FIR filter later in this chapter. The IIR filter is similar, but includes feedback from the output of the filter back to the input. In other fields of math this is known as "Auto-Regressive." In combination these filters are also ARMA filters.

The idea of a transform is simply to show the data from a different viewpoint. For example, the FFT converts the viewpoint of the data set from a time domain perspective to a frequency domain perspective.

All of the above, filters and transforms, demand a lot of multiplies and additions. In DSP terms these are called Multiply/Accumulates (MACs). For example, the simple form of an FIR filter is:

$$y(n) = \sum_{i=0}^{N-1} a(i) * x(n-i)$$

where y(n) is the output sample at time n, a(i) is the ith coefficient or weighting factor, and x(n – i) is the (n – i)th input sample.

Filter lengths (N) can be as long as 50 to 100 taps. This means that during each sample period, there are 50 to 100 multiplies and additions performed. The sample period, of course, depends on the signal being manipulated.

Table 6.1 summarizes the sample periods generally assumed for various signals.

Table 6.1. Typical signal bandwidths and sample periods.

Signal	Frequency band	Sample rate	Sample period
Telecom	4 kHz	8 ks/s	125 us
Audio	20 kHz	48 ks/s	20 us
SD Video (480p)		12 Mp/s	
HD Video (1080p)		120 Mp/s	

s/s means samples per second.
p/s means pixels per second.

Using pixel rate, rather than bandwidth, seems to be a better way to measure the requirement as video has two different concepts involved in its creation. The first is the rate of the sequence of images (i.e., frame rate). Generally the frame rate is between 25 and 60 Hz in a video application. The second is the number of pixels per frame, or the detail in each image (e.g., there are about 400 thousand pixels in a 480p frame and about two million pixels in a 1080p frame). If we then multiply the number pixels per frame by the frame rate we can see the result in pixels per second.

Therefore, a 50 tap FIR filter for an audio signal would require 2.5 million multiplies and adds each second.

It is worthwhile to give the FIR filter some physical significance. An FIR filter is a common technique used to eliminate the erratic nature of stock market prices. For example, when the day-to-day closing prices are plotted it can be difficult to ascertain the trend of the stock due to the large variations in pricing. A simple way of smoothing the data is to generate the average closing values of previous days to represent the current day's value of five, for example. For the new average value each day, the oldest value is dropped and the latest value is added. Each daily average value (Average (n)) would be the sum of the weighted value of the last five days closing price, where the weighting factors (a(i)s) are each 1/5. In equation form, the average is determined accordingly:

$$\text{Average}(n) = (1/5)*d(n) + (1/5)*d(n-1) + (1/5)*d(n-2) + (1/5)*d(n-3) + (1/5)*d(n-4)$$

where d(n−i) is the daily stock closing price for the (n−i)th day.

In the stock market is called the "Moving Average," which are sometimes 50-day or 100-day Moving Averages. In either case, they act as a smoothing filter (low pass) on the data. Figure 6.3 shows a 90 day Moving Average for a particular stock.

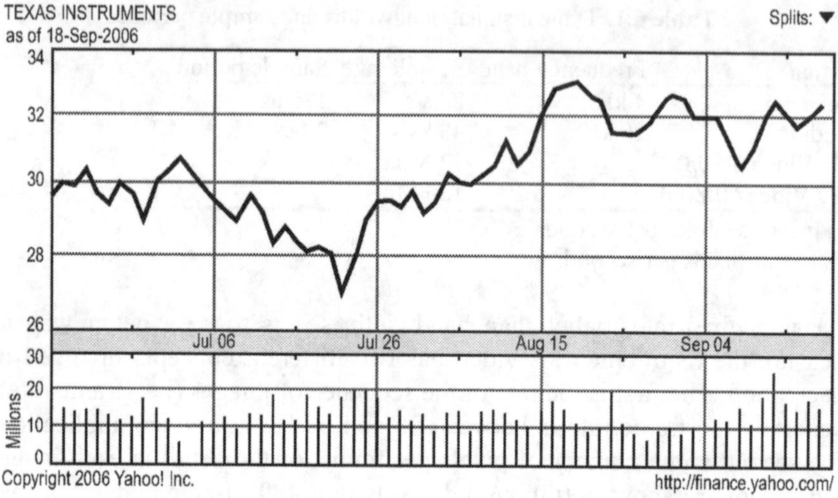

Fig. 6.3. A 90 day moving average plot of TI stock.

Real time. Real time addresses two different aspects. The first is the concept that for each sample period there needs to be a new input accepted and a new output created. This is generally considered the primary aspect of "Real Time." The second concept, latency, is more difficult to understand. Latency is the measure of the delay from the input of the signal to the output of the signal. The easiest example of latency can be described by thinking of a live audio system, such as a recording studio. In this type of application, the musicians have an earpiece to listen to themselves and to the others in the recording session. The delay between the creation of the sound and their listening to it in their earpiece must be less than 5mS or so for it to be considered real time. An example of an application that does not have a latency requirement would be the playback of a movie in a home theater. In this case, there is no issue with the length of time it takes to get the signal from the source (e.g., DVD) to the screen and speakers, as long as the delay is identical for both the video and audio, so the two are synchronized.

Deterministic. Deterministic also relates to two different aspects of digital signal processing. The most important aspect is the concept that each instruction should take the same time to execute independent of the data or state of the execution unit. Maintaining determinism guarantees that the process will always be in real time.

The second aspect addresses the idea of predictability. This aspect tends to be important early in the design cycle when estimates are generated relating to the capability of the DSP to meet the demands of the system.

Interrupt handling. The way DSPs handle interrupts has evolved since their initial development. Initial considerations were that DSPs should not have interrupts at all due to the reasoning that real-time systems cannot be interrupted and at the same time guarantee real time. However, this hard position on interrupts was quickly compromised and interrupt handling was added. The compromise, in some cases, included the concept that the CPU could not be interrupted until it was in a place it could safely accept interrupts without compromising real time. The compromises since then have taken many directions, but the common theme remains that real-time systems can not be interrupted.

Accurate. Accuracy is signal dependent. Each type of signal has different accuracy requirements and each signal type has three different concepts of accuracy. The three concepts are:

- Data accuracy. Data accuracy equates to the signal to noise ratio or to the dynamic range of the data set. The greater the number of bits, the higher the signal to noise ratio (or dynamic range). The simple relationship between dynamic range and the number of bits is 6dB per bit.
- Coefficient accuracy (filter coefficients). Coefficient accuracy becomes important in the use of digital filters, particularly recursive filters (IIR filters) when the coefficients are near the unit circle.
- Internal accuracy. Internal accuracy needs to be larger than the data accuracy to guard against introducing errors into the data. Later in the chapter I will show the block diagrams for several DSPs. You should note that each has a greater internal accuracy than its data accuracy.

Table 6.2 is a chart summarizing the various aspects of accuracy for each of the DSPs. I have added a few other architectures for your information. Notice that the last four examples are those which have been successfully used in audio applications. Three of the four are floating point architectures. The one that is not is the 56K originally from Motorola. I have shown both the accuracy for single and double precision for the last of the devices. I do so as it was specifically designed to easily do double precision math. But, to be fair, all of the devices can do double precision math.

Now you can compare Table 6.3 which shows several signal types with their various accuracy requirements with Table 6.2 on the same aspects for the various architectures.

Special hardware. In addition to the aforementioned general characteristics of a DSP, there is hardware specific to the DSP to help manage the real-time nature of the mathematically intense algorithms. Some of those special pieces of hardware include:

Table 6.2. Accuracy in signal processing architectures.

DSP architecture	Data	Coefficient	Internal
TMS32010	16	12/16	32
TMS320C62x	16	16	32
TMS320C30	24	24	32
Motorola 56K	24	24	32
ADI 21060	32	32	40
TMS320C672x	24	24/53	32/64

Table 6.3. Accuracy requirements of applications.

Signal	Data accuracy	Coefficient accuracy	Internal accuracy
Telecom	8 bits (13 bits)	16 bits	32 bits
Audio	24 bits	32 bits	53 bits
Video	8 bits	16 bits	32 bits

Loop hardware. Many DSP algorithms are repetitive. For example, a 20 tap FIR filter will take the latest 20 data points, multiply each data point by a unique coefficient and add all of the results to generate an output. Each sample period the oldest data input (i.e., n−20) is dropped and the newest input data (n) is added. This new data set is then multiplied by the same set of coefficients and a new out put is generated. This continually repeats and is known as an iterative algorithm that works best with special looping hardware, which allows the algorithm to run with little or no overhead.

Data memory management. One of the other important aspects of DSP theory is that of the delay operator. In a sampled data system not only is the present input data important, but also the previous N data points. As discussed in the previous paragraph, an FIR filter uses the previous N (in this case 20) data point to determine the present output. To make this happen smoothly, the data has to be managed in such a way that the DSP automatically has the right data point at the right time in its calculation. To ensure that this occurs, special memory management schemes are combined with the loop hardware to make the data management relatively transparent.

Saturation logic. Saturation logic simply tests the results of each operation to determine if those results are out of bounds. To best understand this concept, let us consider an analog system. In an analog system, when the signal gets too big, it does not exceed the largest possible value that the analog system allows. Rather, it saturates at this largest value, either positive or negative. Not so in a digital system. When two values are multiplied together and the result is larger than the largest possible value, it wraps to the most negative value. Therefore, the concept of saturation logic was developed. But, before describing the details of saturation logic, it is important to note that there are other concepts used to minimize this possibility of wrapping.

For example, one concept is to scale all of the data and coefficients to be constrained to be between –1 and +1. This guarantees that when two numbers are multiplied together their results will always be between –1 and +1. Thus, no wrapping. While that addresses the multiply function, it does not handle addition, or the accumulator function. If a series of numbers are added together it is very possible that the sum could fall outside of the range of –1 and +1. This is a very complex issue and is generally resolved by the use of guard bits in the accumulator or register. Using a floating point processor is also a good way to overcome the wrapping issue. But, enough about these topics. Let us revisit the concept of saturation logic. It simply tests the results of each operation to determine if those results are out of bounds. If they are, then the maximum positive or negative number is assigned to the result. This can be both a positive and negative occurrence depending on your system and therefore the saturation logic can be turned on or turned off.

Multi-MAC. Prior to the introduction of the DSP, the use of multiplies in an algorithm was minimized. But, with the hardware multiplier the goal changed from the minimization of multiplies to the optimization of the multiply and add function (MAC). It then became obvious that, in a mathematic intense environment, more MACs directly impacted the performance of the DSP. Therefore, the performance of the DSP has grown rapidly over the last couple of decades and most of the growth is the result of the increasing the number of MAC units that can execute in parallel. Figure 6.4 shows a chart on how DSP performance has increased over the last two decades.

Bit reversal (8). This concept was adopted to increase the performance of the FFT. In an FFT, there are two ways to present the data to the transform. The first is to present the data in the correct order. The output of the transformed data will then be in bit reversed order. The second way is to present to the transform the data in bit reversed order and then the transformed data will be in the correct order. Figure 6.5 shows a simple example using an eight-point FFT.

By now, hopefully, you have begun to appreciate the differences between a DSP architecture and other popular microprocessor architectures. And, you have begun to appreciate why these differences are necessary. Simply put, all of the architectural uniqueness exists to guarantee real time is never violated. Remember, real time is extremely unforgiving. The next section will talk about how DSPs will likely continue to evolve so as to understand how they will influence our futures.

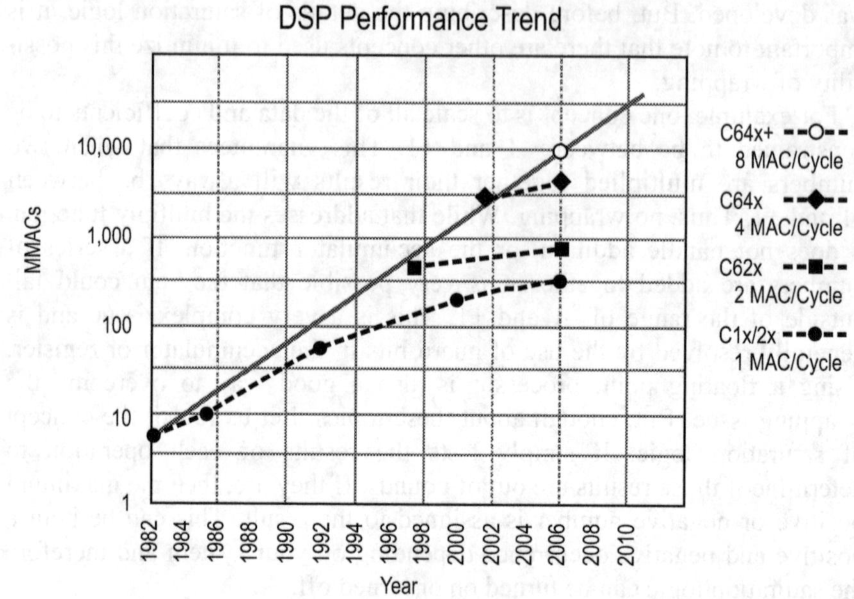

Fig. 6.4. There are two drivers for DSP performance: clock speed and the number of MAC units with the number of MAC units as the dominate driver.

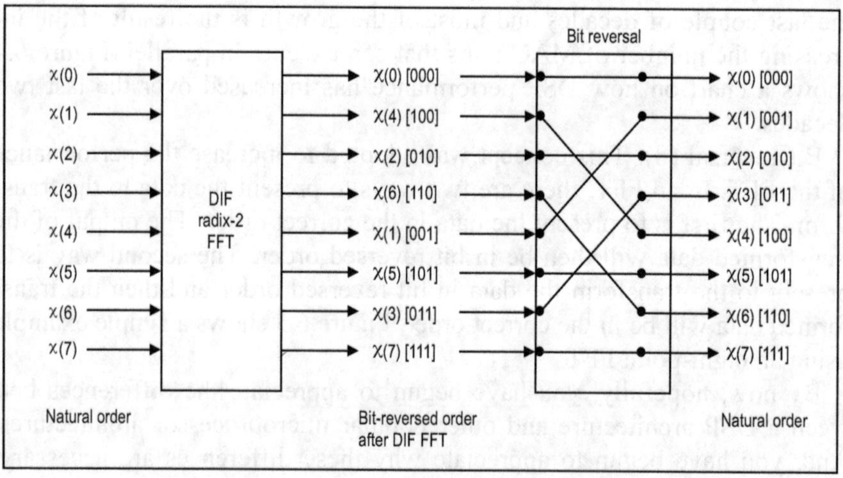

Fig. 6.5. An example of bit reversal in an eight-point FFT.

The evolving architecture of a DSP

As previously discusses, the primary breakthrough that allowed the development widespread adoption of the DSP was the addition of a hardware multiplier to the microprocessor. With this innovation, the objective in digital signal processing was no longer to reduce the number of multiplies an algorithm needed, but to optimize the number of multiplies and adds needed.

Now, a brief review of microcomputer architecture will help in the understanding of the evolving DSP architecture. When DSPs were first architected there were two generally accepted microprocessor/microcomputer architectural styles. One was the Harvard architecture and the other was the von Neumann architecture. The Harvard architecture was a two buss architecture (one for program and one for data) and the von Neumann was a single buss architecture (both program and data were stored in the same memory space). Neither architecture exactly fit the engineering community's needs for a mathematical intense processor. The simple reason is that a multiply has two inputs (multiplier and multiplicand) and one output. To get the two inputs into the multiplier in one instruction cycle required two busses. Of the two, the Harvard architectural style had two busses, so it was what the engineering community gravitated to for real-time DSP. However, it was modified so as to take advantage of both busses when performing multiplications. The modifications allowed both busses and both memory locations to handle data.

Special instructions were also created to perform all of the necessary operations to do a multiply (MPY) in one instruction cycle. The necessary operations were:

- two memory accesses to load the multiplier and multiplicand into the multiplication unit;
- the execution of the multiplication;
- a data shift to manage the binary point;
- one memory access to store the result into memory.

Later, the accumulate operation (addition) was included into the above MPY instruction to allow the full MAC (multiply/accumulate function) to occur in one instruction cycle. The MAC instruction allowed for a single instruction cycle FIR filter tap. So, in addition to the operations necessary for the multiply instruction, there were additional data manipulations in memory to update the data inputs. Figure 6.6 shows the block diagram for an early DSP architecture, the TMS32010.

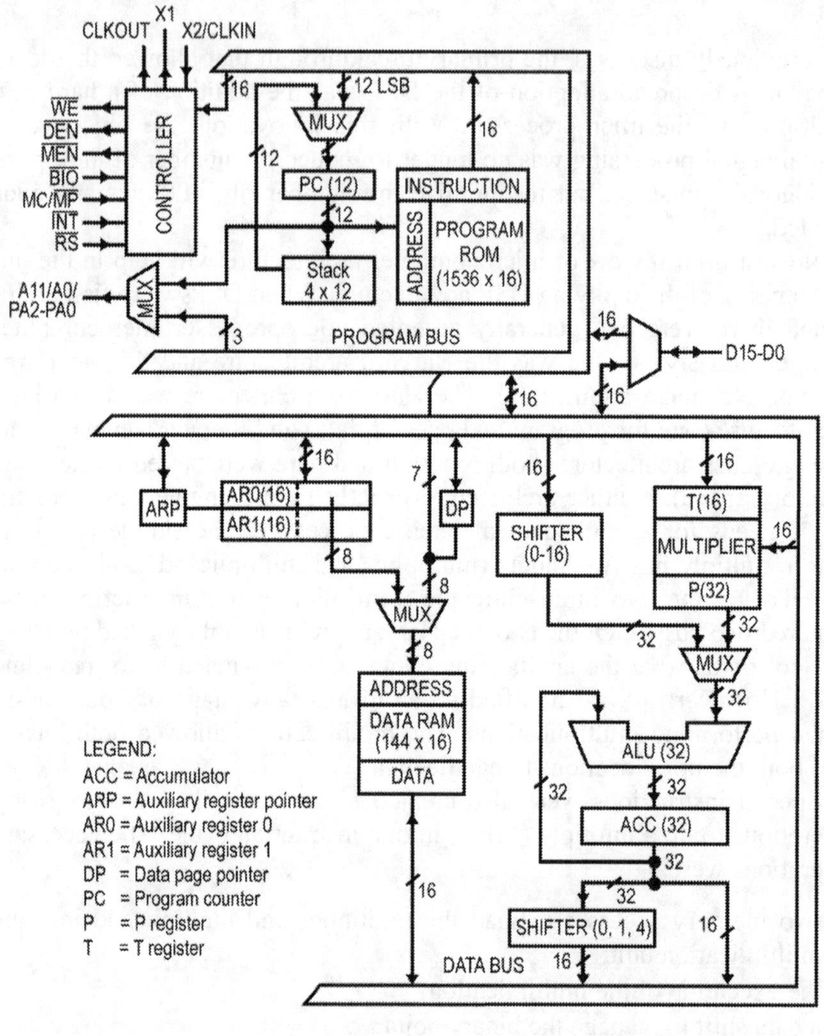

Fig. 6.6. Block diagram of the TMS32010, TI's first DSP offering.

After the success of the initial DSP's using concepts from the Harvard architecture, a new concept was introduced. That is, the combining of the best concepts of both the Harvard and von Neumann architectures thus creating a multiple bus von Neumann style DSP. Figure 6.7 shows an example of such an architectural approach, the TMS320C30.

As the Integrated Circuit technology grew in performance, an old architectural concept, was introduced to the DSP. That old concept is known as

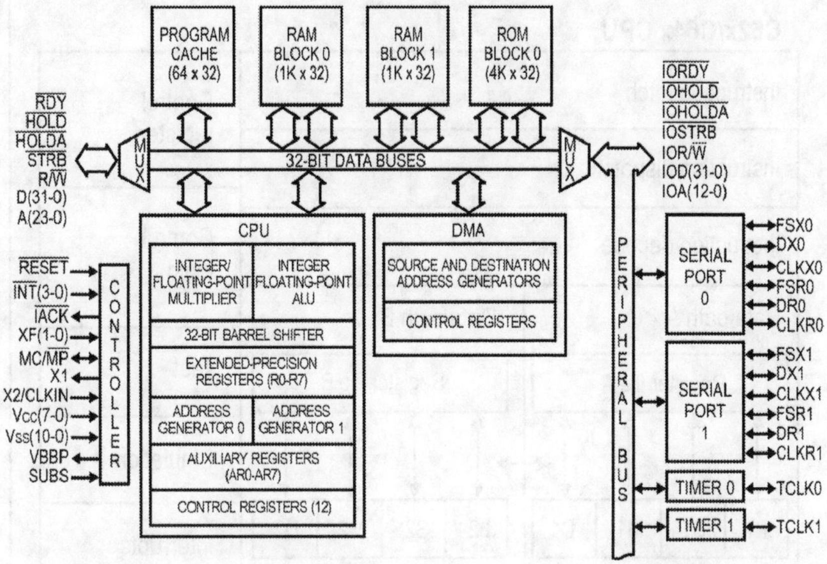

Fig. 6.7. Block diagram of the TMS320C30, TI's first floating point DSP.

Very Long Instruction Word (VLIW) architecture. Figure 6.8 is an example of a VLIW DSP architecture. VLIW allowed significant parallel processing to occur while maintaining the real-time constraints. The VLIW architecture will be discussed in more detail in the next chapter.

As an aside, if you look back at Figure 6.4, you can see how DSP performance has increased through both the raw performance increase from IC technology (i.e., higher clock speed resulting from more advanced technology nodes) and the use of parallel processing units (i.e., more multiply-accumulate units accommodated by the architecture).

What is next in the evolution of the DSP

DSP has made a couple of significant transitions in it brief 40 year history. The first was in the late 1970s when it moved from digital signal processing to the digital signal processor. The second transition is occurring now in the early part of the first decade of the 21st century. This transition is a bit more subtle. It is the transition from DSP as a theory and device to DSP as an enabler. If this sounds a bit confusing, let me give an example of another area of technology that experienced the same transition.

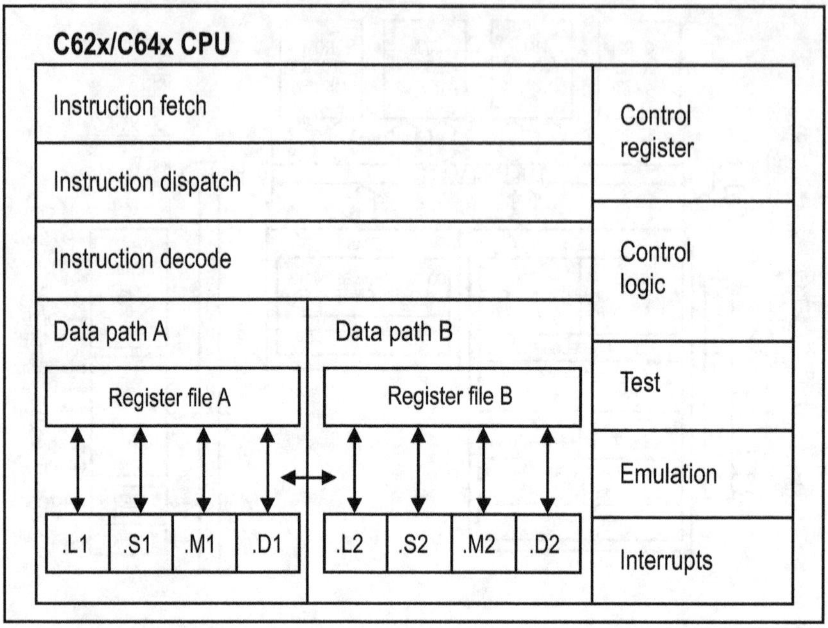

Fig. 6.8. Block diagram of the TMS320C6201, TI's first VLIW DSP architecture.

That area of technology was CMOS technology (or as it was called at Texas Instruments, MOS technology). When I was hired into TI, we had a MOS department within the semiconductor business unit. But, if I look around the company today (and also the industry), I don't find business units named "MOS" or "CMOS." In fact, every business unit in the company uses CMOS to create their products. In other words, "MOS" or "CMOS" have become the enablers for new products rather than the products themselves. In the same way, the industry is in a transition where a DSP is no longer a product in a company's portfolio; but, rather, it is becoming an enabler for many products at the TI. This trend is not unique to TI, but is an industry wide phenomenon.

As integrated circuit technology has advanced over the years it has progressed from technology measured in microns (i.e., micrometers), to submicrons, to nanometers. With these advances, we have shifted from thousands, to millions to hundreds of millions of transistors on a single device. What this means to the industry is that we can now put sophisticated systems, or sub-systems, on one piece of silicon, known as a System-on-a-Chip (SoC). On these SoCs we can integrate host processors, DSPs, accelerators (e.g., signal specific DSPs) and peripherals. What this allows is the ability to now put what we used to include on a board-level product, on a single integrated circuit. We can now integrate not only millions of gates of digital circuits, but also analog circuits and RF circuits. This is all good news. But, along with it comes some bad news. That is, the cost to create new devices is becoming more prohibitive at each new technology node. For example, presently a typical IC design costs in the order of thirty million dollars (or more) which makes an application specific IC unrealistic of most opportunities.

As evidenced above, we are moving to a programmable world. There are four reasons for this movement to a programmable world:

- Integrated circuit technology trends make it possible. The consistent dividend we have received and will continue to receive from advancing IC technology is more transistors. Specifically, at each new technology node the density of transistors doubles. So, every two to three years we get twice the number of transistors for the same, or similar, cost.
- Integrated circuit development cost trends make it necessary. The cost of developing new ICs is becoming prohibitive for all but the largest opportunities. Even the cost of creating the mask set for a new device is millions of dollars. Without programmability, this will continue to reduce the number of companies who can innovate using IC development.
- Integrated circuit design complexity trends make programmability the best solution. The task of designing a new IC can be compared to designing a large city, like New York City, from scratch, and expecting every traffic light to work correctly the first time. Some would argue that it would be easier to design the city than a state of the art high performance processor or SoC.
- Market trends demand it. Time to market is a vital requirement in today's environment. Simply put, the first to market with a new product idea makes an unfair amount of profit. Second to market will probably break even. Third and beyond will certainly lose money even if their product is superior to the other products.

The good news of programmability is that it expands the innovation happening in the world of digital signal processing, which is illustrated by the graph shown in Figure 6.9.

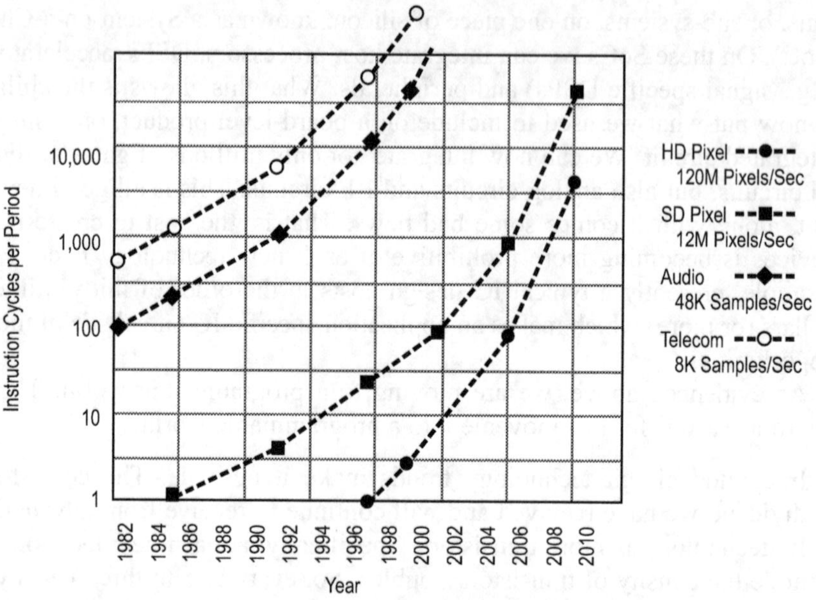

Fig. 6.9. The growth in the number of instructions per sample period for voice, audio, standard and high definition video.

This graph shows how the number of instruction cycles per sample period have increased for four different applications: Telecommunications, Audio, Standard Definition Video and High Definition Video. As we see in Figure 6.9, when the industry introduced their first DSPs, we had about 625 instructions per sample period in a telecommunications system. To guarantee real-time, all of the processing for one sample period had to have fewer than this magic number of 625 instructions. But, notice what has happened over the 20 or so years since those first devices were introduced: Performance has gone up significantly. We are at the point where we virtually have an infinite number of instructions per sample period available to us. So, what have we done with all of those additional instructions? We've innovated. Here are some of those innovations:

- Reduction of vocoder data rates from 64K bps to 1 to 2K bps
- Significantly improved the voice quality of the vocoder in spite of the data rate reduction
- Eliminated echos over the phone line
- Reduced unwanted out of band and in band noise
- Taken modem data rates from 2400 bps to 56K bps

That is all quite impressive.

But look again at Figure 6.9. There is a second line that represents the improvement in instructions per sample in audio. For audio, that first generation of DSPs allowed 100 instruction cycles per sample period. This seemed to be enough to make it interesting. But, once again, we have increased the number of instruction cycles per sample period to the point that we are doing a lot of innovation in this area. We went from monaural audio, to stereo sound, to 5.1 to 7.2. We also included room corrections and room enhancement. Once again, quite impressive. And unlike telecom, we also are increasing the sample rate from 48KHz to 96KHz to192KHz to 284KHz. But, that is another story in itself.

There are two other lines on the figure. Both are for video systems. The first for Standard Definition (SD) video and the second for High Definition (HD) video. If we chose 100 instructions per pixel as the threshold of when a programmable DSP becomes useful in a system then we can determine that this threshold was met for SD in the year 2000, approximately, and HD was met in 2006. The significance of this trend can be demonstrated by looking at the impact DSPs have already made in voice and audio based products. If you think back to the earlier discussion on the four candidates for the next wave of the revolution you will note that all have ties to video, imaging or vision concepts. Just as the explosion of innovation occurred in voice and audio products began when the number of instructions per sample rate grew passed 100 instructions per sample.

Summary

DSP has taken on many roles in our lives. It has gone through some significant changes over its 40+ year history. In this chapter I have discussed several aspects and influences of DSP. First I talked about the early days of discovery (or rediscovery) of DSP theory. As the theoretical basis of DSP was being established, IC technology was also advancing. Once the theory of DSP was combined with the advancing IC technology, the DSP emerged. I next discussed the revolution that the DSP began, is growing and is continuing to grow. Then I discussed the architectural uniqueness of DSPs with some explanation of the significance of the uniqueness. That discussion was followed by a look into the future and the direction DSP architectures will take. Finally I discussed the drivers for those future DSPs.

We have an exciting future ahead of us as new discoveries and new applications of those discoveries are made useful to us as individuals and to us as a society.

7 VLIW DSP Processor for High-End Mobile Communication Applications

Christian Panis

Catena Radio Design, B.V.

Today's mobile communication business is impacted by high data rates, mobility demands and flexibility requirements. High data rates are implicating high processing power, whereas mobility demands have to be covered by ultra-low power dissipation, and fast evolving standards demand for flexibility. The solution to this problem looks like dedicated hardware implementations, but those are lacking in flexibility. Adapting a system to a successor standard or supporting different systems on the same hardware platform is merely impossible.

If only high data rates and flexibility aspects were taken into account, the problem could be tackled traditionally by deeply pipelined high performance processor architectures, providing several GHz of processing power. Executing software on the processor architecture gives the flexibility to develop platforms which support different standards where even some of them can evolve without demanding for a new platform. The benefits of deeply pipelined processors are soon outweighed by control and data dependencies in the application code itself and the high clock rates are conflicting with ultra-low power requirements anyway.

The traditional solution approaches do not seem to fit. The exit strategy out of the dilemma of providing high data rate support at low power dissipation with enough flexibility is *efficient use of resources*. Identifying those parts of the application where flexibility provides an advantage over dedicated hardware implementations requires to handle complex trade-offs between hardware and software partitioning.

This chapter introduces an application-specific adaptable core architecture for SoC platform solutions. The architecture provides sufficient processing power as well as the needed flexibility without violating ultra-low power dissipation constraints. A brief introduction in the varying needs of

J. Nurmi (ed.), Processor Design: System-on-Chip Computing for ASICs and FPGAs, 121–148.

mobile communications is followed by a summary of requirements of DSP processing. Different architectural concepts as introduced in Chapter 2 are briefly evaluated on their suitability. In the second part of this chapter the *3a* architecture is introduced, an RISC based signal processing architecture providing increased efficiency via application specific modifications. The trade-off between kernel benchmarking and application benchmarking is discussed, followed by an introduction of Design space exploration, an application benchmarking method making use of the target application. The chapter ends with discussing challenges of configurability and a brief summary.

Trends in mobile communication

Edholms law introduced in 2004 by Phil Edholm [78], describes the trend of data rate increase in communication, claiming that the increase in data rate is predictable, similar to Moore's law[1] for transistor integration [293].

As illustrated in Figure 7.1, the communication standards are split into three groups: wireline, nomadic, and wireless communication standards. Wireline standards are traditionally dominating the field of high speed communication. The group of nomadic standards is being affected by increased mobility providing lower data rates. The focus of the wireless communication is mobility at the price of lower data rates compared to wireline communication, but this gap is getting closer as illustrated in Figure 7.1.

High data rates implicate increased signal processing requirements and increased memory demands caused by the higher data throughput. The traditional approaches to tackle this problem are deeply pipelined processor architectures with high clock rates up to several GHz, but this approach violates one of the key demands in wireless communication: ultra-low power dissipation.

The target of providing re-programmable/re-configurable platforms to support different standards on the same hardware platform is still suffering

[1] "The complexity for minimum component costs has increased at a rate of roughly a factor of two per year ... Certainly over the short term this rate can be expected to continue, if not to increase. Over the longer term, the rate of increase is a bit more uncertain, although there is no reason to believe it will not remain nearly constant for at least 10 years. That means by 1975, the number of components per integrated circuit for minimum cost will be 65,000. I believe that such a large circuit can be built on a single wafer." Electronics Magazine, 19th April 1965.

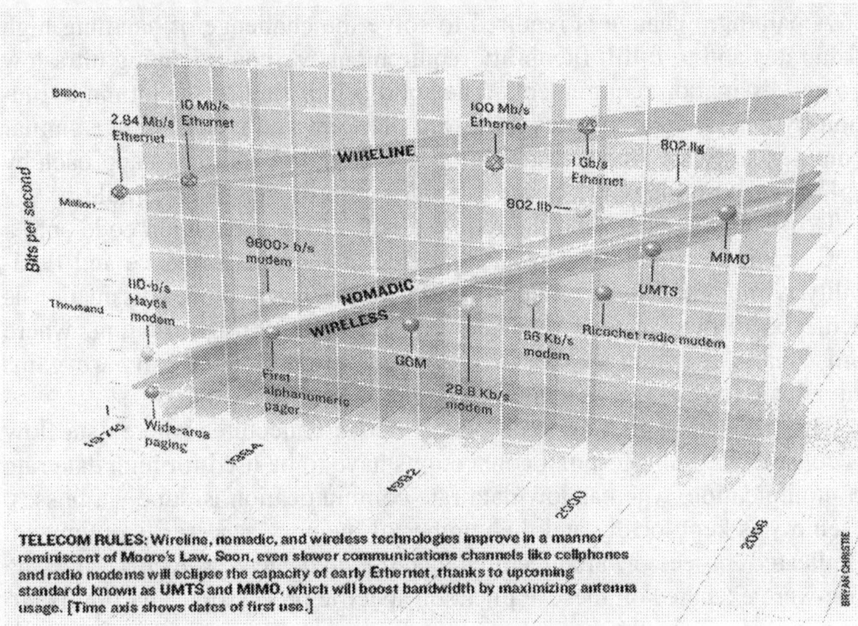

Fig. 7.1. Trends in mobile communication. © IEEE, 2004 [78].

from contradictory requirements of *high data rates*, *flexibility*, and *ultra-low power dissipation*. The focus of the ultra-low power aspect is mainly set in mobile terminal application, whereas the power dissipation requirements for base-stations are still not that severe. For years industry has been working on how to overcome this problem with different approaches. The SDR forum (Software Defined Radio forum) provides a platform for industrial and scientific partners working on solutions in this area [370].

Sandbridge Technologies, Inc. as one example is tackling the problem with high performance multiprocessor SoC architectures [156]. The SDR solution is based on several Sandblaster DSPs [155], where each of them utilizes a complex internal multi-VLIW architecture. Token-triggered multithreading supports the execution of several tasks in parallel. The significant micro-architectural complexity of the chosen architecture makes an efficient programming of the available resources at least questionable if not impossible.

The SDR forum suggests waiting for technologies with smaller feature size as one possible solution to tackle the power dissipation issue. The supply voltage, contributing quadratic to active power dissipation, is not significantly decreasing anymore and the leakage current is getting problematic. Therefore the decrease of power dissipation from one silicon generation to the next one is not that significant.

A paradigm change is required to solve the challenge of handling high data rates and to fulfill flexibility requirements without violating ultra-low power dissipation requirements. The approach of finding the ultimate solution by designing a complex hardware platform and trying to map similar complex software architectures onto it has to fail. A top-down approach by splitting the problem into easier solvable sub-problems is preferable.

How can we understand the requirements of the application code onto a SoC architecture which is able to provide enough performance and flexibility without violating low-power constraints? Low power dissipation is achieved through a well-balanced hardware/software partitioning, where those parts are kept flexible whose implementation in software gains advantage for the overall solution.

The evaluation process shall lead to SoC architectures where data flow is considered. Point-to-point connections have to be used for high data rate communication, whereas low data rate communication is done via busses. Memory is kept locally and the number of memory accesses is optimized. Dedicated processor architectures handle dedicated problems, and the firmware executed on these application specific processors is partitioned to be handled efficiently.

The SoC platform shall then be *power and silicon efficient*, still providing enough flexibility to obtain the requirements of evolving standards and multi-standard solutions. *Design Space Exploration* is a methodology for giving system architects the possibility of quantifying architectural drafts already in an early project phase. The *3a*'s Design Space Exploration methodology is introduced later in this chapter.

DSP-specific requirements

DSP algorithms are the backbone of communication standards as illustrated in Figure 7.1. The mapping of these algorithms onto a target platform requires considering specific aspects. This section covers some of these aspects for giving a basis to check various architectural concepts on their suitability.

- Real-time requirements

Real-time requirements arise when algorithms have to be executed and completed within a pre-defined time frame. Unpredictable delay elements such as data dependent execution time of instructions are violating real-time requirements [128]. Data and program cache misses can increase the

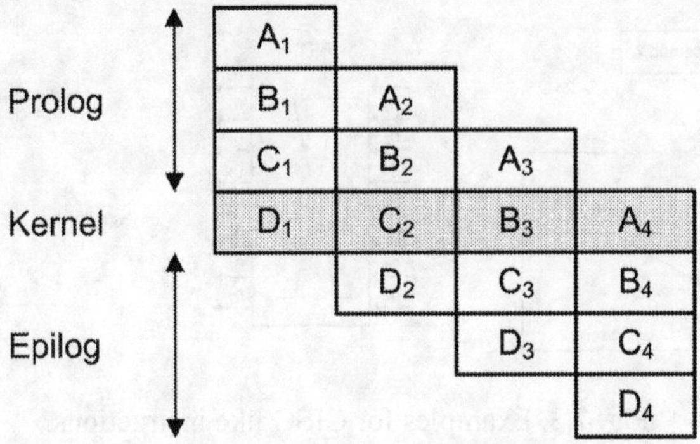

Fig. 7.2. Software pipelining.

execution time of the algorithm and therefore violate the real-time constraints. To guarantee real-time execution, the *worst-case execution time* (WCET) [187] has to be analyzed and considered during task scheduling.

- Hardware loop support

DSP algorithms are often loop-centric. Therefore efficient loop handling significantly increases the execution efficiency of the target platform [336]. The term *zero overhead loop* is used to describe micro-architectures supporting loop execution without loop handling overhead. The loop counter is implicitly inc/decremented and at the end of the loop body, the jump back to loop start is done without interaction of the application program. Otherwise for small loop bodies the overhead for loop handling can get significant. Loop-centric algorithms can make use of the possibility of executing more then one instruction per cycle (ILP, Instruction Level Parallelism greater than one). *Loop unrolling* in combination with *software pipelining* allows executing multiple loop iterations in parallel, therefore decreasing the total number of cycles required for the loop execution [252]. The code density is getting slightly worse through the additional instructions of prolog and epilog as illustrated in Figure 7.2.

- Support of CISC instructions

The execution of DSP algorithms is getting more efficient, by introducing some CISC-like instructions. The most famous one is the multiply-accumulate (MAC) instruction, used as kernel for FIR filter implementations and illustrated on the left side of Figure 7.3. Specialized addressing modes like bit-reversal addressing, as illustrated on the right side of Figure 7.3, are increasing the execution efficiency of arithmetic kernels like FFT [118].

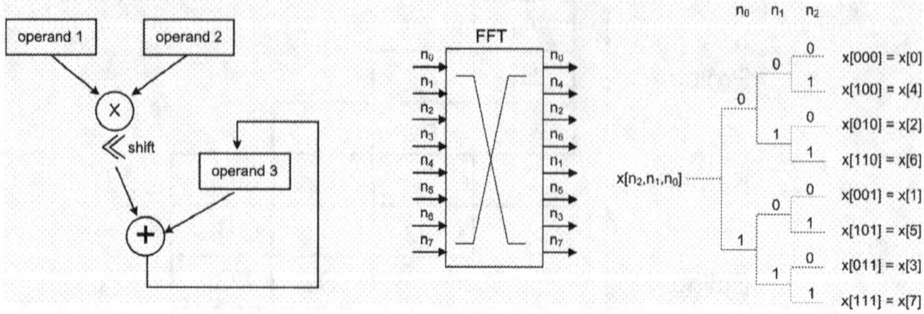

Fig. 7.3. Examples for CISC like instructions.

- Control code support

In current application code for DSP architectures, control code is not that separated anymore as it was 15 years ago and thus automatic code generation through high-level languages and compilers like C/C++ is needed [3,19]. Structural programming techniques result in code sections with increased amount of control dependencies with poor ILP. The development of an optimizing compiler which generates efficient code for a VLIW architecture has to be done in parallel to the design of ISA and micro-architecture. Later in this chapter some of the aspects to be considered are mentioned.

Microarchitectural concepts

In Chapter 2 of this book, several micro-architectural concepts are introduced. Some of them will be now briefly checked regarding their suitability to provide increased ILP, to enable the development of a high-level language compiler and to guarantee the worst-case execution time of algorithms.

- Scalar processor architectures

On scalar processor architectures one instruction per cycle can be executed, which limits the theoretical achievable ILP to one. Dynamic instruction scheduling during runtime enables the minimization of the average execution time, but complicates the calculation of the worst case execution time [377]. To obtain real-time requirements, the minimization of the worst-case execution time is important. Hardware support for resolving dependencies and resource allocation reduces the complexity of compiler development.

- Superscalar processor architectures

  Superscalar processor architectures are using dynamic scheduling, similar as scalar architectures. Therefore the issue of minimizing the worst-case execution time is not resolved. Different to scalar architectures, the execution of more than one instruction per cycle is supported, increasing the theoretical achievable ILP [374]. The combination of increased ILP and dynamic scheduling support requires additional hardware circuits like scoreboards and for example the introduction of Tomasulo's scheme [376], which is based on register renaming in combination with scoreboard elements. Similar as for scalar architectures, the hardware support for resolving dependencies and scheduling of instructions makes compiler development easier.

- VLIW

  VLIW stands for Very Long Instruction Word, meaning that one VLIW instruction consists of several atomic instructions executed at the same time. Different to scalar or superscalar architectures, data and control dependencies are resolved at compile time and instruction scheduling is done statically [126]. The scheduling information, indicating when an atomic instruction has to be executed relative to other instructions, is stored through its location. Static scheduling allows optimizing the worst case execution time. The drawback of static scheduling is the missing support for unpredictable delay elements e.g. caused by cache misses.

  VLIW enables to execute several instructions in parallel and therefore offers an ILP greater than one similar to superscalar architectures. Compiler complexity is increased by missing hardware support for dependency resolution and instruction scheduling. That is probably one of the reasons why VLIW is only successful in the area of embedded DSP architectures, where quite often assembly programming still dominates.

- Dynamic VLIW

  Dynamic VLIW as introduced in [357] removes responsibility from the compiler. There is dedicated hardware support for instruction instantiation. Instruction scheduling is done statically like in VLIW processors whereas issuing of instructions is done dynamically, similar to concepts of super scalar architectures. This decoupling of instruction scheduling and issuing was motivated by having binary compatible versions of the same architecture. For embedded solutions this aspect is not that severe. As analyzed in [357] the increased hardware complexity does not gain advantage in performance.

• EPIC

The EPIC architecture, introduced in 1990s by HP and Intel tries to overcome the drawbacks of static scheduling by introducing a stronger hardware support compared to traditional VLIW architectures [383]. Improving the approach of dynamic VLIW the assignment to functional units is done in hardware, whereas the instruction scheduling is still done statically as for VLIW architectures.

Summarizing the discussed micro-architectural concepts, VLIW allows a minimization of the worst-case execution time and therefore to satisfy real-time requirements. Further improvement steps like EPIC require additional hardware circuits and increase implementation complexity. The advantages of these improvements are a simplified compiler design and a solution for binary compatibility. Apart from the aspect of code density and more complex compiler design, VLIW seems to be a good candidate to execute DSP algorithms efficiently. Later in this chapter a VLIW architecture is introduced. This architecture provides solutions for the code density drawback and the lacking compiler support.

VLIW and SW programmability

One of the major drawbacks of VLIW architectures is still the poor compiler support. For loop-centric DSP applications, that weights not so severe. The achieved ILP and the good predictability at least made it a success for DSP core architectures. Due to increased application complexity, high-level language programmability was getting more and more an issue in the 1990's. To improve compiler support for VLIW architectures micro-architectural features have to be checked on their suitability for an optimizing compiler. Some of these features are discussed in this section. Most of the aspects mentioned here can be found more deeply analyzed in [10,126].

• Load–store architecture

Load–store architectures are decoupling the data memory access from arithmetic operations. Dedicated load–store instructions are transferring data between register file and memory, whereas the operands for arithmetic instructions reside exclusively in the register file. Slower memory can be used consuming more clock cycles than used during execution of arithmetic instructions. Therefore a more efficient use of the hardware resources is possible. If the provided ILP is greater than 1, methods like software pipelining or loop unrolling can make use of the increased parallelism. The decoupling of the arithmetic instructions from the memory accesses increases register pressure. Register pressure is a measure for the balance between intermediate results to be stored in the register file at time *t*, with the number of available registers. The increase is due to the fact that intermediate results have to be kept in registers as well [365]. Therefore a sophisticated liveness analysis and register allocation is required to prevent excessive spill code. The liveness analysis analyzes a Control Flow Graph (CFG) to determine at which places variables are alive or not. Spill code describes code sections, used to free space in the register file by storing intermediate results to memory and restore them, when the values are needed again.

- Large uniform register sets

  Registers are used for keeping as much intermediate results as possible in order to reduce the number of memory instructions. For load–store architectures they play an even more important role because they have to keep the values fetched from data memory to be used as operands for arithmetic instructions. For implementation reasons register files are often split into sub-register sets. Some sub-register sets are assigned to a group of functions which allow reducing the number of read and writing *ports* at the sub-register file. Examples are banked register files, used e.g. at the Motorola 56000 processor for the address registers. Each of the two AGU (Address Generation Units) has its own small register set. The general purpose register file of TMC62xx from Texas Instruments is an example where banked register sets are assigned to a group of data paths. Separate read/write ports are introduced to exchange data between the register file parts, requiring cycles and decreasing code density.

  The advantage of splitting register sets into subsets and making use of banked register files can be outweighed through side-impacts from register allocation and increased spill code effort.

- Mode independent instruction set

  Mode dependent instructions are introduced to improve code density. The same instruction coding can be used for different instructions and the differentiation is indicated by the set mode for that code section.

  The use of mode dependent instructions limits the instruction scheduling procedure. Instructions of the same code section require the same mode setting, as illustrated in Figure 7.4. Instructions requiring a different mode setting cannot be moved into this section, even when data dependencies or hardware resources would allow rescheduling. Additional instructions are required for setting and resetting of modes, which decreases code density and makes the use of mode dependent instructions questionable.

- Orthogonal instruction sets

  Orthogonality refers to the extent to which a processor's instruction set is consistent [252]. In general, the more orthogonal an instruction set is, the easier the processor is to be programmed. For compiler development

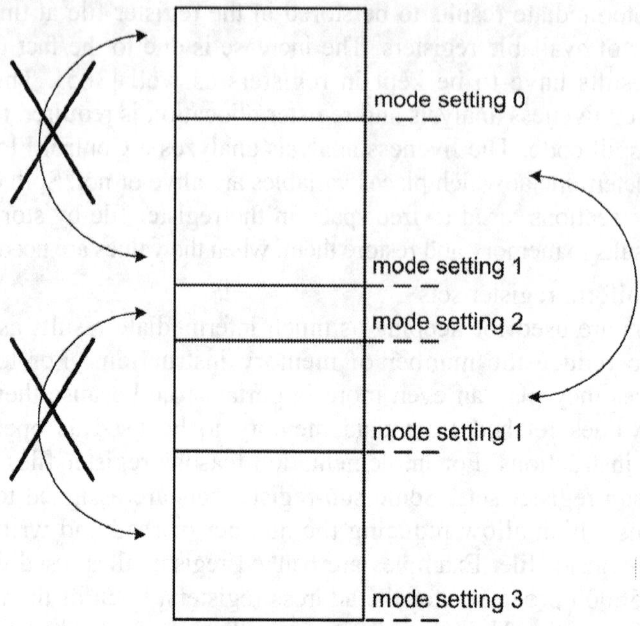

Fig. 7.4. Mode dependent code sections.

this is even more relevant. A higher number of inconsistencies and implicit dependencies increase the complexity of code generation algorithms and reduce the possibility to find a global optimum.

- Simple issue rules, no implicit dependencies

To improve code density and cycle efficiency implicit dependencies are introduced, for instance instructions whose execution duration depends on the size of the operands [21,22], or when registers of a register subset are directly addressed without explicit coding [389]. Further examples are conditional execution instructions depending on only one condition flag, or register sets, whose meaning changes depending on the chosen mode and instruction [389].

All these implicit instructions are causing significant complexity for compiler development, leading to less efficient code and are failing the target to improve code density and cycle efficiency.

3a, an application specific adaptable core architecture

In the first part of this chapter the challenges in communication technology have been introduced and the consequences for SoC solutions with focus on embedded processors have been summarized. Micro-architectural requirements of algorithms in the field of signal processing have been briefly touched and some known micro-architectural concepts have been checked on their suitability. VLIW based architectures thereby give a significant advantage by providing the needed real time requirements and an ILP much greater than one. The major drawbacks are poor compiler support and low code density. The compiler issue has been discussed before by summarizing micro-architectural concepts which facilitate the development of an optimizing compiler.

The second part of this chapter introduces *3a* [323], an application specific adaptable core architecture which enables a power- and area-optimized solution of the core subsystem by efficiently make use of available resources. The main architectural features are introduced, followed by a subchapter discussing benchmarking approaches and the differences between kernel and application benchmarking. Design space exploration is then introduced as an example of application benchmarking.

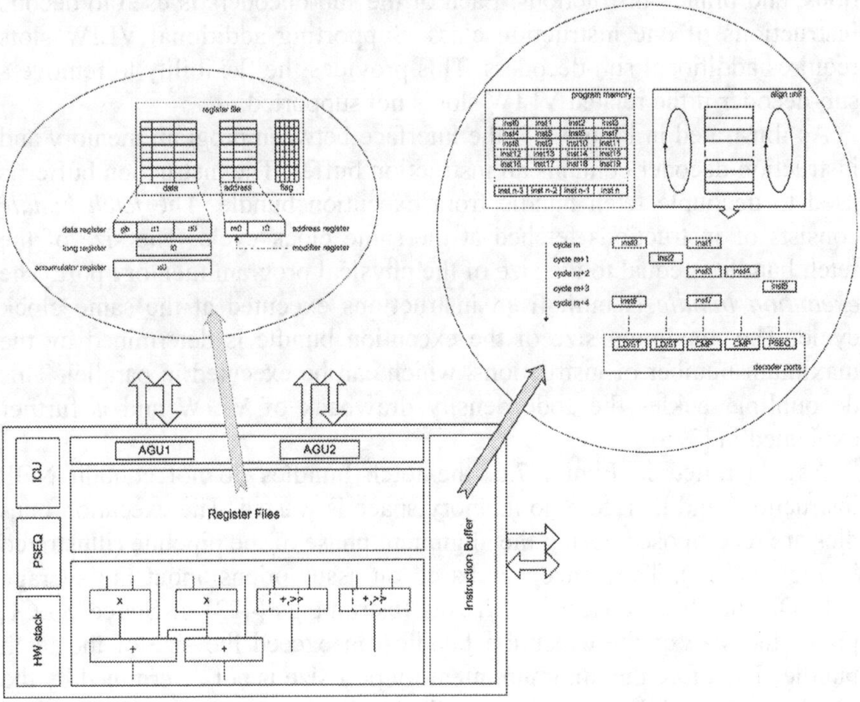

Fig. 7.5. *3a* top-level architectural overview.

Concept

3a is based on an RISC like modified dual-Harvard load–store architecture which additionally supports typical DSP CISC instructions like MAC or specific addressing modes like bit-reversed or modulo addressing. Arithmetic instructions exclusively access register operands, whereas separate instructions are used to transfer data between register file and data memory [238].

The orthogonal register file as illustrated in Figure 7.5 is split into three functional subsets. Registers of the same subset are not restricted to any instruction or groups of instructions. The data register file is used to store operands and intermediate results of the arithmetic instructions. The address register file is mainly used for address calculations. The flag register set mirrors the current status of the registers (data and address) and contains dynamic flag information. Dynamic flags are set based upon the destination register of an operation.

Each of the VLIW slots has its own instruction sub-decoder unit. Three classes of operations are known: arithmetic instructions, load–store instructions, and branch instructions. Each of the sub-decoders is used to decode instructions of one instruction class. Supporting additional VLIW slots requires additional sub-decoders. This provides the flexibility to remove a sub-decoder if the related VLIW slot is not supported.

As illustrated in Figure 7.5, the interface between program memory and instruction decoder contains an instruction buffer. The instruction buffer is used to decouple fetch bundle from execution bundle. The *fetch bundle* consists of instructions fetched at the same clock cycle. The size of the fetch bundle is equal to the size of the physical program memory port. The *execution bundle* is built from instructions executed at the same clock cycle. The maximum size of the execution bundle is determined by the maximum number of instructions, which can be executed in parallel. This decoupling tackles the code density drawback of VLIW and is further explained in [325].

As illustrated in Figure 7.6, the fetch bundles do not contain NOP instructions and therefore no memory space is wasted. The execution bundles are recomposed during the alignment phase of the pipeline (illustrated in Figure 7.12). The concept is based on assumptions about the average ILP. Distinct from concepts of Texas Instruments [422] or Starcore LCC [389], the size of the execution bundle can exceed the size of the fetch bundle. Therefore the minimum memory port size is not determined by the maximum size of the execution bundle. During loop execution, the missing bandwidth of the memory port is compensated by instructions stored in the instruction buffer.

Active power consumption can be reduced by executing instructions already stored in the instruction buffer. The once fetched instructions of the loop body are executed from the buffer without further program memory access. A significant amount of program memory cycles is saved, as long the buffer size fits to the algorithm requirements [325].

If-then-else constructs are essential parts of control code and their implementation in assembly language usually yields branch instructions with small branching distances. Branch instructions at pipelined processor architectures are causing branch delays. Some of the branch delays can be filled with useful instructions, whereas unused branch delays are causing overhead in cycle count and code effort. One possibility to overcome this performance loss is hardware circuits for branch prediction [448].

Another possibility is to prevent the occurrence of branches at all by using *predicated* or *conditional execution*. In predicated execution, instructions

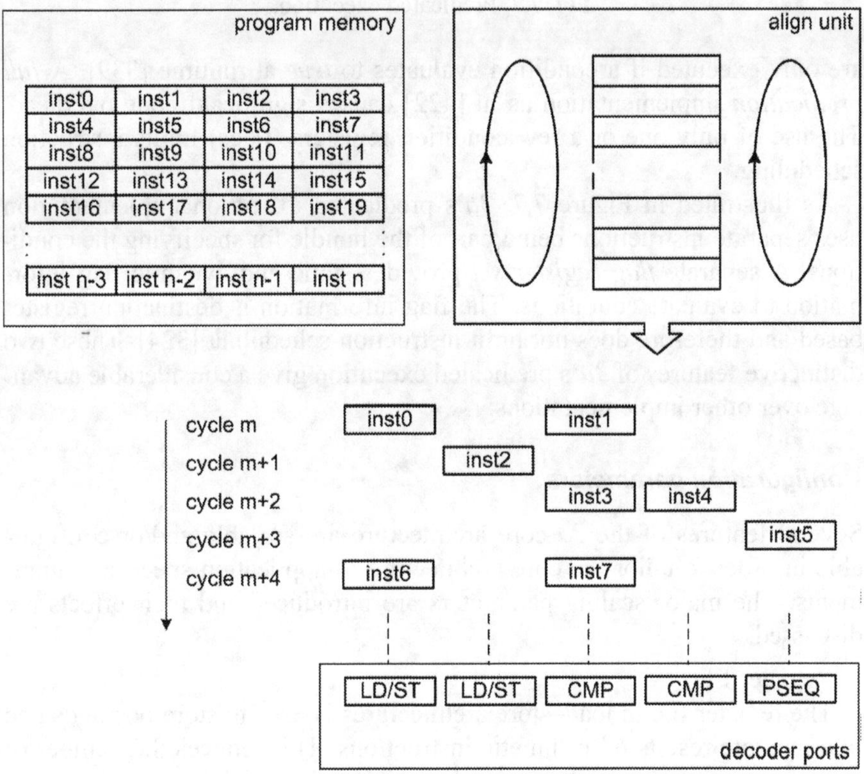

Fig. 7.6. Instruction buffer.

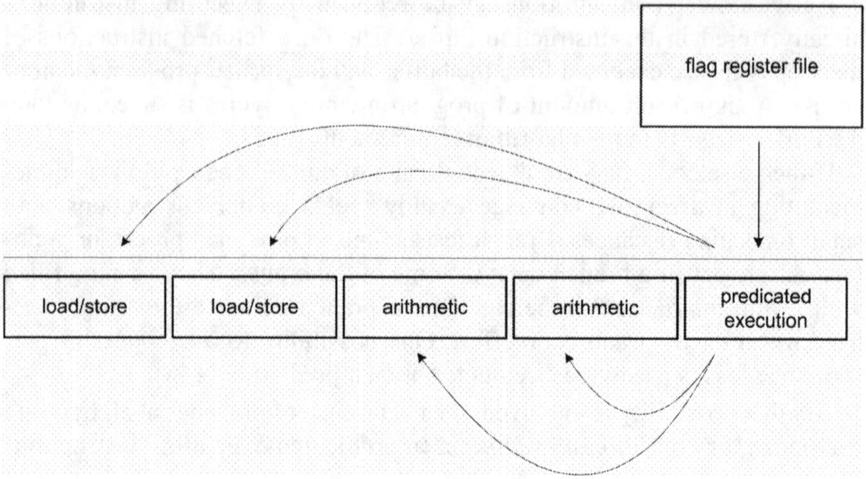

Fig. 7.7. Predicated execution.

are only executed if a condition evaluates to *true* at runtime [339]. A *full predication* implementation as in [422] causes significant code overhead. The use of only one or a few condition registers (flags) limits instruction scheduling.

As illustrated in Figure 7.7, *3a*'s predicated execution implementation uses separate instructions being part of the bundle for specifying the conditions. A separate *flag register set* provides static and dynamic flag information to evaluate conditions. The flag information is destination register based and therefore does not limit instruction scheduling [324]. These two distinctive features of *3a*'s predicated execution give a considerable advantage over other implementations.

Configuration parameters

Several features of the *3a* core architecture are scalable and/or configurable in order to tailor the core architecture to application specific requirements. The major scaling parameters are introduced and their effects are discussed.

- Register file

The register file in load–store architectures is used to store operands and intermediate results of arithmetic instructions. This reduces the number of data memory access and decouples the physical timing of the memory access with the timing of the core architecture. Slower memories can be used if additional pipeline stages for the load–store instructions are introduced.

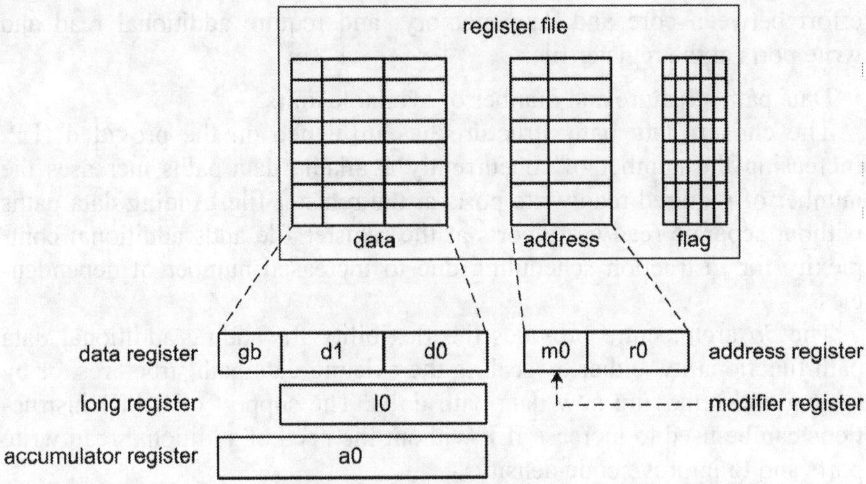

Fig. 7.8. *3a* data register file.

The size of the register file has influence on the number of data memory accesses. More entries permit to store more intermediate results. Less registers in a register file will lead to increased spill code. From this perspective, large register files are favored, but the number of register entries influences code density, because more register entries require an increased space in the instruction encoding. Therefore DSP core architectures introduced banked register files [297,422] and registers dedicated to one instruction or a group of instructions [389]. Both limit the optimization potential during compilation.

The *3a* architecture contains an orthogonal register file split into three functional sub-register sets. The data register file supports different data types, *data* and *long* and *accumulator* register, as illustrated in Figure 7.8. The number of supported registers and the size of the register entries can be scaled and the sharing of registers inside the register file configured.

- Memory interface

The memory interface of the *3a* architecture contains a number of independent AGU units. The generated addresses can be mapped to each of the memory ports. Each AGU has access to all address-registers. The size of the memory port and the number of ports and AGUs can be scaled. Benchmarking of several application examples indicates that two AGUs are reasonable trade-off in terms of implementation feasibility and application requirements. Scaling the size of the memory port has to be carefully adjusted to the size and structure of the registers.

Adding additional memory ports increases the potential memory bandwidth. On the contrary, additional memory ports also increase the wiring effort between core and data memory, and require additional read and write ports at the register file.

- Data path structure and number of available units

The chosen data path structure has influence on the provided ILP. Increasing the number of concurrently available data paths increases the number of required read/write ports at the register file. Adding data paths without separate read/write ports at the register file adds additional complexity for instruction scheduling due to increased number of dependencies.

The *3a* architecture provides the flexibility for adding additional data path functionality, either by scaling the existing data path structures, or by adding and removing new data path units. The support of SIMD instructions can be used to increase ILP without the need of additional read/write ports and to improve code density.

In Figure 7.9 the ALU data paths are reused for the ADD operation of MAC instructions. The advantage of this reuse is a reduced requirement on write ports to the register file. Its drawback is the increased latency for all other ALU operations, which then also take place in the second execution cycle. Similar to MAC instructions they need two clock cycles causing an increased *define-in-use dependency.* Define-in-use dependency describes the number of cycles required between an intermediate result has been calculated until it can be used by consecutive operations. Increasing the number of pipeline stages of the *execution phase* leads to increased define-in-use dependency, which can be partly compensated by loop unrolling/software pipelining mechanism.

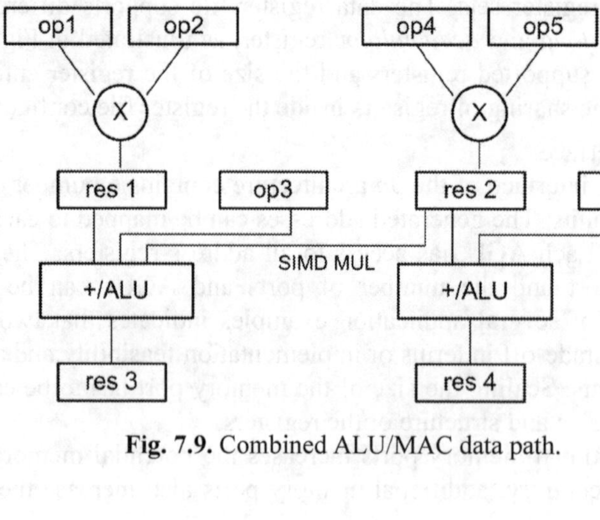

Fig. 7.9. Combined ALU/MAC data path.

In Figure 7.10, separate ALU data paths independent of MAC instructions are added. Additional write ports are required, but the latency of ALU operations is decreased.

Adding new data path structures has to be done considering latency and the chosen pipeline structure. If the latency of the operation significantly differs from the remaining data paths, an implementation as Co-processor is probably better suited. Data transfer can take place via memory mapped I/O.

- ISA and binary coding

The Instruction Set Architecture (ISA) describes the supported functionality of the micro-architecture and the relation between components. All supported instructions must have their hardware equivalent, whereas not all provided hardware needs to be *addressed* by the ISA.

The finally chosen ISA is then assigned to a binary encoding. This assignment process can be optimized to improve code density. Figure 7.11 shows an example of the influence of the chosen native word size. The

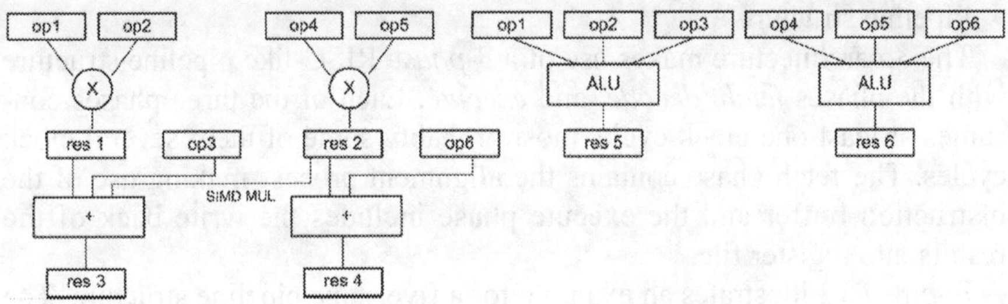

Fig. 7.10. Separated ALU/MAC data path.

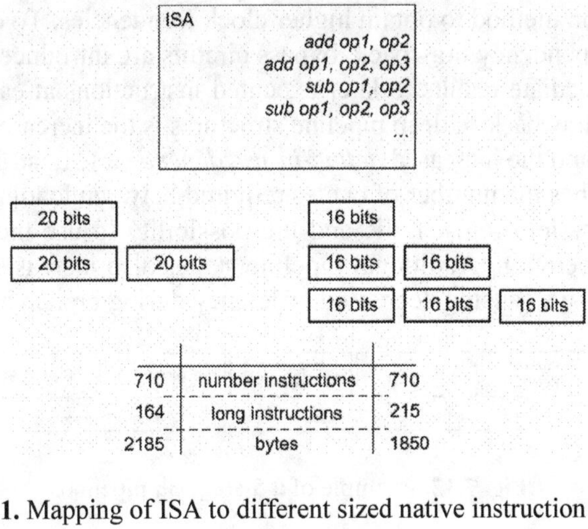

Fig. 7.11. Mapping of ISA to different sized native instruction words.

same ISA is mapped onto different instruction partitioning. On the left side of Figure 7.11, a native instruction word size of 20 bits is assumed, which allows mapping the ISA into a single native instruction word. The second instruction word, mentioned as long instructions, is used for long immediate values only. In the mapping illustrated on the right side of Figure 7.11, a reduced native instruction word size is chosen. This requires more often a second or sometimes even third instruction word for encoding the complete ISA. Depending on the application code the chosen native instruction word size has influence on the code density. The example in Figure 7.11 is done for a synthetic control code benchmark. The reduced native instruction word size of 16 bits requires more long instruction words compared to the 20 bits version, but the overall code size is improved by about 16%. For a different application code it might look different.

Active power consumption can be reduced when the mapping of the ISA to the binary encoding considers the state change frequency at the program memory port. More details can be found in [194,195].

- Pipeline structure

The *3a* architecture makes use of a *3-phase* RISC-like pipeline structure with the phases *fetch, decode*, and *execute*. Each of the three phases consumes at least one clock cycle, most probably some of them several clock cycles. The fetch phase contains the alignment process making use of the instruction buffer and the execute phase includes the write back of the results into register file.

Figure 7.12 illustrates an example for a five stage pipeline structure. The bundle alignment consumes a separate pipeline stage, the execution phase is split over two pipeline stages. Adding pipeline stages allows increasing the maximum clock frequency. Increasing the number of pipeline stages is quite common method to obtain higher clock frequencies. To overcome the increased dependency drawback, bypass circuits are introduced for bypassing of intermediate results to later executed instructions at earlier pipeline stages. The drawback of deep pipeline structures is the increased *load-in-use dependency* and the increased *define-in-use dependency*. Load-in-use dependency describes the number of cycles required between loading an operand from memory into register file and the possibility to use the operand for arithmetic operation. Additional pipeline stages also increases the number of branch delays causing decreased efficiency during branch handling.

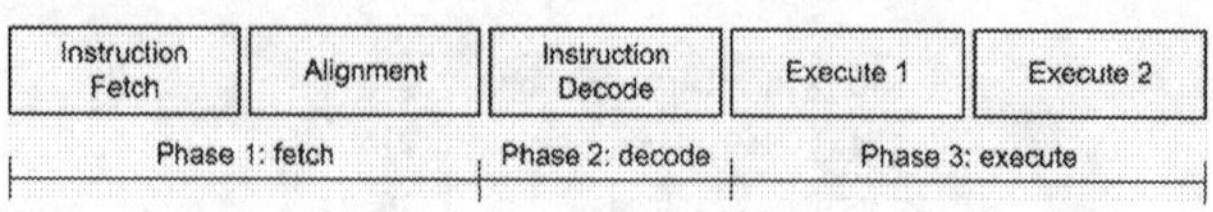

Fig. 7.12. Example of a 5-stage *3a* pipeline.

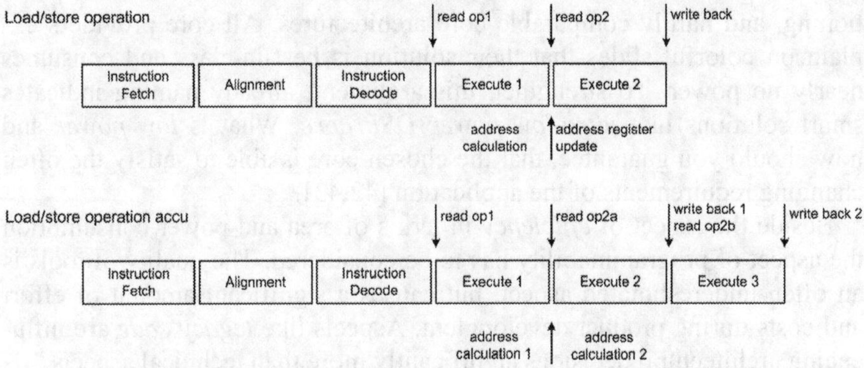

Fig. 7.13. Pipeline structure examples.

Bypass circuits are introduced in order to overcome the drawback of increased dependencies, but bypass circuits themselves introduce another drawback through the increased circuit complexity. In combination with inherent code dependencies, this can even lead to performance degradation [377].

Nevertheless, the *3a* architecture allows introducing additional pipeline stages. Even instructions of the same *instruction class* can use a different number of pipeline stages as illustrated for load–store operations in Figure 7.13. Instruction class describes a group of instructions making use of the same instruction decoder.

The load–store operations for data and long registers can be executed within the pipeline structure as illustrated in the upper part of Figure 7.13. The word length of accumulator registers exceeds the size of the program memory port, therefore an additional load–store and a shift operation is required. In the example given in the lower part of Figure 7.13 the additional instructions are not needed. The load–store operation of an accumulator register consists of two interleaved load–store operations without additional code effort, but in the consecutive cycle of such an instruction, no further load-store operations can be executed. This restriction has to be resolved and obeyed during instruction scheduling.

Benchmarking: kernel versus application benchmarking

This section deals with finding a proper way of identifying key parameters for system partitioning and of choosing suited core architectures.

For embedded solutions the complexity increases. This is due to a wide range of possible system architectures, different hardware/software partitioning, and hardly comparable core architectures. All core providers explain on colorful slides that their solution is best-in-class and consumes nearly no power. To strengthen this argument, already naming indicates smart solutions like *ultra-low power DSP core*. What is *low power* and how should you guarantee, that the chosen core is able to satisfy the often changing requirements of the application [42,43]?

Beside the aspect of *efficiency* in terms of area and power consumption the aspect of programmability has to be considered. The quality of tools is an often underestimated aspect, but causes a significant amount of effort and costs during product development. Aspects like *legacy code* are influencing architectural decisions significantly more than technical aspects.

Comparing embedded core architectures is a non-trivial task. In the next subsection some of the commonly used benchmarking suites are introduced. The benchmarking is not only considering the core architecture itself, but the combination of core, memory concept and tooling [145].

Traditional processor benchmarking

For comparing the processor subsystem architecture without influence of the tooling, the benchmark examples must be coded manually in assembly language. The drawback of this method is the limited size of the possible benchmark examples. The most famous one in the DSP world are the BDTI Benchmarks [44].

- BDTI benchmarking

  BDTI, Berkeley Design Technologies Incorporated, is a spin-off of University of California in Berkeley. Besides a suite of 12 benchmark kernels, a detailed description and implementation limitations are provided. The implementation of the benchmark suite has to be done in assembly language and gets verified by BDTI before any results are being published. For the whole suite, the core has to use a standard mode and the mentioned restrictions during implementation provide the best achievable objectivity with kernel benchmarks. The 12 kernel benchmarks are: Real Block FIR, Single-Sample FIR, Complex-Block FIR, LMS Adaptive FIR, Two-Biquad IIR, Vector Dot Product, Vector Add, Vector Maximum, Viterbi Decoder, Control, 256-Point FFT, and Bit Unpack.

  Companies like Tensilica are making use of these benchmarks to illustrate the advantage of application specific re-configurability, whereas the *DSP community* still questions if this comparison is valid [173]. The perfect *BDTI-benchmark core* provides exactly one instruction per benchmark. In the 1990s, core architectures seem to have been influenced more to achieve a high BDTImark than by requirements of the application code. BDTImark was developed in 1997 by BDTI as signal processing speed metric, improved in 1999 and released as BDTImark 2000.

- DSPstone benchmarking

  The target of the DSPstone benchmarks is to validate a core architecture in conjunction with compiler technology. More than 30 different benchmarks are used to validate the difference between manually coded assembly language and C-Compiler based results. The chosen kernels are divided into three groups, namely *application code*, *DSP kernels*, and *C-kernels*.

- EEMBC benchmarking

  The EEMBC benchmarking suite [117] provides application benchmarks for the different fields of applications like automotive, consumer or telecom. The kernels are C-code based and a balance of synthetic and real-world benchmarks.

- Summary

  The three mentioned benchmark suites provide the possibility to compare core architectures in terms of efficiency. Their methods are slightly different, and the first two examples are dedicated for DSP core architectures.

  Synthetic benchmark kernels are suited to give a first indication and to compare different architectural concepts. However, the comparison is based on algorithms whose significance for the real application is questionable.

Application benchmarking

Application benchmarks are based on real-world application code, mostly in C. The advantage of this benchmark method is that the comparison gives an answer on how well certain core architectures and their tool-chains are suited to fulfill the requirements of a certain application code.

Complete DSP applications like v.90 modem or GSM coder are typically used to compare the suitability of micro-architecture, ISA and tool-chain. Application benchmarking provides metrics like memory and MIPS consumption for the application code. The results can differ to what has been showed by kernel benchmarks. This can get significant when kernel functions are implemented in dedicated hardware to achieve competitive final products.

3a allows application specific adaptations of the main architectural features, as already illustrated in beforehand. The analysis process is based on application benchmarking, giving metrics like memory consumption and cycle count as well as indications on how to improve the core architecture.

Design space exploration

Design space exploration is application benchmarking with the target application. *3a*'s design space exploration is based on an optimizing C-compiler and a configurable ISS (Instruction Set Simulator) [125]. The analyzing methodology is not limited to *3a*.

Design space exploration as illustrated in Figure 7.14 is split into three phases. Not all phases are required to be done in sequence. The depth of analysis effort can be seen as a trade-off between spent effort and result accuracy.

- Application code analysis

During phase one the application code is analyzed to get first quantitative estimations for MIPS and memory requirements. A multi-way VLIW architecture with a large register file serves as a base architecture. This prevents limitations due to the chosen core variation. The target of the analysis is to identify hot spots which are suited to be implemented in separate dedicated entities and their embedding into the overall system. These entities can be separate processor-like structures as well as dedicated hardware implementations, either connected as co-processor to the main processor or connected

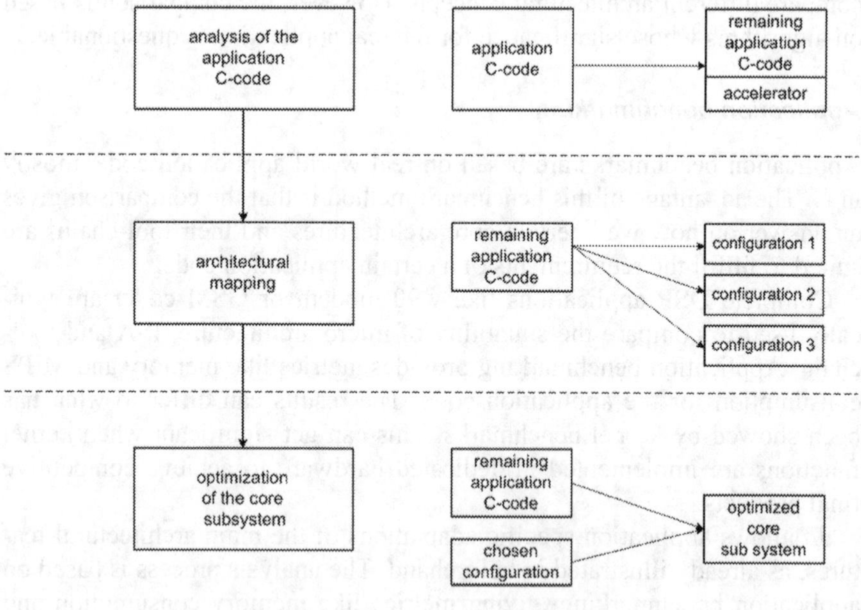

Fig. 7.14. 3-phase design space exploration.

via the system bus. Data exchange rate with the remaining system parts is significantly influencing the separation decision and the way of connection. As illustrated in Figure 7.14, the result of phase one is the application code without those parts which are going to be implemented in dedicated hardware solutions or on separate programmable entities.

- Architectural mapping
  In phase two, the remaining application code is used to identify the key parameters of the core architecture. ILP limitations in the application code are identified and a reasonable trade-off between performance requirements, implementation issues and code density is identified. The results of phase two are the key parameter of a core architecture which is able to execute the remaining application code under performance and resource constraints.
- Core subsystem Optimization
  During phase three the memory footprint and the power consumption of the core architecture is optimized by adaptations of data path width, ISA and binary coding. The result of phase three is an application specific core architecture. The degree of optimization can be varied from case to case.

The analyzing methodology of the application code as described in the three phase approach is making use of a design flow as illustrated in Figure 7.15.

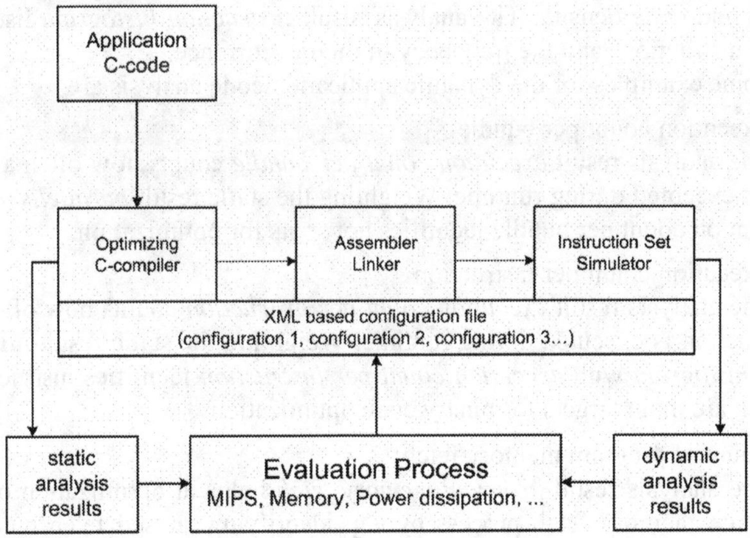

Fig. 7.15. Design space exploration Flow.

The evaluation phase is based upon a configurable tool-chain, which itself consists of an optimizing C-compiler, a linker and an ISS.

The compiler and the linker generate static analysis results, whereas the simulator provides dynamic analysis results. The combination of both result types is basis for the three-stage optimization process. Some examples for static results are

- Code density
  Code density is mainly influenced by the chosen ISA and how well the application code can utilize the core architecture. Code density is simply a measure of the instruction traffic; it is the reciprocal of the number of instruction bits required to execute a program normalized by the dynamic HLL operation count [128]. The result is normalized to bytes which allow a comparison independent from the chosen native instruction word size.
- Parallelism
  Providing the execution of several instructions in parallel increases the theoretical performance of a DSP core, but control and data dependencies in the application code are limiting the actual usage of parallel units. Therefore the analysis result parallelism is used to analyze how efficient the application code can make use of provided hardware resources.
- Instruction histogram
  Mapping of an instruction set architecture to a binary encoding has influence on power dissipation (switching activity at the program memory port) and code density. The analysis result *instruction histogram* lists the used instructions and the frequency of their occurrence.

Some examples for the dynamic application code analysis are

- Execution count per bundle
  The analysis result *execution count per bundle* counts how often a bundle is executed during runtime. Weighting the static result *parallelism* with execution count per bundle identifies hot spots for optimization.
- Execution count per instruction
  The analysis result *execution count per instruction* counts how often an instruction is executed during runtime. Weighting the static result *instruction histogram* with *execution count per instruction* identifies instructions which are main targets for binary code optimization.
- Count of program memory fetch
  The analysis result *program memory fetch* helps at optimization of the program memory fetch process by e.g. identifying unused fetch bundles. Control code with low branch distances leads to a significant number of superfluous fetch cycles when fetching instructions which are never executed.

Evaluation phase

System evaluation is first done for several core configurations (illustrated in Figure 7.15) with the scalable configuration parameters being specified beforehand. The procedure does not care about binary encoding of the ISA, therefore it is called *Evaluation phase*.

Production phase

The production phase is based upon the results of the evaluation phase (e.g. the number of supported registers or data paths). Those key parameters of the core are chosen and fixed. As the next step, the binary mapping of the ISA has to be defined as illustrated in the example of Figure 7.11. This is done by using the *bincode-generator*, a tool that assists in generating and optimizing the binary encoding for the chosen core architecture. Dynamic and static analysis results describing the metrics of the application code are used as additional input and guidance for optimization.

The complexity of configurability

The challenges of modern communication applications have been mentioned in the beginning of this chapter, pointing out that traditional approaches to tackle the issues seem to fail. The exit strategy proposed in the introduction sounds evident: *efficient use of resources*.

Therefore the second part of this chapter was used to introduce the *3a* architecture, which allows adapting the main architectural features of the core architecture to application specific requirements. To identify the requirements of the application code and to optimize the core subsystem to an area and power dissipation optimum, the design space exploration methodology has been briefly mentioned and its advantages compared to traditional kernel benchmarks have been illustrated.

One aspect not tackled yet is the challenges of configurability and scalability. Some of the issues that arise in this context are for instance:

- Consistency of the tools building the tool-chain
- Consistency of tool-chain and hardware description of the core subsystem
- Consistency of chosen configuration and documentation
- Providing tools able to cope with different configurations and modified ISA

- Description of the core architecture, which can be used by the different disciplines like core hardware, tool-chain, documentation

To cope with these challenges, the 3a concept is based on an XML based configuration file.

Configuration file

As already mentioned in Figure 7.15, a single configuration file is used to describe the current core configuration. The configuration file contains the main architectural features like structure of the register file, number of registers, number and kind of supported data paths and the pipeline structure. The ISA is described and which instruction makes use of what core features at which time.

The advantage of a central configuration file is having the core configuration stored at one place. All tools of the tool-chain are configured by the same file, yielding high consistency even for such a configurable and scalable concept.

Besides the tool-chain for code development, some more tools are using the configuration file. In Figure 7.16 the relation of configuration file, tool-chain and additional tools is illustrated. The upper part builds the evaluation

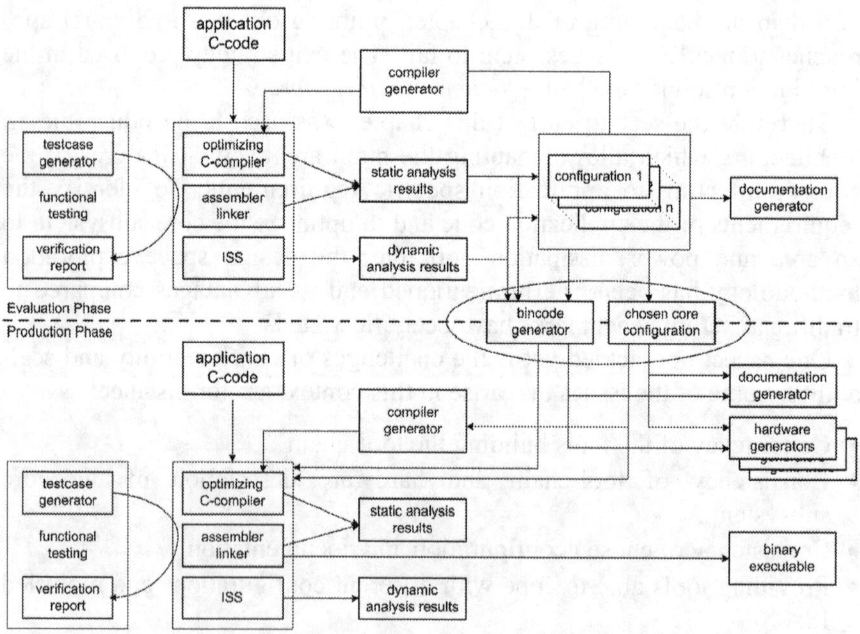

Fig. 7.16. *3a*'s Design flow (evaluation/production phase).

phase, whereas the lower part describes the production phase of the optimization process. The interface of the two phases is the *bincode-generator* and the chosen core configuration.

Static and dynamic analysis results as described before in detail are used to modify the rules for the compilation process and also as input for the *bincode-generator*.

- Hardware-generators

  One example of the HW generators, as illustrated in figure 7.16, is the Decoder Generator. Based on the generated binary coding for a dedicated ISA configuration a VHDL/Verilog description of the decoder structure is provided. The concept assumes to be correct by construction, which allows skipping full exhaustive test suites each time the configuration is changed.

- Functional testing

  A configurable concept inherently struggles by additional verification efforts. Any change in configuration must not have unexpected influence on the functionality. The XML based configuration file contains certain corner test cases for each instruction. The test case generator as illustrated in Figure 7.16 makes use of the corner case tests and automatically generates verification reports.

- Documentation generation

  Based on the XML configuration file and a style sheet, an XSL processor is used for generating a DOCBook File also based on XML. A browsable HTML description can be generated directly from that, whereas generating a PDF file needs an intermediate step via Latex using the *dblatex* tool. The documentation generation in the evaluation phase is missing the binary coding, which is later added by the *bincode-generator*.

Summary

Rapidly changing performance demands in modern communication technologies challenge system engineers to cope with the contradictory requirements of high data rates, mobility demands and flexibility requirements. Traditional approaches like deeply pipelined high performance processors or highly sophisticated SDR solutions are not solving the ultra-low power requirements related to mobility demands. A paradigm shift is required: efficiently make use of available resources while still having flexibility of a programmable solution.

Different architectural concepts are introduced and their suitability as target platform for DSP algorithms is analyzed. VLIW architectures allow minimizing the worst case execution time and therefore to guarantee real-time requirements. Micro-architectural features hampering development of optimizing C-compiler for VLIW architectures are discussed.

3a, an application specific adaptable core architecture is introduced. The main architectural features of *3a* can be configured and/or scaled to application specific requirements. During definition of the ISA, the development of an optimizing C-compiler has been taken into account. An overview of present kernel and application benchmarks is given, followed by an introduction of *3a*'s design space exploration. This methodology enables system architects to quantify different architectural choices and to support a balanced hardware/software partitioning.

Acknowledgment

Defining and developing of an application adaptable core architecture considering hardware, software, and algorithmic aspects can only be done with a highly competent and enthusiastic team. Therefore I would like to express my deepest thanks to the research project team, namely Dr. Ulrich Hirnschrott, Stefan Farfeleder, Wolfgang Lazian, Edwin Steiner, and Volker Gierenz.

I would like to express my thanks to Kave Kianush (CTO and VP of Catena) and Rien Geurtsen (President of Catena) supporting our research work.

Furthermore I would like to thank Prof. Andreas Krall (Vienna University of Technology), Prof. Tobias G. Noll (RWTH Aachen), Prof. Lajos Gazsi (Fellow Infineon Technologies), and Prof. Jari Nurmi (Tampere University of Technology) for their support and advice over the last years.

8 Customizable Processors and Processor Customization

Steven Leibson

Tensilica, Inc.

Introduction

The first commercial microprocessor chip, Intel's 4004, appeared in November, 1971. Since then, most designers have used fixed-ISA (instruction-set architecture) processors in their system designs – first as processor chips in board-level designs and later as processor cores in SOCs. In fact, many system designers cannot envision designing a custom processor for their projects because the use of fixed-ISA machines has become so thoroughly engrained in the conventional design methodology. A small cadre of designers created custom processors based on bit-slice technology in the 1980s for applications with very high performance requirements, but electronic system design has largely evolved into an exercise in adapting standardized processor and DSP architectures to target tasks, often with additional hardware acceleration to bridge the inevitable gap between a task's required computations and the fixed-ISA processor's abilities.

Designers working for conventional processor vendors have always had the ability to extend their own processor ISAs as needed. For example, two features that distinguish DSPs from general-purpose processors are hardware MACs (multiplier/accumulators) and dual load/store units for XY-memory addressing. No intrinsic hardware limitation prevents processor designers from incorporating such features in their general-purpose processor ISAs. However, adding such resources increases the processor's silicon area (and therefore its cost) which may not be used by many of the applications of that processor. Successful DSPs must have these features. Successful general-purpose processors don't. Similarly, many other task-specific features are not good candidates for universal inclusion in general-purpose processor designs.

J. Nurmi (ed.), Processor Design: System-on-Chip Computing for ASICs and FPGAs, 149–175.

However, vendors of customizable processor IP (intellectual property) now offer SOC designers the tools needed to easily customize a processor's ISA for specific applications (and to automatically generate the associated software-development tools needed to program these processors), which produces processor cores that can execute specific, targeted tasks in many fewer clock cycles. This execution efficiency can be used to either increase a processor's execution "reach" without boosting its clock frequency or it can be used to reduce the maximum required clock rate for a given set of tasks, thus lowering the power and energy dissipation required to execute the target tasks.

The flexible silicon resources in high-gate-count SOCs now permit the use of customized processors in a wide variety of products, ranging from low-cost consumer products including MP3 players, video games, and printers to high-end products such as metropolitan-area network routers. Customized processors incorporate task-specific registers and task-specific hardware acceleration in the processor's execution pipeline instead of employing accelerators external to the processor. This move brings the acceleration hardware into intimate contact with the processor's registers, register files, memory-management hardware, and memory interfaces, which greatly improves the efficiency and throughput of data movement into and out of the accelerators and the added task-specific hardware greatly improves the processor's efficiency (measured in clock count) on the task's computations.

For example, MP3 players and other products that play and record audio may benefit from the use of a processor with audio-specific extensions such as one or two audio MACs. Video products such as camcorders and DVD players can benefit from processors that can directly manipulate RGB data as a native data type and networking products including network interfaces, switches, and routers benefit from processors with packet-processing hardware extensions. These abilities are all made possible by the flexibility available in the use of the SOC design methodology and the plastic nature of ASIC silicon.

This chapter is divided into two parts. The first part is a general discussion of the abilities and benefits of using customizable processors. The second part discusses the specifics of one commercial customizable processor: Tensilica's Xtensa family.

A benefits analysis of processor customization

In a generic 90 nm standard-cell foundry process, silicon density routinely exceeds 200K usable gates per mm^2. Consequently, a low-cost chip (measuring 50 mm^2) can carry 10M gates of logic. Simply because it's possible, some designer working on an SOC development team somewhere will find ways to exploit this potential in any given market. Designers face a set of daunting challenges however, because the traditional HDL-based approach to designing SOC hardware is failing:

- **Design effort:** In the past, silicon capacity and design-automation tools limited the practical size of a block of RTL to smaller than 100K gates. Blocks of 500K gates are now within the capacity of today's tools, but existing design methods are not keeping pace with silicon fabrication capacity, which now routinely puts many millions of gates on an SOC.
- **Verification difficulty:** The internal complexity of a typical logic block – hence the number of potential design bugs – grows far more rapidly than does its gate count. Consequently, verification complexity has grown disproportionately. Many teams that have developed real-world SOC designs report that they now spend as much as 90% of their development effort on verification.
- **Cost of fixing bugs:** The cost of fixing an SOC design bug is rising. Higher staff costs caused by growing design teams, bigger NRE fees, and lost profitability and market share make show-stopper design bugs intolerable. Design methods that reduce the occurrence of, or permit painless workarounds for such show-stoppers pay for themselves rapidly.
- **Late hardware/software integration:** All embedded systems now contain significant amounts of software or firmware. Software integration is typically the last step in the system-development process and routinely gets blamed for overall program delays.
- **Complexity and change in standards:** Standard communication protocols are growing rapidly in complexity. The need to conserve scarce communications spectrum plus the inventiveness of modern protocol designers has resulted in the creation of complex new standards such as the IPv6 Internet Protocol packet forwarding, G.729 voice coding, JPEG2000 image compression, MPEG4 video, and Rjindael AES encryption. These new protocol standards have much greater computational demands than their predecessors.

Although general-purpose, firmware-programmable embedded processor cores with fixed ISAs can handle many tasks, they often lack the bandwidth needed to perform complex data-processing tasks such as network packet processing, video processing, and encryption. To meet aggressive performance goals, chip designers have long turned to hardwired logic to implement these key functions. As the complexity and bandwidth requirements of electronic systems increase, the total amount of logic rises steadily.

Even as SOC designers wrestle with the growing resource demands of advanced chip design, they face two additional worries:

- How do SOC design teams ensure that the chip specification really satisfies customer needs?
- How do SOC design teams ensure that the chip really meets those specifications?

Further, a good SOC design team will also anticipate future needs of current customers and potential future customers – it has a built-in road map for the SOC design.

If the design team fails to create an appropriate SOC specification, the chip may work perfectly but it will not sell well enough to justify the design and manufacturing costs. Changes in the chip's requirements may be driven by demands of specific key customers, or may reflect rapid changes that frequently occur in the market such as the emergence of new data format standards or new feature expectations across an entire product category. While most SOC designs include some form of embedded control processor (usually with a fixed ISA), the limited performance of these general-purpose processors often precludes them from being used to meet the performance requirements of essential on-chip data-processing tasks so task-specific hardware is designed, which means that firmware often cannot be used to add or change fundamental new features. If such an SOC's functional abilities need to change, the chip must be redesigned. That is a costly path.

If the SOC design team fails to meet the specifications of the chip's design, additional time and resources must go towards changing or fixing the design errors. This resource diversion delays market entry and causes companies to miss key customer commitments. This sort of failure usually surfaces as a program delay. This delay may come in the form of missed integration or verification milestones, or it may come in the form of hardware bugs – explicit logic errors that are not caught by the relatively limited verification coverage of gate-level simulation. The underlying cause of the error might be a subtle bug in one design element or it might be a miscommunication of requirements – caused by subtle differences in assumptions between hardware and firmware teams, between design and verification teams, or between the SOC designer and SOC library or foundry supplier. In any case, the design team is often forced into an urgent cycle of chip re-design, re-verification and re-fabrication. SOC design "spins" rarely take less than less than six months, causing significant disruption to product and business plans.

Using microprocessor cores in SOC design

There are two closely related problems for system developers. One is to develop system designs with significantly fewer resources by making it much, much easier to design the chips in those systems. The second problem is making SOCs sufficiently flexible so every new system design doesn't require a new SOC design.

The way to solve these two problems is to make the SOC sufficiently flexible so that one chip design will efficiently serve 10, or 100, or 1000 different system designs while giving up none or, at most, a few of the benefits of integration. Solving these problems produces off-the-shelf SOCs that satisfy the requirements of next-generation system designs. This design approach amortizes the costs of chip development over a large number of system designs.

The specialized nature of individual embedded applications creates two issues for general-purpose embedded processor cores executing data-intensive tasks. First, there is a poor match between the critical functions of many embedded applications (e.g. image, audio, and protocol processing) and a processor's basic integer ISA (instruction set and register file). As a result of this mismatch, critical embedded applications require more computation cycles when implemented in firmware running on general-purpose embedded processor cores. Second, specialized embedded devices cannot take full advantage of a general-purpose processor's broad capabilities. Consequently, expensive silicon resources built into the processor core are wasted because they're not needed by the specific embedded task that's assigned to the processor.

Most embedded systems interact closely with the real world or communicate complex data at high rates. The associated data-intensive tasks in these applications could be performed by some hypothetical general-purpose microprocessor running at tremendous speed. This is indeed the approach employed in the personal computer market, with the result that the processors used in PCs cost hundreds of dollars and dissipate tens of watts. For embedded applications, expensive and power-hungry processors are not an appropriate alternative.

Instead, designers traditionally turn to hard-wired circuits to perform these data-intensive functions. In the past 10 years, wide availability of logic synthesis and ASIC design tools has made RTL design the standard for hardware developers. RTL-based design is reasonably efficient (compared to custom, transistor-level circuit design) and can effectively exploit the intrinsic parallelism of many data-intensive problems. RTL design methods can often achieve tens or hundreds of times the performance achieved by a general-purpose processor.

Like RTL-based design using logic synthesis, extensible-processor technology enables the design of high-speed logic blocks tailored to the assigned task. The key difference is that RTL state machines are realized in hardware while logic blocks based on configured processor cores realize their state machines with firmware, which makes these blocks far more flexible than hard-wired RTL designs.

Benefiting from microprocessor extensibility

A fully featured configurable and extensible processor consists of a processor design and a design-tool environment that allows a designer to adapt the base processor design by altering or adding major processor functions, thus tuning the processor core to specific application requirements. Typical forms of configurability include additions, deletions, and modifications to memories, to external bus widths and handshake protocols, and to commonly used processor peripherals. An important superset of configurable processors is the extensible processor – a processor whose functions, especially its instruction set, can be extended by the SOC design team to include features never considered or imagined by processor's original designers.

Changing the processor's instruction set, memories and interfaces can significantly improve the core's efficiency and performance, particularly for the data-intensive applications that represent the "heavy lifting" for many embedded systems. These task-specific features might be too specific to justify inclusion in a general-purpose processor's ISA. General-purpose processors are the result of many design compromises where features that provide modest benefits to all customers supercede features that provide dramatic benefits to a few. All design compromise is necessary because the historic costs and difficulty of processor design mandate that only a few different such cores can be built. However, automatic generation of a configurable processor and its associated software tool suite reduces the cost and development time so that inclusion of application-specific features and deletion of unused features suddenly becomes attractive for SOC design [162].

A *configurable processor* is a processor whose features can be pruned or augmented by parametric selection. Configurable processors can be implemented in many different hardware forms, ranging from ASICs with hardware implementation times of many weeks, to FPGAs with implementation times of just minutes. *Extensible processors* – processors whose functions, especially the instruction set, can be extended by the application developer to include features never considered by the original processor designer – are an important superset of configurable processors.

For both configurable and extensible processors, the usefulness of the configurability and extensibility is strongly tied to the automatic availability of both hardware implementation and the software environment. Automated generation of software tools that support the core's extended features is especially important. Configuration or extension of the processor's hardware without synchronized enhancement of the compiler, assembler, simulator, debugger, real-time operating systems, and other software support tools would leave the promises of performance and flexibility through configurability unfulfilled, because the new enhanced processor could not be programmed very easily.

Table 8.1. Processor configuration and extension types.

ISA	Memory System	Interface	Processor Peripherals
–Extensions to ALU functions using general registers (e.g. population count instruction) –Coprocessors supporting application-specific data types (e.g. network packets, pixel blocks), including new registers and register files with varying word widths –High-performance arithmetic and DSP (e.g. compound DSP instructions, vector/SIMD instructions, floating-point instructions), often with wide execution units and registers –Multiple independent operations per instruction –Selection among function unit implementations (e.g. small iterative multiplier vs. pipelined array multiplier)	–Memory-bus width –Instruction-cache size, associativity, and line size –Data-cache size, associativity, line size, and write policy –Memory protection and translation (by segment, by page) –Instruction and data RAM/ROM size and address range	–External bus interface width, protocol, and address decoding –Direct connection of system control registers to internal registers and data ports –Mapping of special-purpose memories (queues, multi-ported memories) into the address space of the processor –State-visibility trace ports and JTAG-based debug ports	–Timers –Interrupt controller: number, priority, type, fast switching registers –Exception vectors addresses –Remote debug and breakpoint controls

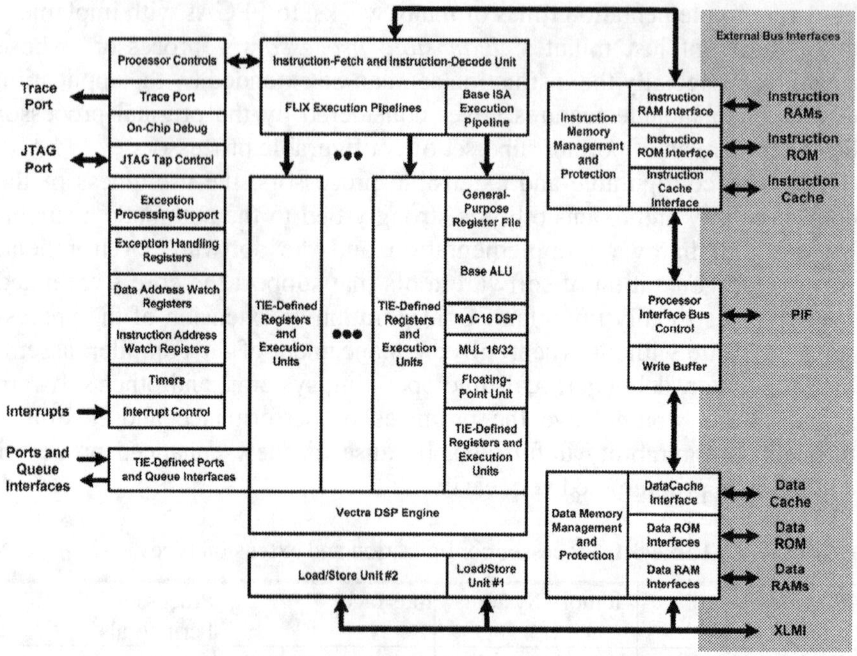

Fig. 8.1. Block diagram of configurable Xtensa processor.

Extensibility's goal is to allow ISA features to be added or adapted in any form that optimizes the cost, power, and application-performance of the processor core. In practice, the configurable and extensible features can be broken into four categories, with examples, as shown in Table 8.1.

A block diagram for Tensilica's configurable, extensible Xtensa processor appears in Figure 8.1. The figure identifies baseline ISA features, scaleable register files, memories and interfaces, optional and configurable processor peripherals, selectable DSP coprocessors, and facilities to integrate user-defined instruction-set extensions.

Processor extensibility serves as a particularly potent form of configurability because it handles a wide range of applications and is easily used by designers with a wide range of skills. Processor extensibility allows a system designer or application expert to directly exploit proprietary insight about the application's functional and performance needs directly in the processor core's instruction-set and register extensions.

How microprocessor use differs between SOC and board-level design

Hardwired RTL design has many attractive characteristics – small area, low power, and high-throughput. However, the liabilities of RTL (difficult design, slow verification, and poor scalability to complex problems) are starting to outweigh the benefits in large SOC designs. A design methodology that retains most of the efficiency benefits of RTL while reducing design time and risk has a natural appeal. Application-specific processors as a replacement for complex RTL fit this need.

An application-specific processor can implement data-path operations that closely match those of RTL functions. Equivalents of the RTL datapaths are implemented using the integer pipeline of the base processor, plus additional execution units, new registers and register files, and other functions added by the chip architect for a specific application. For Tensilica's Xtensa processor cores, those extensions are defined in a Verilog-like language called TIE, which is optimized for the high-level specification of data-path functions in the form of instruction semantics and encoding. A TIE description is much more concise than RTL because it omits all sequential logic, including state machine descriptions, pipeline registers, and initialization sequences. The new processor instructions and registers described in TIE are available to the firmware programmer via the same compiler and assembler that target the processor's base ISA. All sequencing of operations within the processor's datapaths is controlled by firmware, through the processor's existing instruction fetch, decode and execution mechanisms. The firmware is written in a high-level language such as C or C++.

Extended processors used as RTL-block replacements routinely use the same structures as traditional data-path-based RTL blocks: deep pipelines, parallel execution units, problem-specific state registers, and wide data paths to local and global memories. These extended processors can sustain the same high computation throughput and support the same low-level data interfaces as typical RTL designs.

Control of the extended-processor datapaths is very different however. Cycle-by-cycle control of the processor's datapaths is not fixed in hardwired state transitions. Instead, the sequence of operations is explicit in the

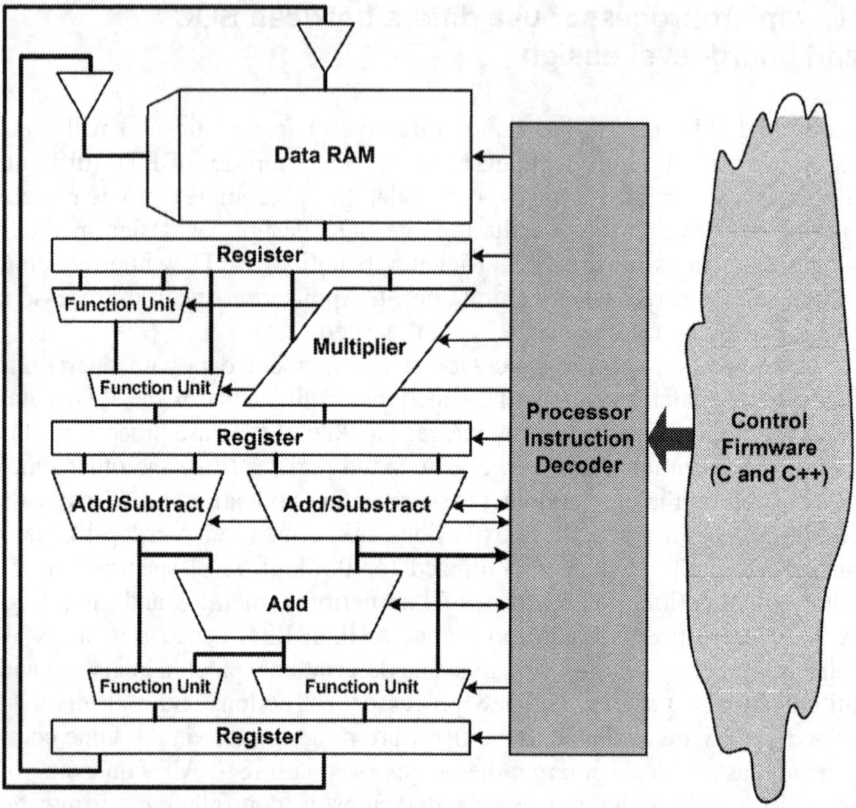

Fig. 8.2. Programmable function: data-path + processor + software.

firmware executed by the processor (shown in Figure 8.2). Control-flow decisions are made explicitly in branches; memory references are explicit in load and store operations; sequences of computations are explicit sequences of general-purpose and application-specific computational operations.

This design migration from hardwired state machine to firmware program control has important implications:

- **Flexibility:** Chip developers, system builders, and end-users (when appropriate) can change the block's function just by changing the firmware. The need for silicon respins is greatly reduced.
- **Firmware-based development:** Developers can use sophisticated, low-cost software-development methods for implementing most chip features.
- **Faster, more complete system modeling:** RTL simulation is slow. For a 10-million-gate design, even the fastest software-based logic simulator may not exceed a few cycles per second. By contrast, firmware

simulations for extended processors run at hundreds of thousands or millions of cycles per second on instruction-set simulators.

- **Unification of control and data:** No modern system consists solely of hardwired logic. There is always a processor and some software. Moving functions previously handled by RTL into a processor removes the artificial distinction between control and data processing.
- **Time-to-market:** Moving critical functions from RTL to application-specific processors simplifies the SOC design, accelerates system modeling, and pulls in finalization of hardware. Firmware-based state machines easily accommodate changes to standards, because implementation details are not "cast in stone."
- **Designer productivity:** Most importantly, migration from RTL-based design to the use of application-specific processor cores boosts the engineering team's productivity by reducing the engineering resources needed to manually code and verify RTL hardware. A processor-based SOC design approach sharply cuts risks of fatal logic bugs and permits graceful recovery when (usually not if) a design bug appears or a specification change occurs.

Three examples, ranging from simple to complex, serve to illustrate how data-path extensions allow extensible processors to replace RTL hardware in a wide variety of situations.

The first example, from the cellular telephone world, is the GSM audio codec used in cell phones. Profiling the codec code using an unaugmented RISC processor revealed that out of more than 200 million processor cycles, 80% of the cycles were devoted to executing multiplications. The simple addition of a hardware multiplier therefore produces a substantial acceleration of this codec implementation. Tensilica's Xtensa processor offers a multiplier as a configuration option. That means that a designer can add a hardware multiplier to the processor's data path and multiplication instructions to the processor's instruction set simply by checking a box in a configuration page of the Xtensa processor generator.

The addition of a hardware multiplier cuts the number of cycles needed to execute the GSM audio codec code from 204 million to 28 million cycles, a 7x improvement in execution time. Adding a few more gates to the processor pipeline by selecting a MAC instead of a multiplier further reduces the number of cycles needed to execute the audio codec code to 17.9 million. These configuration options coupled with code profiling allow an SOC designer to rapidly explore a design space and to make informed cost/benefit decisions for various design approaches.

A more complex example, Viterbi decoding, also comes from GSM cellular telephony. GSM employs Viterbi decoding to pull information symbols

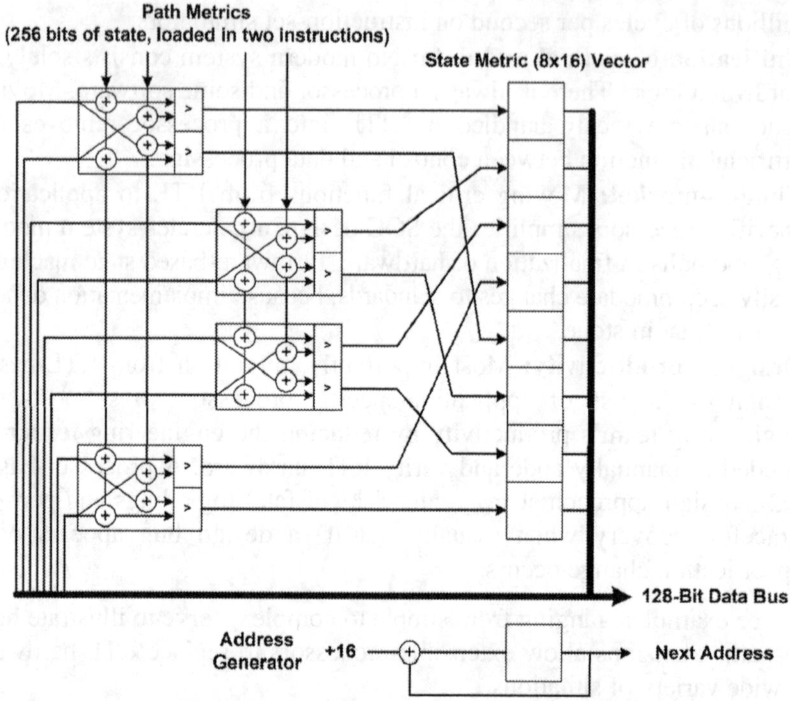

Fig. 8.3. Viterbi butterfly pipeline hardware.

out of a noisy transmission channel. This decoding scheme employs "butterfly" operations consisting of eight logical operations (four additions, two comparisons, and two selections) and uses eight butterfly operations to decode each symbol in the received digital information stream. Typically, RISC processors need 50 to 80 instruction cycles to execute one Viterbi butterfly. A high-end VLIW DSP (TI's 320C64xx) requires only 1.75 cycles per Viterbi butterfly. The TIE language allows a designer to add a Viterbi butterfly instruction to the Xtensa processor's ISA, which uses the processor's configurable 128-bit I/O bus to load data for eight symbols at a time, adds the pipeline hardware shown in Figure 8.3, and results in an average butterfly execution time of 0.16 cycles per butterfly. An unaugmented Xtensa processor executes Viterbi butterflies in 42 cycles, so the butterfly hardware adds only 11,000 gates and yet it achieves a 250x speed improvement over the out-of-the-box Xtensa processor.

The third example, MPEG4, is from the video world. One of the most difficult parts of encoding MPEG4 video data is motion estimation, which requires the ability to search adjacent video frames for similar pixel blocks. The search algorithm used for motion estimation employs a SAD

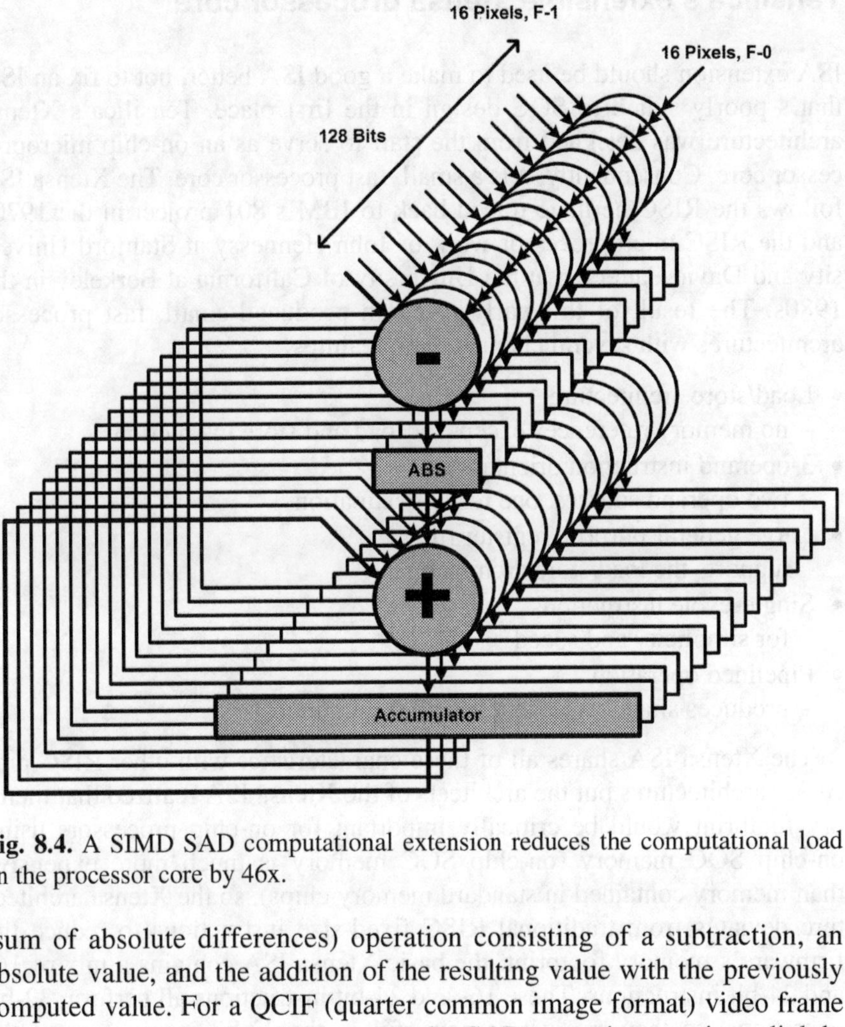

Fig. 8.4. A SIMD SAD computational extension reduces the computational load on the processor core by 46x.

(sum of absolute differences) operation consisting of a subtraction, an absolute value, and the addition of the resulting value with the previously computed value. For a QCIF (quarter common image format) video frame and a 15 frames/second image rate, the SAD operation requires slightly more than 641 million operations/second. As shown in Figure 8.4, it is possible to add SIMD (single-instruction, multiple-data) SAD hardware capable of executing 16 pixel-wide SAD instructions per cycle using TIE. (Note: Using the Xtensa processor's 128-bit maximum bus width, it is also possible to load all 16 pixels worth of data in one instruction.) The combination of executing all three SAD component operations in one cycle and the SIMD operation that computes the values for 16 pixels in one clock cycle reduces number of cycles required to execute the algorithm from 641 million to 14 million operations/second – a substantial reduction.

Tensilica's extensible Xtensa processor core

ISA extension should be used to make a good ISA better, not to fix an ISA that's poorly suited to SOC design in the first place. Tensilica's Xtensa architecture was designed from the start to serve as an on-chip microprocessor core. Consequently, it is a small, fast processor core. The Xtensa ISA follows the RISC heritage traced back to IBM's 801 project in the 1970s and the RISC microprocessor work of John Hennessy at Stanford University and David Patterson at the University of California at Berkeley in the 1980s. The result of this early research produced small, fast processor architectures with several characteristic features:

- Load/store architecture
  – no memory references except for load and store instructions
- 3-operand instruction orientation
  – two operand sources, one result destination
- Large general-purpose register file
  – supports the load/store architecture
- Single-cycle instructions
  – for simplicity and speed
- Pipelined operation
  – produces single-cycle instruction throughput

The Xtensa ISA shares all of these characteristics with other RISC processor architectures but the architects of the Xtensa ISA realized that memory footprint would be critically important for on-chip processors using on-chip SOC memory (on-chip SOC memory is much more expensive than memory contained in standard memory chips), so the Xtensa architecture deviates from traditional RISC fixed-size instructions to reduce the firmware's memory footprint: the basic Xtensa ISA contains a mix of 16- and 24-bit instructions. These 16- and 24-bit instructions all perform 32-bit operations, so they are just as powerful as the 32-bit instructions of the older RISC architectures – they're merely smaller, which reduces program size and therefore reduces on-chip memory costs.

The original RISC processors employed fixed-size instructions of 32 bits to simplify the processor's fetch/decode/execute circuitry but the mechanism that converts incoming instruction words into 16- and 24-bit instructions in the Xtensa processor does not require complex logic. The amount of memory saved through the use of 16- and 24-bit instructions more than compensates for the gates used to implement the processor's mixed-size instruction-fetch and instruction-decode unit.

Because an Xtensa processor's instruction-fetch cycle retrieves more than one instruction per cycle (including fractions of an instruction word) and because a single Xtensa instruction can cross aligned fetch (word) boundaries, the Xtensa processor stores fetched words from the instruction stream in a FIFO holding/alignment buffer. For the base Xtensa ISA, the instruction buffer is 32-bits wide and two entries deep. It can be deeper and wider for certain Xtensa configurations. In addition to supporting the 16- and 24-bit instructions in the Xtensa processor's base ISA, this mixed-size instruction-fetch and instruction-decode unit also supports extended 32- and 64-bit, multi-operation instructions that can be added to the configurable Xtensa processor.

Xtensa configurable registers and register files

The Xtensa processor's base ISA incorporates the following register files and registers:

- A 32-bit, general-purpose register file that employs register windows
- A 32-bit program counter
- Various special registers

Figure 8.5 shows the processor's general-purpose 32-bit register file. This file, called the AR register file, has either 32 or 64 entries (a configurable attribute). Xtensa instructions access this physical register file through a sliding 16-register window. Register windowing allows the processor to have a relatively large number of physical registers while restricting the number of bits needed to encode a source or destination operand address to 4 bits. Thus a 3-operand Xtensa processor instruction needs only 12 bits to specify the registers holding the instruction's three operands.

Register windowing is a key ISA feature that allowed the Xtensa architects to achieve the small instruction sizes needed to minimize the application firmware's footprint while allowing for a large general-purpose register file, which boosts compiler efficiency and improves the performance of compiled code. Xtensa function calls and call returns slide the register window up and down the physical register file. Because window movement is restricted to a maximum of 12 register entries per call, some register entries in the physical general-purpose register file are shared between the current register window and the previous window. This overlap provides a set of register entries that can be used to automatically pass argument values and return function values between a calling function and the called function. XCC – the Xtensa C/C++ compiler – automatically uses the features of this register-windowing scheme to minimize the number of instructions used in function calls.

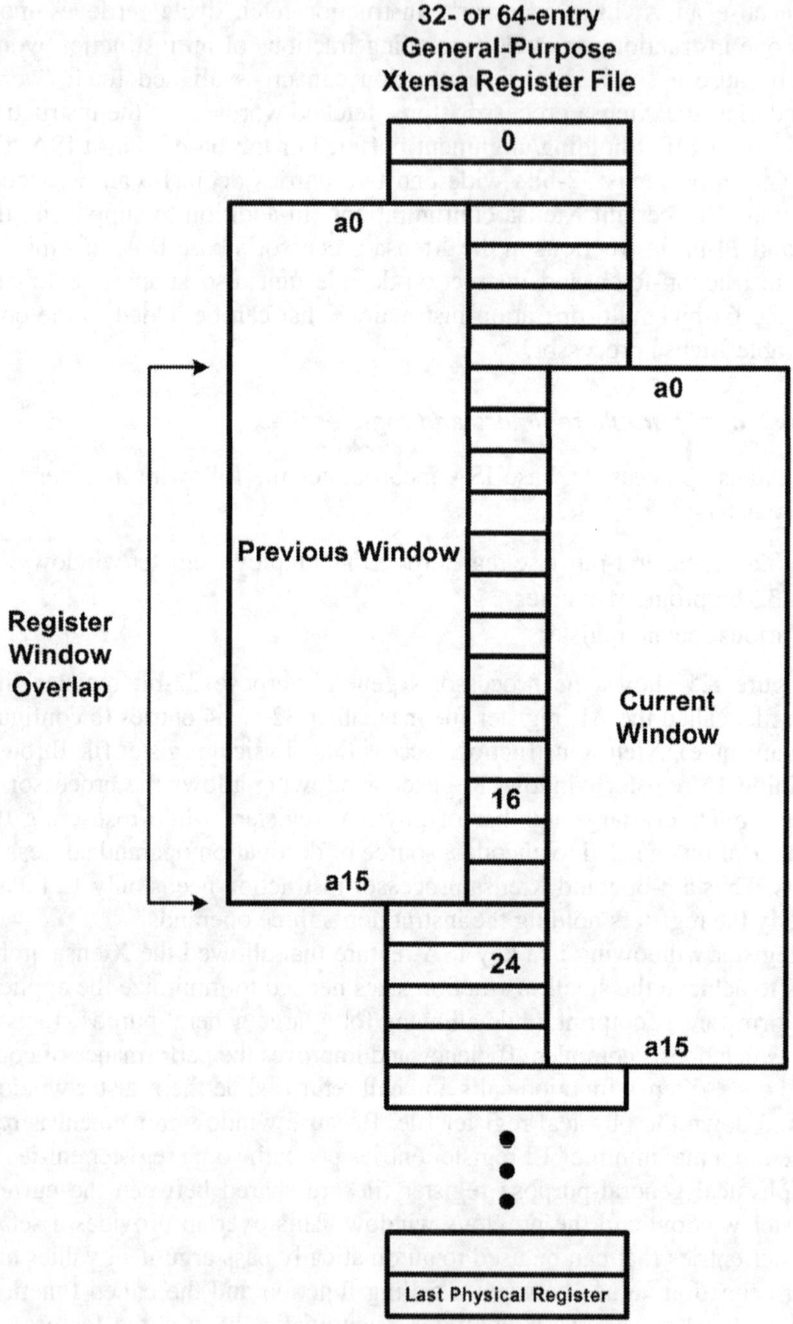

Fig. 8.5. The Xtensa ISA incorporates a 32- or 64-entry register file with register windows. Each entry in the file is a 32-bit, general-purpose register.

The Xtensa program counter

The Xtensa ISA employs a 32-bit program counter, which encompasses a 4-Gbyte address space. During a function call, only the lower 30 bits of the return address are saved, which restricts nested subroutines using function calls to a 1-Gbyte address space. (The other two address bits are used to store the amount of register-window translation: 0, 4, 8, or 12 entries.) A function-call return restores the saved 30-bit return address and leaves the upper 2 bits of the program counter untouched. Thus related sets of function calls can operate in any of the four 1-Gbyte address spaces that comprise the overall 4-Gbyte space. Jump instructions, which use 32-bit target addresses stored in register-file entries, shift program execution across the 1-Gbyte boundaries. This scheme reduces the number of memory-accesses required to implement function calls, which improves the overall performance of the Xtensa ISA.

The resulting 1-Gbyte address restriction that the Xtensa ISA imposes on linked subroutines puts no real practical limits the design of current-generation SOCs, which almost universally employ per-processor code footprints much smaller than 1 Gbyte. Per-processor SOC code footprints are unlikely to approach 1 Gbyte for many, many years. That's simply too much code to run on any one embedded processor core in an SOC.

Memory address space

Unlike practical code-size limits, data-space requirements for SOCs seem to grow without limit, especially for media-oriented chips. Consequently, the Xtensa ISA provides full, unrestricted 32-bit data-space addressing for its load and store instructions. The Xtensa ISA employs a Harvard architecture that physically separates instruction and data spaces, although they share the same 4-Gbyte address space.

An Xtensa processor's local memories are divided into instruction and data memories and the processor employs separate instruction and data caches. The existence of local memories and caches, the address spaces allocated to the local memories, and the width of the bus interfaces to these local memories are all configuration options for Xtensa processors. Load, store, and fetch operations to addresses that are not allocated to local memories (as defined in the processor's configuration) are directed to the Xtensa processor's main bus interface, called the PIF (processor interface). The existence and width of a processor's PIF bus is another Xtensa configuration option.

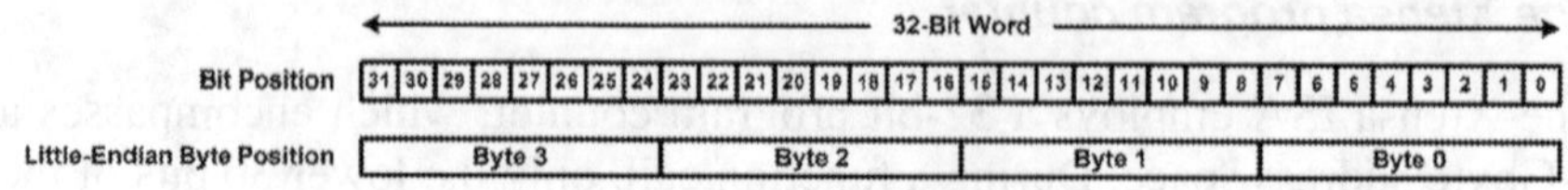

Fig. 8.6. Little-endian byte ordering for a 32-bit word.

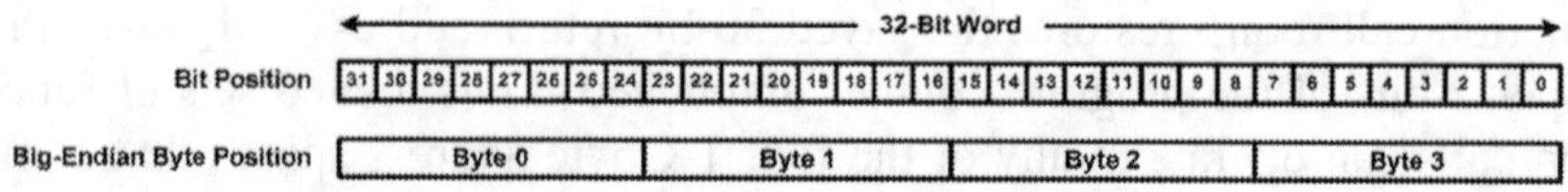

Fig. 8.7. Big-endian byte ordering for a 32-bit word.

Configurable byte ordering

As a configuration option, Xtensa processors support either big- or little-endian byte ordering, shown in Figures 8.6 and 8.7. Little-endian byte ordering stores a number's low-order byte in the right-most byte of the 32-bit word and the high-order byte in the left-most byte. (The least-significant or "little" end comes first.) For example, a 4-byte integer is stored using the little-endian format in a 32-bit memory word as shown in Figure 8.6.

Big-endian byte ordering stores a 4-byte integer's high-order byte in a memory location's right-most byte and the low-order byte in the left-most byte. (The most-significant or "big" end comes first.) The long integer would then appear as shown in Figure 8.7.

Fast endian conversion with one instruction extension

There are a seemingly endless number of arguments regarding the relative merits of big- and little-endian byte ordering. Many of these arguments stem from religious discussions about the merits of the IBM PC vs. the (pre-Intel) Apple Macintosh computers because the underlying Intel and Motorola processors in those computers used opposite byte orderings.

In reality, both byte-ordering formats have advantages. Little-endian assembly language instructions pick up a multi-byte number in the same manner for all number formats. Because of the one-to-one relationship between address offset and byte number (offset 0 is byte 0), multiple precision math routines are correspondingly easy to write for little-endian processors. Numbers stored in big-endian form, with the high-order byte first, can easily be tested as positive or negative by looking at the byte value stored at offset zero, no matter the number of bytes contained in the number format. Software doesn't need to know how long the number is nor does it need to skip over any bytes to find the byte containing the sign

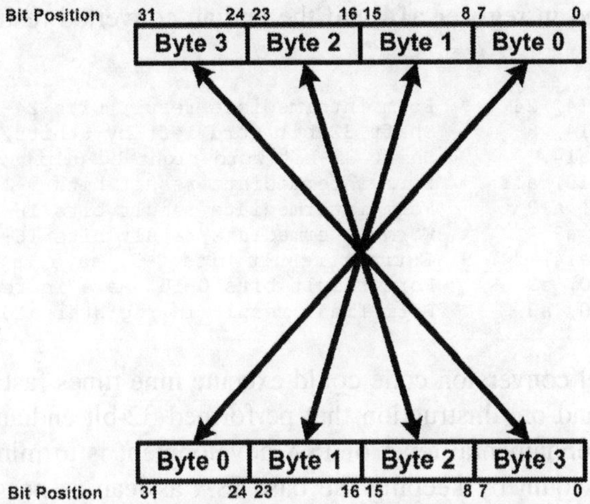

Fig. 8.8. Conversion from one endian format to the other for a 32-bit word.

information. Numbers stored in big-endian format are also stored in the order in which they are printed out, so binary-to-decimal conversion routines are particularly efficient.

The existence of "endian" issues means that whenever multi-byte values are written to a file, the software must know whether the underlying processor is a big- or little-endian machine and whether the target file format is big- or little-endian. For example, if a big-endian machine operates on a standard-format, little-endian file, the software must first reverse the byte order of all the multi-byte values in the file or the file format will not adhere to the file-format standard. SOC designers developing chips that must deal with standard file formats must be concerned with byte ordering because many of these formats have an established byte ordering. Any software that deals with these file formats must correctly deal with the format's byte ordering. Figure 8.8 illustrates how bytes are swapped in a 32-bit word to convert from one endian format to the other.

Audio and video media streams come in both endian formats as well so nearly any SOC dealing with multimedia files will need to work with both byte-ordering formats. Consequently, the choice of a processor's endian orientation is usually not critical because the processor will need to make byte-ordering conversions in any case. The following line of C code converts from one endian format to the other:

```
unsigned ss = (s<<24)|((s<<8)&0xff0000)|((s>>8)&0xff00)|s>>24);
```

The Xtensa C/C++ compiler (XCC) translates the line of C above into nine base Xtensa assembly-language instructions as follows (the input word is passed in register a14 and the endian-converted result ends up in register a10):

```
slli a9, a14, 24        Form intermediate result bits 24-31 in a9
slli a8, a14, 8         Shift 32-bit word left by 8 bits, save in a8
srli a10, a14, 8        Shift 32-bit word right by 8 bits, save in a10
and a10, a10, a11       Form intermediate result bits 9-15 in a10
and a8, a8, a13         Form intermediate result bits 16-23 in a8
or a8, a8, a9           Form intermediate result bits 16-31 in a8
extui a9, a14, 24, 8    Extract result bits 0-7, save in register a9
or a10, a10, a9         Form result bits 0-15, save in register a10
or a10, a10, a8         Form final result in register a10
```

The C-level conversion code could execute nine times faster if the base Xtensa ISA had one instruction that performed 32-bit endian conversion. However, a fundamental tenet of ISA development is to minimize a processor's gate count by keeping the base ISA as lean as possible and not larding it with instructions that might only be useful in some applications. Not all processors need to perform endian conversions. While such specialized instructions may improve the processor's performance for a few application programs, they increase the processor's size for every SOC design using that processor. ISA design for *configurable* processor cores avoids the addition of specialized instructions to the processor's base ISA by allowing each SOC design team to add task-specific instructions needed to meet project-specific design and performance goals.

Specialized instructions easily can be added to an Xtensa configurable processor by the SOC designer using TIE. For example, defining a BYTESWAP instruction that takes a 32-bit word from one of the processor's general-purpose register-file entries, converts the word from one endian format to the other, and stores the result in another register-file entry is a simple task. The TIE description to create this new instruction is remarkably short:

```
operation BYTESWAP {out AR outR, in AR inpR} { }
{
    wire [31:0] reg_swapped =
             {inpR[7:0],inpR[15:8],inpR[23:16],inpR[31:24]};
    assign outR = reg_swapped;
}
```

The interface to the BYTESWAP instruction is defined by the information contained in the first set of curly braces of the operation section. Within the first set of braces, the argument *outR* specifies the destination entry in the AR register file for the result of the instruction. Argument *inpR* specifies the AR register-file entry that provides the source operand for the instruction. The second set of braces in the operation statement can be used to specify additional internal states for this *operation* extension, but this TIE feature is not used in this example.

The behavior of the BYTESWAP instruction is defined within the next set of curly braces. The first line in this group defines how the new machine instruction should compute the byte-swapped 32-bit value and assigns the result of the operation to a temporary variable named *reg_swapped*. Note that the values of the intermediate wires, states, and registers will be visible in the tailored Xtensa debugger for a processor incorporating this BYTESWAP instruction. This feature greatly facilitates the debugging of TIE instructions in a software environment. The second line above assigns the byte-swapped value to the output argument *outR*.

From this single instruction description, the TIE Compiler within the Xtensa processor generator builds the necessary execution-unit hardware, adds it to the processor's RTL description, and adds constructs in the software-development tool suite so that the new BYTESWAP instruction can be used as an intrinsic in a C or C++ program.

The BYTESWAP instruction is an example of a fused instruction. Nine *dependent* instructions (each instruction in the sequence depends on results from previous instructions) have been fused into one. In this example, the circuitry required to implement the function is extremely simple. The execution unit for this instruction needs little more hardware than a few additional wires to scramble byte lanes, yet this new instruction speeds endian conversion by a factor of 9x. This example demonstrates that a small addition to a processor's hardware can yield large performance gains. In addition, the new BYTESWAP instruction doesn't use the intermediate-result registers that are used in the 9-instruction byte-swap routine. Some gates are added in the processor's instruction decoder to decode the new instruction, but these few gates do not make the processor core noticeably larger than the base processor.

The TIE language

In general, most new instructions described in TIE implement more complex operations than BYTESWAP. In addition to the *wire* statement used in the above example, the TIE language includes several operators and built-in functions to describe new instructions. These operators and function modules appear in Tables 8.2 and 8.3, respectively.

Table 8.2. TIE operators.

<table>
<tr><th>Operator Type</th><th>Operator Symbol</th><th>Operation</th></tr>
<tr><td>Arithmetic</td><td>+</td><td>Add</td></tr>
<tr><td></td><td>–</td><td>Subtract</td></tr>
<tr><td></td><td>*</td><td>Multiply</td></tr>
<tr><td>Logical</td><td>!</td><td>Logical negation</td></tr>
<tr><td></td><td>&&</td><td>Logical and</td></tr>
<tr><td></td><td>||</td><td>Logical or</td></tr>
<tr><td>Relational</td><td>></td><td>Greater than</td></tr>
<tr><td></td><td><</td><td>Less than</td></tr>
<tr><td></td><td>>=</td><td>Greater than or equal</td></tr>
<tr><td></td><td><=</td><td>Less than or equal</td></tr>
<tr><td></td><td>==</td><td>Equal</td></tr>
<tr><td></td><td>!=</td><td>Not equal</td></tr>
<tr><td>Bitwise</td><td>~</td><td>Bitwise negation</td></tr>
<tr><td></td><td>&</td><td>Bitwise and</td></tr>
<tr><td></td><td>|</td><td>Bitwise or</td></tr>
<tr><td></td><td>^</td><td>Bitwise ex-or</td></tr>
<tr><td></td><td>^~ or ~^</td><td>Bitwise ex-nor</td></tr>
<tr><td>Reduction</td><td>&</td><td>Reduction and</td></tr>
<tr><td></td><td>~&</td><td>Reduction nand</td></tr>
<tr><td></td><td>|</td><td>Reduction or</td></tr>
<tr><td></td><td>~|</td><td>Reduction nor</td></tr>
<tr><td></td><td>^</td><td>Reduction ex-or</td></tr>
<tr><td></td><td>^~ or ~^</td><td>Reduction ex-nor</td></tr>
<tr><td>Shift</td><td><<</td><td>Left shift</td></tr>
<tr><td></td><td>>></td><td>Right shift</td></tr>
<tr><td>Concatenation</td><td>{ }</td><td>Concatenation</td></tr>
<tr><td>Replication</td><td>{ { } }</td><td>Replication</td></tr>
<tr><td>Conditional</td><td>?:</td><td>Conditional</td></tr>
<tr><td>Built-in modules</td><td><module-name>(...)</td><td>See Table 4.2</td></tr>
</table>

Improving application performance using TIE

As demonstrated above, TIE extensions improve the execution speed of an application running on an Xtensa processor by enabling the creation of single instructions that perform the work of multiple general-purpose instructions. Several techniques can be used to combine multiple general-purpose operations into one instruction. Three common techniques available through TIE are:

- Fusion
- SIMD/vector transformation
- FLIX

Table 8.3. Built-in TIE function modules.

Format	Description	Result Definition
TIEadd(a, b, cin)	Add with carry-in	a + b + cin
TIEaddn(a0, a1, ... an-1)	N-number addition	a0 + a1 + ... + an-1
TIEcmp(a, b, sign)	Signed and unsigned comparison	{a < b, a <= b, a == b, a >= b, a > b}
TIEcsa(a, b, c)	Carry-save adder	{a & b | a & c | b & c, a ^ b ^ c}
TIEmac(a, b, c, sign, negate)	Multiply–accumulate	Negate ? c - a * b : c + a * b where sign specifies how a and b are extended in the same way as for TIEmul
TIEmul(a, b, sign)	Signed and unsigned multiplication	{{m{a[n-1] & sign}}, a} * {{n{b[m-1] & sign}}, b} where n is size of a and m is size of b
TIEmulpp(a, b, sign, negate)	Partial-product multiply	negate ? - a*b : a*b
TIEmux(s, d0, d1, ..., dn-1)	*n*-way multiplexer	s==0 ? d0 : s==1? d1 : ... : s ==n-2 ? dn-2 : dn-1
TIEpsel(s0, d0, s1, d1, ..., sn-1, dn-1)	*n*-way priority selector	s0 ? d0 : s1 ? d1 : ... : sn-1 ? dn-1 : 0
TIEsel(s0, d0, s1, d1, ..., sn-1, dn-1)	*n*-way 1-hot selector	(size{S0} & D0) | (size{S1} & D1) |... (size{Sn-1} & Dn-1) where size is the maximum width of D0 ... Dn-1

To illustrate these three techniques, consider a simple example where an application spends most of its execution time computing the average of two arrays in a loop:

```
unsigned short *a, *b, *c;
...
for (i=0; i<n; i++)
      c[i] = (a[i] + b[i]) >> 1;
```

The above C code adds two short data items and shifts the sum right by 1 bit in each loop iteration. Two base Xtensa instructions are required for each computation, not counting the instructions required for loading and storing the data. These two operations can be fused into a single TIE instruction:

```
operation AVERAGE{out AR res, in AR input0, in AR input1} {} {
      wire [16:0] tmp = input0[15:0] + input1[15:0];
      assign res = tmp[16:1];
}
```

This fused TIE instruction, named *AVERAGE*, takes two input values (*input0* and *input1*) from entries in the AR register file, computes the output value (*res*), and then saves the result in another AR register-file entry. The semantics of the instruction, an add followed by a shift, are described above. A C or C++ program uses the new *AVERAGE* instruction as follows:

```
#include <xtensa/tie/average.h>
unsigned short *a, *b, *c;
...
for (i=0; i<n; i++)
     c[i] = AVERAGE(a[i], b[i]);
```

Assembly code can also directly use the *AVERAGE* instruction. The entire software tool chain recognizes *AVERAGE* as a valid instruction for processors built using this TIE extension.

Instruction fusion is not the only way to form new instructions for Xtensa processors. Two other techniques are SIMD (single-instruction, multiple-data) and FLIX (flexible-length instruction extensions), the Xtensa version of VLIW (very-long instruction word). SIMD instructions gang multiple, parallel execution units that perform the same operation on multiple operands simultaneously. This sort of instruction is particularly useful for stream-processing applications such as digital audio and video.

In the fused-instruction example shown above, one TIE instruction combines an add and a shift operation, cutting the number of instruction cycles for the overall operation in half. Other types of instruction combinations can also improve performance.

Adding SIMD instructions using TIE

The C program in the above example performs the same computation on a new data instance during each loop iteration. SIMD (single-instruction, multiple-data) instructions (also called vector instructions) perform multiple loop iterations simultaneously by performing parallel computations on different data sets during the execution of one instruction. TIE instructions can combine fusion and SIMD techniques. Consider, for example, a case where a TIE instruction computes four *AVERAGE* operations in one instruction:

```
regfile VEC 64 8 v

operation VAVERAGE{out VEC res, in VEC input0, in VEC input1} {} {
     wire [67:0] tmp = {input0[63:48] + input1[63:48],
                        input0[47:32] + input1[47:32],
                        input0[31:16] + input1[31:16],
                        input0[15:0] + input1{15:0]};
     assign res = {tmp[67:52], tmp[50:35], tmp[33:18], tmp[16:1]};
}
```

Computing four 16-bit averages simultaneously requires that each data vector be 64 bits wide (containing four 16-bit scalar quantities). However, the general-purpose AR register file in the Xtensa processor is only 32 bits wide. Therefore, the first line in the SIMD TIE example above creates a new register file, called *VEC*, with eight 64-bit register-file entries that hold 64-bit data vectors for the new SIMD instruction. This new instruction, *VAVERAGE*, takes two 64-bit operands (each containing four 16-bit scalar quantities) from the *VEC* register file, computes four simultaneous averages, and saves the 64-bit vector result in a *VEC* register-file entry. To use the instruction in C/C++, simply modify the original example as follows:

```
#include <xtensa/tie/average.h>
VEC *a, *b, *c;
...
for (i=0; i<n; i+=4) {
    c[i] = VAVERAGE(a[i], b[i]);
```

The C/C++ compiler generated for a processor built with this TIE description automatically recognizes a new 64-bit C/C++ data type called *VEC*, which corresponds to the 64-bit entries in the new register file. In addition to the *VAVERAGE* instruction, the Xtensa processor generator automatically creates new load and store instructions to move 64-bit vectors between the *VEC* register file and memory. The XCC compiler uses these instructions to load and store the 64-bit vectors of type *VEC*. Compared to the fused instruction *AVERAGE*, the SIMD vector-fused instruction *VAVERAGE* requires significantly more hardware (in the form of four 16-bit adders) because it performs four 16-bit additions in parallel. The four 1-bit shifts do not require any additional gates.

The performance improvement gained by combining vectorization and fusion is significantly larger than the performance improvement from fusion alone. The addition of SIMD instructions to an Xtensa processor nicely dovetails with Tensilica's XCC C/C++ compiler, which has the ability to unroll and vectorize the inner loops of application programs. The loop acceleration achieved through vectorization is usually on the order of the number of SIMD units within the enhanced instruction. Thus a 2-operation SIMD instruction approximately doubles loop performance and an 8-operation SIMD instruction speeds up loop execution by about 8x.

Adding FLIX instructions using TIE

Although multiple operations occur simultaneously in a SIMD instruction, they are *dependent* operations. VLIW instructions bundle multiple *independent* operations into one machine instruction. Xtensa FLIX instructions are multi-operation instructions that allow a processor to perform multiple, simultaneous, *independent* operations by encoding the multiple operations into a wide instruction word. Each operation within a FLIX instruction is independent of the others. The XCC compiler for Xtensa processors bundles these independent operations into a FLIX-format instruction as needed to accelerate code. While TIE-defined fused and SIMD instructions are 24 bits wide, FLIX instructions are either 32 or 64 bits wide, to provide enough instruction-word bits to fully describe the multiple independent operations. Xtensa instructions of different sizes (base instructions, single-operation TIE instructions, and multi-operation FLIX instructions) can be freely intermixed.

Consider again the *AVERAGE* example from above. Using base Xtensa instructions, the inner loop contains the *ADD* and *SRAI* instructions to perform the actual computation. Two *L16I* load instructions and one *S16I* store instruction move the data as needed and three *ADDI* instructions update the address pointers used by the loads and stores. A 64-bit FLIX instruction format with one operation slot for the load and store instructions, one slot for the computation instructions, and one slot for address-update instructions can greatly accelerate this code as follows:

```
format flix3 64 {slot0, slot1, slot2}

slot_opcodes slot0 {L16I, S16I}
slot_opcodes slot1 {ADDI}
slot_opcodes slot2 {ADD, SRAI}
```

The first declaration creates a 64-bit instruction and defines an instruction format with three operation slots. The last three lines of code list base ISA instructions that are to be available in each slot defined for this FLIX configuration. Note that all the instructions specified are existing core instructions in the processor's base ISA, so their definition need not be provided in the TIE code because the Xtensa processor generator already knows about base Xtensa instructions.

For this example, the C/C++ program need not be changed at all. The processor generation creates a C/C++ compiler that will compile the original source code while automatically exploiting the FLIX extensions. The generated assembly code for this processor implementation would look like this:

```
loop:
{addi a9,a9,4;      add a12,a10,a8;      l16i a8,a9,0        }
{addi a11,a11,4;    srai   a12,a12,1;    l16i a10,a11,0      }
{addi a13,a13,4;    nop;                 s16i   a12,a13,0    }
```

A computation that requires eight cycles per iteration on a base Xtensa processor now requires just three cycles per iteration, which is nearly a 3x performance increase. It took only five lines of relatively simple TIE code to specify a FLIX configuration format with three instruction slots. Instruction fusion, SIMD, and FLIX techniques can be combined to further reduce cycle count.

Conclusion

Configurable processors are not the right choice for every hardware block on an SOC. Three cases where application-specific processors are not the right choice for a block's design stand out:

- **Small, fixed state machines:** Some logic tasks are too trivial to warrant a processor. For example, bit-serial engines such as simple UARTs fall into this category.
- **Simple data buffering:** Similarly, some logic tasks amount to no more than storage control. A FIFO controller built with a RAM and some wrapper logic can be emulated via memory operations within a processor but a simple hardware FIFO is faster and simpler.
- **Very deep pipelines:** Some computation problems have so much regularity and require very little state-machine control. For these tasks, a single very deep pipeline is the ideal implementation. The common examples – 3D graphics and magnetic-disk read-channel chips – sometimes have pipelines hundreds of clock stages deep. Application-specific processor cores could be used to control such deep pipelines, but the benefits of instruction-by-instruction control would be of less help in these applications, if the algorithms are well known and not subject to change.

The migration of functions from software to hard-wired logic over time is a well-known phenomenon. During early design exploration of pre-release protocol standards, processor-based implementations are common even for simple standards that clearly allow efficient logic implementations. Some common protocol standards that have followed this path include popular video codecs such as MPEG2, 3G wireless protocols such as W-CDMA, and encryption and security algorithms such as SSL and triple-DES. The speed of this migration, however, has been limited by the large gap in performance and design ease between software-based and RTL-based development.

The emergence of configurable and extensible application-specific processors creates a new design path that is quick and easy enough for the development and refinement of new protocols and standards yet efficient enough in silicon area and power to permit very high volume deployment.

9 Run-Time Reconfigurable Processors

Fabio Campi[1] and Claudio Mucci[2]

STMicroelectronics[1]

ARCES, University of Bologna[2]

The emergence of Systems-on-Chip (SoC) has generated a tremendous development in application demands. Continuous algorithmic innovation imposes the need for flexibility, intended as the capability of a given processing engine to adapt to new computation patterns after fabrication. In spite of the boost offered by sheer technology development, programmable architectures can hardly meet requirements. The conventional solution of gaining performance through application-specific circuits is severely affected by design and verification costs, and the concurrent shortening of time-to-market. Design-time programmable processors provide a solution only valid for those products that can afford the cost of providing a new set of masks for each upgrade of the target application field.

The term run-time Reconfigurable Processor (RP) indicates a processor architecture that takes advantage of some form of run-time (dynamically) configurable hardware to provide adaptive instruction set modification, in order to meet application requirements. Such processors hold the promise to couple software flexibility with performance comparable to application-specific hardware. Many open issues remain especially related to the programming environment and the ideal dimensioning of the configurable hardware segments. This section provides an overview of the state of the art in the field. Starting from the formalization of the instruction set metamorphosis concept, it describes its evolution in the research environment up to these days and the first commercial offers that are appearing on the market, underlining classical advantages and main architectural options related to RP design and utilization.

J. Nurmi (ed.), Processor Design: System-on-Chip Computing for ASICs and FPGAs, 177–208.

Embedded microprocessor trends

It is only fair to state that one of the most interesting and unexpected results of the "system-on-a-chip" has been a renaissance in the field of processor architecture. Rather than being a pure "board-on-a-chip", the tight integration of computation and communication, combined with the redefined metrics of embedded applications, has led to a wide range of new (or revised) architectural approaches that attempt to exploit the specific properties of a limited application domain. Examples of these are the configurable instruction-set processor, the embedded very-long instruction set processor (VLIW), the very-short instruction set processor (or vector processor), and the reconfigurable processor.

Jan Rabaey, "Silicon Platforms for Next Generation Wireless Systems – What Role Does Reconfigurable Hardware Play?", 2000 [342]

It is well-accepted that the emergence of SoC has created a significant boost in the field of processor architecture design. Although the definition of SoC in itself does not necessarily imply the utilization of a programmable architecture, it has soon appeared evident that the inclusion of one (or more) processors in the design is unavoidable. Due to their complexity, the time-to-market, and product lifetime that characterize the business segments that they aim at, SoCs *must* feature a high degree of flexibility. Flexibility, intended as the capability to tune the computation, before or after fabrication, to different flavours of the chosen application field is strictly necessary to target market segments and product lifetime wide enough to justify integration costs.

The complexity and the challenges related to the very concept of SoC, such as technology and integration costs, design time, verification issues, have imposed to hardware designers, mostly in spite of their own will, the need for the methodology of IP-re-use. Microprocessors, or better said software programmable architectures in general, are the perfect case of IP-re-usable items: they are highly optimized and qualitatively very complex designs, quite technology independent in principle, but also very general purpose, in that they offer huge benefits on a very diverse range of applications. Moreover, they require significant expertise and complex toolsets to be programmed, but the same programming model, once acquired, can then be utilized for all application domains. As the concept of SoC is now more than 10 years old this tendency appears manifest, to the point that several voices have claimed that microprocessors (and, one could add, processor utilities and peripherals) are the only kind of hardware IP that effectively allows re-use, both for they diffusion and their specificity. Consequently, the idea of SoC has always been associated to that of a microprocessor-based system somehow controlling application-specific logic dedicated to the elaboration of the task for which the SoC was conceived. In particular, this design pattern has proved very successful for the deployment of computation-intensive tasks: telecommunication protocols, multimedia and image processing applications, data encryption algorithms have found great benefit in the above described solutions, opening the way for a new generation of applications. A good example of the tendency described above is the huge popularity of products like MIPS, ARM, and recently Tensilica and ARC that have achieved impressive commercial success marketing processor cores for integration in SoC.

From the algorithmic point of view, the introduction of the SoC has brought an outstanding increase in computational complexity. The applications where algorithm development has been more active in recent years are exactly those that are more closely related to the SoC market: telecommunication protocols (especially in the Wireless field), and multimedia applications. The constraints imposed by the real-time nature of such applications have imposed a heavy price in terms of sheer computational loads. The increase of algorithm complexity over time has been formalized in literature with the so-defined *Shannon's law*. The qualitative graphs in Figure 9.1 [342], depict the increase of algorithmic demand compared with the increase of computational capability offered by technology improvement. It can be observed that algorithm complexity, driven by Shannon's law, can not be tackled by technology alone, bound to Moore's law. For this reason, standard programmable architectures appear increasingly insufficient to handle computational demands.

In conclusion, on one side technology issues are strongly underlying the importance of product flexibility. But on the other hand, unless the design community is ready to renounce to the software programmable processor model for handling high-end algorithmic computations, some kind of architectural breakthrough is needed, to establish new computational patterns that can fill the gap, reverting the relationship between Shannon's and Moore's laws. Another application-related issue that is heavily influencing design is the transition in market shares from supply-wired computing devices to battery-operated portable applications that took place in the late nineties. This market shift has introduced power consumption as a very

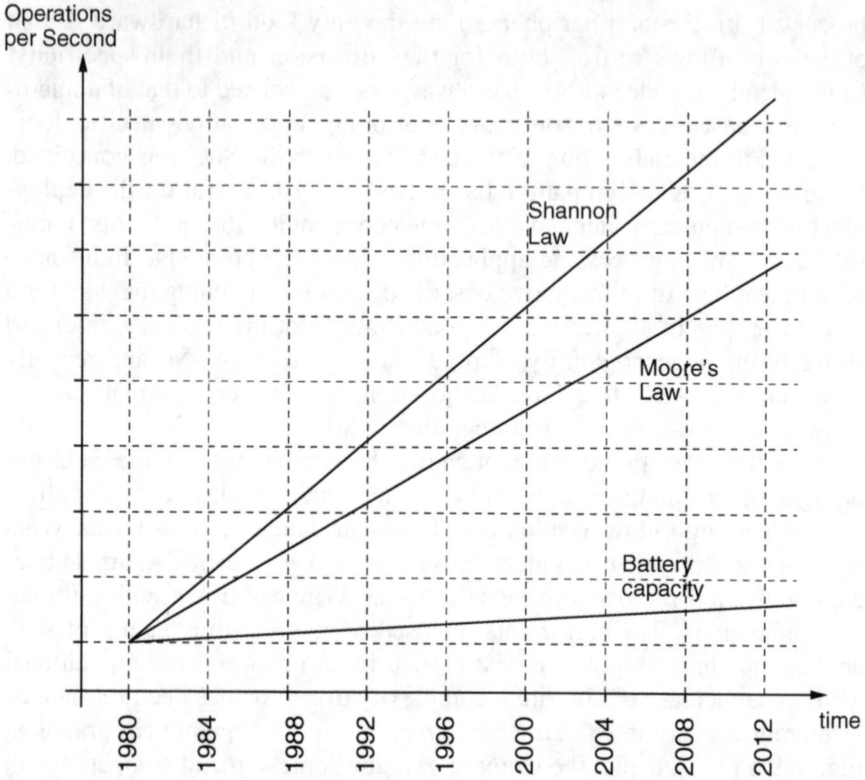

Fig. 9.1. Computational requirements versus Moore's law and battery storage.

stringent constraint in digital architectures design. A third line in Figure 9.1 describes the computational capability allowed by a typical portable application battery. As it can be seen, such increase is almost negligible when compared to both Moore's and Shannon's law. Again, this stalemate can only be broken with the development of a new architectural pattern in computation, that may provide a different relationship between energy consumption and application complexity.

Instruction set metamorphosis

If an application requires more computational power than a general-purpose platform can achieve, users are often driven to an application-specific computer architecture in which fundamental machine capabilities are designed for a particular class of algorithms. Tasks suited to a given application-specific machine perform well, but tasks outside the targeted class usually perform poorly. Computationally intensive applications typically spend most of their execution time within a small portion of the executable code. A general-purpose machine can substantially improve its performance in many of these applications by adapting the processor's configuration and fundamental operations to these frequently accessed portions of code. Segments of the processing platform can be reconfigured to add new capabilities that customize the architecture to individual tasks. Such an architecture retains its general-purpose nature, while reaping the performance benefits of application-specific architectures.

P. Athanas, H. Silvermann, "Processor Reconfiguration through Instruction Set Metamorphosis", 1993 [29]

The main feature of processors in embedded systems is that they are *application-specific* devices. Embedded processors are not general-purpose machines designed to perform reasonably for a very large range of applications; in fact, the service that they are supposed to perform is known a-priori, at least in terms of application context. This fundamental feature makes embedded processors especially suitable for *hardware–software co-design*, the cooperative and concurrent design of both hardware and software components of the processor. The task to be performed is known, even roughly, in advance by the designer. Thus it is possible to partition the computational load between hardware and software, trying to exploit the intrinsic advantages of both.

Figure 9.2 describes qualitatively the possible architectural options for the deployment of computational embedded tasks in SoC design. A conventional

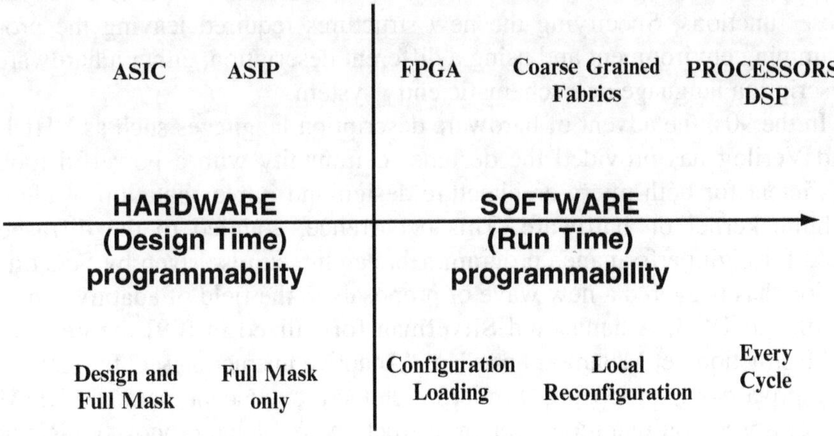

Fig. 9.2. Technology options: deployment of computationally intensive task.

solution, very popular all through the nineties, is to exploit co-design techniques to couple standard processor architectures to application specific circuits (ASIC) utilized as hardware accelerators. This solution, still very common, is becoming increasingly unpractical for two opposite reasons. On one side, the increase of design and verification costs for ASIC circuitry makes the redesign of the acceleration logic too expensive to be re-challenged at every new release of a given product family. On the other hand, the continuous refinement in application standards is forcing the design community to challenge innovative protocol platforms well ahead of their standardization, imposing to SoC designers significant re-engineering of their application portfolio late in the design phase or even after fabrication.

A significant alternative to the concept of hardware acceleration through ASIC circuitry is that of *instruction set metamorphosis*, intended as the extension of the instruction set architecture (ISA) of a given processor architecture in order to match the needs and requirement of a specific application environment. The idea of incorporating some means of adaptation into processor architectures is much older than the concept of embedded processor itself: in 1960 Estrin [122] proposed a machine based on a fixed, general-purpose core augmented by application-specific high speed computational elements defined substructures. The real difficulty in making adaptive architectures practical resided at the time in the difficulty to define an universal language for the description of the ISA extensions, filling the gap between the application developer and the architecture designer: once identified candidate patterns for extension, the programmer could not to have the expertise to define hardware structures at transistor level for these functions. Specifying the new structures required leaving the programming environment and using a different description, either a hardware description language or a schematic entry system.

In the 90s, the advent of hardware description languages such as VHDL and Verilog has provided the designer community with a powerful tool, proficient for both micro-architecture design and implementation of algorithmic kernel on hardware. This occurrence, coupled to the diffused need for high performance programmable architectures driven by SoC diffusion, has triggered a new wave of proposals in the field of adaptive computers. In 1993, Athanas and Silverman formalized in [29] the concept of "Instruction Set Metamorphosis", or "Adaptive instruction set" for embedded processors, proposing a micro-architecture concept named PRISM (Processor Reconfiguration through Instruction Set Metamorphosis). PRISM is based on a fixed RISC processor augmented by customizable segments, identified by application profiling and defined through hardware co-design, in order to tailor the machine computation capabilities to the application environment.

After a profiling step, the program is "co-compiled" in a *software image*, that is traditional program code, and a *hardware image* that is used to manufacture the ISA extension. The concept was proved on a system composed of two boards, one hosting a Motorola M68010 processor and a second featuring a Xilinx FPGA for implementation of the ISA extension. Although similar in concept, this approach differs from dynamically programmable microcode because, rather than composing differently elementary datapath functionality, in this case entirely new primitive operations are added to the processor datapath.

The "Adaptive computation" approach is the starting point for a set of successful trends that have taken root in SoC design. The critical parameters in the definition of adaptive micro-architectures are essentially two:

- the formalization of the compilation/synthesis environment for the extension segments
- the choice of the technology support on which to map them

Companies like ARC technologies and Tensilica have offered very successful commercial implementations of the ISA-metamorphosis concept, exploiting the technology independent design flows offered by the emergence of hardware description languages to provide *design-time configurable* (sometimes mentioned as mask-level configurable) definition of the extension segments. These kind of devices are defined Application-Specific Instruction Set Processors (ASIPs) [211]. In particular Xtensa (see Chapter 8) is a very aggressive RISC architecture specifically tailored for SOC design [356]. Its commercial offer is completed by an advanced design exploration tool that exactly implements the adaptive machine paradigm suggested by Athanas/Silverman. The user is put in condition to analyze and carefully profile its application, and consequently determine candidate kernels for instruction set extension. Such kernels are then described by a proprietary verilog-like language (TIE, Tensilica Instruction Extension) and synthesized into function unit that are then seamlessly plugged in the processor pipeline [261]. Referring to the landscape described in Figure 9.2, products like Xtensa implement instruction set adaptivity at mask level, to provide an original paradigm for SoC design. ASIP design and verification costs are minimized by the re-use of a fixed core and development environment, while application-specific computation capabilities are added at design time to the micro-architecture in order to guarantee performance without altering the computation model.

On the other side of the spectrum top FPGA vendors such as Altera and Xilinx provide in their products microprocessor cores (either soft cores mapped on the field-programmable logic or hard cores hardwired on silicon) that can be coupled to surrounding programmable logic to provide design-time extension of the processor capabilities. In particular, the Altera NIOS II [14] soft processor and toolset are designed in order to allow the user to enhance the standard instruction set with user defined instructions, similarly to the Xtensa model. The adaptive core is then mapped on the Altera FPGA fabric; on one side, this allows dynamic, post-fabrication reconfigurability of the ISA. On the other hand, the core of the processor is also mapped on FPGA fabric and this leads to limitation of the overall performance.

Reconfigurable computing

Configurable computing systems combine programmable hardware with programmable processors to capitalize on the strengths of hardware and software. Often these systems must also address the difficulties of both hardware and software, because they mix the technology. [...]

The configurable computing community is divided into two camps, according to the level of abstraction provided by the programmable hardware. The majority of current research efforts use commercial FPGAs and manipulate digital circuits through logic gates and flip-flops. In the second camp are the newer architectures, which are based on "chunky" function units such as complete ALUs and multipliers. These architectures limit the programmable hardware to the interconnect among the function units, but implement those units in much less IC area.

W. Mangione-Smith et al., "Seeking Solutions in Reconfigurable Computing", 1997 [274]

Even though the mask-level programmable approach is very competitive for large sections of the SOC landscape, there exists a market segment that shows the need for post-fabrication configurability. Mask-level programmable processors provide a solution only valid for that digital processing whose volumes can afford the cost of providing a new set of masks for each upgrade of the target application field. Since hardware for the new instructions is synthesized with an ASIC-like flow and the processor cannot be reconfigured after fabrication, high non-recurrent engineering costs may arise when specifications for the application need to change or evolve during the product lifetime. Furthermore, run-time reconfiguration, as opposed to design-time reconfiguration, may offer a higher degree of flexibility over the same silicon area, allowing to dynamically reprogram the same logic for different applications rather than having to add new silicon for every defined extension segment.

The term *Reconfigurable Computing* (RC) is broadly intended as the capability to couple software based programmability with dynamic hardware programmability. The term is a large umbrella including a lot of different technology implementations: discrete and embedded FPGAs, coarse/fine/mixed grain fabrics, VLIW processing, systolic arrays, processor networks and so on. RC has long been considered [38,50,97,98,179,274,342,461] a feasible alternative to tackle the requirements described in section 0. The utilization of space-based computation patterns allows RC a density in computational power per silicon area largely unmatched by software programmable architectures, while maintaining the flexibility offered by run-time programmability. In [97], this is defined as the "*Density Advantage*" of run-time configurable hardware over software programmable processors. This synergy between dynamic programmability and computational power makes, at least on a theoretical standpoint, reconfigurable hardware the ideal technology option to deploy run-time instruction set metamorphosis.

As described in [38,179,180,274,380], reconfigurable architectures are classified depending on their *grain*, that is a qualitative metric of the bit-width of their interconnect structure and the complexity of their reconfigurable processing elements (PEs). Field Programmable Gate Arrays (FPGAs) are typically island-style architectures where PEs based on lookup tables (LUTs) are merged in a bit-oriented interconnect infrastructure. Featuring small LUT cells and 1-bit interconnect FPGAs are typically described as *fine-grained.* Their very symmetrical and distributed nature makes FPGAs very flexible and general purpose, and they can be used to tackle both computation-intensive and control-oriented tasks, to the point that large commercial FPGAs are often used to build complete Systems-on-Programmable-Chip (SoPCs) [49]. For highly specialized, low volume market segments this represents a viable alternative to SoCs. On the other hand, their fine grain leads to redundancy in the computation fabric, and in particular in the routing architecture. Arithmetic-oriented datapaths feature regular structures, so when targeting computation-intensive applications it is possible to achieve higher efficiency designing PEs composed of hard-wired operators such as ALUs, multipliers or multiplexers. Reconfigurable units relying on such multiple-bit width, custom operators are defined *coarse-grained*. These devices trade part of the flexibility of FPGAs in order to provide higher performance for specific computations. There exist also a set of devices that fall in between the above two classifications, featuring bit widths of 2 or 4 bits, and small computational blocks that are either large LUTS or small arithmetic blocks as 4-bits ALUs. These can be classified as *medium-grained.*

There are a lot of different possibilities to merge programmable architectures and reconfigurable hardware into efficient computing machines. No model is probably the "absolute" answer to embedded systems requirements, but depending on the application some approach proves particularly appealing. All this broad category of digital architectures fall under the cumulative name of "*Reconfigurable Architectures*" (RAs), underlining their capability to reconfigure at execution time part of their hardware structure to support more efficiently the running application. Those parts are defined here "*Reconfigurable Function Units*" (RFUs), and are normally composed by a regular mesh of "*Processing Elements*" (PEs). In this broad domain, we use the definition "*Reconfigurable Instruction Set Processor (RISP)*" for those reconfigurable architectures that are tightly integrated in order to compute as a single adaptive processing unit according to the Athanas/Silverman paradigm regardless of their hybrid nature. The more general term "*Reconfigurable Processor*" includes all architectures that perform processor-oriented computation taking advantage of hardware reconfiguration, elaborating sets of data according to instructions provided in some form of programming code. The next two sections will describe the evolution of the RP concept in the last 10 years, through the description of the most significant contributions in the field by both industry and academia.

Run-time reconfigurable instruction set processors

In the last decade, we have been witnessing several changes in the embedded processors design fuelled by two conflicting trends. First, the industry is dealing with cut-throat competition resulting in the need for increasingly faster time-to-market times in order to cut development costs. At the same time, embedded processors are becoming more complex due to the migration of increasingly more functionality to a single embedded processor in order to cut production costs. This has led to the quest for a flexible and reusable embedded processor which must still achieve high performance levels. As a result,embedded processors have evolved from simple microcontrollers to digital signal processors to programmable processors. We believe that this quest is leading to an embedded processor that comprises a programmable processor augmented with reconfigurable hardware.

S. Wong, S. Vassiliadis, S. Cotofana, "Future Directions of Embedded Processors", 2002 [461]

The first significant attempt at deploying instruction set metamorphosis to embedded systems taking advantage of run-time configurable hardware is P-RISC (PRogrammable Instruction Set Computer), proposed by Razdan/Smith in 1994 [348]. The architecture is depicted in Figure 9.3. The P-RISC micro-architecture is described as a fixed RISC (a MIPS core) that is adaptively extended by instructions mapped on a standard FPGA embedded in the core and defined as a PFU (Programmable Function Unit). In fact, the authors never address explicitly the embedded domain and retain their work on the micro-architecture point of view. But, with respect to the work of Athanas/Silverman they define a straightforward and efficient interface between the core and the PFU, and focus on fitting the PFU into the core pipeline. For this reason, their effort can be considered the first reference for the definition of a run-time adaptive embedded architecture.

They also propose a compilation model for the ISA extension, starting from C specification. A significant drawback of this approach is that the proposed extensions are confined to combinatorial 2-inputs 1-output functions. This is done in order to ease the physical interface between core and extension and, most of all, to define a clear programming pattern for the compiler-based extraction of "interesting" extensions, parts of the source code that would benefit from FPGA mapping. Also, a maximum 15/20 levels of logic are suggested in order to fit in the cycle time of the processor pipeline. The most interesting concept in P-RISC is that the PFU is considered as a function unit of the datapath, just like an additional ALU

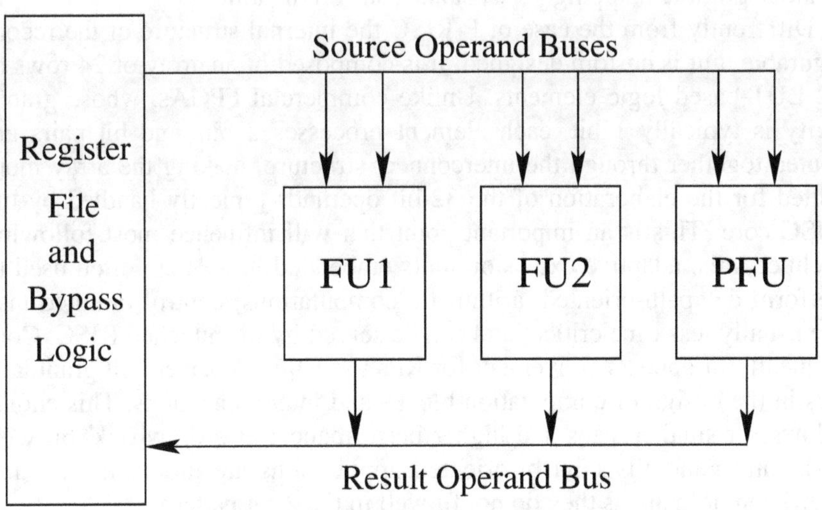

Fig. 9.3. P-RISC architecture.

PFU operands are read and written through the core register file thus providing a low-overhead and tightly coupled model. This is very friendly from the compiler and programmer point of view.

Later attempts try to overcome the shortcomings of P-RISC, while maintaining its significant strong-points. In *Chimaera* [201] the FPGA is divided in eight *rows* (defined *Reconfigurable Function Units* or RFUs) that can be programmed and computed independently, in order to minimize the latency for reconfiguration. *One Chip* [458] is a reconfigurable RISC processor that is also based on the function unit model, but differently from P-RISC and Chimera the reconfigurable units allow the implementation of finite state machines (FSM).

A significant novel step is represented by the "GARP" processor, which is shown in Figure 9.4. GARP [65] couples a MIPS with a custom designed reconfigurable unit, addressed as a coprocessor with explicit *Move* instructions. As in P-RISC, configuration and computation on the reconfigurable unit are triggered by specific assembly instructions. Unlike previously described machines, the granularity of tasks mapped on the unit is quite coarse, to fully exploit the potentiality of the space-based computation approach. Another interesting aspect is that the coprocessor features direct access to memory. This approach allows a larger data bandwidth to the extension unit than that allowed by the core register file, although it raises relevant issues regarding memory access coherency. A significant drawback of GARP is the explicit data movement required to handle the extension unit, that cause a communication overhead that can only be made negligible mapping very coarse tasks on the unit.

Differently from the case of P-RISC the internal structure of the reconfigurable unit is custom designed. It is composed of an array of 24 rows of 32 LUT-based logic elements. Unlike commercial FPGAs, whose granularity is typically 1 bit, each element processes 2 bits and bit pairs are routed together through the interconnect structure, making the array more suited for the elaboration of the 32-bit operands typically handled by the RISC core. This is an important point that will influence most following architectures: adaptive extension units embedded in RPs are often used to perform datapath-oriented, arithmetic computations; control oriented task are usually less time critical and can be served by the attached RISC. Consequently, it appears convenient for RPs to feature "coarser" bit granularities in the design of computation blocks and interconnections. This choice allows for smaller sizes and higher performance; as a drawback, bit-wise tasks and generally control oriented applications are more difficult and inefficient to map, as they do not fit well in the 2-bit pattern.

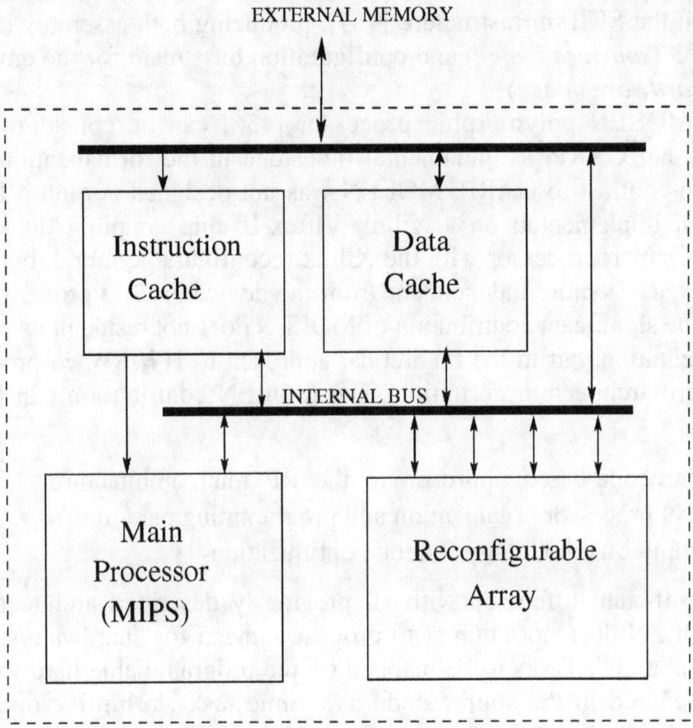

Fig. 9.4. Schematic view of the GARP processor.

The GARP reconfigurable unit is *organized in rows*. Fast carry chains are implemented row-wide to provide efficient 32-bit arithmetical/logical operations on a single row. Each row can be approximated to a 32-bit ALU. An embedded hardware *sequencer* is added to the unit, in order to activate operators (each row can map a single operator) with appropriate timing to build a customized pipeline.

This structure provides a significant enhancement in the programming pattern of the instruction set extension. Candidate kernels are not any more translated in HDL and synthesized over FPGA logic: they are described at C-level and decomposed in Data-Flow-Graphs (DFG), determining elementary operators and their data dependencies, and then mapped over the existing LUT resources. GARP *renounces to logic synthesis* and to a hardware oriented approach of instruction set metamorphosis. The sequencer embedded in the configurable hardware allows for an *imperative* computing pattern that matches very well with C language and the GARP C compiler, thus easing a lot the gap between software and hardware programming indicated by Athanas/Silverman as the key issue in the deployment of instruction set adaptivity. GARP is supported by a compiling tool-chain based on the SUIF infrastructure [457], producing both assembly code for the MIPS (*software image*) and configuration bit-stream for the embedded array (*hardware image*).

The MOLEN polymorphic processor [436] can be considered, with P-RISC and GARP, a fundamental milestone in the formalization of the RISP. In contrast to GARP, MOLEN was not designed at circuit level. It has been implemented on a Xilinx Virtex-II chip coupling the on-chip PowerPC microprocessor with the Xilinx reconfigurable fabric, but in fact the approach is quite independent from the device used to prove its feasibility. The significant contribution of MOLEN does not reside in its physical implementation, but in the theoretical approach to HW/SW co-processing and micro-architecture definition. The MOLEN contribution can be described as

- A microcode-based approach to the RP microarchitecture
- A novel processor organization and programming paradigm
- A compiler methodology for code optimization

A significant difference with all previously described architectures is that MOLEN does not attempt to propose a mean for "hardware/software co-compilation". Tasks to be mapped on the programmable hardware unit are considered in the source code as atomic tasks, primitive operations *Microcoded* in the processor architecture. Instruction set extensions are not determined by co-compilation of the source code, but are defined separately as libraries with an orthogonal HDL-based flow. This could be considered a drawback in the compilation approach, as it requires the user of the RP to be fully responsible of the extension segment design (*hardware image*). Designing microcode for the adaptive extensions (*$\rho\mu$-code*) consists in HDL design, synthesis and place and route of the extension functionality over third-party tools without any assistance from the RP compilation environment. This could raise issues for algorithmic developers not proficient with hardware design. On the other hand, this choice allows a large degree of freedom in the implementation; in fact, the MOLEN paradigm can be extended to any technology support both from the processor and the reconfigurable unit point of view, taking advantage of the development in both fields while retaining the architectural framework. Also, as described in [436], the microcoded approach allows MOLEN-based RISPs to achieve speedups that are almost 100% of the theoretically achievable speedup according to Amdahl's law, much higher than speedups achieved by hybrid compilation.

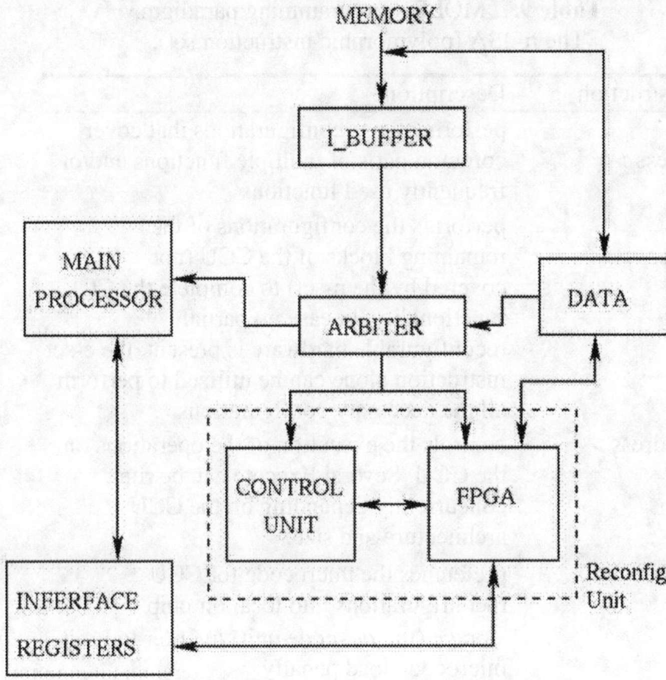

Fig. 9.5. Schematic description of the Molen Architecture.

The MOLEN micro-architecture, shown in Figure 9.5, is structured as follows: instructions are decoded by an arbiter determining which unit is targeted. "Normal" instructions are computed by the *Core Processor* (CP) while instructions targeting the reconfigurable hardware are computed on the RP. This, in turn, is composed of a computational unit called *Custom Configured Unit* (CCU) and a reconfigurable microcode control unit (*ρμ-code unit*). The control unit allows partial reconfiguration of regions in the array, and includes a caching mechanism for the configuration bits. Exchange of data between GPP and RP is performed via specific exchange registers (XREGs). In order to target the *ρμ-coded* processor, a sequential consistency programming paradigm is defined. The paradigm allows concurrent hardware execution: several extension segments can be run concurrently on the CCU depending on its size and internal structure. Table 9.1 describes the list of required instructions, denoted as polymorphic Instruction Set Architecture (π-ISA): six instructions are required for controlling the reconfigurable hardware, while two are used for handling GPP/RP data transfers.

Table 9.1. MOLEN programming paradigm.
The π–ISA (polymorphic instruction set).

Extension Instruction	Description
partial set (p-set < address >)	performs those configurations that cover common parts of multiple functions and/or frequently used functions
complete set (c-set < address >)	performs the configurations of the remaining blocks of the CCU (not covered by the p-set) to complete the CCU functionality. In case no partial reconfigurable hardware is present, the c-set instruction alone can be utilized to perform all the necessary configurations.
Execute < address >	controls the execution of the operations on the CCU. Several Execute can be run concurrently depending on the CCU architecture and size.
Set prefetch < address >	prefetches the microcode for CCU reconfigurations into local on-chip storage (the $\rho\mu$-code unit) in order to limit microcode load penalty.
Execute prefetch < address >	same as the set prefetch instruction, but related to microcode for CCU executions.
Break	used as a synchronization mechanism to complete the parallel execution.
movtx XREGa Rb	move the content of general-purpose register Rb to XREGa.
movfx Ra, XREGb	move the content of exchange register XREGb to general-purpose register Ra.

It should be observed that both in the case of GARP and MOLEN extension segments are handled as coprocessors rather than function units, and data are explicitly transferred from the core to the extensions. In order to guarantee enough data bandwidth direct access from the extension segments to data memory is allowed, although there is no specific handling for multiple access consistency with respect to the core processor. It is interesting to note that, as technology advance in RC leads to the design of denser and denser extension segments, those segments also become more data hungry. One issue that emerges from these later efforts is that the P-RISC function unit paradigm, albeit more elegant and compiler friendly, hardly provide enough computation parallelism to justify the reconfigurable hardware inclusion.

P-RISC, GARP, and MOLEN represent from a theoretical standpoint the milestones that have brought to the formalization of the RISP concept, in terms of micro-architecture, programming paradigm, and compilation approach. From the physical implementation side, an interesting attempt is described in [51], where a RISP is fabricated in CMOS 0.18 μm technology coupling an Xtensa core with an embedded FPGA device to implement essentially the P-RISC concept in one silicon die. The considerable difficulties reported in the eFPGA/core interface in terms of pipeline synchronization and frequency domain adjustment suggest that the GARP approach of designing a specific hardware accelerator rather than relying on existing FPGA technology may prove a good investment in order to design a portable and reliable RP that can be safely included as IP in the embedded domain.

XiRisc [66,270] can be considered the first silicon implementation of a custom designed embedded RISP. The design was performed at circuit level both for what concerns the core and the reconfigurable unit. XiRisc, which is shown in Figure 9.6, is composed of a Very Long Instruction Word (VLIW) core based on a five-stage pipeline. It includes two hardwired computation channels and an additional pipelined run-time configurable datapath (defined PiCoGA) acting as adaptive repository of application-specific functional units. The VLIW core determines two symmetrical separate execution flows. The reconfigurable unit dynamically implements a third concurrent flow, extending the processor instruction set with *multi-cycle* pipelined functionalities of variable latency, according to the instruction set metamorphosis pattern. Unlike GARP or MOLEN, extension segments are tightly integrated in the processor core receiving inputs and writing back results from/to the register file. As in P-RISC, the extension segments are fit in the processor pipeline and there is no direct access to memory. Differently from P-RISC, complex multi-cycle tasks can be mapped on the extension. In order to provide sufficient data bandwidth to the extension segments, PiCoGA features four source and two destination registers for each issued computation. Moreover, it can hold an internal state across several computations, thus reducing the pressure on connection to the register file. Synchronization and consistency between program flow and PiCoGA elaboration is ensured by a register locking mechanism, which handles read-after-write hazards.

According to the MOLEN paradigm, computation and dynamic reconfiguration are handled by a small set of specific assembly instructions Configuration is loaded inside the array reading from an on-chip *configuration cache* that is fed from the on-chip main memory or from an off-chip Flash

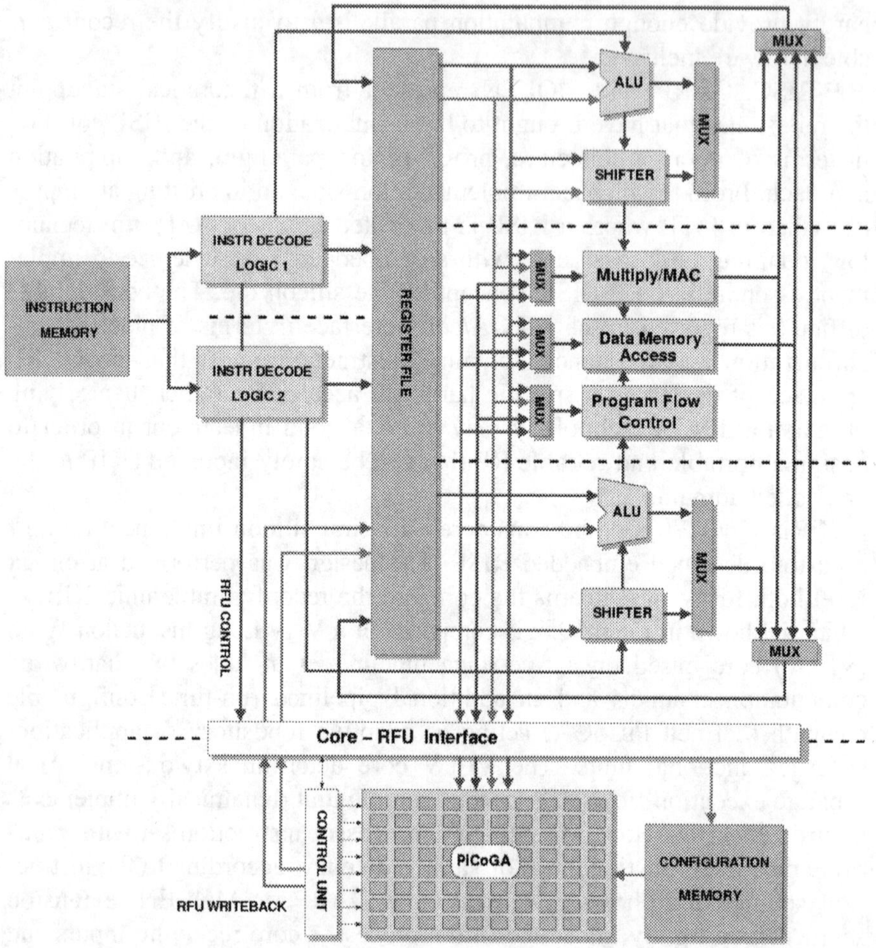

Fig. 9.6. XiRisc internal architecture.

or DRAM repository. PiCoGA is a multi-context device: in order to avoid stalls due to reconfiguration when different configurations are required in a short time span, several extension segments may be stored inside the array (up to 16) and are immediately available.

Similarly to GARP, PiCoGA is organized in rows. A hardwired *pipeline management unit* is embedded in the array, and it is used to activate row elaboration according to the chosen computation pattern. This enables the user to program the extension segments as application-specific datapath implementing *customized pipelines* determined by configuration. Each row represents a stage in the pipeline. This solution imposes some limitation in the definition of extension segments with respect to an HDL-based approach; on the other hand, from the programmer point of view, a direct implementation of a Data Flow Graphs (DFG) becomes quite straightforward significantly shortening development costs. Furthermore, computational resources on the array are utilized according to a heavily pipelined pattern that can maintain high area and energy efficiency while achieving a significant data throughput. DFG handling is quite more sophisticated than in the case of GARP: multiple fan-out nodes are acceptable, and it is possible to fill the pipeline with successive issues only depending on the DFG issue delay (i.e. the pipeline need not to be empty before the insertion of a new issue). Furthermore, DFGs can be cyclic, so a second issue of a given computation may read a state from the first (e.g. to implement accumulators). More than a single node/operator can fit in a given row provided they belong to the same pipeline stage. Finally, up to four pipelined extensions can be computed concurrently on the gate-array, and dynamic hardware resolution logic is used to handle each pipeline flow independently to avoid conflicts on the register file write-back channels.

One of the most interesting aspects of XiRisc is its application development flow [299]. Candidate kernels for implementation on reconfigurable hardware are identified manually by application profiling, as suggested by Athanas/Silverman. Kernels are typically smaller than those utilized for acceleration in GARP or MOLEN, as XiRisc much like P-RISC exploits instruction-level parallelism at assembly level, rather than task-level parallelism. Extracted kernels are written in single-assignment C syntax, and compiled by a proprietary tool that extracts parallelism between the Data Flow Graph (DFG) nodes (C operators) and consequently builds a pipelined implementation of the DFG and map it over the available hardware resources. The proprietary tool, defined *GriffyC compiler* generates the *hardware image* (bitstream) as well as back-annotation information for the compiler scheduling [253] (latency, issue delay, load penalty for each extension).

The pipelined nature of the hardware extension unit allow *every* extension segment to work at the fixed frequency of the main processor (complexity of the kernel is traded in terms of additional latency cycles at fixed speed), thus avoiding all synchronization and clock domain crossing issue. Furthermore, the place and route step can not build paths that require different speed. Thus, all application development can be performed at DFG level, and extensions performance does not depend on the physical mapping step. It is thus possible to iterate design space exploration simply working on software profiling and DFG optimization, only resorting to place and route tools for bit-stream generation.

In conclusion, XiRisc is strictly based on the Athanas/Silverman concept of instruction set adaptivity already implemented in P-RISC. The element of novelty in this respect is that it provides multi-context configuration, a heavy pipelining and a dynamic control of concurrent computation on the extension segments. The hardwired control unit, similar in concept to that of GARP, allows for better and denser utilization of the configurable hardware resources, and enables the user to perform algorithm mapping and design space exploration in a C environment and at DFG-level, avoiding the need for hardware knowledge and hardware related design flows. Several aspects of the XiRisc micro-architecture, compilation approach and ISA-extension comply to the MOLEN paradigm. The difference between the two approaches resides in the extension segments size and nature, and in the type of parallelism they tend to exploit. The high computation density offered by XiRisc allow for a relatively small area and impressive energy consumption reduction figures (70% to 90%), thus making XiRisc a good candidate for IP re-use in SOC design. On the other hand, XiRisc remains oriented at instruction-level parallelism. Sheer speedups remain largely smaller than those available by micro-architectures like MOLEN, based on HDL-oriented microcoding of the extension segments that benefit by massive task-level parallelism.

Coarse-grained reconfigurable processors

Reconfigurable Computing mostly stresses the use of coarse grain reconfigurable arrays (RAs) with paths greater than one bit, because fine-grained architectures are much less efficient because of huge routing area overhead and poor routability. Since computational datapaths have regular structure, full custom design of reconfigurable datapath units can be drastically more area-efficient than by assembling FPGA way from single bit CLBs. Coarse grained architectures provide operator level CFB (complex functional blocks), word level datapaths, and powerful and very area-efficient datapath routing switches.

R. Hartenstein, "A decade of Reconfigurable Computing: a visionary retrospective" [180]

FPGAs have historically been used as programmable computing platforms, in order to provide performance effective solutions to challenge NRE costs and time-to-market issues in the implementation of computationally intensive tasks. For this reason, FPGA fabrics were the immediate choice for implementation of extension segments in the first attempts at designing RPs such as PRISM, P-RISC, OneChip and others. Soon, it appeared evident that RPs required computational features different from standard FPGAs: being essentially part of the processor datapath, extension segments are typically computation – rather than control – oriented and data widths are typically larger than one. Another critical issue is data flow synchronization: in FPGAs control is synthesized on LUTs together with computation. In case of datapath-oriented elaboration control is very regular, so that it is possible to explore alternative, hardwired synchronization means allowing for much more compact implementations: the reconfigurable units in GARP and XiRisc, while still FPGA-oriented, represent a first attempt in enlarging the grain of the processing elements and implementing synchronization with an hardwired sequencer. Furthermore, basic arithmetic operations such as multiplications simply do not fit well when synthesized over fine-grain LUT-based technology, leading to unaffordable delays and area occupation. In such cases, application-specific logic such as hardwired multipliers is the only way to achieve the necessary performance.

If extension segments are very arithmetic-oriented and bit-level computation is not necessary, then the above described approach can be taken to extremes renouncing to LUTs and exploiting coarse-grained hardware. We will define RPs based on coarse-grained hardwired operators rather than on fine-grained LUTs as *coarse-grained reconfigurable processors.* A significant benefit of this approach is the massive reduction of configuration memory and configuration time, as well as the reduction of complexity in the place and route step. The obvious drawback is that algorithm mapping and interconnect resolution, if certainly simpler than in the case of FPGAs, is necessarily non-standard, and very architecture-specific. Shifting towards coarser grained RFUs, the definition of the PE internal structure becomes the focal trade-off in the design of the RP. As the number of options and the complexity in the design of the PE increase, its choice is necessarily influenced by the targeted application domain. As a consequence, performance in that application field will be very impressive, but RPs will not scale well to different application environments. Coarse-grained RPs will then be no more general purpose, as it was the case for the fine-grained architectures described earlier, but rather domain oriented.

The literature in the field of coarse-grain reconfigurable hardware is very rich, and a detailed analysis is out of the scope of the present work. Notwithstanding, this chapter will contain a brief overview of those architectures that were more successfully utilized to build RP, or that are related to the perception of RPs in embedded systems.

PipeRench [159] is one of the first and more original run-time reconfigurable datapaths appearing in literature. It is composed by a set of configurable “stripes”. Each stripe maps a pipeline stage of the required computation, and is composed by an interconnect network and a set of PEs. In turn, each PE contains one arithmetic logic unit and a pass register file that is used to implement the pipeline. ALUs are composed of lookup tables (LUTs) plus specific circuitry for carry chains. Multiplication specific logic is not present in the stripes, so multipliers are built out of multiple adder instances. The configuration overhead is minimized as the configuration flows through the pipeline: configuration is not *static*, but is split into pieces that correspond to each pipeline stage, and each stage is loaded, one per cycle, into hardware. Each stripe can then perform a different functionality per each cycle, thus providing an efficient time-multiplexing in the usage of each resource. The granularity of the computation fabric is parametric, but best performance results are obtained with 16 instances of 4- or 8-bit PEs per stripe, so that we can define the datapath as *average-grained*. The PipeRench fabric is quite general purpose, offering good speedups for a large spectrum of applications.

PipeRench features an efficient compilation toolchain; as it is the case with most average and coarse-grained fabrics, a simplified format of C, in this case defined *Dataflow Intermediate Language* (DIL), is used as entry language. The size and specificity of the target operators make possible to avoid logic synthesis, and perform simple mapping of C-level operators directly to hardware resources. As in the case of XiRisc, operators to be mapped on the fabric are described at Data Flow Graph (DFG) level by a single-assignment C-based format, where variable size can be specified by the programmer, and then translated on one or more PEs on the stripe after an automated Instruction-Level Parallelism (ILP) extraction.

The PACT XPP digital signal processor [445] is composed by a set of *Processing Array Clusters* (PACs), each composed by an array of heterogeneous *Processing Array Elements* (PAEs) and a low level *Configuration Manager* (CM). Configuration Managers are organized in a hierarchical tree that handles the bit-stream loading mechanism. Communication between PAEs is handled by a packet-oriented interconnect network. Each PAE has 16-bit granularity and is composed by synchronization register and arithmetical/logical operations, including multiplication. Data exchange is performed by transmission of packets through the communication network, while I/O is handled by specific ports located at the four corners of the array. The PACT XPP architecture is depicted in Figure 9.7.

In normal operation mode, PAE objects are self-synchronizing: an operation is performed as soon as all necessary data input packets are available, and results are forwarded as soon as they are available. As the full exploitation of parallelism at all levels is very critical to fully exploit the relevant computational potential of the architecture, PACT XPP is programmed through the Native Machine Language (NML), a structural event-based netlist description language. As described in [92] XPP can be applied also to

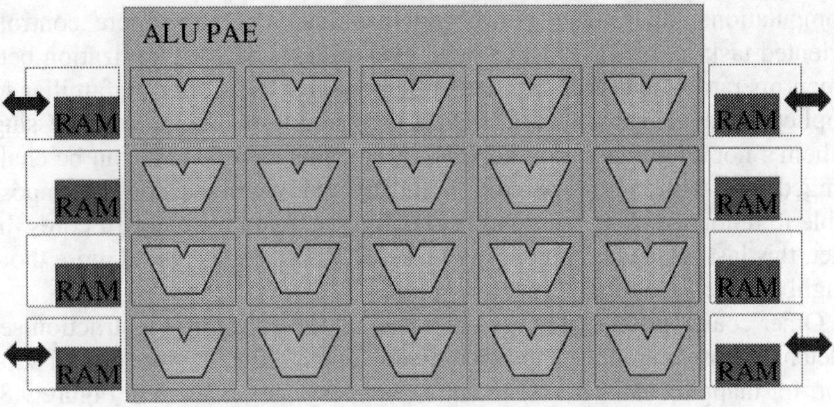

Fig. 9.7. Schematic description of the Pact XPP processor.

control-intensive tasks but the best performance figures are delivered when dealing with data intensive, highly parallel kernels.

We can consider PACT and PipeRench as peculiar RPs, in that they break the classic Athanas/Silverman concept of extending a standard RISC instruction set with new instructions.

Of course, it is possible to use them in cooperation with standard microprocessors to build complete computation systems, but they do not require a main core to handle control, instruction sequencing, and pipeline synchronization. More or less explicitly, they propose to *replace* the concept of instruction sequencing (that is, cycle per cycle instruction fetching) by configuration sequencing (that is, spatial distribution, dynamically pipelined or not, of configuration bits) and they process data streams instead of single random accessed memory words. This concept of communication-centric distributed computation is similar in principle to Transport Triggered Architectures (TTAs) [92], and it is indeed quite promising when applied to reconfigurable hardware because this micro-architectural paradigm, compared to the Von Neumann paradigm, appears more suitable to support a scalable number of function units, each with scalable latency and throughput. On the other hand, this promising approach has three main open issues. One is the communication infrastructure that needs to be large yet flexible enough to allow the necessary throughput between the different function units (PEs). A second open issue is related to tools and programming languages; this concern is certainly true also for adaptive instruction set processors, but the RISC framework in which the latter operate allows the user to be in a more familiar, C-oriented environment for most of the application mapping, focusing on reconfigurable hardware exploitation only for kernels. Non-RISP reconfigurable datapaths can certainly describe computation with C-like syntax, but mapping larger and more control-oriented tasks than RISPs, they need also to describe synchronization between operators, and this requires structures and tools often unfamiliar to application developers. One last issue is related to the memory addressing scheme: not all computation kernels in the embedded domain can be challenged with a streaming paradigm, and for many cases it appears impossible to renounce to the addressing flexibility offered by standard cores. In fact, this last issue is common to all RPs, so it will be discussed more thoroughly in the following.

Other coarse-grained devices are based on the concept of instruction set Metamorphosis introduced above, only utilizing a different architectural support for mapping extension segments: morphosys, also shown in Figure 9.8, [380] is a very successful RP that also been the base for a few successful commercial implementations. It is composed by a small 32-bit RISC core (TinyRisc), coupled to a so-called *Reconfigurable Cell Array*. The array is composed by an 8×8 array of identical Reconfigurable Cells (RCs). Cells are very coarse: each computes 16-bit words and contains multiplier, ALU, shifter, a small local register file and an input multiplexing logic. The architecture comprises a multi-context configuration memory, that is capable to overlap computation and configuration in order to minimize reconfiguration penalty, and a multi-bank frame buffer that is used to overlap computation on one set of data and concurrent transfers on a parallel set to enhance overall data throughput.

Over the RCs, computation is performed in a purely Single Instruction Multiple Data (SIMD) fashion: all cells belonging to the same row receive the same control word, and thus compute the same calculation over extended 128-bit words. It appears thus evident that the Morphosys reconfigurable cell array is very performant and has a much higher area efficiency with respect to FPGA-based solutions described earlier, but it is also rather domain oriented: the machine is conceived for applications with relevant data parallelism, high regularity, and high throughput requirements such as video compression, graphics and image processing, data encryption, and DSP transforms. Its large size and SIMD-oriented structure would make it redundant for application fields that exploit task-level parallelism or instruction-level parallelism.

Several other coarse-grained RPs have appeared in literature, such as REMARC [291] or Butter [57,58], but from the micro-architecture point of view they can all be referred more or less to the concepts described above. In terms of commercially available offers, the coarse-grained RISP concept

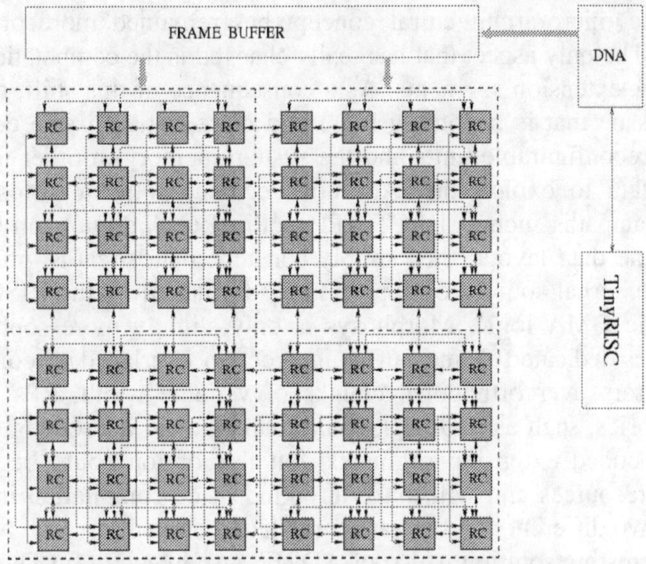

Fig. 9.8. The Morphosys reconfigurable architecture.

is exploited at the moment by two interesting implementations. The S5 [25] core from the US company Stretch consists of a Tensilica Xtensa core whose extensions are dynamically mapped over an Instruction Set Extension Fabric (ISEF). The fabric is provided with 128-bit wide register access and 128-bit wide memory access to feed the extension segments that are described and compiled starting from C/C++ code. ISEF is composed by a *plane* of 4-bit arithmetic/logic elements and a *plane* of multipliers interlinked in a programmable routing fabric. ALUs are cascaded through carry-chains to provide 64-bit arithmetic. ALU elements are provided with embedded registers to implement state variables or pipeline resources. Connection between Xtensa and ISEF is based on the function unit paradigm applied by P-RISC and XiRisc, even though it supports direct access to memory. The DAPDNA-2 core [364] from Japanese IPFLEX is composed by a Digital Application Processor (DAP), a standard Risc core, coupled to a Distributed Network Architecture (DNA). The DNA is a network of coarse-grained processing elements of different nature, processing large bit-sizes such as 32-bit ALUs/Multipliers or performing control tasks such as I/O, address generation, data storage and so on.

In conclusion, RISPs have evolved from the mapping of combinatorial, single cycle functional extension of P-RISC up to the very intensive hyper-parallel SIMD computational pattern of Morphosys-like architectures, but indeed the micro-architectural concept has remained more or less unchanged. The only aspect that has really changed is the computational grain of the ISA extension segments. As a consequence of this shift, one architectural issue that is becoming more and more critical is the connection between reconfigurable units and the system memory in order to provide enough data to exploit the extension segment potential. Most coarse-grained datapaths such as PACT XPP or PipeRench do not actively intervene on the data layout: they simply consume data streams, provided by standard external sources or appropriately formatted by the RISC core or by specific DMA logic. Morphosys is only slightly more sophisticated, featuring a dedicated frame buffer in order to overlap data computation and transfers over orthogonal chunks of variable width. RPs based on FPGA fabrics, such as MOLEN, could map memory addressing as part of the microcoded extension segments, but this option could be costly in terms of resources and will make any kind of co-compilation impossible creating two different and separate compilation domains.

An interesting solution is that of ADRES (Architecture for Dynamically Reconfigurable Embedded systems) [283]. ADRES exploits a RFU similar to that of Morphosys, based on very coarse-grained (32-bit) PEs implementing arithmetical/logical operations or multiplications. Differently from Morphosys, the ADRES RFU is used as function unit in the frame of a VLIW processor. Data exchange with external memory is through the default path of the VLIW processor, and data exchanges take place on the main register file, as it was the case for the XiRisc processor described in section 4. The programming model is simplified because both processor and RFU share the same memory access. ADRES is completed by a compilation environment, DRESC [282], that schedules ISA extensions in order to exploit maximum concurrency with the VLIW core and handles data addressing towards the VLIW register file for both reconfigurable array and hardwired core. A relevant added value of the compiler, that is made possible by the register-file oriented micro-architecture, is that extension instructions are generated by the same compilation flow that produces code for the hardwired core. Even though the RFU has a grain comparable to PACT XPP or MorphoSys, data feed is random accessed and very flexible, and it is not limited to data streaming.

Still, the VLIW register file remains a severe bottleneck for RFU data access. A different solution is provided by Montium [354,382] a coarse-grained RP composed of a scalable set of Tile Processors (TP). A TP is essentially composed by a set of five 16-bit ALUs, controlled by a specific hardwired sequencer. Each TP is provided with 10 1Kbytes RAM buffers, feeding each ALU input; buffers are driven by a configurable *Address Generation Unit* (ATU). Montium can be seen rather as a flexible VLIW than a RP in the context described in this work, but it is affected by the same bottleneck shared by most RP overviewed above: in order to exploit its computational density, it needs to fetch from a repository several operands per clock, and possibly each of them featuring an independent, if regular, addressing pattern. In this respect, automated addressing generation based on regular patterns could be an interesting option: most applications that benefit from hardware mapping are based on loops, and addressing is more often than not generated and incremented with regularity as part of the loop. Automated addressing FSMs could add a new level of configurability to RPs, providing an adaptive addressing mechanism for adaptive units, enhancing potential exploitation of inherent parallelism. As it is the case with adaptive computation, automated addressing can be considered an option only if supported by solid compilation tools that could spare the end user from manual programming. In fact, it appears theoretically possible to automatically extract from a high level (typically C/C++) specification of the algorithm regular addressing patterns to be applied to automated addressing FSM: the same issue has long been discussed for high-end Digital Signal Processors [81] and it is an open research field also for massively parallel systems based on discrete FPGAs [101]. These aspects are only very recently being evaluated in RP architectures.

DREAM [67] is an example of RP that feeds its RFU through automated address generation (Table 9.2). DREAM is an adaptive DSP based on a medium-grained reconfigurable unit. Program control is performed by a standard 32-bit embedded core. Kernel computation is implemented on the RFU, composed of a hardware sequencer and an array of 24×16 4-bit PEs. The RFU accepts up to 12 32-bit inputs and provides 4 32-bit outputs per clock, thus making it impractical to access data on the core register file. For this reason, DREAM is provided with 16 memory banks similar to those of Montium. On the RFU side, an address generator (AG) is connected to each bank. Address Generation parameters are set by specific control instructions, and addresses are incremented automatically at each issue of an RFU instruction for all the duration of the kernel. AGs provide standard STEP and STRIDE [81] capabilities to achieve non-continuous vectorized addressing, and a specific MASK functionality allows power-of-2 modulo addressing in order to realize variable size circular buffers with programmable start point. The reconfigurable architectures are summarized in Table 9.2.

Table 9.2. Schematic description of some RP described in the previous sections.

	Computation model	Computation Grain	Memory Data Access	Entry Language for RFU	Implementation
P-RISC (94) [348]	Function unit	1-bit LUT	Register File	C (Combi natorial functions)	MIPS+Xilinx FPGA (board-level)
GARP (00) [65]	Coprocessor	32-bit row, 2-bit LUT	Passive direct access	C	MIPS + Custom Gate Array
MOLEN (02) [436]	Coprocessor	1-bit LUT	Direct access coded on RFU	HDL	Xilinx Virtex II (PowerPC+ Xilinx FPGA)
(03) [51]	Coprocessor	1-bit LUT	Direct Access coded on RFU	HDL	Xtensa + M2000 eFPGA
XiRisc (03) [66]	Pipelined Function unit	32-bit row, 2-bit LUT	Register File	Single assignment C	Custom VLIW+ Custom Gate array
PipeRench (99) [159]	Pipelined Programmable datapath	128-bit stripe, 8/ 16-bit LUT	Streaming	DIL (Single assignment C)	Custom
Pact XPP (03) [445]	Programmable datapath	16-bit ALU + rfile	Streaming	nML (event-triggered description)	Custom
MorphoSys (00) [380]	Coprocessor	128-bit row, 16-bit ALU	Passive Direct access via Buffer	Assembly at cell level (MorphoAsm)	TinyRISC + Custom Cell Array
ADRES (04) [283]	Function unit	32-bit ALU	Register File	C	Custom
DREAM (07) [67]	Coprocessor	64-bit row, 4-bit LUT + 4-bit ALU	Address Generation FSMs	Single Assigment C	STxp70 + Custom Gate Array

Conclusions

The vision behind this approach is creating and upgrading accelerators by downloading configuration code generated by compilers accepting high level programming language sources onto a general purpose reconfigurable accelerator [...] But the microprocessor will not become obsolete. Its role will be to support future reconfigurable computing platforms by running glue logic, software being not performance critical, and bloatware (software that needs masses of primary memory and hard disk storage space). But with very powerful future reconfigurable accelerator platforms the microprocessor will be the tail wagging the dog.

R. Hartenstein, "The Microprocessor Is No More General Purpose", 1997 [179]

A point common to all literature dealing with RC is that the interaction between microprocessor cores and reconfigurable hardware holds the promise of breaking the gap between computational requirements and programmable architectures capabilities. Many authors have described RC as the main architectural breakthrough in digital processing after the introduction of the microprocessor itself. It is indeed true that the high-end processor/DSP IP market is facing a very consistent offer of processor networks, arrays of processing elements, reconfigurable accelerators for standard processors or RISPs (PicoChip [32], Silicon Hive Moustique [220], NEC Drp [307], IPFlex ADP DNA [209], Stretch S5 [391], MorphICs WSP [295], MorphoTech MS2 [296], Pact XPP [320], Elixent D-Fabrics [119]) that can all be classified under the broad denomination of RPs. Still, despite claims that "The time has come for a reconfigurable revolution" that were regularly iterated over the last 12 years, only few attempts are starting to appear in commercial SoC, very slowly and for very specific application fields. It is probably the huge success of the SoC itself that has slowed down the introduction of reconfigurable hardware in the embedded world, creating such a well accepted legacy that most designers are reluctant to abandon it to explore new design alternatives. The methodology of accelerating computation via ASIC has indeed built the SoC success, and it is so deeply established in our minds that the only metric for evaluating the inclusion of a computational engine in a SoC is currently area occupation. Obviously, the Kgate/mm^2 figures provided by any reconfigurable hardware will never be any way near that of a small microcontroller surrounded by ASIC circuitry, in spite of all the flexibility and configurability it can offer. If this long awaited revolution will finally make its way through in the SoC world, it will be literally imposed by the rate of algorithmic innovation. Applications evolve much faster than standardization, and time to market and cost of SoCs targeting the same algorithms are steadily increasing, so post-fabrication configurability appears a necessity. Still, most of the SoC structure will need to remain unchanged, to preserve the huge legacy that most companies have developed in the field, and because a few general-purpose critical computations will have to be handled by ASIC anyway. For this reason, the revolution will probably consist in a slow and graceful shift of SoCs from application-specific circuits to domain oriented platforms, where different flavours of reconfigurable hardware, each more suited to a given application environment, will be merged with ASIC and general-purpose processors to provide ideal trade-off between performance and post-fabrication programmability.

In this landscape, the fundamental parameters in the evaluation of a candidate RP for inclusion as IP in SoC design can be classified as follows:

a) The design/choice of the reconfigurable fabric (Computation Grain, Interconnect Infrastructure)
b) The application mapping flow and its entry language
c) The interaction between the fabric and the processor core (operand feed, synchronization)

Regarding (a), one point that appears from the analysis performed in the first two sections of this chapter is that fine-grain FPGAs appear mostly redundant for RPs requirements, as the flexibility they provide is rarely exploited by applications and require heavy costs in terms of performance. Moreover, in order to avoid clock domain synchronization issues, it appears advisable to map computation on fixed speed, pipelined fabrics rather than having variable frequency depending on the application mapping. This said, the ideal grain of the processing elements is entirely dependent on algorithmic specifications. Imaging and multimedia applications in general greatly benefit from coarse standard processing elements and large data-widths, and feature relevant data- and task-level parallelism. Telecommunication protocols, especially in the wireless field, feature degrees of instruction-level parallelism, and often require random access to data rather than streaming; also, data widths are normally smaller. Encryption and data security applications normally involve bit-wise computation and parallelism at data and instruction level, but not at task level.

Issue (b) is related to a trade-off between performance and time-to-market. As the orientation towards medium- to coarse-grained architectures appears manifest, C-oriented entry languages are preferred and they are often exploited to map directly operators and their dependencies to hardware through DFG-oriented computation. HDL-based flows are used for targeting FPGA-based RP but this tendency is not suited to SoC design, both for the area penalty of fine-grained RFUs and for the inherent difficulties by application developers in utilizing hardware-oriented programming patterns.

Medium-grain architectures as Piperench or XiRisc exploit DFG-oriented design tools obtaining good mapping results from user-friendly single assignment C descriptions, but their performance is limited by the grain of their PEs. On the other side of the spectrum, the coarser the grain of the reconfigurable fabric, the more difficult it is to provide efficient implementation and good routability starting from automated tools, so that in most practical cases performance is only obtained with manual mapping or event oriented description languages. An exception to this trend is represented by the DRESC compilation environment.

As for point (c), RISPs adaptively extend a standard processor ISA with application-specific instructions, typically with some sort of tightly coupled function unit or loosely coupled coprocessor interface. The coprocessor interface eases synchronization issues and has proved so far very popular especially for market-oriented RP implementation, but the function unit interface as described in ADRES holds the promise for a better and more coherent compilation framework. Other RPs explore innovative patterns, based on data streaming, distributed processing and data driven computation. In all cases, data I/O to reconfigurable hardware appears at the moment a severe bottleneck, and how to program and implement data movements remains an open issue, especially from the compilation point of view. Finally, the choice of the controlling RISC core appears often irrelevant to the definition of the architecture. With the relevant exceptions of ADRES and XiRisc, reconfigurable units have grown to the point where co-computation between the core and the reconfigurable device is not worth exploiting, and only basic control tasks are normally trusted to the RISC. The presence of a RISC core is almost always necessary, but there exist no real constraint on its nature and its computational capabilities.

In conclusion, dynamic instruction set metamorphosis is, after all, finding its way into SoC design. The elegant Athanas/Silverman approach applied in machines like PRISM and P-RISC is still valid. On the practical side though, the level of parallelism that it targeted appears in most cases not massive enough to justify the costs and uncertainties of including reconfigurable hardware in embedded systems, so today's reconfigurable extensions tend to be larger, coarser and more complex than the first embedded FPGA fabrics. Guidelines for the design of the reconfigurable unit can hardly be formalized, as the choice of granularity and internal structure still must be application oriented in order to find the ideal cost/

performance trade-off. RPs are certainly not application specific, but to date they are not general purpose either, as to deliver necessary performance they still need to be domain oriented. The choice of the ideal programming pattern and language is strongly oriented towards C/C++ and its hardware dialects, but this aspect is also related to the targeted reconfigurable hardware, and thus, ultimately, to the application domain. Research is very active in various aspects, but the best approach to RP appears at the moment to couple a small standard core dedicated to program flow, data movement and synchronization tasks to one (or more than one) large, medium to coarse-grained, *domain oriented* configurable devices performing intensive computation. So that, as Reiner Hartenstein colourfully predicted in 1997, embedded systems remain processor-oriented, although microprocessors are becoming the small, smart tail wagging a big dull dog made of run-time programmable hardware.

10 Coprocessor Approach to Accelerating Multimedia Applications

Claudio Brunelli and Jari Nurmi

Tampere University of Technology

In this chapter, we describe a coprocessor approach to accelerating applications. We are concentrating on multimedia applications as an illustrative example of where this approach can be applied. In everyday life we all deal with a number of integrated digital systems which are present into almost all the products for consumer electronics. The heart of those systems is usually a microprocessor core, which is mainly used to control all the other components.

Need for accelerators

Many of the items that feature a microprocessor inside them are handheld devices, thus posing severe constraints both on the area available and on the energy budget and power dissipation. Despite of this, the will for conveniently including more and more complex and smart functionalities even on simplest devices has been pushing system designers to consider new ways of empowering their architectures with special solutions to guarantee the ability of executing the required functionalities while respecting the posed constraints.

In particular, users ask for their portable devices (PDAs) to support heavy multimedia and 3D graphics applications, besides other ones which are commonly referred to as "general-purpose" computations [213].

Besides others which need general-purpose computations, multimedia applications need to squeeze all the possible performance out of the computation system, and standard RISC architectures can hardly provide

J. Nurmi (ed.), Processor Design: System-on-Chip Computing for ASICs and FPGAs, 209–228.

the necessary computational capabilities. For this reason, computer architectures shifted from the usage of conventional microprocessors to entire "Systems on a Chip" (SoC). Inside SoCs there is usually a main core (might be an RISC microprocessor or even a powerful DSP) together with a set of other components ranging from I/O peripherals and DMA controllers to memory blocks and other dedicated computation engines (coprocessors and dedicated processors tailored to accelerate a precise set of algorithms). At the beginning such dedicated accelerators were usually custom ASIC blocks; more recently the fast pace at which applications and standards are changing pushed SoC designers to go for components which are programmable and flexible, while providing at the same time a significant speed-up.

The topic of microprocessor acceleration is so wide, ranging from the diverse metrics which have been proposed to measure performance in a fair and coherent way, going through the different types of algorithmic exploitations, to the analysis of different architectures of accelerators, that even an entire book could hardly cover it in a proper way. This chapter thus will not give an exhaustive dissertation about all the existing typologies of accelerators, but will rather try to give a brief overview of the main issues related to the topic of microprocessor acceleration, to take then a deeper insight into one typology of accelerators in particular which has been used for many years and still is used in several designs thanks to its modularity and ease of use: the coprocessors.

Accelerators and different types of parallelism

Several different architectural approaches were introduced so far to achieve high performance; in principle, each solution tries to guarantee computational efficiency by taking advantage of known properties of application programs running on them. One of the key properties that can be exploited is the *parallelism*, which can be exploited at several levels:

- Instruction level parallelism (ILP), which measures the possibility to execute several instructions at the same time
- Loop level parallelism, which happens when consecutive loop iterations can be executed in parallel
- Task level parallelism, which means that entire procedures inside the program can be executed in parallel
- Program level parallelism, which is present if multiple independent processes can be executed in parallel
- Data parallelism

Task level and program level parallelism can be usually exploited by the operating system (OS) in very complex systems, like multiprocessor SoCs. The first two typologies of parallelism, instead, are suitable for exploitation in a wide range of machines. In particular, ILP can be exploited by VLIW, pipelined, or superscalar architectures; to exploit data parallelism there are instead SIMD, systolic, neural and vector architectures [376,377].

Processor architectures and different approaches to acceleration

Besides general-purpose microprocessors, a special family of microprocessor is the one of DSP processors: they try to guarantee high performance in executing a very special category of algorithms, while keeping some degree of flexibility thank to the fact that they can be programmed using high-level languages.

In the 1990s also RISC microprocessors became very popular, to the point that they replaced some CISC architectures (initially very widely adopted thanks to the fact that they needed smaller program memory to run applications). The success of ASIC and SoCs eased the advent of post-RISC processors, which are usually generic RISC architectures, augmented with additional components.

Several different approaches have been proposed to implement the acceleration of a microprocessor, but in general the main idea consists in starting from a general-purpose RISC core and adding extra components like dedicated hardware accelerators (for example, this approach is followed by Tensilica with their Xtensa architecture [259]). This is due to the fact that it is good to keep a certain degree of software programmability to keep up with the fact that applications and protocols change fast, so having a programmable core in the system is recommendable to guarantee general validity and flexibility to the platform.

One possible way of accelerating a programmable core consists in general into exploiting instruction and/or data parallelism of applications by providing the processor with VLIW or SIMD extensions; another way consists in adding special functional units (for example, MAC circuits, barrel shifter, or other special components designed to speed up the execution of DSP algorithms) in the datapath of the programmable core (usually an RISC core or a DSP) [259]: this way the instruction set of the core is extended with specialized instructions aiming at speeding-up operations which are both heavy and frequent.

This approach is anyway not always possible or convenient, especially if the component to plug into the pipeline is very large. Moreover, large ASIC blocks are now not so convenient in the sense that they usually cost a lot and lack flexibility, so that they become useless whenever the application or the standard they implement changes.

For this reason many microprocessors come with a special interface meant to ease the attachment of external accelerators; there are basically two possibilities:

- using large, general-purpose accelerators to be used for as many applications as possible;
- using very large, powerful, run-time reconfigurable accelerators.

The design and verification issues related to coprocessors can be faced independently from the ones related to the main processor: this way it is possible to parallelize the design activities, saving then time, or (in case which the core already exists before the coprocessors are designed) the coprocessors can be just plugged into the system as black boxes, with no need to modify the architecture of the processor.

The rest of the chapter is organized as follows: the next section explains what are the requirements and constraints posed by applications to the design and the architecture of hardware coprocessors. The following two sections will then provide a general overview about floating-point units (FPUs) and reconfigurable machines, respectively, stating in which application domains they are needed and why they are getting more and more popular; some examples are given to illustrate different practical approaches to the implementation of those architectures. The last section will describe examples of each of those two typologies of accelerators.

Requirements of applications for hardware coprocessors

Thanks to the advances in modern microelectronics technology, today a single chip can host an entire system which is cheap and at the same time powerful enough to run several applications, including demanding ones like image, video, graphics, and audio, which are becoming extremely popular at consumer level even in portable devices.

These applications are then to be carefully analyzed to determine an optimal way to map them to the hardware available: since normally applications are made up of a control part and a computation part, the first stage usually consists in locating the computational kernels of the algorithms. These kernels are usually mapped on dedicated parts of the system (namely dedicated processing engines), optimized to exploit the regularity of the operations operated on large amounts of data, while the remaining parts of the code (the control part) is implemented by software running on a regular microprocessor. Sometimes special versions of known algorithms are set up in order to meet the demand for an optimal implementation on hardware circuits.

Different application domains call for different kinds of accelerators: for example, applications like robotics, automation, Dolby digital audio, and 3D graphics require floating-point computation [131,214,259], making thus the insertion of FPU very useful and sometimes even necessary.

To cover the broad range of modern and computationally demanding applications like imaging, video compression, multimedia, we also need some other kind of accelerator: those applications usually benefit from regular, vector architectures able to exploit the regularity of data while satisfying the high bandwidth requirements. A possibility consists in producing so called multimedia SoC, which usually are a particular version of multiprocessor systems (MPSoCs) containing different types of processors, which meets far better the demands than homogeneous MPSoCs. Such machines are usually quite large, so a very effective way of solving this problem which is widely accepted nowadays is to make those architectures run-time *reconfigurable*. This means that the hardware is done so that the datapath of the architecture can be changed by modifying the value of special bits, named *configuration bits* or *configware*. One first example of reconfigurable that became very popular is given by FPGA processors, which can be used to implement virtually any circuit by sending the right configuration bits to the device.

The idea of reconfigurability was then developed further, leading to custom devices used to implement powerful computation engines; this way it is possible implementing several different functionalities on the same component, saving area and at the same time tailoring the hardware at run-time to implement an optimal circuit for a given application. Reconfigurability is an excellent mean of combining the performance of hardware circuits with the flexibility of programmable architectures.

Accelerators come in different forms and can differ a lot from each others: differences can relate to the purpose for which they are designed (accelerators can be specifically designed to implement a single algorithm, or can instead support a broad series of different applications), their implementation technology (ASIC custom design, ASIC standard-cells, FPGAs), they way which they interface to the rest of the system, and their architecture.

Numeric coprocessors: floating-point units

As stated in the previous section, several of the applications which are nowadays very common require (or anyway benefit a lot) from the usage of floating-point arithmetic. Historically, the drawback related to floating-point arithmetic is its significantly higher complexity when compared to integer arithmetic. Such complexity translates into very long execution time when implemented using software routines; for this reason were introduced hardware circuits able to perform floating-point operations in much less time, commonly known as FPUs. Again, the complexity of floating-point arithmetic lead to the fact that the area of the FPUs is usually quite large; this point usually discouraged designers to include them into their systems, at least until some years ago. Nowadays, thanks to the advances in VLSI technology, that problem is becoming less acute, so FPUs are appearing also in embedded and area-limited devices.

There are different existing typologies of FPU, ranging from proprietary to open-source ones, supporting the IEEE-754 standard or not, able of single-precision or double precision computation, for usage with CISC or RISC machines; they can come either as external coprocessors or internal functional unit.

Considering the world of RISC cores, one of the most important examples is given by FPUs for ARM, called VFP-9, VFP-10 and VFP-11, just to cite some of them [24]. They are very powerful, vector FPUs, supporting also double precision to enhance accuracy in calculation which can be needed in some kind of applications like scientific calculus or precise positioning control systems. They are pipelined and present some functions which can be configured by the user through software configuration.

MEIKO is an FPU developed at SUN, and has been used together with the Leon processor [227], which is an open source RISC core developed at Gaisler Research.

Also an FPU is made available from Gaisler Research, called GRFPU [135], which was as well interfaced with LEON processor. Anyway, the GRFPU is not really open-source: only some post-synthesis netlists are available for personal usage. It also supports double precision arithmetic, is pipelined, and complies with SPARC V8 floating-point instructions.

Generic FPUs are instead available from Opencores, provided by Jidan Al-Eryani and from Usselmann. The FPU from Jidan Al-Eryani is a complete coprocessor, which features a hardware logic to handle denormal operands, even though it does not support parallel execution of the instructions. The device from Usselmann is instead a set of functional units (like GRFPU), and does not feature an independent register file, nor a control and interface logic. The functional units are pipelined and can operate in parallel.

Hitachi created an FPU for CalmRISC32 microprocessor [214]; it features single precision arithmetic and all the common arithmetic operations; it has also load/store operations.

Finally, a particular case is the one of FPUs which are especially optimized for FPGA implementation: Xilinx and Altera have available dedicated FPUs to their soft-cores (Microblaze and Nios) [13,15,340,467,468].

Various types of reconfigurable accelerators

Reconfigurable accelerators appeared quite recently on the scene of embedded systems, but opened quickly a wide set of research paths. They give the advantages of offering high performance using at the same time a small amount of configuration data (configware) and guaranteeing limited energy consumption.

The term "reconfigurable hardware" is used for a wide spectrum of devices. Designers interpreted the term in many different ways, producing items that are somewhat configurable at run time (for which the term "configurable" would be maybe more precise), devices whose instruction set can be dynamically adapted like Tensilica proposed [356], and finally (and more properly) devices whose hardware circuits can be dynamically be modified and adapted at run time to implement a set of given applications.

Again, machines belonging to the last category can be categorized according to some criteria like the way which they are connected to the rest of the system (coprocessors, functional units, I/O processors), to the bit-width of the operands they can process (leading to fine grain or coarse grain architectures) [180,423].

Tensilica interpreted (re)configurability at several abstraction levels, providing automatic tools able to place and connect together many, embedded IP blocks to optimally partition a complex algorithm and get it executed simultaneously by the processing units. This approach can be very efficient in situations where a clear partition of the applications can be carried out, even though sometimes it might lead to the implementation of large and complex systems.

In the case of XiRISC processor, developed at the University of Bologna, the reconfigurable hardware is a functional unit named PicoGA inserted inside the datapath of a VLIW microprocessor [272]. An advantage of this approach consists in the low latency and overhead that it takes to feed the FU with new data; on the other hand a limit must be put on the maximum length of the pipeline in order to prevent inefficient execution due to a high stall rate. PicoGA is a fine grain reconfigurable machine, thus it is namely suitable for cryptography and telecommunication applications.

Many different devices follow instead the coprocessor approach. The reconfigurable devices are then external components which are somehow attached to the core or to the system bus. It is almost impossible to report all the different reconfigurable architectures which have been proposed, but we will now try to report at least some of them.

Montium from University of Twente [193] is a coarse grain reconfigurable machine, mainly targeted to video applications.

Also Morphosys [379] and Adres [283] are coarse grain reconfigurable coprocessors, organized in an 8×8 matrix of processing elements. They are targeted to image processing and applications characterized by a large number of data-parallel operations, but also to applications like Viterbi decoder. ADRES is a scalable *template* of a coarse grain reconfigurable matrix of computing elements, each featuring a simple 16-bit ALU.

Matrix [290], PipeRench [159], Remarc [291], Kress Array [181], are all coarse grain reconfigurable coprocessors. They are mainly targeted to image processing and applications characterized by a large number of data-parallel operations, but also to applications like Viterbi decoder. Usually this kind of machines proved to be able to exploit better the fine grain parallelism available inside applications than multimedia-extended ISA processors.

XPP architecture from PACT [45] is at the moment the only example of coarse grain reconfigurable machine available which is entirely designed using HDL languages, according to our knowledge.

RAA (Reconfigurable Algorithm Accelerator) [251,252] is a generic reconfigurable coprocessor for algorithm acceleration. It implements a MIMD stream processing array, made of 16-bit DSP processor cores and FIFO buffers. Both an RSA encryption algorithm and a GPS correlation algorithm have been implemented on RAA.

Molen [436] is a project in which a dedicated software toolchain is used to map a C code on a platform made up of a PowerPC core coupled with a reconfigurable processor mapped on a Virtex II Pro FPGA.

Table 10.1 summarizes the reconfigurable architectures mentioned. It is clearly visible how most of them are used as a coprocessor for multimedia applications in general.

Table 10.1. Overview of different reconfigurable architectures.

	Grain size	Interfacing style	Application domains
XiRISC	Fine	Functional unit	Cryptography, telecom
Montium	Coarse	Coprocessor	video processing
Morphosys	Coarse	Coprocessor	Image processing, Viterbi decoding
Adres	Coarse	Coprocessor	Image processing, Viterbi decoding
Matrix	Coarse	Coprocessor	Image processing, Viterbi decoding
Piperench	Coarse	Coprocessor	Image processing, Viterbi decoding
Remarc	Coarse	Coprocessor	Image processing, Viterbi decoding
Kress array	Coarse	Coprocessor	Image processing, Viterbi decoding
XPP	Coarse	Coprocessor	Multimedia
RAA	Coarse	Coprocessor	Cryptography, GPS
Molen	Coarse	Coprocessor	Multimedia

Almost all of the architectures mentioned are attached to the system bus, and are coarse grain machines, thus suitable for imaging or streaming applications, characterized by a large number of data-parallel operations. Indeed, for applications like DSP algorithms, multimedia processing and 3D graphics it has been shown how coarse grain reconfigurable architectures allow a significant advantage in terms of ease of mapping of basic functionalities, in terms of latency and energy consumption [102,138,284].

A number of intermediate solutions between opposite approaches are possible: ranging from fine-grain to coarse grain cells, and from external devices to functional units present on internal datapath of microprocessors, a set of compromises can be found.

Milk coprocessor and Butter accelerator

In this section we present two designs which exemplify the interpretation that we gave to the design of FPUs and reconfigurable accelerators.

Milk coprocessor

The FPU we developed is named Milk and was designed to be as much reusable, flexible, and customizable as possible; one first target was to make it fully compliant with the IEEE Std754-1985 standard for floating-point arithmetic, to guarantee general validity.

To enhance the flexibility and portability of our design, we wrote a parametric and technology-independent VHDL model which makes use of so called *generics*, that is a set of constants which allow to specify the values for a set of key parameters of the architecture like the bus width, the polarity of control signals, the value of opcodes, the functional units which should be inserted or removed, and so forth.

Key features of our FPU:

- Hardware circuits that elaborate the denormal operands. This avoids the degradation of performance due to software emulation which is necessary when such circuits are not included. On the other hand, if denormal operands are expected to be not so frequently encountered in the application domain, we provided the possibility to exclude from the implementation the normalization logic to save some area and to reduce the latency of each functional unit by one clock cycle.
- High parallelism, because functional units can operate independently from each other. When different functional units commit their eloboration simultaneously, a multi-port register file allows the concurrent write-back of their results.
- Scalability and adaptability: the functional units can be inserted or removed from the architecture in an immediate way, just setting the value of dedicated VHDL generics. Also many key parameters of the architecture can be tuned to taste of the user (width of the bus, latency of the functional units, opcodes, etc.)
- Modularity of the functional units: each functional unit is dedicated to implement an elementary arithmetic operation in particular. It can be removed from the architecture and also be used as a stand-alone computational element inside other designs.
- Hardware logic for "register locking" and to stall the core, to guarantee the consistency of program execution with no need to rely on the compiler.
- Hardware implementation of division and square root.
- Support by the GCC compiler.

Milk is totally open source and not bounded to any technology in particular; our VHDL model can be mapped either to ASIC standard-cells or FPGA technologies, even though (to preserve generality) we did not make explicit usage of any FPGA-specific optimization. This choice leads to results for FPGA implementations which are less optimized than they could be, but keeps the VHDL model of general validity and portability.

Milk interface

Milk communicates with the main core using a simple interface made up of a 32-bit bidirectional data bus, an address bus (that specifies the register of our FPU the current operation refers to) and some control signals (used for instance to enable the coprocessor and also to specify whether the current data on the bus should be read from or written to Milk).

This straightforward interface allows plugging our FPU easily into different systems; as a proof of this we successfully and easily attached our coprocessor to two different microprocessors: Qrisc from University of Bologna, and Coffee RISC from Tampere University of technology [56].

On the other hand, the simplicity of the interface also introduces a slight penalty to the performance by introducing some data transfer overhead: indeed the core first needs to move the operand(s) of a given instruction to the register file of Milk, then the core must write the instruction word to the control register of Milk, and finally (at least in some cases) the core has to read back the result from the register file of Milk, where it has been written after that the computation is over.

Despite of this, benchmarking results showed that the penalty introduced was not so large. Then, to ease portability to other platforms, we decided not to modify the interface.

It could maybe be possible to improve the situation providing our coprocessor with the capability of accessing directly the main memory in a way invisible to the core. On the other side, this would imply a non-trivial modification of the architecture, which should be provided both with memory access hardware and an arbitration mechanism to guarantee consistent and conflict-free operations.

Architectural choices

We made some architectural choices to provide high computational power while keeping the design also immediately understandable and easily adaptable to diverse implementations. To achieve this goal we had to study the trade-offs between different design choices and the features that our design should support.

For example, we could achieve computational power and efficiency by adopting a parallel nature of the data path, and also using compact tri-state buffer arrays to allow fast switching of internal buses.

We used pipelined functional units to be able of issuing a new instruction every clock cycle when needed. We designed the pipeline stages so that the latencies are as low as possible, while at the same time the maximum clock frequency achievable is high enough to compete with other similar devices.

In addition we tried to maximize the exploitation of parallelism allowing the simultaneous elaboration of more functional units in parallel, allowing this way optimizing compilers to schedule the instruction in an efficient way (for example, issuing several simple operations after that a heavy one has been issued and has still to be completed). This process may involve some kind of reordering of instructions, which then must deal with issues related to instruction dependencies.

Some compilers can take care of solving all these problems; to guarantee the highest possible portability of our design we decided anyway to introduce an hardware mechanism for register locking and processor stalling, which detects the hazards and stalls the processor whenever it tries to access data from a register which is not ready, or tries to execute floating-point instructions dependent on previous ones which are still in execution. This way the consistency of the execution is guaranteed, with no need to rely on special functionalities of the compiler; still it is true that an optimizing compiler which is aware of the hardware can schedule instructions so that the amount of conflicts (and thus the number of time the core is stalled) is minimized.

Allowing parallel execution of any combination of functional unit implied some area redundancy, since a few hardware components had to be replicated inside all the functional units; on the other hand, the global computation efficiency is increased, as well as the modularity of the design: each functional unit can thus be freely removed or inserted in the datapath, without affecting the rest of the system.

The functional units get their operands from a multi-port register file, where they also store the corresponding results when their elaboration is finished. The register file is multi-ported because this way we could allow an arbitrary number of functional units to write their results simultaneously, provided that they don’t have also to write to the same destination register.

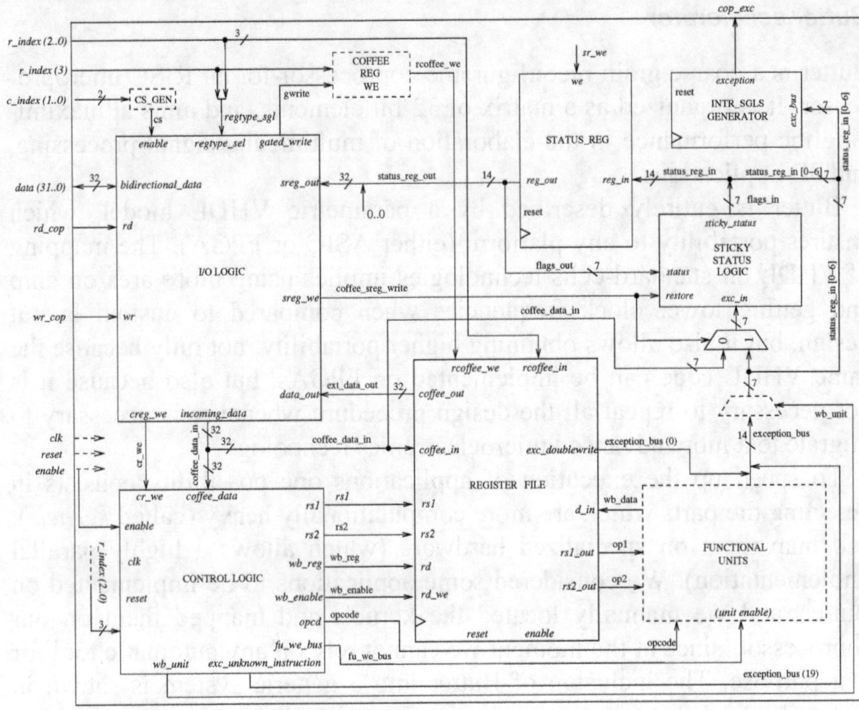

Fig. 10.1. Block scheme of the internal architecture of Milk FPU.

The operands of arithmetic operations are sampled inside each functional unit whenever its execution is triggered, limiting unnecessary switching activity that would lead to useless energy consumption.

The user can select whether or not to mask interrupt signals like “illegal instruction” or the exceptions mentioned by the IEEE Std754-1985 standard (underflow, overflow, division by zero, inexact result, and invalid operation).

The internal architecture of Milk is described by Figure 10.1, where it is depicted at logic block level.

Milk requires 105 Kgates and runs at 400 MHz on a 90 nm standard cells technology, and requires 20K Logic Elements running at 67 MHz on an Altera Stratix FPGA. It is capable of completing instructions in a very small number of clock cycles: 3 for multiplications, 5 for additions, 8 for square root, 11 for divisions, 2 for conversions and 1 for all the other ones (like absolute value, negations, comparisons, etc.).

Butter accelerator

Butter is a coarse grain reconfigurable coprocessor for an RISC microprocessor. It is organized as a matrix of 32-bit elements, and aims at maximizing the performance in the elaboration of multimedia, signal processing, and 3D applications.

Butter is entirely described by a parametric VHDL model, which ensures portability to any platform (either ASIC or FPGA). The mapping of VHDL on standard-cells technologies implies using more area on chip and getting lower clock frequencies when compared to custom layout design, but it also allows obtaining higher portability, not only because the same VHDL code can be implemented on FPGAs, but also because it is not necessary to repeat all the design procedure whenever is necessary to migrate to a more advanced microelectronics technology.

To speed up the execution of applications one possibility consists in detecting the parts which are more computationally heavy (called *kernels*), and map them on specialized hardware (which allows a highly parallel implementation). We considered some applications to be implemented on Butter, and we manually located the kernels and mapped them on our coprocessor, since at the moment we cannot rely on any automatic tool for that purpose. The inclusion of Butter into a generic system is shown in Figure 10.2.

As can be seen in the picture, Butter is a coprocessor attached to the system bus; this might put some constraints to the kind of applications which can benefit from mapping on our architecture, since the bandwidth between the core and the coprocessor is limited. For this reason we decided to provide the possibility of using our machine also as a kind of I/O processor: once that the array has been configured it is ready to process incoming data, which can fed for example from a high-speed input peripheral like a camera. The results are produced as output through another high-speed port which (for instance) could drive directly a display, since Butter is mainly targeted to applications like image and video processing.

Some dedicated *configuration bits* are used to specify the elaboration that the cells of Butter must perform and from where their operands come (thus defining the topology of interconnections between cells). Configuration bits are stored in a dedicated memory inside Butter, and can be written by the core or via DMA transfers.

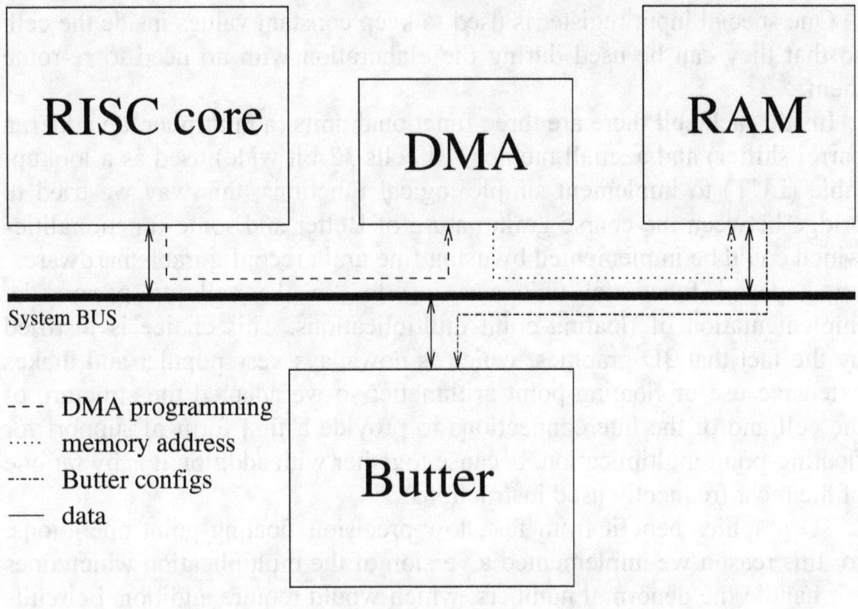

Fig. 10.2. Including Butter accelerator inside a system.

Butter is organized as a matrix of processing elements called *cells*; the number of rows and columns of the matrix is determined by constants included in the VHDL model of Butter (generics).

Internal architecture of Butter: cells and interconnections

Each cell inside Butter has two inputs ports to read 32-bit wide operands; a 6-bit wide input port brings inside the cell the configuration bits, used to specify the kind of operation that the cell should perform on the operands. Also reset and an enable input are used to control the internal registers of the cell.

There are two 32-bit output ports for each cell: they can be used either to bring out the 64-bit result of a 32-bit multiplication, or a generic 32-bit result coming from another functional unit, together with one of the operands which has been fed through the cell (used to simplify the routing).

Input registers inside the cells are used to sample the operands, creating this way a sort of pipeline; these registers can be also disabled to avoid useless dynamic power consumption.

One special input register is used to keep constant values inside the cell, so that they can be used during the elaboration with no need to re-route them.

Inside each cell there are three functional units (a multiplier, an adder, a barrel shifter) and a small memory (4 cells 32-bit wide) used as a lookup-table (LUT) to implement simple logical functions: this way we tried to bridge between the coarse grain nature of Butter and some functionalities which could be implemented by using fine grain reconfigurable hardware.

A special functional unit was inserted in the cells to permit the implementation of floating-point multiplications. This choice is justified by the fact that 3D graphics, which is nowadays very popular and makes extensive use of floating-point arithmetic; so we adapted the structure of the cell and of the interconnections to provide a first form of support for floating-point multiplication, because together with addition it is by far one of the most frequently used instructions.

3D graphics benefit from fast, low precision floating-point operations; for this reason we implemented a version of the multiplication which does not handle the denormal numbers, which would require additional circuits and complicate the mapping procedure. We also decided to get the whole instruction done internally by each cell, so that one multiplication will require just one clock cycle.

Considering the architecture of a simple floating-point multiplier it is apparent how the integer multiplier and the integer adder (which are necessary to build a floating-point multiplier) come at no cost, since they are present anyway inside the cells.

The packing logic that prepares the results produced by the adder and the multiplier, rounding them to be stored in the floating-point format, is too complex to be fit inside the LUT alone, and cannot be mapped on the remaining functional units, so we had to introduce a dedicated block inside the cells: it consists of a portion which calculates the amount of leading zeros for each of the operands, the sign of the result, and packs the internal number into the final format.

The structure of each Butter cell is described in Figure 10.3. The first row of cells read their operands from global vertical interconnections; the results of the elaboration are put as output accessible from the underlying rows. The final result can be read externally of Butter either from its last row at the bottom of the device, or from the rightmost column: this way results can be accessed as soon as they are produced, with no need to wait that they go through all the rows.

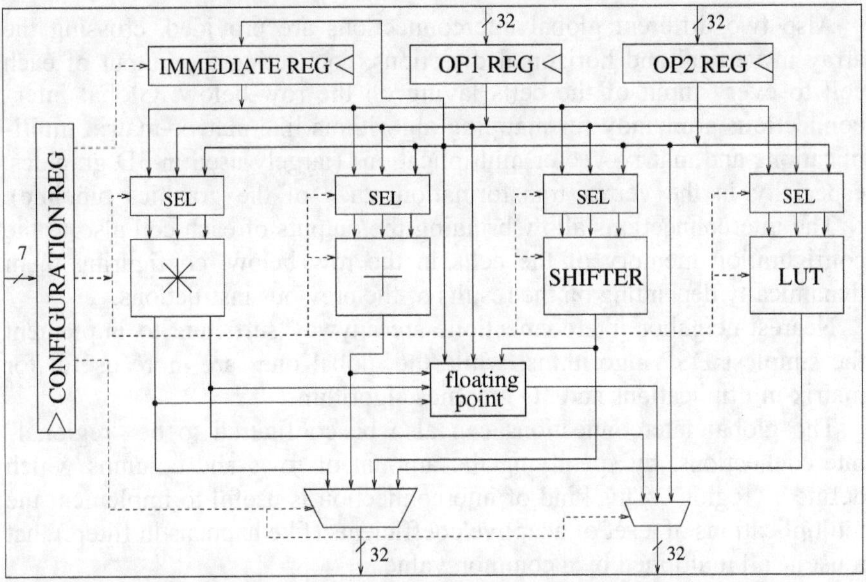

Fig. 10.3. Internal architecture of a cell of Butter accelerator. ©IEEE, 2006 [57].

Both the input ports of cells inside the array can be connected to the outputs of the cell straight above them, or instead to the cell which lays two rows above (interleaved connection). The interleaved interconnection is useful (for example) to propagate the 64-bit result of multiplications splitting their processing over two adjacent rows, for instance like happens in the FIR algorithm: the input operands have to be multiplied by different coefficients, and after that the results must be added together. Thanks to the interleaved connections it is possible to implement the FIR algorithm using only three rows of the array: the first row executes the multiplications, the second row the additions of the least significant bits of the products, and the third row the addition of the most significant bits.

Another possibility provided by the interconnection network consists in enabling the input ports of a cell to receive the results of the cell on its left-side, as well as the ones lying in diagonal direction (upper left and upper right). They are useful in easing and enhancing the mapping of some algorithms, and in reducing the amount of cells used: for example, to elaborate some additions in parallel, using diagonal interconnections we use seven cells instead of eleven (which would be necessary using only vertical and horizontal connections).

Also two different global interconnections are provided, crossing the array in vertical and horizontal directions, connecting the output of each cell to every input of the cells laying on the row below. Global interconnections are handy for mapping algorithms like matrix–matrix multiplications and matrix–vector multiplications (largely used in 3D graphics, especially in the vertex transformation stage of the graphics pipeline).

The interconnections allow bringing the outputs of each cell also to the configuration memory of the cells in the row below, configuring them dynamically depending on the results of the previous instructions.

Nearest-neighbor interconnections are anyway sufficient to implement the simplest DSP algorithms, while the global ones are more useful for matrix-multiplications and 3D graphics algorithms.

The global interconnections can also be configured to be "regional" interconnections, by specifying the amount of rows and columns which define a "region". This kind of interconnection is useful to implement the multiplications of a set of adaptive coefficients (like happens in filters) that must be all multiplied by a common value.

A summary of the local interconnections between two Butter cells is shown in Figure 10.4.

Butter was synthesized on FPGA and a 90 nm Standard-cells technology, achieving, respectively, an operating frequency of 57 and 280 MHz. Thanks to its wide datapath, high parallelism and pipelined nature Butter can run algorithms using a very limited amount of clock cycles; for example, an FIR filter takes 16 cycles, a matrix–vector multiplication takes 4 cycles, and a 2D IDCT 54 cycles.

Conclusions

Considering the diverse requirements dictated by modern applications and the set of constraints mentioned at the beginning of this chapter we came up with the description of a series of possible solutions which are commonly adopted to tackle the problems and to deliver efficient solutions, ranging from programmable processors (RISC, DSPS, etc.) to multiprocessor systems, going through VLIW and SIMD architectures, and heterogeneous systems hosting a set of different types of computation engines, which usually are microprocessors, coprocessors and dedicated accelerators.

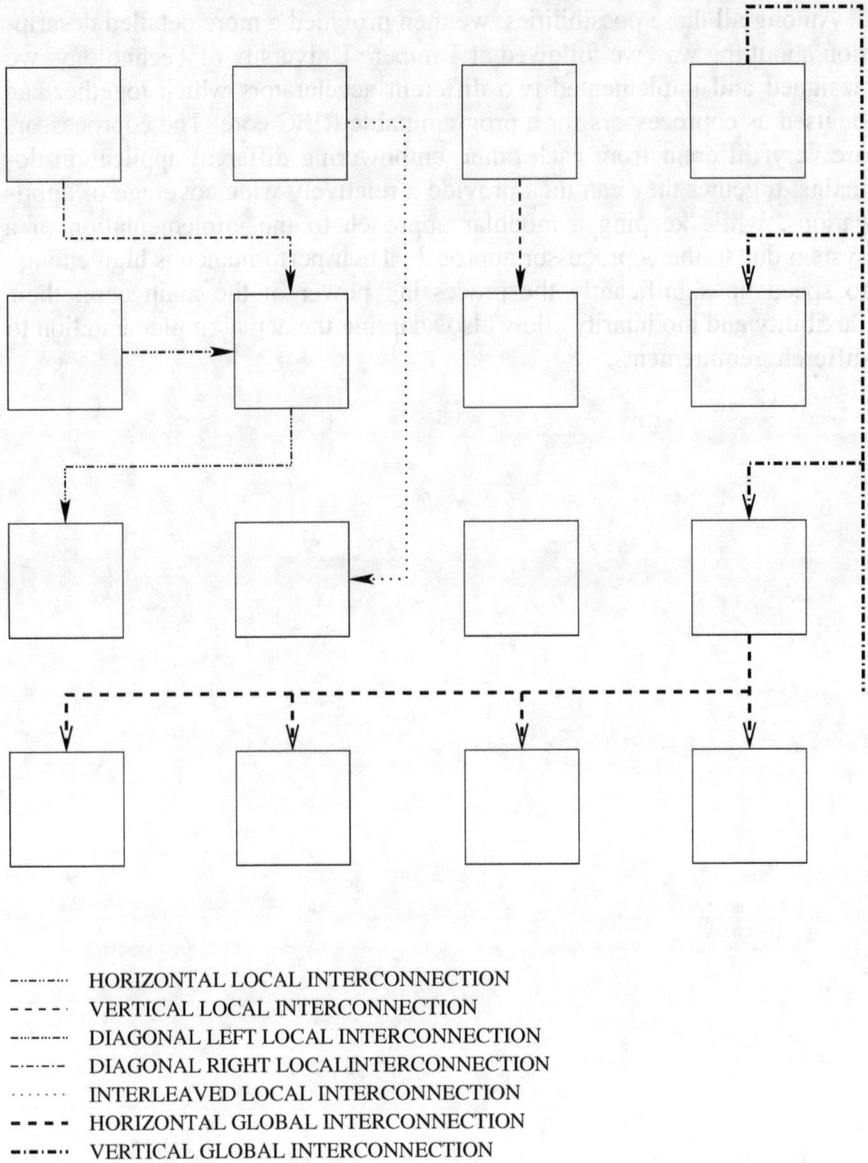

Fig. 10.4. Different kinds of interconnection inside Butter. © IEEE, 2006 [57].

Among all these possibilities, we then provided a more detailed description about the way we followed at Tampere University of Technology: we designed and implemented two different accelerators which together can be used as coprocessors for a programmable RISC core. The coprocessors are very different from each other, empowering different application domains; together they can then provide a relatively wide coverage of applications, while keeping a modular approach to the implementation of a system due to the coprocessor approach. Their performance is high enough to speed up significantly the processing power of the main core; their flexibility and modularity allow also adapting the actual implementation to different requirements.

11 Designing Soft-Core Processors for FPGAs

James Ball

Altera, Inc.

In the mid to late 1990s, soft-core FPGA processors were of interest only to the academic community due to their high cost and low performance. A soft-core FPGA processor might have been able to fit in an 1990's FPGA but it occupied the majority of the device. An off-the-shelf processor chip was cheaper, faster, and readily available.

With today's modern FPGAs, soft-core FPGA processors are now mainstream products. A soft-core processor occupies less than 1% of a high-capacity FPGA device and exceeds 200 DMIPS of performance. A soft-core FPGA processor is so small that in some cases it fits in the unused resources of an FPGA so is essentially free.

All RAM-based FPGA vendors provide soft-core FPGA processors optimized for their FPGAs. Altera [11] has the Nios II [14], Xilinx [465] has the MicroBlaze [466], and Lattice [255] has the Mico32 [256]. All of the soft-core FPGA processors follow the RISC principles developed by Hennessy and Patterson [187].

None of the major FPGA vendors provide soft-core implementations of established processor architectures (e.g. PowerPC, ARM) although some provide hard-core implementations. Soft-core implementations of established processor architectures are avoided partly due to licensing costs but mainly due to their low efficiency in an FPGA.

This chapter discusses the unique aspects of designing soft-core FPGA processors in contrast to designing soft-core ASIC processors. A brief overview of FPGA architecture is followed by a discussion of instruction set and design issues, a comparison of FPGA processor instruction sets, and a case study of the Nios II FPGA processor. The remainder of this chapter assumes that all references to processors are soft-core implementations.

J. Nurmi (ed.), Processor Design: System-on-Chip Computing for ASICs and FPGAs, 229–256.

Note that efficiency is considered to be the ratio of performance to cost. The cost of a circuit in an FPGA is the number of logic elements (LEs). The cost of a circuit in an ASIC is the number of 2-input NAND gates. A logic element is the basic building block of an FPGA. A logic element either consists of a lookup table and a 1-bit flip-flop. Some FPGAs have 4-input lookup tables but some newer FPGAs have 6-input lookup tables. The area of a logic element and a 2-input NAND gate can be converted to silicon area when comparing the efficiency of a circuit in an FPGA and an ASIC.

Configurable processors

FPGA processors and ASIC processors support generation-time configuration options to allow designers to trade off performance and cost. Examples of generation-time configuration options include pipeline implementation, cache size, multiplier implementation, divider implementation, barrel shifter implementation, and tightly coupled memories.

The re-configurability of FPGAs gives FPGA designers an advantage over ASIC designers. FPGA designers tune their FPGA processor configuration on an FPGA whereas ASIC designers tune their ASIC processor configuration on a simulation. Using a simulation to tune the processor configuration takes longer and is less accurate than using an FPGA.

Because FPGA designers can easily tune their designs, FPGA designers can easily switch between FPGA processor pipelines to meet design requirements. This encourages FPGA vendors to provide multiple pipeline implementations of a given instruction set. The Altera Nios II processor provides an example of an FPGA processor that supports multiple pipelines. A detailed overview of the Nios II family of processors is available by Balll [35].

The Nios II processor is available in three pipelines: Nios II/f (fast), Nios II/s (standard), and Nios II/e (economy). The pipelines range in area by a factor of 3 and in performance by a factor of 9. The Nios II/f pipeline is a six-stage pipeline with an optional instruction cache, optional data cache, and dynamic branch prediction. The Nios II/s pipeline is a five-stage pipeline with an optional instruction cache, no data cache, and static branch prediction. The Nios II/e pipeline is six-stage serial pipeline that only executes one instruction at a time.

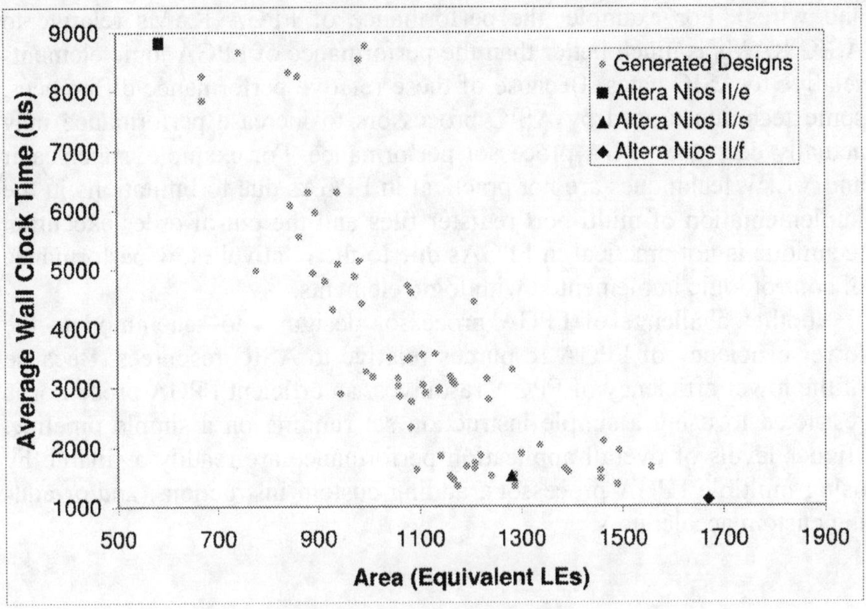

Fig. 11.1. FPGA processor performance vs. area.

Figure 11.1 shows the wide range of performance and area characteristics possible for FPGA processors. The figure includes three Nios II pipelines each in their default configuration as well as several generated designs. The generated designs are different configurations of five pipelines of an academic FPGA processor. The academic processor shown in Figure 11.1 is based on a subset of the MIPS-I instruction set and is designed by University of Toronto researchers Yiannacouras, Steffan and Rose [469]. The figure comes from their research by into FPGA processors.

Challenges of FPGA processor design

FPGA processor design is similar in many respects to ASIC processor design. Indeed, ASIC processors are regularly prototyped in FPGAs during their development. However, these prototypes are not intended to be commercially viable products. FPGA processor design requires re-thinking processing requirements to develop solutions appropriate for FPGAs and not blindly adopting solutions created for ASIC processors.

One challenge of FPGA processor design is to accommodate the different relative performance of FPGA resources (logic elements, RAMs, multipliers, and routing) to the relative performance of ASIC resources (gates, RAMs, and wires). For example, the performance of FPGA RAMs relative to ASIC RAMs is much better than the performance of FPGA logic elements relative to ASIC gates. Because of these relative performance differences, some techniques used by ASIC processors to increase performance may actually decrease FPGA processor performance. For example, superscalar and VLIW techniques are not practical in FPGAs due to limitations in the implementation of multi-port register files and the out-of-order execution technique is not practical in FPGAs due to the relatively low performance of control logic implemented with logic elements.

Another challenge of FPGA processor design is to accommodate the lower efficiency of FPGA resources relative to ASIC resources. Because of the lower efficiency of FPGA resources, an efficient FPGA processor is restricted to using a simple instruction set running on a simple pipeline. Higher levels of overall application performance are readily available by using multiple FPGA processors, adding custom instructions, and/or adding custom accelerators.

Opportunities of FPGA processor design

Even though FPGAs have performance and cost disadvantages relative to ASICs, the flexibility of FPGAs provides unique opportunities in FPGA processor design. An FPGA designer can change their FPGA processor configuration whenever design requirements change. An ASIC designer cannot change their ASIC processor configuration without creating a new ASIC. An ASIC processor compensates for this lack of inherent flexibility in ASICs by increasing the flexibility of the ASIC processor albeit at additional cost.

ASIC processors are typically configured to provide more performance than required. This performance margin accounts for potential inaccuracies in estimating performance requirements and potential increases in future performance requirements. Providing performance margin tends to increase-the cost of the ASIC processor (e.g. larger caches, more complex pipeline). Because an FPGA is inherently flexible, an FPGA processor can avoid this additional cost by being configured with minimal or no performance margin.

ASIC processors tend to have many run-time parameters. Run-time parameters are configured by system software writing to control registers. ASIC resources are provided to support all possible behaviors and a control register selects the desired behavior. An example of a run-time parameter is endianness. Because some ASICs might be used in big- or little-endian systems, an ASIC processor typically provides a control register bit to select the desired endianness. An FPGA processor avoids the additional cost of supporting run-time parameters by converting them into generation-time parameters. Generation-time parameters eliminate the extra FPGA resources required to support multiple behaviors and the associated control register. In the endianness example, an FPGA processor recognizes that endianness is typically a static system setting so it is appropriate to implement it as a generation-time parameter. Caches and debug facilities are two areas with many opportunities to take advantage of converting run-time parameters into generation-time parameters.

Caches

ASIC processors have cache run-time parameters such as enabled, write policy, and line locking. An FPGA processor can convert these run-time parameters into generation-time parameters to avoid the overhead of run-time parameters.

ASIC processors with caches typically have run-time parameters to control whether the caches are enabled or disabled. FPGA processors can convert this run-time cache enabled/disabled parameter into a generation-time cache present/not-present parameter. This implies that if a cache is present, it is always enabled. Having caches always be enabled is acceptable to software as long as the caches are properly initialized when coming out of reset.

Data caches have a write policy of write-back or write-through. ASIC processors typically support a run-time parameter to select between these two modes. It is rare for an application to switch between these modes dynamically in an embedded application. An FPGA processor can make the write policy a generation-time option.

Having the ability to lock cache lines is common in ASIC processors with multi-way caches. This is an inefficient technique to guarantee low-latency access to memory. Instead, the flexibility of an FPGA allows an FPGA designer to add tightly coupled memories to the FPGA processor as needed.

Debug facilities

Processors provide debug facilities to allow software developers to control the processor and observe its state. Typical debug facilities include single stepping, breakpointing, watchpointing, tracing, and examining/modifying memory and registers. The extent of the debug facilities for an ASIC processor is fixed once the ASIC is produced. If a facility desired by a software developer is not available to assist in debug (e.g. trace), the software developer must find some other technique to debug the software.

The flexibility of an FPGA allows an FPGA designer to add/remove debug facilities as required. The debug facilities may be removed after the software is fully debugged. An FPGA designer might even prototype their design in an FPGA larger than required to provide extra resources for debug facilities.

FPGA architecture overview

An FPGA processor designer must have a good understanding of FPGA devices in order to make good design decisions. A brief overview of FPGA architecture is provided to set the context for the remainder of this chapter.

FPGAs are composed of logic elements, RAM blocks, multiplier blocks, and routing. Lewis et al. [263] provide details of the Altera Stratix FPGA architecture which is representative of modern FPGA devices.

FPGA resources are configurable through on-chip configuration SRAMs. Upon power-up a configuration file (typically stored in external non-volatile memory such as flash) is written into the FPGA configuration SRAMs to implement the desired circuit.

FPGA vendors provide devices in a wide variety of sizes. Logic elements range from a few thousand to hundreds of thousands. RAM blocks range from a total chip capacity of several kilobits to several megabits. Multiplier blocks range from zero to dozens. Routing is a property of the FPGA architecture and not typically quoted by FPGA vendors although it can occupy 80% or more of the die area.

Altera and Xilinx both have low-end and high-end FPGA families. Table 11.1 shows examples of FPGA resources for low-end and high-end FPGA families. A small device and large device is shown for each family. The large device family in this example has 6-input logic elements so the table shows the number of equivalent 4-input logic elements. Note that the low-end family has fewer RAM bits per logic element than the high-end family which makes RAM resources relatively scarce in these devices.

Logic elements

Figure 11.2 shows a typical logic element. It consists of a 4-input lookup table, carry-chain logic, and a flip-flop. Some newer FPGA devices have 6-input lookup tables. The lookup table computes any 1-bit function of its inputs. A small configuration SRAM is used to hold the contents of the lookup table. The inputs are connected to the address of the SRAM.

Table 11.1. Altera 90 nm FPGA examples.

	Low-end 90 nm FPGA		High-end 90 nm FPGA	
	Small Device	Large Device	Small Device	Large Device
Altera device Name	Cyclone II 2C8	Cyclone II 2C70	Stratix II 2S30	Stratix II 2S180
4-input logic elements	8,256	68,416	33,880	179,400
Small RAM blocks (512 bits + parity)	0	0	202	930
Medium RAM blocks (4096 bits + parity)	36	250	144	768
Large RAM block (64 Kbytes + parity)	0	0	1	9
Total RAM bits (including parity)	165,888	1,152,000	1,369,728	9,383,040
RAM bits per logic element	20	17	40	52
18-bit multiplier blocks	18	150	64	384
Multiplier blocks per logic element	2.2×10^{-3}	2.2×10^{-3}	1.9×10^{-3}	2.1×10^{-3}

The carry-chain logic provides dedicated circuitry to support faster adders and subtractors than possible just using the lookup table. The flip-flop stores the output of the lookup table or carry-chain logic.

RAM blocks

Designs that require RAMs are common so FPGAs provide dedicated RAM blocks. It is possible but inefficient to use logic element flip-flops to implement RAMs. Some FPGAs do allow the SRAM-based lookup tables in logic elements to be combined to implement efficient but small RAMs.

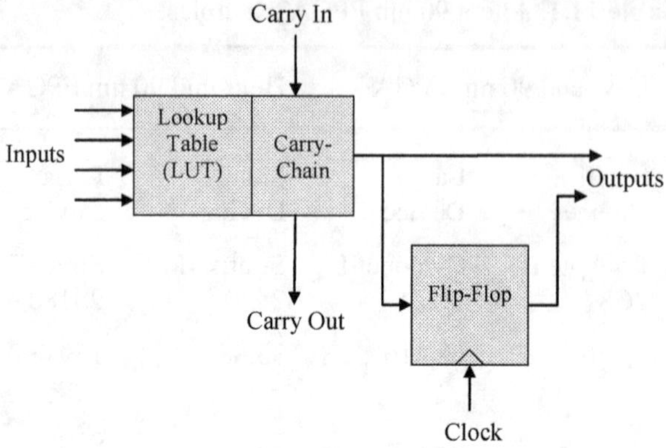

Fig. 11.2. Typical 4-input FPGA logic element (LE).

FPGA RAMs typically support simple dual-port (one read port, one write port) and may also support true dual-port (two read/write ports). The width of the RAM is configurable from just a few bits wide (in some cases as small as 1 bit) up to as many as 128 bits wide. RAMs are typically only available as synchronous SRAMs with registered inputs and optionally registered outputs.

Multiplier blocks

Designs that require multipliers are common (although less common than those that require RAMs) so most FPGAs provide dedicated multiplier blocks. It is possible but inefficient to use logic elements to implement multipliers but not as inefficient as using logic elements to implement RAMs.

Multiplier blocks are composed of several small multipliers (typically 9 bits or 18 bits each) that are combined to create larger multipliers. Some FPGAs have multiplier blocks that provide dedicated circuitry to combine small multipliers. The alternative is to use logic elements and programmable routing resources to combine small multipliers. FPGAs with dedicated circuitry to combine small multipliers offer higher frequency because the delay of the dedicated circuitry is much smaller than the delay of logic elements and programmable routing. The performance of large multipliers is important for FPGA processor design because FPGA processors typically require 32-bit multipliers.

Some multiplier blocks provide features such as saturated arithmetic, accumulators, or special support for barrel shifters. The behavior of the multipliers is configurable including whether the inputs and/or outputs are registered.

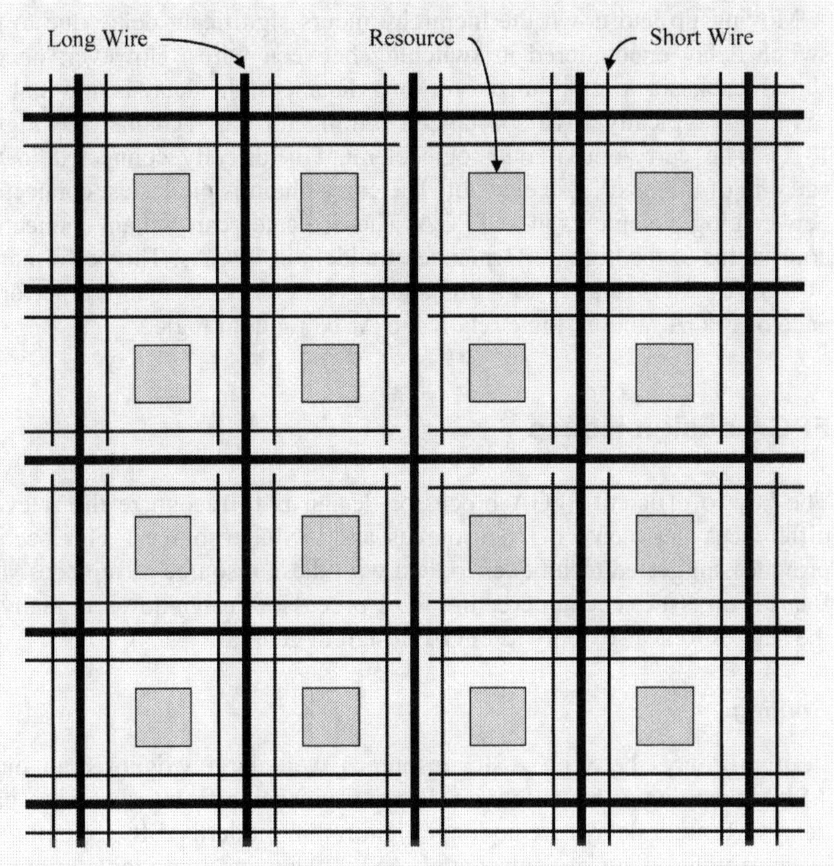

Fig. 11.3. FPGA routing.

Routing

Figure 11.3 shows a simplified representation of FPGA routing. The figure shows that there are short and long wires that connect resources organized in a two-dimensional array. An FPGA typically has more types of wires of different lengths and speeds organized into a hierarchy. Resources have connections to the routing which are not shown for clarity. Configurable switches composed of muxes connect the wires that make up the routing.

Resources are connected with different kinds of routing with different speeds. The routing is organized as a multilevel hierarchy so that resources close to each other have smaller delays than resources further apart. Resources that are further apart are connected through multiple wires and switches so experience longer delays.

Moving up and down the hierarchy incurs significant delay due to the switch delay encountered in switching between wires. However, once a signal reaches a wire optimized for long distances, the delay is minimal.

FPGAs typically have specialized routing for carry-chains (see Figure 11.2). The carry-out of one logic element is directly connected to the carry-in of the next logic element. The carry-chain is the fastest connection between logic elements in an FPGA. The dedicated carry-chain connection contributes to the high performance of adders in FPGAs. The performance of FPGA adders relative to ASIC adders is much better than the performance of FPGA random logic relative to ASIC random logic.

FPGA design issues

The key to efficient FPGA processor design is to recognize the relevant differences between FPGA resources and ASIC resources. This section provides suggested techniques to deal with these resource differences. The focus is on issues encountered in FPGA processor design however many of the suggestions also apply to FPGA design in general.

Routing

Routing delays between ASIC resources were largely ignored in older ASIC technologies because wire delays were insignificant relative to gate delays. Routing delays are no longer ignored in modern ASIC technologies because wire delays are substantial. ASIC place and route tools minimize routing delays by choosing the optimum wire performance and placement customized to the ASIC. FPGA routing is fixed in performance and placement so FPGA place and route tools have less opportunity to minimize routing delays. This fixed nature of FPGA routing in combination with the extra delays of the switches that connect wires in the routing cause large routing delays in FPGA circuits.

Routing delays to and from RAM blocks and multiplier blocks are particularly large due to the relative scarcity of these resources relative to logic elements. The fixed location of these resources tends to increasing routing distance. The routing delays are the largest for the large RAM blocks and the multiplier blocks because they are the scarcest resources.

Suggested techniques

- Minimize the number of logic elements between registers. Try to use all the inputs of the lookup table in each logic element as much as possible.

- FPGA registers are abundant so take advantage of them wherever possible. ASIC registers cost about 10 gates each so they are used sparingly in ASIC designs.
- Provide as much slack as possible to and from the multiplier block and large RAM blocks. Configure the multiplier block to have registered inputs and outputs.
- Ensure that the pipeline stall signals are driven early in the cycle (ideally directly from a register). Stall signals tend to have high fan-out so can experience long routing delays.

Control logic

Control logic consists of combinatorial logic and flip-flops. FPGA control logic has an area penalty and performance penalty relative to ASICs. These penalties encourage FPGA processor designers to employ simple control structures eliminate critical timing paths. performance. Simple control structures do increase average CPI (Cycles Per Instruction) but often overall performance is higher due to a higher frequency of operation.

Adders

An FPGA processor can use adders liberally because they are fast and inexpensive. The performance of FPGA adders relative to ASIC adders is much better than the performance of FPGA random logic relative to ASIC random logic. FPGA adders are also relatively inexpensive because each logic element supports a 1-bit adder. A 32-bit FPGA adder only requires 32 logic elements whereas a 32-bit ASIC adder requires hundreds of gates.

It might make sense for an ASIC processor to share an adder. In an FPGA the additional cost of sharing an adder is comparable to the cost of the adder itself so sharing FPGA adders should be avoided.

Multiplier blocks

FPGA multiplier blocks are configurable just like other FPGA resources. An FPGA processor typically registers the input and output of the multiplier block to avoid critical paths to and from the multiplier. Registering the multiplier input and output creates a multiplier with two cycles of latency and a throughput of one cycle. Some FPGA multipliers also provide optional internal pipeline registers to increase frequency at the expense of increased latency.

Suggested techniques

- In high-performance FPGA processors, multiply instructions achieve a throughput of one result per cycle by delaying the availability of the result by two cycles. This technique requires a pipeline long enough to absorb the latency of the multiply operation.
- In medium-performance FPGA processors with fewer pipeline stages, multiply instructions achieve a throughput of one result every three cycles by stalling multiply instructions for two cycles.
- Some FPGAs only provide multipliers smaller than 32 bits (e.g. 8 bits or 16 bits). A full 32-bit by 32-bit multiplier (with a 64-bit result) is composed of four 16-bit multipliers. Combining these small multipliers using logic elements can create a critical path. If a pipelined multiplier is required, an FPGA processor can insert logic element registers between the small multipliers but this increases latency. If a non-pipelined multiplier can be used, an FPGA processor can create a 32-bit by 16-bit multiplier and use it multiple cycles to compute a 32-bit result.

Equality comparison

An ASIC implements a 32-bit equality comparison between two 32-bit values as 32, 2-input XOR gates followed by a 32-input NOR function. An FPGA with logic elements containing 4-input lookup tables implements this equality comparison with 21 logic elements organized in a tree $\log_4 64 = 3$ levels deep. Each logic element in the tree reduces four inputs to one output in the following manner:

```
64 -> 16 -> 4 -> 1
```

An ASIC implements a 32-bit equality comparison between a 32-bit value and a 32-bit constant as a 32-input AND function. An FPGA with logic elements containing 4-input lookup tables implements this equality comparison with 11 logic elements organized in a tree $\log_4 32 = 3$ (after rounding up) levels deep. Each logic element in the tree reduces four inputs to one output in the following manner:

```
32 -> 8 -> 2 -> 1
```

Equality comparisons are commonly used to evaluate conditional expressions (e.g. conditional branches and compare instructions) and in tag comparisons for caches and TLBs.

Suggested techniques

- Change the RTL to implement an equality comparison as an XOR followed by a subtraction instead of using the == operator. For an *n*-bit equality comparison, an *n*-bit subtraction is faster than an *n*-bit NOR for some values of *n* (device specific). A single FPGA logic element can implement both an XOR and a subtraction. Using an *n*-bit subtraction instead of an *n*-bit NOR would never be used in an ASIC processor because an *n*-bit NOR is faster and uses far fewer gates in an ASIC. An equality comparison of `in1` to `in2` using the XOR/subtraction technique is obtained by examining the sign bit of the following Verilog expression:

```
{1'b0, in1 ^ in2} - 1
```

- Use direct-mapped caches because the tag comparison is usually a critical path. Set-associative caches do perform multiple tag comparisons in parallel but then require extra logic to combine those comparisons.
- Instead of using a fully associative TLB (e.g. 64 entries) which requires 64 equality comparators, use a microTLB with just a few fully associative entries (typically 4–8) that cache a larger RAM-based TLB.

Instruction decoding

Processors decode instructions to create pipeline control signals. RISC processors typically use combinatorial logic to decode instructions. Pipeline control signals used in the same stage that the instruction is decoded can create critical paths. These critical paths are especially severe in FPGA processors due to the low performance of combinatorial logic implemented with logic elements.

Suggested techniques

- FPGA RAMs are typically multiples of 9 bits wide. For a 32-bit instruction, this means there are four unused bits for each instruction in the instruction cache. These four unused bits can be used to provide pre-decoded control signals to reduce critical paths. The pre-decoded control signals are computed when an instruction is fetched from memory and written into the instruction cache.
- Consider using a ROM to assist in the instruction decoding. Using a ROM tends to be slower than using logic elements to decode instructions but uses fewer logic elements. Using a ROM can be a generation-time parameter to allow FPGA designers to trade off FPGA processor memory usage and logic element usage.
- Decode instructions earlier in the pipeline than required and then pipeline them to the required stage. Decoding instructions early gives the routing tools more flexibility in laying out the processor so it tends to improve frequency. The extra pipelining does consume flip-flops but those are abundant in FPGAs.

- Create as many pipeline control signals as possible in the same pipeline stage to allow the synthesis tool more opportunities to create optimal decoding logic.

Multiplexers

ASIC muxes are composed of gates which are fast and inexpensive. Circuits that contain a high concentration of muxes benefit from the ability of an ASIC to have high wire density where required. Having a high wire density helps to pack muxes closely together and minimize routing delays.

FPGA muxes are composed of logic elements which are relatively slow and expensive. FPGA muxes are relatively sparsely packed because the routing is fixed so circuits with a high concentration of muxes experience high routing delays.

Processors naturally contain circuits with high concentrations of muxes. Metzgen [285] reports that 30% of the Nios II/f logic elements are used for muxing. Muxes are commonly used in bypass circuits (a.k.a. forwarding logic), general-purpose register write circuits, control register read circuits, barrel shift circuits, load data align circuits, and next PC (program counter) circuits.

Pipelined processors use muxes to bypass results from later pipeline stages into earlier pipeline stages. The number of mux inputs is proportional to the length of the pipeline and the number of mux bits equals the datapath width. There are typically two such bypass muxes in an RISC processor (one for each input operand).

Processors use muxes to select among all possible instruction results to be written to the general-purpose register file. For example, a typical FPGA processor instruction set provides several arithmetic instructions (e.g. add, subtract), several logical instructions (e.g. and, or, xor), shift/rotate, load, and other instructions that all write the general-purpose register file. All of these instruction results are muxed together to create the register write value. All this muxing results in a 32-bit wide mux with several inputs.

The control registers of a processor are typically read with a control register read instruction. This instruction reads a specified control register and writes a specified general-purpose register. The control register read instruction uses a mux to select the desired control register value.

Suggested techniques

- Reduce the size of the bypass muxes by omitting some stages from the bypassing. This increases CPI because stalls are required to prevent pipeline hazards.

- Only bypass one of the 2-input operands. Compilers tend to only create pipeline hazards on one of the input operands so the bypass muxes for the other operand are rarely used.
- Implement a multi-threaded version of the processor that performs round-robin context switching every cycle between threads. This technique was employed by Fort et al. [130] on a Nios II-compatible implementation. It eliminates the bypass logic entirely without a performance penalty assuming that the application has enough threads.
- Provide early signals to the bypass muxes to allocate most of the cycle for the muxing and routing delays.
- Add a pipeline stage before the general-purpose register file write to provide extra time to mux the write value.
- Consider reducing the performance of the control register read instruction to allow more time to mux between the control registers. This instruction is typically not performance critical.
- An Altera FPGA 4-input logic element is able to implement 2:1 priority mux. However, if the output of the mux is registered and the mux is wider than a few bits, a logic element is able to implement a 3:1 priority mux. Designing an FPGA processor with this in mind produces considerable increases in efficiency.

Constants

A logic element is much more efficient at storing a constant value than a variable value. Each logic element only has one flip-flop bit but also includes a 4-input lookup table that can store 16 1-bit constants and mux between them.

Suggested Techniques

- Whenever possible, convert run-time parameters to generation-time constants. For example, an ASIC processor typically stores the exception vector address in a control register. Storing an exception vector address in a register allows the most flexibility for software but isn't always required. Instead, the exception vector address can be specified by the FPGA designer as a generation-time constant and the overhead of a control register is saved.

Barrel shifters

Barrel shifters perform a shift or rotate by any amount from 0 to 31. An ASIC implements a barrel shifter using several levels of multiplexers. For example, a 32-bit barrel shifter that is composed of 2:1 muxes has $\log_2 32 = 5$

levels of muxes. Using 2:1 muxes to implement a barrel shifter in an FPGA is inefficient.

Typically each logic element only implements one 2:1 mux so a 32-bit barrel shifter requires 160 logic elements for muxing and is five logic elements deep. This is a large number of logic elements for a relatively infrequent operation. Also, given the low density of muxes possible in an FPGA, the barrel shifter is slow due to long wire delays between the logic elements.

Suggested techniques

- An Altera FPGA 4-input logic element is able to implement a 4:1 mux when used in barrel shifters and the result of the mux is registered. A 32-bit barrel shifter implemented using this technique requires 96 logic elements for muxing and is only $\log_4 32 = 3$ (when rounded up) logic elements deep. The negative edge of the clock may be used to reduce the barrel shift latency from three to two cycles.
- A 32-bit barrel shifter may be implemented with a 32-bit multiplier block. Using a multiplier block takes advantage of the property that multiplying a number by 2^n is equivalent to shifting it left by n bits. The 5-bit shift value in the shift/rotate instructions is converted to 2^n using logic elements and provided to the multiplier as one of the inputs. The other multiplier input is provided the value to be shifted or rotated. All combinations of shift, rotate, left, right, arithmetic, and logical can be implemented using this technique. More details on using a multiplier to perform barrel shifts are provided by Metzgen [285].

RAM blocks

ASICs contain RAMs optimized for the design. An ASIC designer chooses the size, number of ports, and power vs. speed characteristics of each RAM. An FPGA contains a fixed set of RAMs. An FPGA designer uses those RAMs as best as possible. Because the FPGA RAMs are fixed, it enables the FPGA designer to make some design choices not practical for ASIC designers.

Suggested techniques

- FPGA RAMs are typically dual-ported so an FPGA designer can take advantage of having more than just a single read/write port. An ASIC designer typically tries to utilize single port RAMs as much as possible at the expense of additional control logic and muxing to share ports. For example, if a design needs a RAM that is rarely written and read at the same time, an ASIC designer can use a single port RAM. However, in an FPGA the RAMs already have a separate read and write ports so adding additional logic to share one read/write port is not efficient.

Register files

A typical single-scalar RISC processor requires a register file with two read ports and one write port. The processor uses the two read ports to read the 2-input operands and the write port to write back the instruction result.

ASICs implement a register file with either flip-flops and muxes or RAMs. Implementing a register file using flip-flops and muxes on an FPGA is not practical due to the size and low speed of such a solution. For example, a register file with 32-bit registers requires 1024 logic elements just for the flip-flops and about the same number of logic elements to implement a 32-bit wide 32:1 mux. Given the low performance of muxing on an FPGA, the read time for this register file is unacceptably slow. The only practical implementation of a register file on an FPGA is to use RAMs.

Suggested techniques

- Use multiple RAMs to implement a register file with multiple read ports. Each RAM is written with the same value at the same time so they have the same contents. Use two simple dual-port RAMs to implement a register file with two read ports and one write port.
- Avoid circuits that require multiple write ports. The technique of using multiple RAMs to obtain more read ports doesn't work to increase write ports.

Power

Power is not typically a large concern for FPGA processors because FPGAs tend not to be used in power sensitive applications. This is due to the relatively large power consumption of FPGAs over ASICs.

Suggested techniques

- ASIC processors optimized for low-power consumption only access internal RAMs for caches and register files when absolutely necessary because RAM accesses are power intensive operations. Reducing RAM accesses to reduce power consumption should be avoided in FPGA processors because the overhead to do this tends to add logic in critical paths.

Instruction set issues

Many of the original RISC philosophies created when ASIC technology was less advanced lend themselves well to FPGA processors. RISC advocates using a load/store architecture, simple addressing modes, simple instruction formats, and only providing instructions usable by a compiler or those required to support the operating system.

It is important to choose an instruction set that is efficient for a wide range of pipelines. Instruction set features that expose the pipeline such as branch delay slots or load delay slots are discouraged because they tend to increase the efficiency of some pipelines at the expense of others.

This section provides examples of techniques to create an efficient instruction set for FPGA processors.

Instruction encoding

- Keep all the instructions the same length (e.g. 32 bits). Some ASIC processors support multiple instruction lengths to reduce code size. An FPGA processor could support multiple instruction lengths but the increased cost and reduction in performance outweigh the savings from reduced code size for most applications.
- Provide only the minimum number of instruction formats. The practical minimum is two formats: register/register and register/immediate. The register/register format has two source operands consisting of general-purpose registers. The register/immediate format has one source operand consisting of a general-purpose register and another operand consisting of an immediate value. Immediates are typically 16 bits. Adding another instruction format to support a larger immediate for subroutine call instructions is useful to reduce subroutine call overhead.
- Keep the instruction set extremely regular to simplify instruction decoding and hazard detection.
- Arrange opcodes to make good use of don't cares to improve instruction decoding.
- Avoid optimizations in the instruction set that utilize alignment requirements to obtain the maximum range of immediate fields. For example, if load instructions include a 16-bit immediate as an offset, it is tempting to treat the offset as a word offset for load word instructions and byte offset for load byte instructions. To do this requires a mux on the offset field. This mux may create a critical path so it is best to always treat the offset as a byte offset.

- Detecting branch instructions can be a critical path because they affect the early stages of the pipeline. Make it easy to detect branch instructions by minimizing the number of bits of the instruction required to detect a branch and by minimizing the number of bits of the instruction required to differentiate between conditional and unconditional branches.
- Computing the branch target can be a critical path. Make it easy to compute the branch target by making the branch target address be relative to the PC (program counter) of the branch instruction and always having the offset be located in the same bits of all branch instructions.
- Minimize the number of instructions that are optional. Typically only multiply and divide should be optional. Make sure that omitted instructions can be emulated by a trap handler.
- Don't combine barrel shifts with arithmetic/logical instructions. Barrel shifts tend to be substantially slower than arithmetic/logical instructions on FPGAs. Barrel shifts should be their own instructions.
- Provide support for designers to extend the instruction set with custom instructions. Custom instructions can dramatically increase performance of some applications.
- Avoid instructions with dependencies on other instructions such as branch delay slots or instruction pairs designed to create large immediate values. Branch delay slots decrease the taken branch penalty but increase the control logic complexity and may introduce delays in critical paths. Branch delay slots also increase the design complexity and testing burden and tend to be a source of bugs. Branch delay slots are only a good fit for some pipelines; for other pipelines they are a burden to support.

Registers

- FPGAs use RAMs to implement the general-purpose register file. Even though RAM bits are relatively inexpensive on an FPGA, limit the number of registers to 32. More than thirty-two registers provide little performance improvement and increase the size of the register number field in instructions. A larger register number field means there are fewer bits in the instruction for other purposes and slows down the hazard detection logic.
- Register windows are generally considered inefficient even for ASIC-based RISC processors. They do reduce memory traffic but increase control logic complexity which tends to reduce frequency and increase

cost. Register windows are particularly inefficient for FPGA processors. Register windows can also cause very long (thousands of cycles) worst-case interrupt response times.

- Avoid special registers such as condition code registers or registers used by multiply and divide instructions. These registers complicate hazard detection, require extra muxing, and require extra instructions to manage them.
- Minimize the number of control registers and control register bits. Adding control registers increases muxing. Adding control register bits uses logic element flip-flops. Generally flip-flops are readily available but they can increase the size of pipelines optimized for very small size. Reserve registers in the general-purpose register file for things like the exception return address instead of using dedicated control registers.
- Don't define configuration registers to provide information about the pipeline such as cache sizes and which instructions are present. This static information is better provided to software by utilizing a BSP (Board Support Package) that provides this information in a C header file. The main reason to have configuration registers is for debuggers to detect the characteristics of your processor. In this case, put this information in the debug facilities of the processor instead of making them part of the base instruction set.

Operating system

- Provide a simple exception model. Most exceptions aren't performance critical so can share a single exception vector address. Sharing reduces hardware costs at a modest increase in code size to process exceptions.
- Don't burden the base instruction set with support for low-latency interrupts such as shadow registers and interrupt vectors because many applications don't require them. Provide optional add-ons such as an interrupt vector custom instruction to provide this functionality.
- Provide simple cache management instructions. Supporting many variations of ways to flush a cache increases control complexity and isn't required by most applications. Minimally, there should be instructions to flush a single line in the instruction and data cache. Flushing the entire cache is rarely performed so can be handled by a software loop that flushes each line individually.

FPGA processor instruction set comparison

Table 11.2 provides information about the FPGA processor instruction sets from the leading FPGA vendors. All instruction sets are quite similar to the original Stanford MIPS RISC instruction set. The MicroBlaze is the oldest instruction set of the three followed by Nios II and then Mico32. The MicroBlaze is notable for its support of branch delay slots.

Table 11.2. FPGA processor instruction sets.

	Nios II	MicroBlaze	Mico32
Architecture	RISC	RISC	RISC
Instructions (excluding floating-point)	82	113	62
Instruction formats	register/register register/imm16 imm26	register/register register/imm16	register/register register/imm16 imm26
Instruction size	32 bits	32 bits	32 bits
Datapath size	32 bits	32 bits	32 bits
General-purpose registers	32	32	32
Minimum control register bits	4	88	86
Delay slot	No	Yes	No
Number of interrupts	32	1	32
Vectored interrupts	Optional via custom instruction	No	No
Custom instructions	Up to 256	No	No
Floating point	Optional via custom instructions	Optional	No
Integer multiply	Optional	Optional	Optional
Integer divide	Optional	Optional	Optional
Signed load	Yes	No	Optional

Case study – Nios II

The Nios II was introduced by Altera in 2004. It is a replacement for the original Nios processor which featured an instruction set with 16-bit instructions and register windows. The Nios II is representative of FPGA processors and is described in detail in this case study.

This section provides an overview of the Nios II instruction formats, an overview of the Nios II/f processor, a description of the Nios II/f processor pipeline, a discussion of FPGA-related Nios II/f design decisions, and Nios II/f instruction performance.

Nios II instruction formats

The Nios II instruction set has the following instruction formats: immediate, register, and call. These formats are shown below.

Immediate Format

31 30 29 28 27	26 25 24 23 22	21 20 19 18 17 16 15 14 13 12 11 10 9 8 7 6	5 4 3 2 1 0
SRC REGNUM	DST REGNUM	IMM16	OP

Register Format

31 30 29 28 27	26 25 24 23 22	21 20 19 18 17	16 15 14 13 12 11	10 9 8 7 6	5 4 3 2 1 0
SRC1 REGNUM	SRC2 REGNUM	DST REGNUM	OPX	SHIFT COUNT	OP

Call Format

31 30 29 28 27 26 25 24 23 22 21 20 19 18 17 16 15 14 13 12 11 10 9 8 7 6	5 4 3 2 1 0
IMM26	OP

The immediate format provides a 5-bit source register number, a 16-bit immediate value, and a 5-bit destination register number. The immediate format is used by arithmetic/logical instructions with immediates, load/store instructions, and branch instructions. The OP field specifies the instruction.

The register format provides two 5-bit source register numbers, a 5-bit destination register number, and a 5-bit shift count. The register format is used by arithmetic/logical instructions without immediates, shift/rotate instructions, and subroutine call/return instructions that jump indirectly to the address in a register. The OP and OPX fields specify the instruction.

The call format provides a 26-bit immediate. The call format is used by the subroutine call instruction that jumps to an immediate address. The OP field specifies the instruction.

The destination register number is either implicit (implied by the instruction) or in one of two locations (depending on the instruction format). The source register number(s) are always explicitly specified in the instruction and are always in the same location when present.

Nios II/f overview

The Nios II/f processor achieves high performance and high efficiency in Altera FPGAs. A detailed analysis of the Nios II/f processor is available by Metzgen [285]. The Nios II/f processor has the following characteristics:

- Six-stage pipeline
- Requires 1800 4-input logic elements
- Runs over 200 MHz and over 225 D-MIPS in an Altera Stratix II FPGAs
- Optional instruction cache
    - Configurable size (up to 64 Kbytes)
    - 32-byte line
    - Direct-mapped
    - Critical word first
- Optional data cache
    - Configurable size (up to 64 Kbytes)
    - Configurable line size (4, 16, or 32 bytes)
    - Direct mapped
    - Writeback with write allocate
    - One line writeback buffer
        - Allows line fill to occur before writeback on dirty miss
- Dynamic branch prediction
    - 2-bit gShare algorithm developed by McFarling [13]
    - 256 entries in BHT (Branch History Table)

Nios II/f pipeline

Figure 11.4 shows the Nios II/f pipeline in its default configuration. The dashed lines show the boundaries of the pipeline stages. A rectangle whose top touches a pipeline boundary contains a register. Arrows that cross pipeline boundaries have implicit registers at the pipeline boundary. Each stage of the Nios II/f pipeline is described in detail.

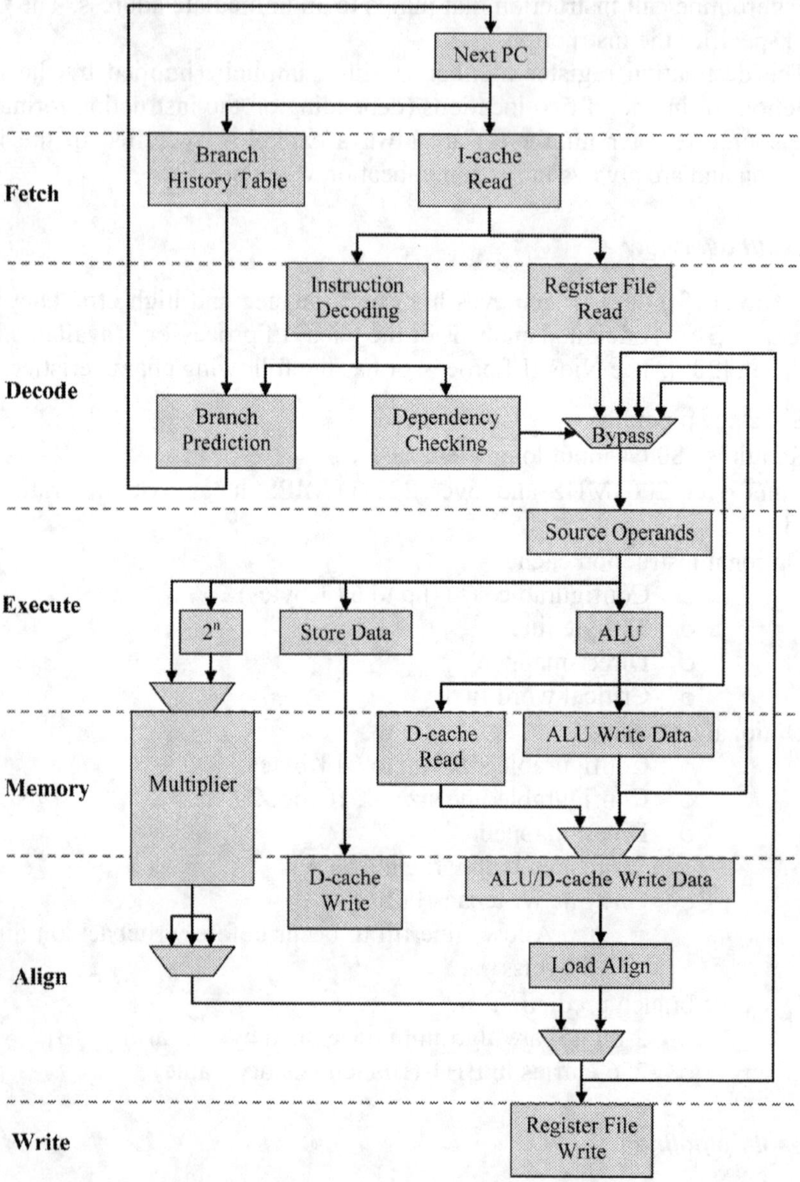

Fig. 11.4. Nios II/f pipeline.

Fetch stage

The next instruction is fetched from the instruction cache using the next PC.

The branch history table is read to obtain the 2-bit branch prediction for the next instruction in case it is a conditional branch instruction.

Decode stage

The instruction is decoded to provide inputs to control logic (not shown).

The two source operands are read from the register file RAM and sent to the bypass logic. The 5-bit source operand register fields are fed from the output of the instruction cache RAM into the register file RAM address inputs.

The dependency checking logic determines if the decode stage needs to stall due to a dependency on an older instruction in the pipeline. It does this by comparing the 5-bit register number field of the decode-stage instruction against older instructions in the pipeline. The register number comparisons are also used to control the bypass muxes to choose the correct version of the input operands.

If the decode-stage instruction is a conditional branch, the branch prediction information is consulted to determine if the branch is predicted taken or not taken. If predicted taken, the next PC logic computes the branch target address and uses that as the next address.

Execute stage

The ALU uses the two source operands to compute the result for arithmetic, logical, and comparison instructions. The comparison is also used to resolve conditional branch instructions. The ALU also computes the memory address for load/store instructions.

If the execute-stage instruction is a shift/rotate instruction, 2^n of the least-significant 5 bits of one source operand is computed and the result sent to the multiplier. Otherwise, the source operands are sent directly to the multiplier.

Memory stage

The data cache is read to obtain the data for load instructions and the multiplier computes its products. The multiplier is a hard-macro provided by the FPGA and doesn't use logic elements. The multiplier has registered inputs and registered outputs.

Align stage

The data cache load data is aligned and sign-extended. The data cache is written for store instructions.

The multiplier produces a 64-bit product from its two 32-bit source operands. A 3:1 mux selects the least-significant 32 bits, the most-significant

32 bits, or the bitwise OR of them. This handles all cases required by the multiply, shift, and rotate instructions.

A mux produces the final write data for the register file.

Write stage

The write data from the align-stage is written into the register file.

Nios II/f design decisions

In many respects, the Nios II/f pipeline is a classic RISC implementation. However, there are several key design decisions made to achieve an efficient FPGA implementation.

- The instruction cache RAM data out to register file RAM address is a critical path (largely due to long wire delays between RAMs). The instruction cache is direct mapped to avoid the additional delay of comparing tags to control a mux to select the appropriate instruction. Notice that the pipeline takes advantage of the Nios II instruction set properties related to the fixed location of source operand register numbers so that that the register file RAM address is extracted directly from the 5-bit source register fields in the instruction without any additional logic. Eliminating the need for each logic to decode source register numbers helps avoid a critical path.
- Because RAM bits a relatively inexpensive on FPGAs, the instruction cache has one valid bit for each instruction instead of one valid bit for each line. This simplifies the instruction cache fill logic.
- The multiplier block is used as a barrel shifter for shift/rotate instructions. This avoids the overhead of using logic elements to implement the shift/rotate instructions.
- The BHT is read in the fetch stage but isn't used until the decode stage. Using the BHT output in the decode stage avoids a long path from the output of the BHT RAM to the next PC logic and the instruction cache/BHT RAM inputs.
- The load data is read in the memory stage from the data cache RAM but isn't sent to the bypass muxes. This could be done for load word instructions which need no alignment or sign extension. However, this would create a long path from the data cache RAM output to the bypass muxes.
- The instruction cache hit and data cache hit signals are computed in the fetch and memory stages (respectively). These signals aren't used in their stages; instead they are pipelined into the next stage. This prevents a long path from the cache hit signal into the stall logic. Instead, the stall logic uses a registered version of the hit signal one stage later.

  - o When an instruction misses in the instruction cache, it isn't stalled in the fetch stage; instead it progresses to the decode stage and then continuously recirculates back to the fetch stage until the instruction hits in the instruction cache.
  - o When an instruction misses in the data cache, it still proceeds to the align stage and stalls while miss processing is active (line fills and/or spills to memory).
- Minimal logic is in the path of signals sent to/from the multiplier block. This helps allow time for the typically long wire delays to/from the multiplier block.
- Dynamic branch prediction using the gShare algorithm is used. The gShare algorithm achieves a prediction accuracy greater than 90% according to McFarling [280]. Static branch prediction schemes such as predicting taken on backward branches and not-taken on forward branches provide only 80% accuracy according to Ball and Larus [37] so are not used due to the resultant decrease in performance.
- Dynamic branch prediction typically requires a BHT (Branch History Table) and a BTB (Branch Target Buffer). The Nios II/f BHT is a 256 entry RAM with 2 bits per entry. These small RAMs are readily available on Altera FPGAs. The BTB is typically a much wider RAM because it stores the predicted target address. The Nios II instruction set encodes a branch target as a 16-bit PC-relative immediate offset. Given the relatively high performance of adders in FPGAs, the branch target is calculated faster than using a BTB (Branch Target Buffer) RAM to lookup the target address. This allows the BTB to be omitted in the Nios II/f.

Nios II/f instruction performance

Table 11.3 shows the number of cycles each Nios II instruction takes to execute on the Nios II/f processor. Only instructions which have a significant performance effect are listed.

Closing comments

FPGA processor design in many ways is simpler than ASIC processor design because FPGA processors have to be simple to be efficient. Higher levels of performance on an FPGA are achieved by using custom instructions, custom accelerators, or multiple FPGA processors.

Mapping a processor designed for ASIC implementation onto an FPGA produces an inefficient result due to the unique characteristics of FPGAs relative to ASICs. FPGA processors are their own design specialty.

FPGA processors are an active area of research. Greater efficiencies for commercial offerings are possible using techniques such as round-robin threaded implementations. FPGA processors may also help drive the requirements for future FPGA devices to help reduce some of the area and speed penalties of current devices.

Table 11.3. Nios II/f instruction performance.

Instruction Type	Cycles
Arithmetic/logic instructions	1
Branch (correctly predicted taken)	2
Branch (correctly predicted not taken)	1
Branch (mispredicted)	4
Function call to immediate address	2
Function call to address in a register	3
Function call return	3
Load/store (assuming hit in data cache)	1
Shift/rotate	1
Multiply	1
Divide	4–67

Acknowledgments

I'd like to thank Kerry Veenstra of Altera for his work on the Nios II instruction set. Most of this chapter's section on instruction set issues is a direct result of his work.

12 Protocol Processor Design Issues

Dragos Truscan,[1] Seppo Virtanen,[2] and Johan Lilius[1]

Åbo Akademi University[1]

University of Turku[2]

Introduction

Addressing the conflicting requirements of network-enabled embedded systems has become, in recent years, an important challenge for designers. Balancing the scale between the need for short time-to-market and the requirement for cost-efficient development cycles of new products is a demanding goal that has traditionally been approached by using *general purpose processors* (GPPs). However, general purpose microprocessors are no longer an appealing alternative for networking hardware due to their lack of optimized execution units for network processing. Using general purpose processors all networking functionality must be implemented in software. This in turn leads to very high CPU clock frequency requirements. General purpose processors that operate in a suitable frequency range are often too expensive, consume too much power or occupy physically too much space in the target system with all their required external circuitry. Also, many general purpose processor features, like *floating point arithmetic units* (FPUs), can usually not be taken advantage of in networking applications.

The increasing demands for functionality and performance of the current networking applications require the use of dedicated hardware circuits to boost the performance of the system. Consequently, extremely complicated *application specific integrated circuits* (ASICs) started to be developed. Being a hardware-based solution, they can provide higher performance, and lower power consumption and blueprint area than a general purpose

J. Nurmi (ed.), Processor Design: System-on-Chip Computing for ASICs and FPGAs, 257–285.

processor implementation. The drawback is that the ASIC design process is demanding and expensive, and the time-to-market tends to be long. Also, ASICs are usually minimally programmable and thus need to be redesigned for updated or new network protocols, which makes them inflexible in dynamic market segments.

As suggested for instance by Comer [86] and by Henriksson [189], the challenge in designing processors for network or protocol processing systems is to find an architecture that is as good a compromise as possible between a general purpose processor and a protocol-application-specific custom chip (i.e., an ASIC implementation). Network processors are often *system-on-chip* (SoC) or *network-on-chip* (NoC) type devices that typically contain a general purpose microprocessor core, and possibly also programmable parallel general purpose engines. In comparison, protocol processors are better categorized as coprocessors or *intellectual property* (IP) blocks used as part of a complex SoC or NoC device.

An ideal protocol processor would harness both the programmability of general purpose processors and the application-specific hardware optimization of ASICs. As a compromise solution, *application specific instruction-set processors* (ASIPs) have emerged as a flexible, high performance, and cost effective alternative for these applications, tailored to process application-specific tasks in an optimized manner. Typically, the architecture should be optimized for a family of applications, such that more than one application can be served by the same ASIP. ASIPs are biased to meet specific performance requirements by using dedicated processing elements implemented in hardware, to support the tasks demanding high performance. A GPP (also referred to as *controller*) is used to drive the activity of the processing elements. Using the program code running on this controller, ASIPs may be programmed to serve several applications. Thus, they provide a good solution for complex applications, where flexibility is needed not only to accommodate design errors, but also to upgrade the specifications [80].

Being an optimized solution for a given application or a set of applications, ASIPs provide improved performance as compared to the general purpose processors. According to [226], this increase can be 2 to 100 times. Nevertheless, since ASIPs are not fully hardware-based solutions they can be several times slower as compared to a corresponding ASIC solution. Even with this drawback, the payoff is far greater in terms of flexibility and upgradeability. Programmable architectures bring important benefits like rapid time-to-market, flexibility of design, and consequently, an increased product lifetime and upgradeability. Looking at the currently available solutions for embedded systems targeted at processing network data or protocols, it can be observed that there is a continuing need to better understand the exact needs of protocol processing in terms of designing and implementing both hardware and software architectures. Thus, it has become obvious that in addition to developing new architectures with optimized processing elements, effort is also needed in developing application-domain-specific processor design methodologies for protocol processing. Such methodologies should support the designer in analyzing the application domain (and preferably the particular application in question), as well as in exploring and evaluating different hardware/software configurations for performing the target application. This chapter provides an analysis of the design space for programmable protocol and network processors, and makes an effort to point out characteristics and methodological requirements of this application domain that need to be dealt with when designing programmable protocol processors.

This chapter has its foundation in our research originally presented in [428,439,440].

Domain and application analysis for optimized protocol processing hardware

In computer networks the communication tasks are usually too complex to be implemented using monolithic protocols. Instead, modular, or layered, protocol architectures are preferred. The protocols form a stack of layers, in which each layer communicates with the one above it and the one below it, by passing information through predefined *service access points* (SAPs), using protocol-specific logical service primitives. The advantage of a layered protocol architecture is that each layer abstracts away some technical functions from the layer above it. For instance, a programmer designing a new networking application (working in the topmost layer) does not have to worry about voltage levels and their corresponding logic states in the lowest layer. Another important benefit of this construction is the possibility to use many different physical mediums and well designed standard protocols for each medium type to perform the same high level task. The best known protocol stack reference model is without a doubt the Open Systems Interconnection (OSI) reference model [96], defined by the International Organization for Standardization (ISO).

Protocols encapsulate the information to be exchanged into *protocol data units* (PDUs). PDUs are protocol-specific, i.e., the PDUs of a certain protocol are understood only by stations supporting the same protocol. PDUs are constructed of a header, a payload, and a trailer. The payload is the actual data being carried. The header and trailer contain protocol-specific control information organized into fields defined in the protocol specification. The fields may contain information such as protocol version, payload length, traffic class, receiving station address, error-checking checksum etc. Many protocols do not have a trailer part in their PDUs, but carry all the necessary control information in the header.

Characteristics of protocol processing

In order to develop devices for the protocol processing application domain, careful studies on protocols and protocol processing applications are needed. Emphasis should be on finding functionality that varies very little or not at all from one protocol to another. As a starting point, Jantsch et al. have identified three typical characteristics of protocol processing in [212]: (1) pattern matching and replacement in bitstrings (especially in frame or cell header analysis); (2) control dominated operation (large finite state machines and nested if-then-else and case structures); and (3) the need for irregular memory accesses (managing tables and buffers of various sizes). In [439,440], we have reviewed six widely used protocols in a search for additional characteristic functionality that could be taken advantage of in protocol processing hardware design. The emphasis in these studies has been on analyzing protocols that can be regarded as layer 1–3 protocols (physical, data link, and network layer) in the OSI reference model. The protocols in these layers are not end-to-end protocols, but require intermediate stations (e.g., repeaters, bridges, switches, routers, etc.) between the source and destination devices. Thus, these protocols present a clear need for application-specific hardware systems in addition to application software, whereas the end-to-end protocols in OSI layers 4 and above are often completely implemented as software running on networked workstations. Table 12.1 summarizes some characteristic functionality for the protocols analyzed in [439,440].

As can be seen in this table, most of the detected characteristic functionality for a particular protocol was found to actually be characteristic to several other protocols as well. A processor with optimized hardware support for the found common functionality should be easily and clearly able to outperform a similar processor with general purpose processing units. Also, the power consumption and area use of such an optimized processor can be expected to be less than those of a general purpose implementation.

This is on the one hand due to a reduced clock speed requirement (the optimized processor is likely to provide equal processing performance at a lower clock speed), and on the other hand due to the fact that the optimized processor needs to implement only the required subset of operations (general purpose execution units may also implement extra functionality that is not needed for the desired protocol processing functionality). The last row in Table 12.1 displays a very important finding in terms of processor design: protocol processing can be implemented using only unsigned arithmetic (i.e., there is no need for managing negative values), which makes hardware implementations considerably simpler.

Specifying and analyzing the target application

General knowledge of inter-protocol similarities as discussed previously may not be adequate when optimizing an architecture for a particular protocol processing application. For this reason, the protocol processor design process should be such that it makes it easy to identify frequently used operations within the particular target application. This in turn leads to the need for providing easy ways to integrate application-specific operations into hardware.

The specifications of applications targeted to programmable architectures have been traditionally written either in machine language or in a high-level programming language (typically C). In the latter case, the resulting application specification is mapped onto the architecture using specialized tools. Currently, this approach cannot cope with the increasing complexity of specifications anymore, thus more elaborated methods for the application specification are needed. Not only the use of higher levels of abstraction is required, but also a systematic application analysis process.

Table 12.1. Summary of typical and essential functions found in commonly used protocols. Bitwise manipulation means bit-pattern matching and/or masking inside data words, and/or *n*-bit shifting. High bitrate means speeds above 500 Mbps.

Processing Characteristic	SDH	IEEE 802.3	IEEE 802.11	ATM	IP	IPv6
High bitrate	Yes	Yes	No	Yes	Yes*	Yes*
Boolean evaluations	Yes	Yes	Yes	Yes	Yes	Yes
Bitwise manipulation	Yes	Yes	Yes	Yes	Yes	Yes
Counter functions	Yes	Yes	Yes	Yes	Yes	Yes
Timer functions	Yes	Yes	Yes	No	No	No
Checksum calculation	No**	Yes	Yes	Yes	Yes	No***
Random numbers	No	Yes	Yes	No	No	No
Buffering	Yes	Yes	Yes	Yes	Yes	Yes
Unsigned arithmetic	Yes	Yes	Yes	Yes	Yes	Yes

* The required processing speed in the network layer depends on lower layer protocols used.
** SDH uses parity bits, the calculation of which does not classify as a checksum.
*** Checksums are needed in some cases, e.g., in routers when processing control messages.

On the one hand, such techniques allow the designer to focus on the relevant details of the specification at different development stages. On the other hand, they reduce the risk of missing functionality at later phases of the development process, which would delay the development cycle considerably.

Optimized hardware is one of the key aspects of ASIPs. The hardware resources of the architecture are the ones providing an increased performance as compared to a software-based approach. Hence, one of the most important issues in ASIP design is that from the application specification one can identify complex and frequently used processing tasks to be implemented using dedicated hardware. Furthermore, these processing tasks may be optimized at application level by optimizing the application specification towards optimal performance.

In recent years, efforts have been put in improving the application specification process by employing techniques and concepts specific to the software engineering domain. Among these techniques, the use of object-oriented principles and more recently of the model-driven ones caught ground, by providing the designer with higher levels of abstraction and a visual modeling environment for application specification.

Using the mechanisms provided by the model-driven philosophy, the application specification process is spread out over several abstraction levels, each modeling the application at a specific level of detail. The starting point in the application specification process is extracting the requirements of the application from informal specifications like textual documents, customer discussions etc. into a given formalism. The resulting specification is used as a starting point for the application analysis step, where the main features of the application are identified. This step is performed iteratively, at each iteration round new information is added to the specification. The main goal of the application analysis is identifying the functionality that has to be supported by the implementation platform, namely by the protocol processor under design. Several modeling languages have been proposed for modeling the application specification, typically tailored for specific processor architectures [372] or general purpose architectures [316].

Hardware abstraction to handle the complexity of specifications

Managing the complexity of hardware specifications requires the development of abstract views of hardware implementations. As for software, abstractions have to be defined at several levels, starting from logical circuits and continuing with the components of the architecture, each capturing only details relevant to a given step of the development. Similarly to software specifications, the hardware specification techniques gradually increased their level of abstraction over the years. If initially hardware has been designed in terms of transistors and logic gates, nowadays, *hardware description languages* (HDLs) (e.g., VHDL, Verilog) and even object-oriented system-level specification languages (e.g., SystemC) have become popular in industrial environments.

Several studies have pointed out the need for abstraction levels to address the complexity of hardware specification [229,363].

Platform-based design

The concept of *Platform-based Design* (PBD) was initially formulated by Sangiovanni-Vincentelli [363] as a solution for addressing the complexity of hardware specifications. PBD is defined as a conceptual framework to be applied to specific solutions and it was gradually adopted by the industry.

The main concept of PBD is the *platform*, a collection of concepts present on a given level of abstraction of a specification. A platform on a given abstraction layer is obtained as a refinement of a platform on a higher abstraction layer. Similarly, a platform on a given layer serves as an abstraction for the platforms specifying the system in the subsequent design steps. The evolution of the specification from one platform to the next is assumed via predefined transformers and specialized tools. A given pair of platforms in combination with the transformers forms a *platform stack*.

As such an architecture platform represents a collection of predefined components that are designed interdependently to support a specific family of architectures, customized for a specific problem. The approach promotes the use of component libraries, thus enabling automation and reuse. A subset of an architecture platform components selected for a specific application is referred to as an *architecture platform instance*.

The architecture platform is abstracted even more to a level where software primitives are describing the architecture. These software primitives are referred to as the *programming model* or the *application programming interface* (API) of a given architecture.

Programming models

In recent years, several programmable processors have been developed, also in the area of protocol processing [190]. Soon after, the difficulty in programming them, due to their complex architecture and the variable instruction set, became an obstacle to using them in practice [276]. Consequently, the use of a *programming model* of the architecture has been suggested [229] in order to provide not only an abstraction of the hardware details, but also a functional view of the architecture. Such a model facilitates programming the architecture by allowing the designer to focus on the functionality that the architecture provides, rather than on the hardware implementation of this functionality.

A programming model for a protocol processor architecture needs to be defined for the following purposes: (a) to provide an abstraction of the hardware architecture, enabling the designer to focus on the functionality of the architecture, rather than on its physical details; (b) to bridge the gap between the hardware architecture and the application during the mapping process.

Custom design frameworks

The problem with the constantly increasing level of abstraction is that a small change at a high level may result in an unacceptable result in terms of circuit complexity in lower levels. For this reason, the target implementation architecture must support the system designer with adequate tools for reliable high level design, including simulators, physical characteristics estimators, and tools for the transition from the high abstraction level description to synthesizable processor models. Emphasis should be on being able to obtain reliable results rapidly from simulations and estimations, and on providing a precise and reliable synthesis model that correctly reflects the characteristics of the simulated model.

The initial problem to be solved in protocol processor hardware mapping is selecting the target hardware platform or ASIP design environment for the implementation. The choice can be one from the numerous available commercial solutions like, for instance, the LISATek suite from CoWare [197] or the Chess/Checkers suite from Target Compiler Technologies [435]. The LISATek suite uses the LISA language for constructing machine descriptions from which software tools and a synthesis model can be generated, whereas the Chess/Checkers suite includes a retargetable compiler and an instruction-set simulator generator that operate on an ASIP processor model called the *instruction-set graph* (ISG) [435].

ASIP design methodologies aim to find sequences of general purpose operations in the target application, and group these sequences into a hardware implementation [26,52,166,434]. The recurring command sequences that are chosen for hardware implementation are often quite short, usually less than 10 and often only 2–3 general purpose processor commands of length. In ASIP methodologies, often a general purpose processing core is enhanced with hardware execution units for the detected command sequences. In [26], this kind of an approach provides a performance increase of no more than 30% when compared to a general purpose processor designed with equal area and power constraints. Also, typical protocol processing operations are more complex than 2–3 sequential instructions, which seriously limits the usability of traditional ASIP methodologies for optimizing protocol processor performance: complex application-domain specific operations (e.g., cyclic redundancy checking in the protocol processing application domain) are likely to not be well optimized in ASIP approaches due to the shortness of the detected command sequences. If larger, yet frequently occurring protocol processing operations could be detected for hardware implementation, greater increases in processing speeds could be expected.

An optimal tuning of the architecture, with respect to the application requirements can be realized by employing collections of customized tools needed to configure, compile, analyze, optimize, simulate, explore, and synthesize architectural configurations. Protocol processors are usually accompanied by custom design frameworks that enable the designer to squeeze the optimal performance out of a given architecture. Such a design framework should provide several features: out of them, the *architectural estimation* is one of the key ones. Embedded systems in general, and hardware in particular, imply high cost of designing and manufacturing. Therefore, it is important to be able to estimate the characteristics of the final product as early as possible in the development process. System-level estimation is of particular importance to embedded systems, since, by their definition, they are systems that must comply with tight constraints in terms of size, energy consumption, and cost. Based on the estimation results, the architectural design space exploration is performed. During this process, the designer evaluates several architectural configurations and selects the one(s) complying best with the requirements of the application. *Simulation* represents an equally important technique in developing programmable architectures. To prevent and detect inherent errors in the specifications of both the application and the architecture, the simulation has to be performed at different levels of abstraction, with respect to both functional and non-functional requirements of the system. Finally, once the configuration of a given programmable architecture is chosen to implement an application, the *hardware synthesis* enables the specification of the configuration in a HDL and also its synthesis, using specific synthesis tools. Furthermore, an important feature of such a design framework is its capability of *rapid generation of different models* (e.g., simulation, synthesis, estimation, etc.) of the system.

Design processes

There are two important categories of methodological approaches that could be used and taken advantage of in protocol processor design. In the *top-down* approach [137,287,372], a high level system specification is gradually refined towards an implementation. When enough detail is gathered, the refined specification is partitioned into hardware and software, and co-designed. Such approaches are also known as hw/sw co-design. The *top-down* approaches are typically based on a *model of computation* (MOC), in which notions of formal semantics for communication and concurrency are defined. The methods in this category focus on the properties of the application and on the simulation of the system as a whole, but generally the implementation obtained is less efficient for families of applications and for applications using several distinct algorithms. An overview of top-down approaches may be found in [471].

In the *meet-in-the-middle* approaches, the application and the architecture are developed independently. A top-down design flow is used to specify the application. The architecture is developed following a *bottom-up* flow, in which the functionality it provides is identified starting from the hardware layer of the architecture. When both specifications are complete, the application is mapped onto the architecture. Such an approach enables the reuse of both software and hardware, cutting down development times and design effort. The price to pay is the significant effort in designing complex libraries, since any new hardware component has to be designed from scratch. Here we adopt the second category to develop programmable architectures.

Our choice is justified by two reasons: the architectures that we address promote the reuse of IP components at different levels of abstraction, and such architectures are intended to be used as an implementation platform for several applications of the same family.

The Y-chart approach

The *Y-chart* approach [228,229] is a generic framework for designing programmable architectures, in which the architecture is tuned to provide the performance required by a set of applications. Being based on a meet-in-the-middle flow, the Y-chart approach promotes the idea of a clear distinction between the application and the architecture, each being developed independently. The implementation of the application onto an *instance* of the architecture (i.e., of a *configuration*) is done through a *mapping* process.

The general view of the approach is shown in Figure 12.1. The application running on a given architecture instance is obtained through the mapping process and the performance of the resulting implementation is evaluated in the *Performance Analysis* step.

All three main artifacts of the approach (i.e., application, architecture, and mapping) are considered to be equally important in achieving an optimal configuration of the architecture. Therefore, although the *performance numbers* aim mainly at suggesting improvements of the proposed configuration(s), the other two artifacts may also be targeted. For instance, at application-level, some of the algorithms may be optimized, or the mapping process modified based on certain heuristics. The mapping process is performed iteratively until satisfactory performance numbers are obtained. The process of trying out different architectural configurations to find "the optimal" one is also known as *design space exploration.*

Exploring the design space is very important to find a good, close to optimal, combination of hardware and software for the given protocol processing application. The exploration requires that several processor architecture candidates for implementation are specified, and that the application software is tuned for each candidate.

A very important aspect of any design framework or design process is tool support and automation. The importance is clear when considering the need for creating and editing the artifacts of the design flow at different levels of abstraction. In design frameworks that support multiple levels of abstraction, there are many tasks in refining the specifications at each step

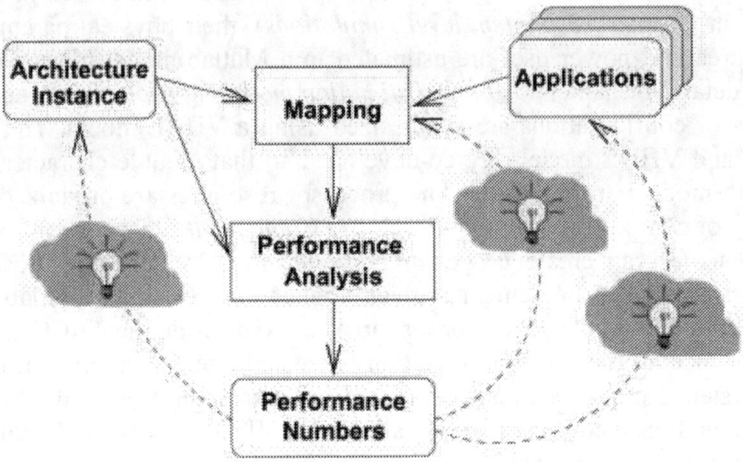

Fig. 12.1. The Y-chart approach. © Springer, 2002 [229].

of the design flow that benefit greatly from tools that reduce the need for manual design work or completely automate a certain part of the design process. Automated or semi-automated tools are also involved in generating different artifacts, for example a simulation model or a synthesis model from an abstract specification, or application software for a generated optimized protocol processor architecture.

Another important feature appreciated by system designers is the possibility of using component library based approaches. Such approaches allow the designer to conveniently reuse components from previous designs in new ones instead of designing new components every time, thus achieving shorter design and test time (the components in a library have already been tested for correct operation in the previous project). This kind of approach is especially useful in developing application-optimized protocol processor architectures as noted earlier in this chapter, many protocols and protocol processing applications exhibit similar requirements from processing functionality.

The TACO framework for protocol processor design

Tools for application-specific hardware/software CO-design (TACO) [440,443] is an integrated design framework for fast prototyping, simulation, estimation, and synthesis of programmable protocol processors based on the *transport triggered architecture* (TTA) concept [92]. TACO processors are simulated using a SystemC [418] model (discussed in more detail in Chapter 18, *System level simulations*), their physical parameters (like area and power use) are estimated in a Matlab model (discussed in more detail in Chapter 17, *Early Estimation models of processors*), and the processor configurations are synthesized using a VHDL model. The SystemC and VHDL models are co-developed so that module characteristics in both models are the same. The processor resources are organized in a library of components, namely the *TACO Component Library*, from which the designer can create *processor configurations* by selecting those resources needed to implement a given application. SystemC, Matlab, and VHDL models of the processor resources coexist inside the TACO framework. Once a processor configuration is created one can simulate it using the SystemC model, estimate its physical characteristics using the Matlab model, and once configuration is validated, a VHDL model can be built to be synthesized in hardware.

In this section we discuss how the TACO Framework addresses the previously discussed design issues of protocol processors. We start by introducing the TACO processor, then we discuss the abstraction layers in TACO and the TACO Component Library used to support IP-reuse and automation. After that, we discuss the TACO design methodology and exemplify the main steps with excerpts from an IPv6 router specification.

The TACO processor

Hardware considerations. The TACO processor is a programmable and configurable protocol processor, which provides dedicated hardware to deal efficiently with protocol processing tasks. The processor is based on the TTA architecture, a processor architecture proposed for application specific processors. The architecture is modular, has a scalable performance, provides flexibility, and makes it easy to control the processor cycle time. A TACO processor (Figure 12.2) consists of a set of *functional units* (FUs) connected by an *interconnection network* (IN). The FUs may be of different types, each implementing its function(s) independently. Each FU performs an application-domain specific (i.e., protocol processing) operation or a group of parameterizable operations, for example Checksum calculation or bitstring matching. Typically the operation is rather complex, especially in comparison to traditional ASIP approaches. There may be more

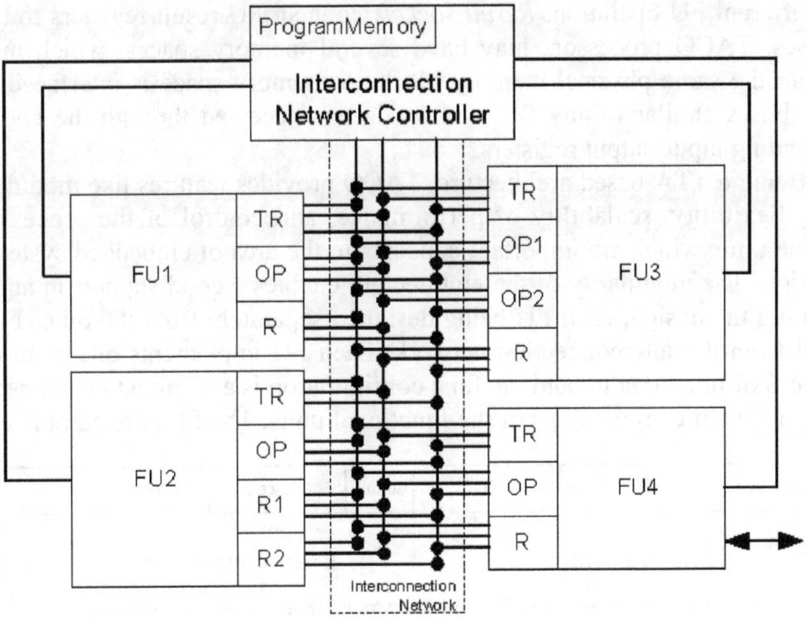

Fig. 12.2. Generic architecture of the TACO processor. An architecture instance with three buses and four FUs is shown.

than one FU of the same type in a TACO processor, allowing parallel execution of multiple instances of the same protocol processing task. In turn, the interconnection network is composed of one or many *buses*, which are controlled by an *interconnection network controller*.

A protocol processing FU is basically composed of an interface to the interconnection network and an internal logic. The interface consists of several registers that are used for storing input and output values. Input registers are of two types: *operand* and *trigger*, respectively. *Operand registers* are used to store input values received from the buses. *Trigger registers* are a special kind of operand registers that, once written with data, trigger the computation of the FU. There may be zero, one or more operand registers and exactly one trigger register in an FU. In addition, *output registers* are used for providing the result of the computation to the buses.

Some FUs may also have a *result signal* that provides a Boolean result of the computation to the interconnection network controller.

The FUs of TACO are connected to the buses via *sockets*. There are three types of sockets: *operand*, *trigger*, and *result* sockets, respectively. *Operand sockets* connect one bus to an input register of an FU. Trigger sockets are a special kind of operand sockets, which connect a given bus to the trigger register of an FU. There may be several trigger sockets connected to the same trigger register of an FU, each socket corresponding to a different FU operation. *Result sockets* connect FU result registers to the buses. TACO processors may have several memory spaces, which may share the same physical memory block. A memory space is interfaced to the buses similar to any FU, and its data is accessed through the corresponding input/output registers.

Being a TTA-based architecture, TACO provides features like modularity, flexibility, scalability of performance, and control of the processor cycle time, which are important concepts in the area of embedded systems design. The modularity of the architecture enables a good support in automating the design, each FU being designed separately from the other FUs and from the interconnection network. Each FU implements one or more pieces of functionality and the final configuration is assembled by connecting different combinations of the functional units. The FUs are completely

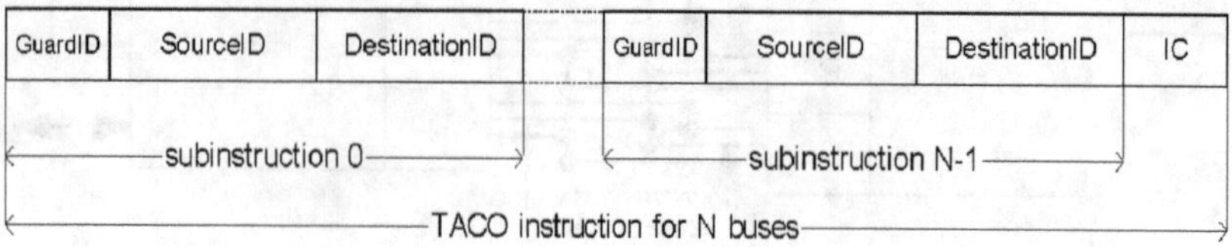

Fig. 12.3. TACO instruction format.

independent of each other and, at the same time, of the interconnection network, all of them being interfaced in a similar manner. The performance of the architecture can be scaled up by adding extra FUs, buses, or by increasing the capacity of the data transports and storage. The functionality of the architecture may be enhanced, by adding new FU types to provide computational support for the application.

Software considerations. In TACO, data transports are programmed and they trigger operations, in contrast to the traditional processors, where the operations of the processor trigger data transports. The architecture is programmed using only one type of instruction, the move, which specifies data transports over the buses. An operation of the processor occurs as a side-effect of the transports between functional units. Each transport has a *source*, a *destination*, and *data* to carry from one FU register to another. The parallelism level in TACO can be exploited not only by increasing the number of FUs of the same type, but also by adding more buses. This allows for executing several bus transports in the same processor cycle.

TACO has a variable-length instruction format based on the number of buses present in a given configuration. An *instruction* (Figure 12.3) is composed of several *subinstructions* (i.e., *moves*), which specify data transports on individual buses. In addition, an *IC* field is used to specify immediate integers on the buses. In turn, a *subinstruction* consists of three fields: *GuardID*, *SourceID*, and *DestinationID*.

The *GuardID* field enables the conditional execution of the subinstruction. Upon evaluation of this field by the interconnection network controller, the subinstruction is either dispatched on its corresponding bus or ignored. GuardID values are configurable, based on combinations of result signal values received from FUs. For instance, 64 guard combinations can be defined, provided that a GuardID on 6 bits is used. The process of defining the GuardIDs for a given processor configuration is done at configuration-time based on the user experience and also on the application requirements. Combinations of the FU result signals may be obtained using the negation (*not*) and the conjunction (*and*) boolean operations. For instance, to implement the following hypothetical example:

```
if (x>2 or y<3 or z<>0) then
     do_something1();
else
     do_something_else();
```

on a TACO processor configuration with three comparator FUs, one could define a GuardID based on the result signals of each comparator FU. Let these result signals be a (true for $x > 2$), b (true for $y < 3$), and c (true for $z = 0$), respectively. A guard that replaces the conditional statement can be written as `myGuardID=!a.!b.c`. As such, a TACO-like implementation of the previous code would be as below:

```
compare(x>2);compare(y<3);compare(z=0);
!myGuardID:do_something1;
myGuardID:do_something_else;
```

For brevity, the TACO subinstructions in the previous example have been replaced with a textual description of the operation performed (e.g., *compare*). We will discuss in the following sections, how the TACO operations are implemented, in practice, in terms of bus transports.

Finally, the *SourceID* and *DestinationID* fields are used to specify the source and target logical addresses (i.e., sockets) between which a transport takes place.

The interconnection network controller is the "brain" of the processor, being in charge of implementing data transports on the buses. The controller uses a program memory from which it fetches instructions, splits them into subinstructions, and dispatches each subinstruction on the corresponding bus. It is important to mention that, when adding FUs, one does not have to change the instruction format, as long as the existing FUs are addressable by the length of source and destination addresses.

A *program counter* (PC) is maintained by the interconnection network controller. A built-in trigger socket is used to load the PC with a desired value (either absolute or relative), making possible to implement program jumps. More details on the actual implementation of the controller may be found in [440].

Visibility of data transports at the architectural level is an important feature of TACO, allowing compilers to optimize and schedule these transports. Since all the operations of the processor are side-effects of the data transports on the buses, the processor cycle-time depends on the availability of the FU results. Furthermore, since the main emphasis of the TTA architecture is moving data, it provides an important platform for data-intensive applications, like protocol processing.

Programming for TACO. From the programmer's point of view, programming TACO processors is a matter of moving data from output to input registers, identified using the address spaces of the corresponding sockets. An example of an instruction composed of three subinstructions is given below:

```
g1:src1 > dst1; !g2:src2 > dst2; +02 > dst3;
```

Using registers for interfacing FUs allows one to apply TTA-specific optimization techniques [92], like moving operands from an output register to an input register without additional temporary storage (bypassing), using the same output register or general purpose register for multiple data transports (operand sharing), or removing registers that are no longer in use, etc. All these techniques help in reducing code size and consequently, in reducing the number of bus transports. Some general compiler optimizations may also be performed on the TACO assembler code, like sinking, loop unrolling, etc. The necessary allocation and scheduling, along with transforming the assembler code into hexadecimal code is left as a task for the TACO compiler which is currently under development.

Abstraction layers in TACO

In this subsection we discuss the abstraction layers that are used in designing the TACO processors.

The TACO complete model. At the lowest level of detail lays the *TACO Complete Model* providing an accurate representation of the hardware details of each component of the architecture. This view of the architecture is specified using VHDL and serves for synthesis purpose and for obtaining an exact measure of the physical characteristics of the system.

The TACO complete model is implemented as a library of module descriptions in VHDL. The first version of the synthesis model was written using Alcatel's 0.35 μm technology libraries. The current version is implemented using 0.18 μm standard CMOS technology. The TACO VHDL module library includes descriptions for all hardware blocks of the TACO hardware platform, including functional units, sockets and the interconnection network controller.

All VHDL descriptions of functional units as well as other modules in the hardware platform have already been individually simulated and synthesized at the time of adding these descriptions into the TACO library. Thus, the individual VHDL module descriptions have already been verified to meet the functional specifications given in the corresponding SystemC module descriptions at the time of writing or generating the top-level file. The top-level VHDL file defines exactly the same architecture as the architecture simulated in SystemC. Once the top-level VHDL file for a processor architecture has been generated, the architecture defined by it is simulated using VHDL simulation tools. In these simulations, the synthesis model is appended with a non-synthesizable part that is used for simulating the different memory blocks needed in the specified architecture. For example, program memory is simulated by reading the application code used in the SystemC simulations from a file. A similar solution is used for simulating the inbound and outbound network buffers. By using identical application code in both SystemC and VHDL simulations, identical functionality and execution scheduling is expected: in TACO processors, the application code has been scheduled already at the time of application software implementation. Thus, the synthesis model is verified by comparing the VHDL simulation results to the SystemC simulation results running the same application code and the same network data. Successful completion of the VHDL simulations permits proceeding to synthesis.

The TACO cycle-accurate model. As an abstraction of the previous level, the *Cycle-Accurate Model* of TACO specifies the TACO architecture with respect to the timing characteristics of each component. This specification serves for simulation purposes, in order to estimate the performance of the processor at early stages of design. Currently, this view of the architecture is specified using the SystemC language. We will discuss the TACO SystemC simulation environment in detail in Chapter 18, *System Level Simulations*.

The TACO component model. The *Component Model* specifies the TACO processor as a collection of components and their interconnections without giving details about their implementation. This model is used during the configuration process of the TACO processor. Both the *qualitative* and the *quantitative* configurations of the TACO processor are specified at the Component Model layer. Qualitative configurations take into account functional requirements of the application, whereas quantitative requirements address non-functional requirements. UML [316] is employed as a modeling language for the component model. A UML profile for TACO [427] is used to model not only the physical architecture of the TACO processor, but also the Matlab estimations (area, gate delay, power consumption) of the TACO components.

The TACO programming model. On the highest level of abstraction, the *Programming Model* of the TACO processor provides an additional abstraction layer of the hardware architecture, where a set of *programming primitives* are defined and used to specify the functionality provided by a given TACO configuration. The TACO programming model is composed of several programming primitives, split into two categories: *functional primitives* and *control primitives*.

Functional primitives (FPs) are the programming constructs providing computational support for a given application. Their main characteristic is that their presence in a TACO configuration varies with the types of TACO resources (FUs) included in a configuration. Each FU provides one or more processing functions, whereas buses are only used to support data transports between FUs. Each processing function is associated with a specific trigger socket. When data is written through this socket into the trigger register of an FU, the function is executed. For executing an FP, several bus transports may be required. In the first stage, the operand registers of the FU are set up, whereas in the second stage, the trigger register is written and, consequently, the operation is executed. This implies that an operation is equivalent with a number of TACO bus transports and, even more, every time the operation is invoked, the exact same sequence of transports is used. Therefore, we can abstract the operations of the processor as macros containing TACO bus transports. The benefit from this approach is that, every time an FP is used, it can be automatically translated into bus transports. For instance, an addition functional primitive can be expressed using three TACO bus transports, each of them writing or reading one of the registers of the FU (e.g., *COUNTER FU*) implementing the primitive.

```
addition(a:int, b:int, c:int){
//COUNTER FU
a > TSC;
b > TIC;
RC > c;
}
```

In turn, *control primitives* are used to support the sequencing of the FPs in a given application, and they are typically present in all TACO configurations. Three types of control primitives are defined in TACO: *unconditional jumps*, *conditional jumps*, and *labels*. Using these types of operations more complex programming structures like loops, cases, subroutines, etc. can be defined.

The set of functional and control primitives provided by a given processor configuration form the API of that configuration. Adding or removing FUs modifies accordingly the API of the configuration. The programming interface can be used not only as a language for programming TACO, but also as a bridge between the application specification and the architecture.

The TACO component library

To provide prerequisites for automation and reuse, the TACO processor resources are organized in a library of components, namely the *TACO Component Library*. Each resource included in this library is specified from three perspectives. A *simulation model* provides SystemC executable specifications, enabling the simulation of the processor configurations, in order not only to check its functional correctness, but also to evaluate its performance. A *synthesis model* provides implementations of each processor resource in VHDL. The synthesis model of TACO targets synthesizable off-the-shelf ASIC components, thus enabling the designer to generate synthesizable processor configurations at system-level. The simulation model is developed in concordance with the performance characteristics provided by the synthesis model, such that the simulation of the system provides cycle accurate results, with respect to the synthesis model. An *estimation model* allows one to obtain high level estimations of different physical characteristics like area, power consumption, and gate delay of the components. The estimates are based on a mathematical model (see Chapter 17) built in Matlab [312, 443].

Building the TACO Component library is an iterative process that relies both on the analysis of the processing needs of different protocols and on the TACO resources identified in previous applications. The SystemC and VHDL modules corresponding to each TACO resource are designed independently. Each new FU is simulated, synthesized, and validated before being added to the library. TACO configurations are created using top-level files, which specify the required modules and their interconnection, as we will discuss in the following sections. Based on the results obtained from the simulation and estimation models, optimizations may be suggested for improving the provided performance, with respect to a given application family.

Furthermore, in order to provide a "unified" component library, in which all the library information is available in a single model, we adopt an UML-based approach. The TACO UML Library not only conjoins, but also abstracts the information of the four different libraries, in order to facilitate the generation of the various TACO models. As such, four categories of information are included in the library:

- *structural* – internal structure of components, like registers, result signals, etc.
- *physical* – characteristics estimates of area, power use, and gate delay; simulation and synthesis specifications pointers to the SystemC and VHDL implementations.
- *functional* – FPs (ie., the API) provided by each component, and their implementations in terms of TACO bus transports, as well as additional implementations (e.g., using the C language).

The TACO UML Library is modeled using the TACO Profile definition, in order to benefit from support in UML tools. The approach allows one to graphically create and maintain such a library, and moreover, to store it as a UML model. Figure 12.4 provides a snapshot of the TACO UML Library for IPv6 routing using the Coral UML tool.[1]

[1] Available for download at *http://mde.abo.fi*

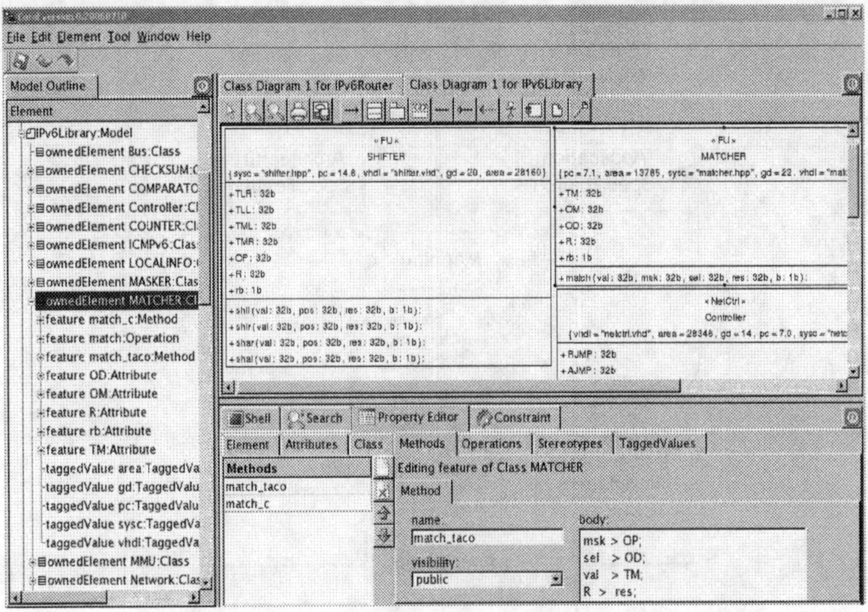

Fig. 12.4. Caption of the TACO library for IPv6 in Coral.[2]

At the bottom of the screen, a property editor allows one to edit the properties of TACO elements. In this particular example, a property editor for editing the methods of the MATCHER FU is shown at the bottom of the screen. The left-hand side panel provides a list of library elements, while the main editor lets one to graphically create and edit the library components.

The TACO Library is currently built manually by the TACO hardware designer and complemented with estimation information extracted from the libraries of the TACO framework. The process of building the library might seem tedious, but the number of library elements is relatively small (e.g., 15 for IPv6 routing) and new additions may occur rather seldom. Nevertheless, once we have the TACO library built, we are able to quickly create processor configurations from which can further generate different artifacts of the development process.

[2] The authors wish to thank Tero Nurmi, University of Turku for providing the physical estimates included in the TACO UML library.

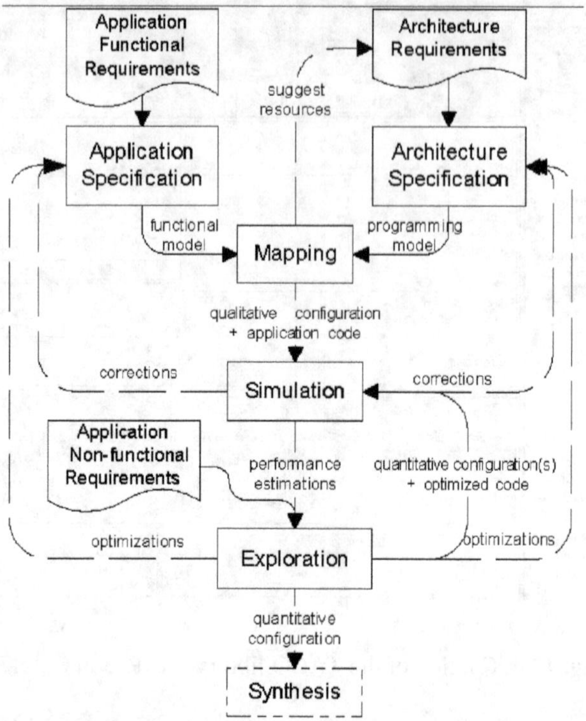

Fig. 12.5. Development flow for programmable architectures.

The TACO design methodology

In this section we discuss the TACO design methodology. During our presentation small excerpts from an IPv6 router application implementation on TACO are provided as example. As mentioned previously, we follow the Y-chart approach [229] as the general framework to develop applications targeted to TACO. A general view of the employed methodology is given in Figure 12.5. Our methodology differs from the Y-chart method in the fact that only the *Application Functional Requirements* are taken into consideration during the application specification process. The impact of such an approach is that the mapping process of the architecture is performed only with respect to the application functionality. Consequently, a qualitative configuration of the architecture is obtained. The *Application Non-functional Requirements* (e.g., performance and physical characteristics) of the application are taken into account only later in the architecture exploration phase of the methodology, where a quantitative configuration of the architecture is obtained.

Application specification. In the *Application Specification* phase, the target application is analyzed with respect to the functional requirements in a top-down manner. Several subphases are used, for instance requirements analysis, application analysis, etc. At each subphase, more details are gathered into the specification. The main goal of this phase is to identify the pieces of functionality of the application that have to be supported by the architecture.

The *Unified Modeling Language* (UML) [316] is used as a graphical notation during this phase. The result of the *Application Specification Phase* is a collaboration diagram (see Figure 12.6), in which the objects (i.e., components of the system) communicate via a message-based scheme, while their internal behavior is modeled with activity graphs.

Architecture specification. The *Architecture Specification* phase starts from the *Architecture Requirements* that capture the functionality that the architecture has to provide. The approach may be seen as a combination of a top-down and a bottom-up approach. In the former, hardware resources are identified from requirements of the TACO architecture; in the latter, the functionality supported by the architecture is extracted from its hardware resources, i.e., starting from the TACO Complete Model towards the TACO Programming Model. Several subphases are used for specifying the system at various levels of abstraction. The end result of the architecture specification is a collection of domain-specific components, modeled at the three hardware abstraction layers of TACO and included in the UML model of the TACO library.

Mapping. The specifications resulting from the application and architecture specification processes provide a functional view of the application and of the architecture, respectively, which in turn form the input to the *Mapping* process. During this phase, the functionality required by the application is mapped to the functionality provided by the architecture. Two artifacts are obtained: a *qualitative configuration* of the architecture to support the functional requirements of the application, and the *application code* to run on this configuration. If some functionality of the application is not supported by the architecture, new resources of the architecture may be suggested. The mapping process is based on three subphases:

A. Express the application specification with TACO programming primitives. During this subphase the application is modeled using UML activity diagrams [316] to model the computations (represented as *activity* and *subactivity* states) that are applied on each PDU. The subphase is based on two steps. In the first one, the subactivity states are hierarchically decomposed into less complex subactivity states, until they match the granularity of the TACO FPs (see activity 'G' of Figure 12.6). This step is heavily based

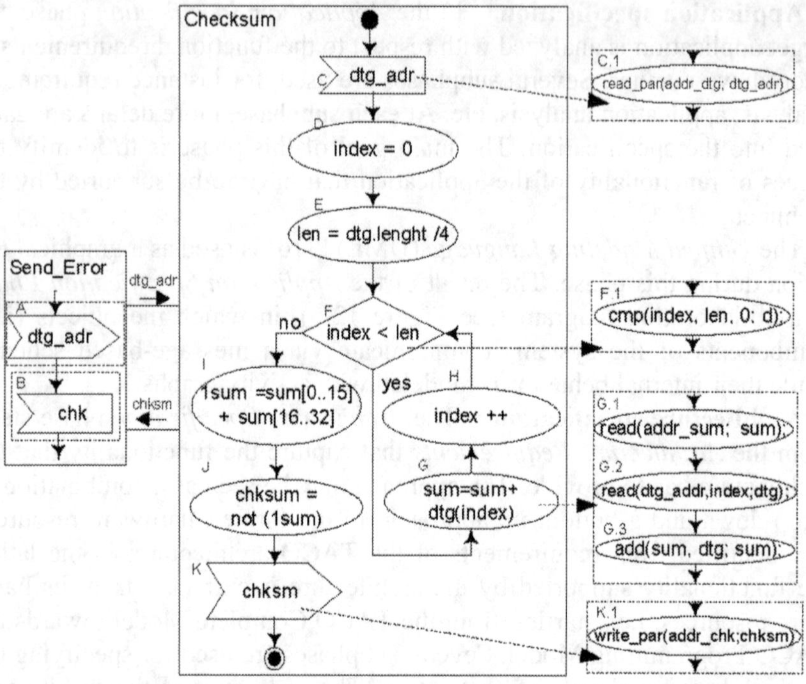

Fig. 12.6. Decomposing activity states to match the complexity of TACO operations.

on the experience of the TACO domain expert and thus, performed manually. A certain level of tool support could be assumed in decomposing the subactivity states, though. In the second step, each "leaf" subactivity state (e.g., 'G.3') is implemented by one of the FPs (e.g., `add()`) provided by the TACO programming model. We remind the reader that TACO FPs are available in the TACO UML Library, as discussed previously.

B. Create qualitative configuration. The process consists of interrogating the TACO Component library in order to identify which functional unit supports each required programming primitive. In certain situations, the same programming primitive may be provided by several TACO FUs, in which case the designer chooses manually the desired FU. The result of this step is a qualitative configuration (see Figure 12.7) of the processor in which one FU of each required type and one bus are included.

C. Create the application code. The process consists of two steps. Firstly, the control flow of the application is transformed into control primitives of the TACO architecture, as follows:

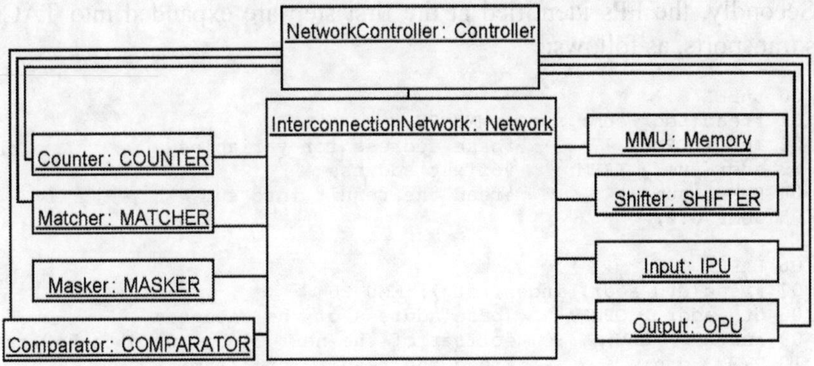

Fig. 12.7. Qualitative configuration of TACO for the IPv6 router.

1. each *activity state*, *branch state*, *send signal*, *receive signal state* is transformed into a TACO subroutine in which the first instruction is a *label*;
2. a *transition* between two blocks is transformed into an unconditional *jump* instruction, where the *address* of the jump corresponds to the *label* of the target activity;
3. *guarded transitions*, only allowed in combination with branch states, are transformed into *conditional jumps* pointing at the label of the target subroutine.

The result of applying the transformation to the example in Figure 12.6 is shown below. In this example, the TACO FP implementing each action state has been replaced with the tag (e.g., 'A.1') of the corresponding activity states, in order to provide a better view of the structure of the presented code.

```
label A.1;     label D.1;       label G.1;     label I.1;
 A.1;           D.1;             G1;            I.1;
 JUMP C;        JUMP E.1;        JUMP G.2;      JUMP J.1;
 JUMP B;
               label E.1;        label G.2;     label J.1;
 label B;       E;               G.2;           J.1;
 B.1            e: JUMP F.1;     JUMP G.3;      JUMP K.1;
 JUMP ...;                      label G.3;
 ........       label F.1;       G.3;           label K.1;
                F.1;             JUMP H.1;      K.1.
 label C.1;     d: JUMP G.1;                    JUMP B.1;
 C.1;           !d: JUMP I.1;    label H.1;
 JUMP D.1;                       H.1;           label B.1;
                                 JUMP F.1;      B.1;
                                                JUMP B.1;
```

Secondly, the FPs identified at the first step are expanded into TACO bus transports, as follows:

```
LABEL G.1;
#read(addr_sum; sum); MMU FU
+00 > OPMM;            #base address for variables
addr_sum > TRMM;       #offset address
RMM > sum;             #read the result into sum
JUMP G.2;

LABEL G.2;
#read(dtg_addr, index; dtg); MMU FU
dtg_addr > OPMM;       #base address of the datagram
index > TRMM;          #offset of the next 32-bit word to be read
RMM > dtg;             #read the result into 'sum'
JUMP G.3;

LABEL G.3;
#add(sum, dtg; sum); COUNTER FU
sum > TSC;             #initialize counter with the 'sum' value
dtg > TIC;             #increment counter value with 'dtg'
RC  > sum;             # read result of the addition
JUMP H.1;

LABEL H.1;
#inc(index); COUNTER FU
index > TSC;           #initialize counter with the 'index' value
+01 > TIC;             #increment counter value by 1
RC  > index;           #read result of the addition into 'index'
JUMP H.2;

LABEL H.2;
#write(addr_index); MMU FU
+00 > OPMM;            #base address for variables
addr_index > TRMM;     #offset address of the 'index'
JUMP F.1;

LABEL F.1;
#cmp(index, len, 0; d); COMPARATOR FU
len > OPC;
index > TLTC;          # index < len, d is guard signal
d: JUMP G.1;           # conditional JUMP
!d: JUMP I1;
.....................................
LABEL K.1;
#write_par(addr_chk, chksum); MMU FU
+00 > OPMM;            #base address for variables
addr_chk > TWMM;       #offset address
RMM > sum;             #write the result into memory
JUMP B.1;

LABEL B.1;
#read_par(addr_chk, chksum; chk); MMU FU
+00 > OPMM;            #base address for variables
addr_chk > TRMM;       #offset address
RMM > sum;             #read the result into sum
JUMP ...;
```

We remark that the *send signal* and *receive signal* activities constitute a special case. A shared memory location is used for passing the message from the sender to the receiver. Basically, the send signal activity writes the value of the message into a given memory location (lines 35–40) and the receive signal activity reads (lines 41–46) the value of the message from the same memory location.

System-level simulation. In the *Simulation* phase, the functionality of the resulted system is validated, with respect to the requirements of the application. The validation is performed based on the input/output behavior of the system. In the situation that errors in the specification (of both the application and the architecture) occur, corrections are suggested and once they are attended the mapping process is performed again. It is important to mention that, by having already designed SystemC specifications for all the TACO components, allows that only the top-level configuration file of the processor need to be generated from the TACO component models. Issues related to the system-level simulation of TACO processors will be discussed in Chapter 18.

Design space exploration. The *Exploration* phase deals with tailoring the architecture towards an optimal implementation of the application, in terms of performance requirements and physical constraints. This phase implies performing estimations of the architecture from several perspectives. For instance, from the simulation process the designer may collect performance estimates with respect to the throughput of the application. Additionally, an estimation of the physical characteristics like occupied area, power use, etc. of the configuration is performed. Based on this data, mainly optimizations of the architecture, but also of the application specification are suggested. The exploration process is performed iteratively until a satisfactory configuration (i.e., *quantitative configuration*) is obtained. For each quantitative configuration, the application code is optimized such that it takes into account the parallelism of the configuration.

Synthesis. The designer selects from the exploration process one or many TACO processor configurations suited for implementing the application and, subsequently, synthesizes them in hardware. To build a VHDL description of a TACO architecture, a top-level file specifying all needed modules and their interconnections have to be created. Again, having already built VHDL specifications stored in the TACO Component library enable us to automatically generate the TACO Complete model; i.e., the top-level configuration file corresponding to a given configuration. The code below presents as example the declaration of the Matcher functional unit of Figure 12.7.

```
signal m1 op1 load, m1 od2 load, signal m1 trg load : std ulogic;
signal m1 OP1 : unsigned(datawidth-1 downto 0);
signal m1 OD2 : unsigned(datawidth-1 downto 0);
signal m1 TR : unsigned(datawidth-1 downto 0);
signal m1 ResultR : unsigned(datawidth-1 downto 0);
signal m1 guard bit : std ulogic;
matcher operand : input_socket
   generic map (socket_address => 16#OPM1#)
       port map (clk => clk, reset => reset, dst_address =>
       icnw_dst_address, net_data_in => icnw_data, load =>
       m1_op1_load, socket_data_out => m1_OP1);
matcher data : input_socket
   generic map (socket_address => 16#ODM1#)
       port map (clk => clk, reset => reset, dst_address =>
       icnw_dst_address, net_data_in => icnw_data, load =>
       m1_od2_load, socket_data_out => m1_OD2);
matcher trigger : input_socket
   generic map (socket_address => 16#TM1#)
       port map (clk => clk, reset => reset, dst_address =>
       icnw_dst_address, net_data_in => icnw_data, load =>
       m1_trg_load, socket_data_out => m1_TR);
matcher result : output_socket
   generic map (socket_address => 16#RM1#)
       port map (src_address => icnw_src_address, socket_data_in
       => m1_ResultR, net_data_out => icnw_data);
matcher fu : matcher
   port map (clk => clk, reset => reset, op1_load =>
   m1_op1_load, od2_load => m1_od2_load, trg_load => m1_trg_load,
   OP1 => m1_OP1, OD2 => m1_OD2, TR => m1_TR,
   ResultR => m1_ResultR, guard bit => icnw guard());
```

Conclusions

Several issues have to be addressed in the process of designing protocol processors. The initial problem to be solved is the target hardware platform. The alternatives vary from custom design frameworks with excellent application-domain optimization through traditional ASIP methodologies with a smaller level of domain optimization to fixed hardware implementations with minimal programmability (ASIC). The available tools for automating the design process at different abstraction levels depends on this choice. In any case, in order to evaluate the design processor models for simulation, physical characteristics estimation and logic synthesis are needed. A key issue in the development of such processor models is that they need to provide a simple enough API through which models for a given protocol processor architecture can be generated. This is especially important for system level simulations of architectures: in addition to functional verification, simulations need to provide hardware-accurate results like cycle-by-cycle simulation and register transfer statistics. Whilst providing such precise information, the simulations are also required to be very fast (i.e., to run a test bench in a matter of seconds) to facilitate rapid architectural exploration at the system level. In Chapter 18, we discuss these issues in more detail and also reflect on the use of object-oriented programming techniques and a heterogeneous level of abstraction in the simulator implementation.

Another problem to be addressed is defining rigorous application specification processes which are able to capture those characteristics of the application required to tune up the implementation platform for optimal performance. Such specification processes not only have to cope with the increasing complexity of applications through abstraction layers and tool support, but also through employing application-domain-specific methodologies and specification languages/models that address the specifics of each application family.

Last but not least, the application specification processes should be able to facilitate the mapping of the application on selected architectures naturally and with as much tool support and automation as possible.

13 Java Co-Processor for Embedded Systems

Tero Säntti, Joonas Tyystjärvi, and Juha Plosila

University of Turku

Introduction

Java is very popular and portable, as it is a write-once run-anywhere language. This enables coders to develop portable software for any platform. Java code is first compiled into bytecode, which is then run on a Java Virtual Machine (hereafter JVM). The JVM acts as an interpreter from bytecode to native microcode, or more recently uses just-in-time compilation (JIT) to affect the same result a bit faster at the cost of memory. This software's only approach is quite inefficient in terms of power consumption and execution time. These problems rise from the fact that executing one Java instruction requires several native instructions. Another source for inefficiency is the cache usage. As the JVM is the only part of software running natively, it occupies the instruction cache, whereas the Java bytecode is treated as data for the JVM, hence being located in the data cache. Also the actual data processed by the Java code is assigned to the data cache. This clearly causes more memory accesses missing the cache. When the execution of the bytecode is performed on a hardware co-processor this is avoided and the overall amount of memory accesses is reduced.

This work is a part of the REALJava project, which aims to design a Java co-processor that is easily integrated to various systems. We have chosen to use asynchronous techniques in this project because then we can achieve good performance with reasonable power consumption and very easy integration with existing systems, since no clock limitations need to be considered. Asynchronous self-timed circuit technology [386], where timing is based on local handshakes between circuit blocks instead of a global clock signal, provides a promising platform for obtaining a highly modular low-power and low-noise Java accelerator implementation.

J. Nurmi (ed.), Processor Design: System-on-Chip Computing for ASICs and FPGAs, 287–308.

The rest of the chapter is organized as follows. In the next section we shortly describe the structure of any JVM, and show how the REALJava co-processor fits into the specifications. The following section describes the hardware co-processor design, and the software portion of the virtual machine is presented in another section. Yet another section details the current status of the system. Finally, we draw some conclusions and describe the future efforts related to the REALJava co-processor.

Generic virtual machine architecture

Advantages of virtual machines

One of the most important reasons to use a virtual machine is that the code is "write once, run anywhere". This means that the code needs to be compiled only once, and then it can be run on any platform, even over the net. Another important advantage is that new versions of hardware need only a new virtual machine to run all existing software. This reduces development costs of a new generation of devices.

Even though the reasons listed above are very attractive, especially to industry, if not the consumer, they are not the only reasons to use virtual machines. Virtual machines provide improved security features, the additional layer between the code and the executing hardware can be used to increase security. While making it hard for you to shoot yourself in the foot, it also makes it harder for others to shoot you in the foot. Software downloaded from the internet can be verified to ensure it is original and is not malevolent.

The security advantages are also present in some fully interpreted languages, such as TCL and JavaScript. The difference here is in the execution time. Virtual machines get some kind of precompiled input files (Java bytecode for example) thus they need less run-time interpreting and manipulation of the source code resulting in better performance. According to [286] TCL has been (informally) measured to be up to 200 times slower than fully compiled C++, whereas semi-compiled languages fall in the range of 10 to 20 times slower than C++. With modern technologies, such as JIT, the difference is dropping below "only" five times slower. The penalty for JIT is increased memory requirements. The performance is still significantly below C++, due to the fact that a C++ compiler can optimize register allocation, whereas a JIT compiler has to work starting with Java bytecode operating on a stack.

JVM implementations

The traditional way to implement a JVM is to use a software interpreter. This approach takes the precompiled Java bytecode and executes it by interpreting one instruction at a time. The execution speed of this way is rather poor, partly due to stack emulation and partly because each bytecode instruction is processed using an indirect jump from the main loop to the implementation of a given instruction with the current instruction as a key. Recent developments in this area have led to use of JIT, which means that the software executing bytecode takes pieces of the code and compiles, them to the underlying hardware's native instruction set. The most advanced systems using this approach do not recompile or optimize everything, but focus more attention on code segments that are used often (loops, etc.). JIT is not suitable for resource limited environments, as the recompiled code segments require extra memory space at run-time, and the dynamic compiler required for the JIT takes both memory space and permanent storage space. Another downside of JIT is the unpredictability of execution time, one (usually) cannot know in advance if a given code segment is already compiled or not. Also resource limited systems might purge old recompiled segments out of the memory to make room for new segments. Both interpreting and JIT compiling run on the highest path in Figure 13.1.

Java bytecode can also be transformed to some other virtual machine architecture (XVM) using a transcoder. This transcoding can be done at execution time or during software download. The execution time transcoding moves along the highest path in Figure 13.1 breaking down at the last fork, whereas the download time option breaks down at the previous fork. This approach also allows other languages to be compiled for the same XVM, even all the way from the source code. However there are drawbacks, such as increased storage space requirements to store the original code and the XVM code, longer download delays for non-local software that needs to be transcoded and successive optimizations performed by compilers with different targets producing inefficient code.

Our virtual machine is of the first type, that is 100% bytecode compatible, but it transfers the execution away from the CPU to a co-processor designed to execute simple Java bytecode instructions. The CPU still maintains control of all system specific operations (such as file system, network, I/O), complex instruction (class loading and verifying) and memory management (especially garbage collection). This partitioning provides easy integration to multiple systems, since the underlying host architecture is

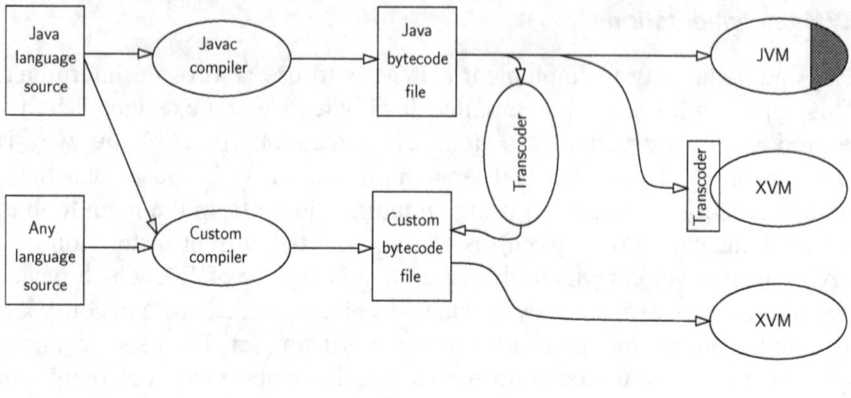

Fig. 13.1. Possibilities to develop the Java virtual machine.

irrelevant to the co-processor. The use of a co-processor can be seen as implementing the highlighted part of JVM in Figure 13.1 in hardware.

Using hardware systems in virtual machine implementations

A stand-alone solution implements the whole virtual machine in hardware, and thus needs no CPU. Examples of this approach include Sun's PicoJava, JOP and aJile. With this strategy complex instructions (class loading and verifying, etc.) and garbage collection are hard to implement, and the resulting virtual machine is not easily integrated to an existing system. Out of these three example systems only PicoJava implements the full J2SE (standard edition of Java) [396]. The other two implement J2ME (microedition for small systems) [397] and for instance the garbage collection is completely left out from JOP, due to real-time performance goals set for the system.

Using a co-processor for the execution of Java bytecode provides easier integration to existing systems, as platform dependent features, such as GSM stack for mobile phones and I/O devices, can be coded in software for the CPU. Also the physical size of the Java core is reduced, as complex functionality is performed in the CPU, leaving the co-processor to deal with the instructions that are suitable for direct hardware implementation. This approach has been used with inSilicon JVXtreme, CCL Java coprocessor and the REALJava co-processor discussed in this chapter.

Co-processors typically use either autonomous or parallel execution model. Parallel model is similar to Intel x87 floating point unit architecture, where each instruction is routed to both the CPU and the co-processor and the unit whose instruction set a given instruction falls into executes it. On the other hand in autonomous execution model the CPU just configures the co-processor and lets it handle execution on its own. During this time the CPU is free for other tasks. The REALJava co-processor uses the autonomous model. It should be noted, that the autonomous model allows using several co-processors at once, unlike the other hardware schemes presented here. Using several co-processors in a JVM is well justified, as Java supports multithreading at language level.

Solutions using a hardware interpreter have also been used. These are exclusively targeted to a certain CPU, as the interpreter translates the bytecode instructions to the CPU's native instruction set. ARM Jazelle [23] is a well-known example of this approach.

Structure of the co-processor

The pipeline structure [403] of the Java co-processor differs from the structure normally used for general-purpose processors. This is due to the fact, that normally the instruction set of a processor is engineered with hardware implementation in mind, but this is not the case for Java. The Java bytecode has been designed to be executed in software, resulting in several significant differences. Additionally the bytecode instructions are based on a stack, instead of the normal RISC approach of using several registers. This calls for optimizations not seen in modern general-purpose processor design.

General-purpose processor pipeline architecture

The normal strategy for pipelining a general-purpose processor involves five stages, namely:

1. Instruction fetch
2. Instruction decode/register access
3. Execute/ALU
4. Memory access
5. Write back

This approach has been used in several processors and is also presented in several textbooks, such as the DLX processor presented in [187]. This strategy is based on the assumption that the processor has internal registers for temporary or working data storage. Usually these registers can be accessed in parallel, and there are several registers available. As an example the DLX processor has 32 32-bit general-purpose registers. Some processors also include separate registers for storing floating point numbers. The DLX processor provides 32 32-bit floating point registers, which can be used as even–odd pairs to hold 16 64-bit double-precision values. Several register access optimization strategies have been developed, including operand forwarding and splitting register accesses to writes in the first half of the clock cycle and reads in the second half.

The modified architecture for the Java co-processor

The Java Virtual Machine Specification [267] states that the JVM has no internal registers, instead the temporary and working data is stored in a stack. Normally the software coder can improve performance by reordering the register accesses to keep the pipeline flowing, but in Java this is not possible, since all instructions manipulate data which is located at the top of the stack. This situation is somewhat comparable with normal processor architecture with only one register available to the programmer, or the old accumulator architecture. This would keep the pipeline stalled for a large portion of the time, because of data dependency issues. To keep our pipeline in effective use, we have modified the normal pipelining strategy to better suit the stack based operation.

As shown in Figure 13.2, the modified architecture begins with instruction fetching, we just use a FIFO inside this unit to provide the folding unit with fast access to the instruction stream. The instruction decoder is the next unit. A technique called instruction folding, which will be explained in more detail in another section, is used to reduce unnecessary stack accesses, and the folding is also included in the decoder stage. After that we have an intermediate buffer level to store the folded instructions before execution. This buffer also performs minor operations, such as extending literal data items to 32 bits.

The next stage performs operand fetching, if necessary. Then comes the ALU, which contains the write back stage. The write back stage is included to the ALU because the bytecode instructions are based on the stack. One might wonder what this has to do with selecting the pipeline stages, but the answer is rather simple. In Java bytecode the instructions take the operands from the stack and write the result back to the stack. This would cause the "normal" pipeline structure to generate excessive stalls to move the data to and from the stack. Thus the execution in the ALU would

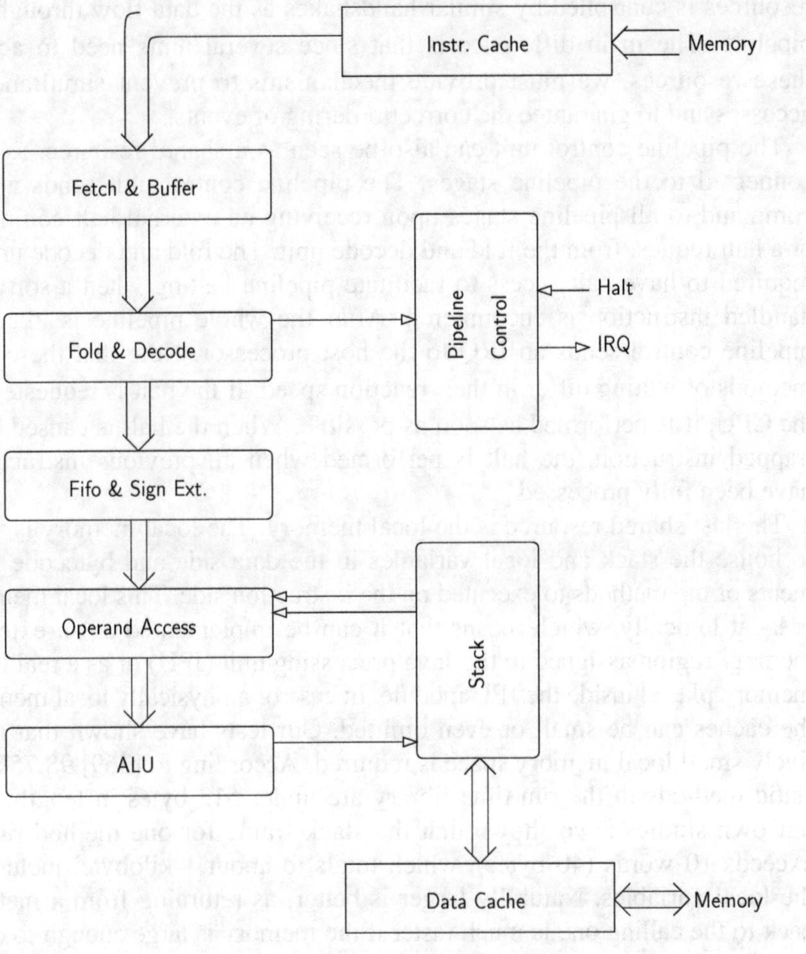

Fig. 13.2. A simplified view of the pipeline.

be often halted while the data is moved back and forth. Actually we will also describe two other methods to alleviate this problem, but they will be presented later in sections.

Shared resources

Several pipeline stages need to access shared resources. These include the stack, the control registers and the program counter. Access to these resources is controlled by similar handshakes as the data flow through the pipeline. The main difference is that since several units need to access these resources, we must provide mechanisms to prevent simultaneous accesses and to guarantee the correct ordering of events.

The pipeline control unit can also be seen as a shared resource, as it is connected to the pipeline stages. The pipeline control unit sends a halt command to all pipeline stages upon receiving an external halt command or a halt request from the fold and decode unit. The fold and decode unit is required to have halt access to facilitate pipeline halting when a software handled instruction is encountered. After the whole pipeline is idle, the pipeline control sends an IRQ to the host processor. Note that these two methods of halting differ in their reaction speed. If the halt is requested by the CPU, it is performed as soon as possible. When the halt is caused by a trapped instruction, the halt is performed when all previous instructions have been fully processed.

The last shared resource is the local memory. The local memory is used to house the stack and local variables in the data side and bytecode segments of the methods to executed on the instruction side. This local memory is local logically, which means that it can be implemented as an external memory region assigned to the Java processing unit (JPU) or as a real local memory placed inside the JPU module. In case of a physically local memory the caches can be small or even omitted. Our tests have shown that relatively small local memory space is required. According to [369] 98.75% of static methods in the run-time library are under 512 bytes in length, and our own studies have shown that the stack frame for one method rarely exceeds 10 words (40 bytes), which totals to about 1 kilobyte, including the local variables. Naturally larger is better, as returning from a method back to the calling one is much faster if the memory is large enough to contain the stack frame and code segment for several methods at the same time.

Instruction preprocessing

This block starts after the instruction cache. The cache handles all communication with the physical memory, regardless of whether the memory is physically local or external. This partitioning of responsibilities allows using different memory technologies without modifications to the instruction fetching unit. The physical addresses are generated at the instruction fetching buffer, using the program counter (PC) and CODE_OFFSET registers. The CODE_OFFSET register holds the starting address of the current code segment in the memory. The instruction folding process is described later.

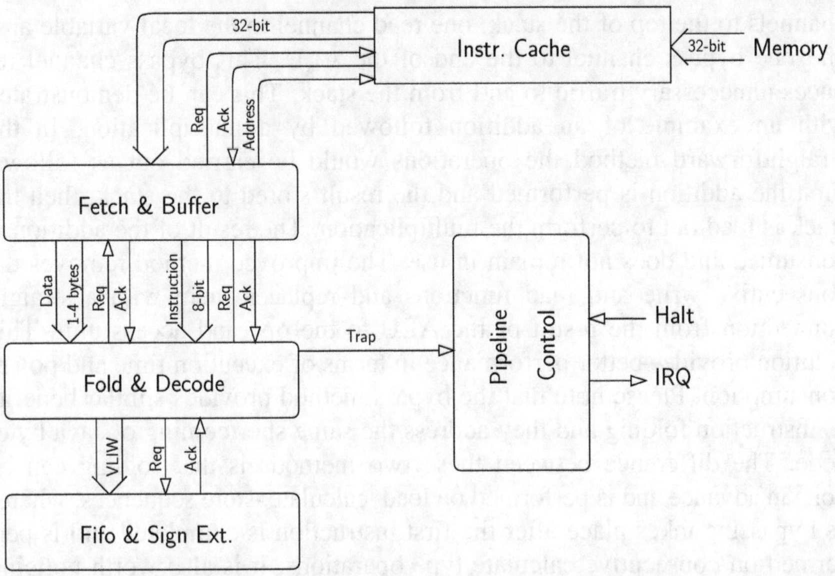

Fig. 13.3. The instruction preprocessing pipeline.

The pipeline control unit is connected to the folding and decoding unit with two way communication. The folding unit needs to request a halt when it encounters an instruction to be handled in software. Of course the pipeline control unit must be able to stop the processing in this segment, so there needs to be bidirectional channel. The control unit also connects to all other pipeline stages, with a halt signal. The CPU can also request a halt, for thread switching or setting new values to internal registers.

As shown in Figure 13.3, the folding and decode unit has two communication channels to the instruction buffer, one for actual instructions and one for literal data. After an instruction has been decoded and folded, the VLIW (very long instruction word) is sent to the FIFO in then main pipeline. The instruction folding unit is covered in more detail later. The FIFO is only a few levels long and provides timing margin for folding and performs sign extension.

Operand access, ALU and result storing

The operand access unit takes care of providing the ALU with the actual operands, which may come from the local variable area, the stack or as literal data from the bytecode stream. The operand access has two read channels to the top of the stack, one read channel to the local variable area and one bypass channel to the end of the ALU. This bypass channel reduces unnecessary traffic to and from the stack. This can be demonstrated with an example of an addition followed by a multiplication. In the straightforward method the operations would be carried out as follows. First the addition is performed and the result stored to the stack, then the stack is read out to perform the multiplication. The result of the addition is consumed and does not remain in use. The improved method removes the consecutive write and read functions and replaces them with a straight connection from the result of the ALU to the operand access unit. This solution provides better performance in terms of execution time and power consumption. Please note that the bypass method provides similar benefits as instruction folding and they address the same shortcoming of Java bytecode. The difference between these two methods is that folding can be done in advance and is performed on load–calculate–store sequences, whereas bypassing takes place after the first instruction is completed and is performed on consecutive calculate type operations. It is also worth noticing that the bypass method is quite similar to operand forwarding in general-purpose processors.

Figure 13.4 shows the data connections in the execution part of the pipeline. The request and acknowledge signals are not shown, in order to keep the figure readable. The result of the ALU usually goes to the top of the stack. In some cases the result is directed to a local variable. The third possibility is to intercept the result and direct it straight to operand access unit. This happens when the current instruction in ALU pushes its result to the top of the stack and the next instruction pops it away. The state of the stack remains as if the first result had never been pushed. The interception thus saves power and time, at the cost of slightly more complex logic.

Caches, stack and registers

The JPU contains two caches, namely the data cache and the instruction cache. The instruction cache is (quite naturally) read-only, whereas the data cache can be written and read. The instruction cache is less complex also because it is connected to only one unit, namely the instruction buffer. The data cache, on the other hand, is connected to the stack and to the local variable control. The writing to the data cache is implemented using the write-through strategy, in order to keep the memory consistency during traps and context switches easier to manage. Both caches are also giving state information to the pipeline control unit, to notify the controller when the current operations are finished.

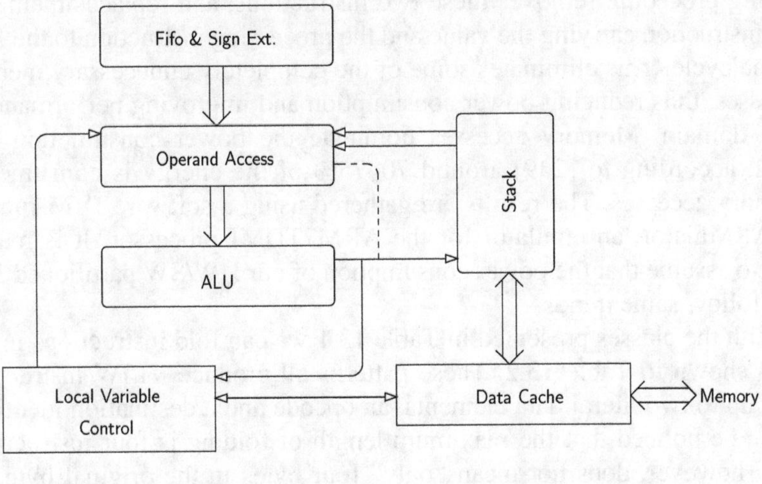

Fig. 13.4. The execution pipeline and data transportation.

The stack is implemented as ring buffer with memory roll-back. The buffer holds the top of the stack. When the buffer is close to full, the bottom of the ring is rolled to the memory via data cache. Naturally if the buffer is close to being empty more data is retrieved from the memory.

The stack performs these transactions automatically, and no direct commands are required during normal execution. However a command for flushing the stack to the memory is required, since jumping to a method causes a new stack-frame to be initialized, with its own local variables, etc.

The internal registers of the JPU are all addressable from the CPU. This is required in order to be able to configure the JPU in the beginning of the execution as well as during thread switching. The internal register file also contains configuration data from the JPU to the CPU. The most important piece of information delivered here is the size of the local memory. The system also supports multiple instruction sets so this information needs to be delivered to the software. Currently two instruction sets have been designed, one with floating point instructions in hardware and one with software emulation.

Instruction folding in more detail

The instruction folding [404] is performed in order to remove unnecessary cycles in ALU and also to minimize redundant stack accesses. These performance hindrances are caused by bytecode instructions first pushing a value to the stack and immediately popping it out for processing. The folding procedure removes these two instructions, and replaces them with one instruction carrying the value and the processing instruction to the ALU in one cycle. This eliminates some of the completely unnecessary memory accesses, thus reducing power consumption and improving performance in time domain. Memory accesses dominate the power consumption of a JVM, according to [249] around 70–75% of the energy is consumed in memory accesses. The results are gathered using a software JVM running on ARMulator, an emulator for the ARM7TDMI processor. It is reasonable to assume that the power consumption of our HW/SW partitioned JVM will follow same trends.

With the classes presented in Table 13.1 we can fold instructions in patterns shown in Table 13.2. These patterns all produce VLIW instructions with up to two literal data elements, an opcode and a destination identifier. It can be noticed that the maximum length of folding is four instructions. This, however, does not mean "only" four bytes in the original bytecode stream. The original stream may have had some literal data included, and these are also placed in the VLIW, as shown later in Figure 13.5.

The fact that the whole co-processor is asynchronous helps us in the folding. In asynchronous circuits the blocks can run at independent speeds. This means that the folding unit can perform for instance a maximum of n foldings per second, whereas the ALU may be significantly slower, say $n/2$ operations per second. The negative effects of independent speed, such as waiting for one long operation halting all other pipeline segments, can be reduced using an intermediate FIFO. The timing marginal for folding is

Table 13.1. Instruction classes.

Mnemonic	Description
LV	A local variable load, a load from a global register or a push constant
OP	An operation that uses the top two entries of the stack and produces a one word result which is stored on the top of the stack
OP1	An operation that uses the topmost element of the stack and breaks the group
OP2	An operation that uses the top two entries of the stack and breaks the group
MEM	A local variable store or a global register store
NF	Non-foldable instruction
TRAP	An instruction which is trapped by the hardware and is executed in software instead

Table 13.2. Possible patterns.

Pattern	Instructions
LV LV OP MEM	4
LV LV OP	3
LV LV OP2	3
LV OP MEM	3
LV OP	2
LV OP1	2
LV OP2	2
LV MEM	2
OP MEM	2

increased because with asynchronous techniques all units exhibit average case performance. This means that the ALU may complete some instructions (bit-wise OR, etc.) in very short time, whereas some instructions (32-bit multiplication) take a lot more time. Since folding may produce new VLIW instructions at the rate of 1/1 to 1/4 in comparison to the original bytecode stream, the FIFO balances the effects of both folding and the average case performance of the ALU. In our architecture the FIFO also performs minor tasks, such as sign extension and address calculation for local variable accesses.

The folding unit receives data from the instruction buffer. The instruction cache handles the actual memory accessing, so the instruction buffer needs only to access the cache. The address is generated at the instruction buffer. The fold and decode unit has two communication channels to the instruction buffer. This is required because instructions may be followed by data, such as literal operand or an address. The amount of data can be found out only by decoding the instruction first. After the decoding is completed, the correct amount of data bytes is read in parallel. The amount of data is between 0 and 4 bytes. If it is 0 bytes, no request is sent to the data read port. Since we read the data items in parallel to the fetch data module shown in Figure 13.5, the instruction buffer can move the next instruction to the output end of the buffer without unnecessary delays.

After the instruction has been decoded and the data related to that instruction is read in, the next instruction is checked to see if it can be folded with the previous one. If it can be, then the procedure is repeated to see if the third instruction can be folded. If at any point the instructions cannot be folded together, the previous instructions are sent out, and the procedure starts over with the current instruction as a base for new foldings.

Figure 13.5 shows the internal structure of the folding unit. The FSM stands for Finite State Machine, which controls the operation of the unit. The ROM table approach is chosen, because Java bytecode is not optimized for hardware decoding. The instructions of Java bytecode are just listed in seemingly random order and given the order number as an opcode. This would lead to a very complicated decoder, if implemented directly using standard logic elements. The ROM approach is also further validated by the fact that we can easily store microcode and other metadata in the same table, as well as instruction classes and the number of literal data bytes related to a given instruction. This keeps our FSM simple and fast. All the entries in the ROM table are coded with one-hot scheme and the table is implemented as a precharged MOS NOR ROM matrix. The precharging is done when request is low, so the response time is minimal.

The output format register stores partial foldings, until they are completed. If a folding pattern is not terminated with a valid instruction for that pattern, the partial folding is executed one by one, and folding of the next instruction will be attempted. The register keeps record of which fields in it are valid at any given time. When a pattern is completed, the register pushes its contents to the FIFO in the main pipeline, and prepares for a new folding autonomously.

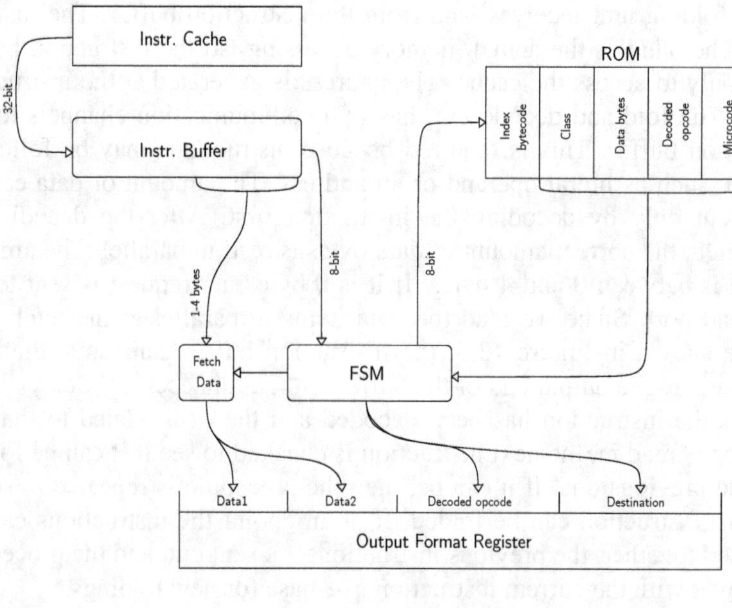

Fig. 13.5. The internal structure of the folding unit. © IEEE 2005 [404].

Software support

Because the JPU executes only a subset of the instructions in the JVM instruction set, executing actual Java programs with it requires supporting software written for a general-purpose CPU. This supporting software needs to do most of the things that a generic virtual machine does, but it also needs to control bytecode execution and memory usage on the JPU. In this section, we briefly discuss the operation of a generic JVM and the software components required to support a JPU.

Virtual machine software

A typical JVM implemented purely in software loads Java classes, manages resources such as memory and threads, provides an interface to the virtual machine for native code, and most importantly, controls bytecode execution. The part of a virtual machine that executes bytecode is called its execution engine. This is the part that typically uses the largest amount of CPU time.

The simplest software implementation of a bytecode execution engine is a bytecode interpreter. In an interpreter, the software fetches one instruction at a time, branches according to its opcode and executes the native instructions corresponding to the Java bytecode instruction. This loop is continued until the interpreter encounters an instruction that requires special processing. For example, a method invocation instruction may require calling a native function.

Although an interpreter is simple to implement, it is not very efficient. Since the JVM is entirely stack-based, interpreting bytecode requires a large amount of memory accesses even for relatively simple operations. For example, the following sequence of instructions, which multiplies a local variable with 5, requires 6 stack accesses:

```
ALOAD_0       ; push 1
SIPUSH 5      ; push 1
IMUL          ; pop 2, push 1
ASTORE_0      ; pop 1
```

There is also a per-instruction overhead in an interpreter caused by the pointer access used to fetch the instruction and by the instruction dispatch itself.

Many optimizations have been developed to reduce this overhead. Direct- and inline-threading [63] seek to reduce the instruction fetch and instruction dispatch time by converting opcodes into the corresponding native instructions before execution. JIT [94] further optimizes the generated code and reduces stack accesses as well by using the host CPU's registers to store intermediate results. Because the bytecode instructions are replaced with native instructions, all of these optimizations also remove the bytecode program counter.

The structure of the execution engine needs to be changed to support a bytecode co-processor. Rather than executing every instruction in software, when the virtual machine encounters a sequence of instructions that the co-processor can execute, it delegates execution to the co-processor. The co-processor executes instructions until it encounters an instruction that it cannot execute or the software commands it to halt.

Since most of the processing is done on the co-processor, many improvements to the software part of the execution engine become unnecessary and impractical to implement. Because the virtual machine needs to be able to update the stack and the internal registers of the co-processor when it resumes execution, optimizations that reduce stack accesses or replace the program counter become unusable as such. However, they also become largely unnecessary, because the co-processor takes care of most of the bulk stack manipulation.

We implement a simple JVM in C++ with support for JNI [395] and the standard edition of the Java 2 platform [396]. The current version of our virtual machine only works on Windows and Linux on x86 computers. Our virtual machine also contains a simple emulator of the hardware's capabilities, and can be used for testing new functionality on software.

The structure of our virtual machine is shown in Figure 13.6. Like a generic virtual machine, our virtual machine contains a native interface, a heap memory manager and a class loader. Our execution engine, however, is split between the software and the co-processor. Our virtual machine also manages memory on the co-processor for java stack and method usage and implements a simple thread scheduler for allocating co-processors to separate threads.

Bytecode execution and trap handling

The virtual machine needs to do some preparation before it starts executing code on a co-processor. First, it has to acquire the lock on the co-processor. Co-processor locking is discussed in the next section. Second, it has to check that the current thread's stack frame and the current method are loaded in the co-processor's memory. Memory management will be discussed further in a later section. Finally, it has to update the internal registers

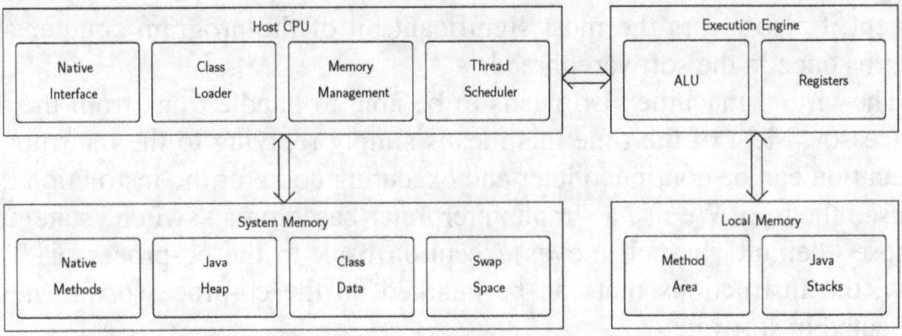

Fig. 13.6. Logical layout of the REALJava virtual machine.

of the co-processor. While the co-processor is executing bytecode, the software is free to do other tasks. For example, it could optimize methods by converting instructions to "fast" versions in advance or load and verify classes that will be needed soon.

When the virtual machine calls a new method on the co-processor, it has to do some extra work on the stack. First, it checks if there is enough space for the method invocation in the currently allocated stack page. If there is not, or a stack page is not currently allocated for the thread, it allocates a new stack page on the co-processor. Once there is enough space for the method invocation, the method parameters are popped from the stack and the current registers are pushed to the stack top in reverse order. If the previous stack frame was swapped out by the allocation or there is none, a magic number (we use a bit pattern of all ones) is pushed instead of the current program counter. After the registers have been pushed, the local variable pointer is set to the current stack top pointer, the stack top pointer is incremented by the amount of local variables, and the method parameters are stored in the local variables. Once this is done, execution can proceed on hardware.

Sometimes, a return from a function must be executed in software when a method returns in the bytecode. For example, when calling a Java method from native code, the software needs to be able to return to the proper native function after the call. For this reason, the software must be able to force a trap in one of the return instructions. The most significant bit of the program counter register is used for this purpose. A limitation in Java's exception handler implementation practically limits the Java program counter to 16 bits [267], so the upper 16 bits can be used to store data required by the virtual machine. If a software return is required, the bit is set to 1 when storing the registers of the previous stack frame. When the return instruction traps, the software handles it like the co-processor would, except it also clears the most significant bit of the program counter and returns once in the software thread.

The virtual machine also needs to be able to handle traps from the co-processor. Most of the time this means simply replying to the interrupt so execution can be continued later and executing code for the instruction that caused the trap. We use a simple interpreter based on a "switch" statement whose "default" branch moves execution back to the co-processor. This way, the instructions that can be handled on the co-processor are never executed in software.

Co-processor allocation with multithreading

In a single-threaded environment, controlling co-processor usage is simple. Since there is only one thread using the co-processor, no locking or pre-empting is necessary. When the single thread needs to execute code on the co-processor, it always knows that the co-processor is in the state that it was when it made the last interrupt request. Therefore, it can send "continue" commands to the co-processor any time it wants.

In a multithreaded environment, the virtual machine must control co-processor usage for two reasons. First, to prevent invalid behavior, two threads must be prevented from accessing a single co-processor simultaneously. For this reason, each co-processor has to have a lock that the threads acquire before using it. Second, a single thread must be prevented from holding the lock on a co-processor indefinitely, because it could lead to starvation and possibly deadlocks.

We implement a simple time slice-based pre-empting system on software. To implement pre-empting, we have each thread's "resume" routine occasionally poll the virtual machine's access control system to see if it has exceeded its time slice. If it has, it sends the co-processor a "halt" command, stops execution and releases its lock on the co-processor. Once it can acquire the lock again, it can resume execution. We also poll the access control system whenever the co-processor traps normally. Another way to implement pre-empting, which requires hardware support, is to have the hardware trap after a predefined number of clock cycles has passed since execution was last resumed. This removes the need for active polling.

A thread must also release its lock on a co-processor if it does something that could potentially take a long time without needing the co-processor. Otherwise, other threads might starve or even become deadlocked. The most common possibly lengthy operations during normal execution are monitor acquisitions and native method invocations. For monitor acquisitions, it is usually possible to atomically test whether a monitor can be acquired and acquire it if possible, so a co-processor release is only necessary if the monitor is acquired by another thread.

Native methods, on the other hand, are essentially "black boxes" for the virtual machine, since it cannot know what a dynamically loaded chunk of native code will do, and how long it will take to execute the code. A native function might, for example, initiate a native I/O operation and block until input arrives. Therefore, for most native method calls, the lock on the co-processor must be released. As an exception, certain known "safe" functions in the standard library can be assumed to execute quickly. The method Math.sin, for example, takes a constant, short amount of time to execute regardless of input, and therefore should not require releasing the lock. A simple list of such methods can be implemented in the virtual machine.

Co-processor memory management

In our current implementation, memory on the co-processor is used for two purposes: thread stacks and method bytecode. Memory on both is allocated in a similar way. The memory is split to a stack region and a method region when the virtual machine starts. We currently split the available memory simply in half, but in practice, the amount of memory required for stacks is not very large. Programs that do not use heavy recursion usually do not require stacks larger than 10 kilobytes [63].

Memory is allocated in fixed-size pages. This makes reclaiming and swapping out memory easier. The virtual machine must swap out memory from the co-processor if it runs out of memory allocated for stack pages. Methods are never swapped out, because they are always stored in the host CPU's memory. If the virtual machine has to allocate a new method page, it simply overwrites the least recently used page.

We store some information in software for each page. First, we use a bit vector in which one indicates a page that is currently in use. Second, we use an array to store the time that the page was last used and certain information used when swapping out pages. The time is updated every time the page is used by the software.

In order to allocate pages, the virtual machine first checks if there are unused pages in the region required. If there is, this page is returned. Otherwise, the virtual machine finds the least recently used page, swaps it out to the host CPU's memory if it is a stack page and returns that. We use a simple "sliding window" algorithm to allocate multiple pages.

If a stack page refers to another page that gets swapped out, the referring page must be updated to prevent return statements from returning to a page used by another thread. Therefore, if there is a stack frame with a pointer to the swapped out page, the previous frame's program counter in that frame is set to the magic number mentioned in an earlier section. When this return is executed, the page is swapped back in. Swapping out method pages is easier, because the only thing that needs to be done is to remove the mapping from the swapped out methods to the pages.

Garbage collection considerations

Garbage collection means finding the set of objects that are reachable and reclaiming the memory used by unreachable objects. Reachable objects are ones that are referred to from the thread stacks, the local variables, static member variables or other reachable objects. Most JVMs implement garbage collection to ensure that unreachable objects will not cause the virtual machine to run out of memory.

Although the garbage collector can be modified to run in parallel with normal execution, the rest of the system is usually stopped for garbage collection. Garbage collection is started when the virtual machine runs out of heap memory. Since the virtual machine is often running multiple threads, the garbage collector has to wait until each of these has stopped running. The individual threads therefore need to poll the garbage collector at certain points during execution to check if garbage collection is starting. These points usually include method invocations and backwards jumps. Polling at backwards jumps is important because without it, an infinite or very long loop could prevent garbage collection and stall the whole virtual machine.

When the code is being executed on a co-processor, polling for garbage collection at backwards jumps becomes impractical and time-consuming. However, since the co-processor thread scheduler is guaranteed to halt any execution at some point after the thread's time slice runs out, this is not a problem. Polling can be done every time a thread releases the lock on the co-processor or interpreter it owns. This way, every thread is guaranteed to stop at some point.

The virtual machine also needs to be able to locate the current stack frames for garbage collection. We use a simple array to store for each stack page the information of whether the page is loaded in, what its hardware address is and if it is not loaded in, what its contents are. All stack frames are found by traversing backwards from the top stack frame until the stack bottom is reached.

Current status and future work

The virtual machine has been implemented with an FPGA-based co-processor and PC-based software portion. This system provides all the basic functionality of the final system, and runs at 100 MHz. Currently the system does not provide reasonable execution speed due to massive communication overhead. This overhead is caused by the link between the FPGA board and the PC. The link utilizes standard parallel port, and achieves maximum data rate of about 80,000 bytes per second in burst mode. The main reasons for building this prototype include validating the co-processor concept and facilitating early testing of hardware and software cooperation. These goals have been met, and the system passes several test programs, including Embedded Caffeine Mark [62]. The limitations in the system are mostly due to the GNU Classpath implementation, which is not complete at this time. Two different local memory strategies have been tested with this system. The first version used a 32 megabyte external SDRAM chip for local memory whereas the second used only 49,152 bytes of physically local memory. The smaller memory was implemented using single clock cycle Block RAMs in the FPGA chip, providing massive throughtput advantage over the 70 ns external memory. In all our tests so far the smaller memory size has been adequate.

Our next step is moving the co-processor and the software portion to the same chip. This will be done using a Xilinx ML310 demonstration board housing a Virtex II Pro FPGA chip with an embedded PowerPC 405 core. This PowerPC core will be executing the software portion, and the co-processor logic will be implemented in the FPGA fabric. With this setup we will see the true performance of the system more accurately, as the communication data rate will be more appropriate. This system also models the whole target domain more accurately, as the raw calculation power of a modern PC is far superior in comparison to the relatively slow processors used in embedded systems and portable devices.

In the software portion we plan to develop a smaller memory footprint version of both the libraries and the software portion of the JVM. The target would be somewhat similar to the J2ME [397] provided by Sun for small embedded devices. Also garbage collection and possibilities for hardware assistance in it will be investigated further. We will take a closer look on the real-time performance [420] in this context. Since current embedded systems are required to run several application programs at the same time, it makes sense to study the effects of creating a multitasking JVM using several co-processors. Please note that multitasking means having several applications running at the same time on the same system, while multithreading means having several threads of execution in a single task.

As the final step, we will design an asynchronous implementation of the co-processor using the Haste design language and Timeless Design Environment (TiDE) toolset for asynchronous design by Handshake Solutions [175]. The goal will be to achieve a sufficient operation speed with very low-power consumption. To demonstrate performance of our asynchronous Java core in its intended environment, we will establish an NoC case study in which a number of Java co-processors and general-purpose CPUs, and possibly some other functional units, are integrated into a single chip. We plan to use the communication platform described in [266], as it provides support for both synchronous and asynchronous cores.

Summary

This chapter shortly described the strategies used in JVM implementations, along with some examples of using hardware solutions. To address the issues that were discovered to be sources of inefficiency, a co-processor architecture was presented. In order to make use of the co-processor, the software portion of the JVM was also discussed. The current status of implementing the co-processor and resulting virtual machine were described, with final notes outlining the future efforts related to the REAL-Java co-processor.

14 Stream Multicore Processors

Michael Bedford Taylor, Walter Lee, Jason Eric Miller, David Wentzlaff, Ian Bratt, Ben Greenwald, Henry Hoffmann, Paul Johnson, Jason Kim, James Psota, Arvind Saraf, Nathan Shnidman, Volker Strumpen, Matt Frank, Rodric Rabbah, Saman Amarasinghe, and Anant Agarwal

Massachusetts Institute of Technology

Introduction

The physical realities of wire delay and power consumption seriously challenge the ability of microprocessor designers to continue designing monolithic architectures with centralized resources. Materials and process changes have proven insufficient to solve the fundamental physics problems, and it is increasingly challenging for existing architectures to turn chip resources into higher performance, at tractable costs. Fast moving VLSI technology will soon offer tens of billions of transistors, massive chip-level wire bandwidth for local interconnect, and a modestly larger number of pins. Processors need to convert the abundant chip-level resources into power-efficient application performance, while mitigating the negative effects of wire delays.

This chapter discusses the architecture of the Raw Microprocessor, an early multicore processor developed at MIT [411,447]. Raw is a *tiled multicore architecture* containing 16 homogeneous tiles arranged in a grid. Each tile contains a processor, caches, and several mesh routers. Raw is a general-purpose multicore architecture in that it supports various models of computation including instruction-level parallelism (ILP), streaming, data-level parallelism (parallelism:DLP), and thread-level parallelism (TLP). Raw's point-to-point interconnection networks between tiles support these models by routing both scalar operands and streams with extremely low latency between architecturally exposed function units. The Raw chip was

J. Nurmi (ed.), Processor Design: System-on-Chip Computing for ASICs and FPGAs, 309–338.

Fig. 14.1. The Raw board. The Raw multicore chip is in the center. DRAMs, PCI slots, and a support chipset implemented in FPGAs surround the Raw chip.

successfully fabricated and demonstrated in 2002. Figure 14.1 shows a photograph of the Raw system prototype.

The Raw processor is an early prototype in a growing trend for processor design toward chip multiprocessors, or as commonly called today: multicore processors. For example, the IBM/Toshiba/Sony Cell processor has 9 cores [199], the Sun Niagara has 8 cores [235], the Cavium Octeon has 16 cores [472], RMI XLR732 has 8 cores [347], the IBM/Microsoft Xbox 360 CPU has 3 cores [17], and most vendors are shipping dual-core chips. In addition, Intel has demonstrated an 80 core prototype code named Polaris [215], and Cisco has described a next-generation network processor containing 192 Tensilica Xtensa cores [111].

Multicore architectures address the physical challenges of power and wire delay by favoring several simpler cores over a large monolithic processor. Tiled multicore processors further arrange the abundant on-chip resources – including logic, wires, and pins – in a scalable tiled pattern. Tiled multicore architectures replace global-access centralized structures (such as giant register files, buses, and centralized caches), where wire delay and power efficiency scale poorly, with small distributed structures (such as mesh-based on-chip networks) that facilitate efficient local accesses.

The tiled multicore architecture methodology is equally effective at improving power efficiency because it replaces a large monolithic core by several smaller voltage-scaled cores. Small decentralized structures are known to be significantly more power efficient compared to a large monolithic structure. Thus, while Raw was not specifically designed as a low-power architecture, it is 80% more efficient in terms of ops/Watt compared to a monolithic (centralized) processor with equivalent resources.

Design philosophy

Just as VLSI advances have created an opportunity for massively parallel multicore processors, they have also expanded the number of applications that are implementable in application-specific integrated circuits (ASICs). Early studies [31,447] to determine the factors responsible for the significantly better performance of application-specific VLSI chips revealed four main factors: specialization; exploitation of parallel resources (gates, wires, and pins); management of wires and wire delay; and management of pins. The goal of multicore processors is to leverage these four factors, and yet implement the gamut of general-purpose features such as functional unit virtualization, unpredictable interrupts, instruction virtualization, and data caching. The processor also needs to exploit ILP in sequential programs, and space and time multiplex (i.e., context switch) threads of control for thread and task-level parallelism.

1. **Specialization:** ASICs specialize each "operation" at the gate level. In both the VLSI circuit and microprocessor context, an operation roughly corresponds to the unit of work that can be done in one cycle. A VLSI circuit forms operations by combinational logic paths, or "operators", between flip-flops. A microprocessor, on the other hand, has an instruction set that defines the operations that can be performed. Specialized operators – for example, implementing an incompatible floating point operation, or implementing a linear feedback shift register – can yield an order of magnitude performance improvement over an extant general-purpose processor that may require many instructions to perform the same one-cycle operation as the VLSI hardware. As an example, customized Tensilica ASIC processors take advantage of specialization by augmenting a general-purpose processor core with specialized instructions for specific applications.
2. **Exploitation of parallel resources:** ASICs further exploit plentiful silicon area to implement enough operators and communications channels to sustain a tremendous number of parallel operations in each clock cycle. Applications that merit direct digital VLSI circuit implementations typically exhibit massive, operation-level parallelism. While an aggressive VLIW implementation like Intel's Itanium II [303] executes six instructions per cycle, graphics accelerators may perform hundreds or thousands of word-level operations per cycle. Because they operate

on very small word operands, logic emulation circuits such as Xilinx-II Pro FPGAs can perform hundreds of thousands of operations each cycle. The addition of MMX and SSE-style multigranular instructions that operate on multiple subwords marks an effort to improve the efficiency of microprocessors by exploiting additional parallelism available due to smaller word sizes.

The Itanium II die photo reveals that less than two percent of the die area is dedicated to its 6-way issue integer execution core. Clearly, the ALU area is not a significant constraint on the execution width of a modern-day wide-issue microprocessor. On the other hand, the presence of many physical execution units is a minimum prerequisite to the exploitation of the same massive parallelism that ASICs are able to exploit.

1. **Management of wires and wire delay:** ASIC designers can place and wire communicating operations in ways that minimize wire delay, minimize latency, and maximize bandwidth. In contrast, it is now well known that the delay of the interconnect inside traditional microprocessors limits scalability [2,196,321,361,412]. Itanium II's 6-way integer execution unit presents evidence for this – it spends over half of its critical path in the bypass paths of the ALUs. ASIC designers manage wire delay inherent in large distributed arrays of function units in multiple steps. First, they place close together operations that need to communicate frequently. Second, when high bandwidth is needed, they create multiple customized communication channels. Finally, they introduce pipeline registers between distant operators, thereby converting propagation delay into pipeline latency. By doing so, the designer acknowledges the inherent tradeoff between parallelism and latency: leveraging more resources requires signals to travel greater distances. The Alpha 21264 is an example of a microprocessor that acknowledges this tradeoff on a small scale: it incurs a one-cycle latency for signals to travel between its two integer clusters.
2. **Management of pins:** ASICs customize the usage of their pins. Rather than being bottlenecked by a cache-oriented multilevel hierarchical memory system (and subsequently by a generic PCI-style I/O system), ASICs utilize their pins in ways that fit the applications at hand, maximizing realizable I/O bandwidth or minimizing latency. This efficiency applies not just when an ASIC accesses external DRAMs, but also in the way that it connects to high-bandwidth input devices like wide-word analog-to-digital converters, CCDs, and sensor arrays. There are currently few easy ways to arrange for these devices to stream data into a general-purpose microprocessor in a high-bandwidth way, especially

since DRAM must almost always be used as an intermediate buffer. In some senses, microprocessors strive to minimize, rather than maximize, the usage of pin resources, by hiding them through a hierarchy of caches.

The Raw processor

The Raw processor addresses the challenge of whether future general-purpose multicore architecture could be built to run a wide range of applications with reasonable performance in the face of increasing wire delays. This chapter evaluates the Raw microprocessor and discusses its success in achieving these goals. Raw takes the following approach to leveraging the four factors behind the success of ASICs.

1. Raw implements the most common operations needed by ILP or stream applications in specialized hardware mechanisms. Most of the primitive mechanisms are exposed to software through a new ISA. These mechanisms include the usual integer and floating point operations, specialized bit manipulation operations, scalar operand routing between adjacent function units, operand bypass between function units, registers and I/O queues, and data cache access (i.e., data load with tag check).
2. Raw implements a large number of these operators which exploit the copious VLSI resources, including gates, wires and pins, and exposes them through a new ISA, such that the software can take advantage of them for both ILP and highly parallel applications.
3. Raw manages the effect of wire delays by exposing the wiring channel operators to the software, so that the software can account for latencies by orchestrating both scalar and stream data transport. By orchestrating operand flow on the interconnect, Raw can also create customized communications patterns. Taken together, the wiring channel operators provide the abstraction of a scalar operand network [48] that offers very low latency for scalar data transport and enables the exploitation of ILP.
4. Raw software manages the pins for cache data fetches and for specialized stream interfaces to DRAM or I/O devices.

The software crisis

The proliferation of multicore architectures has given rise to a software crisis. This crisis is two-fold. First, while multicore architectures excel at parallel processing, they seldom support legacy single-threaded applications unless the applications are recompiled to take advantage of the on-chip parallelism. Recent work [452], however, has shown that the parallel resources in multicore processors facilitate the execution of legacy codes (with acceptable performance) through the use of dynamic binary translation engines. The engines can spatially implement the components of traditional superscalar processors across the distinct cores in a multicore processor.

The second factor contributing to the software crisis is the fact that very few programmers know how to program the massive on-chip parallelism afforded by multicore systems. Existing programming models are largely inadequate for parallel programming by non-expert end-users who are, for the most part, trained in classical von Neumann imperative languages such as C or Java. The von Neumann model provides a simple abstraction of a single thread of control reading and writing data sequentially from memory. With parallel processing, there are multiple threads of control that requires thoughtful orchestration to deal with non-determinism, and avoid deadlocks and data races.

Several new programming models have emerged (or re-emerged) in response to the parallel programming software challenge. Some of these models (e.g., Fortress [9], X10 [73], UPC [430]) are designed to target the general-purpose programming population, while other models are domain-specific and target important application areas.

The stream programming paradigm is a promising example. Stream programming breaks the von Neumann language barrier by encapsulating computation in actors. Each actor has its own thread of control, as well as its own address space to avoid the pitfalls of a shared address space. Data is streamed between the actors (or I/O devices and peripherals) in a producer–consumer fashion, often using FIFO channels (first-in-first-out), much as in power-efficient ASICs. An actor reads data from its input stream, operates on the data, and outputs the results to a new stream that is consumed by another actor. Architectures like Raw support the streaming programming model by allowing streams to be efficiently carried over the interconnection network between tiles for processing by the actors resident on the cores.

Streaming offers an approach for exposing parallelism and communication in a manner suitable for mapping programs to multicore architectures without the heroic efforts needed to extract parallelism from von Neumann languages. The stream programming model is motivated by trends in the application space toward network processing, image, voice, and multimedia programs, cryptography, and security. Stream processing is increasingly crucial to a plethora of embedded systems, including handheld computers,

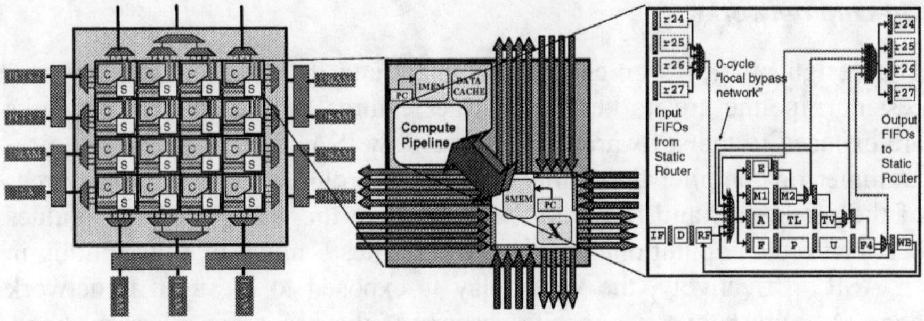

Fig. 14.2. The Raw microprocessor comprises 16 tiles. Each tile has a compute processor, routers, network wires, and instruction and data memories.

cell phones, and DSPs. And there is already growing evidence that streaming applications consume a substantial fraction of the computation cycles on consumer machines [355].

Raw architecture overview

The Raw architecture supports an ISA that provides a parallel interface to the gate, pin, and wiring resources of the chip through suitable high level-abstractions. As illustrated in Figure 14.2, the Raw processor exposes the copious gate resources of the chip by dividing the usable silicon area into an array of 16 identical, programmable tiles. A tile embodies a single core, with an 8-stage in-order single-issue MIPS-style processing pipeline, a 4-stage single-precision pipelined FPU, a 32KB data cache, two types of communication routers (static and dynamic) and 32KB and 64KB of software-managed instruction caches for the processing pipeline and static router, respectively. Each tile is sized so that the amount of time for a signal to travel through a small amount of logic and across the tile is one clock cycle. Future Raw processors may have hundreds or even thousands of tiles.

The tiles are interconnected by 4 32-bit full-duplex on-chip networks, consisting of over 12,500 wires (see Figure 14.2). Two of the networks are static (routes are specified at compile time) and two are dynamic (routes are specified at run time). Each tile is connected only to its four neighbors. Every wire is registered at the input to its destination tile. This means that *the longest wire in the system is no longer than the length or width of a tile*. This property ensures high clock speeds, and the continued scalability of the architecture.

On-chip networks

The design of Raw's on-chip interconnect and its interface with the processing pipeline are its key innovative features. These on-chip networks are exposed to the software through the Raw ISA, thereby giving the programmer or compiler the ability to directly program the wiring resources of the processor, and to carefully orchestrate the transfer of data values between the computational portions of the tiles – much like the routing in an ASIC. Effectively, the wire delay is exposed to the user as network hops. A route between opposite corners of the processor takes six hops, which corresponds to approximately six cycles of wire delay. To minimize the latency of inter-tile scalar data transport (which is critical for ILP) the on-chip networks are not only register mapped but also integrated directly into the bypass paths of the processor pipeline. The register mapped ports allow an instruction to place a value on the network with no overhead. Similarly, instructions using values from the network simply read from the register mapped ports. The programmable switches bear the responsibility of routing operands through the network.

Raw's on-chip interconnects are examples of scalar operand networks [412], which provide an interesting way of looking at modern day processors. The register file used to be the central communication mechanism between functional units in a processor. Starting with the first pipelined processors, the bypass network has become largely responsible for the communication of active values and the register file is more of a checkpointing facility for inactive values. The Raw networks, and in particular the static networks, serve as 2D bypass networks between tiles.

The static router in each tile contains a 64KB software-managed instruction cache and a pair of routing crossbars. Compiler generated routing instructions are 64 bits and encode a small command (e.g., conditional branch with/without decrement) and several routes, one for each crossbar output. Each Raw static router, also known as a switch processor, contains a 4-way crossbar, with each way corresponding to one of the cardinal directions (north, east, south, and west). The single-cycle routing instructions are one example of Raw's use of specialization. Because the router program memory is cached, there is no practical architectural limit on the number of simultaneous communication patterns that can be supported in a computation. This feature, coupled with the extremely low latency and low occupancy of the in-order inter-tile ALU-to-ALU operand delivery (three cycles nearest neighbor) distinguishes Raw from prior systolic or message passing systems [18,164,245].

The switch processor instructions may route operands already on the network (for example, reading from the north and sending south), inject

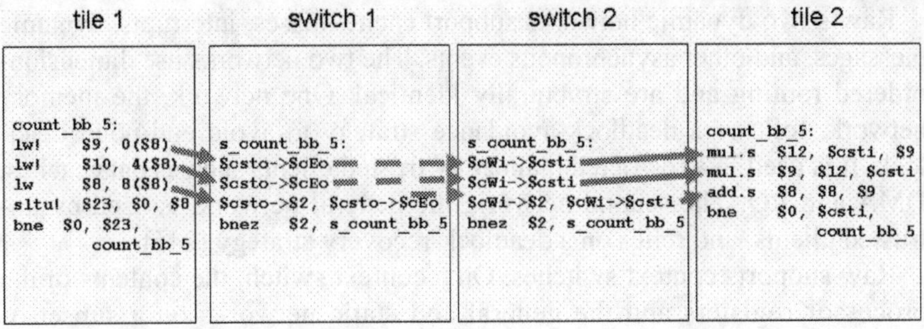

Fig. 14.3. Example of a 2-tile Raw program. Dashed edges represent dependences between switch instructions and solid edges represent dependences between a processor and its switch.

operands from the processor into the network, or drain operands from the network and send them to the processor or to a local register. If the switch wants to pull the incoming value from the north and send it west, the instruction looks like *$cNi→$cWo*, where the *i* and *o* represent input and output, and the *N* and *W* represent north and west, respectively. The input ports (*$cNi, $cWi, $cEi, $cSi*) each have 4-element FIFOs to buffer incoming data. To send a value from the network to the processor, the switch must read from one of the input ports or a local register and write to *$csti*. The FIFO between switch and processor allows the switch to write four words of data before the processor reads any values. To read a value from the switch, the processor simply reads from *$csti*.

Figure 14.3 shows an example of a 2-tile Raw program where the two tiles are located next to each other in a horizontal row. The program contains processor code and router code for each tile. Both tiles and switches have the same control flow, with one tile calculating the branch condition and using the switches to propagate the condition to both switches and the other tile. Solid edges represent data dependences between a processor and the local switch. The dashed edges represent data dependences between the two switches. In the figure, instruction opcodes containing a "!" symbol write to the register mapped network port called *$csto*. Physically, *$csto* is implemented via a queue. The processor inserts words into the queue and the switch reads words from the queue. The switch may place values read from the *$csto* queue onto the network or into a local switch register. The queue contains eight words of storage space allowing the processor to write up to eight words to the register mapped port before the switch reads any values. The Raw ISA provides a *route* opcode for moving operands around. For brevity and space, this opcode is omitted from the example.

Raw's two dynamic networks support cache misses, interrupts, dynamic messages, and other asynchronous events. The two networks use dimension-ordered routing and are structurally identical. One network, the memory network, follows a deadlock-avoidance strategy to avoid end-point deadlock. It is used in a restricted manner by trusted clients such as data caches, DMA and I/O. The second network, the general network, is used by untrusted clients, and relies on a deadlock-recovery strategy [245].

Raw supports context switches. On a context switch, the contents of the processor registers and the general and static networks on a subset of the Raw chip occupied by the process (possibly including multiple tiles) are saved off and the process and its network data can be restored at any time to a new offset on the Raw grid.

Direct I/O interfaces

On the edges of the network, the network channels are multiplexed down onto the pins of the chip to form flexible I/O ports that can be used for DRAM accesses or external device I/O. To toggle a pin, the user programs one of the on-chip networks to route a value off the side of the array. The package is a 1657-pin CCGA (ceramic column–grid array) and provides 14 full-duplex, 32-bit I/O ports. Raw implementations with fewer pins are made possible via logical channels (as is already the case for two out of the 16 logical ports), or simply by bonding out only a subset of the ports.

The static and dynamics networks, the data cache of the compute processors, and the external DRAMs connected to the I/O ports comprise Raw's memory system. The memory network is used for cache-based memory traffic while the static and general dynamic networks are used for stream-based memory traffic. Systems designed for memory-intensive applications can have up to 14 full-duplex full-bandwidth DRAM banks by placing one on each of the chip's 14 physical I/O ports. Minimal embedded Raw systems may eliminate DRAM altogether: booting from a single ROM and executing programs entirely out of the on-chip memories. In addition to transferring data directly to the tiles, off-chip devices connected to the I/O ports can route data through the on-chip networks to other devices in order to perform glueless DMA and peer-to-peer communication.

ISA analogs to physical resources

By creating first class architectural analogs to the physical chip resources, Raw attempts to minimize the ISA gap – that is, the gap between the resources that a VLSI chip has available and the amount of resources that are

Table 14.1. How Raw converts increasing quantities of physical entities into ISA entities.

Physical Entity	Raw ISA Analog	Conventional ISA Analog
Gates	Tiles, new instructions	New instructions
Wires, Wire delay	Routes, network hops	none
Pins	I/O ports	none

usable by software. Unlike conventional ISAs, Raw exposes the quantity of all three underlying physical resources (gates, wires and pins) in the ISA. Furthermore, it does this in a manner that is backwards-compatible – the instruction set does not change with varying degrees of resources.

Table 14.1 contrasts the ways that the Raw ISA and conventional ISAs expose physical resources to the programmer. Because the Raw ISA has more direct interfaces, Raw processors can have more functional units and more flexible and efficient pin utilization. High-end Raw processors will typically have more pins, because the architecture is better at turning pin count into performance and functionality. Finally, Raw processors are more predictable and have higher frequencies because of the explicit exposure of wire delay.

This approach makes Raw scalable. Creating subsequent, more powerful, generations of the processor is straightforward, and is as simple as stamping out as many tiles and I/O ports as the silicon die and package allow. The design has no centralized resources, no global buses, and no structures that get larger as the tile or pin count increases. Finally, the longest wire, the design complexity, and the verification complexity are all independent of transistor count.

Related architectures

Raw distinguishes itself from others by being a modeless architecture and supporting all forms of parallelism, including ILP, DLP, TLP, and streams. Several other projects have attempted to exploit specific forms of parallelism. These include systolic (iWarp [164]), vector (VIRAM [237]), stream (Imagine [219]), shared-memory (DASH [262]), and message passing (J machine [309]). These machines, however, were not designed for ILP. In contrast, Raw was designed to exploit ILP effectively in addition to these other forms of parallelism. ILP presents a difficult challenge because it requires that the architecture be able to transport scalar operands between logic units with very low latency, even when there are a large number of highly irregular communication patterns. A recent paper [412] employs a 5-tuple to characterize the cost of sending operands between functional units in a number of architectures (see Table 14.5 for a list of the components in this 5-tuple). Qualitatively, larger 5-tuple values represent proportionally more expensive operand transport costs. The large values in the network 5-tuples for iWarp <1,6,5,0,1>, shared memory <1,18,2,14,1>, and message passing <3,7,1,1,12>, compared to the low numbers in the 5-tuples of machines that can exploit ILP (e.g., superscalar <0,0,0,0,0>, Raw <0,1,1,1,0>, Grid <0,0,½,0,0>, and ILDP <0,1,0,1,0>) quantitatively demonstrate the difference. The low 5-tuple of Raw's scalar operand network compared to that of iWarp enables Raw to exploit diverse forms of parallelism, and is a direct consequence of the integration of the interconnect into Raw's pipeline and Raw's early pipeline commit point. We will further discuss the comparison with iWarp here, but see [412] for more details on comparing networks for ILP.

Raw supports statically orchestrated communication like iWarp or NuMesh [375]. iWarp and NuMesh support a small number of fixed communication patterns, and can switch between these patterns quickly. However, establishing a new pattern is more expensive. Raw supports statically orchestrated communication by using a programmable switch that issues an instruction each cycle. The instruction specifies the routes through the switch during that cycle. Because the switch program memory in Raw is large, and virtualized through caching, there is no practical architectural limit on the number of simultaneous communication patterns that can be supported in a computation. This virtualization becomes particularly important for supporting ILP, because switch programs become as large or even larger than the compute programs.

Processors like Grid [304] and ILDP [230] are targeted specifically for ILP and propose using low latency scalar operand networks. Raw shares in their ILP philosophy, and implements a static-transport, point-to-point scalar operand network, while Grid uses a dynamic-transport, point-to-point network, and ILDP uses a broadcast based dynamic-transport network. Both Raw and Grid perform compile time instruction assignment to compute nodes, while ILDP uses dynamic assignment of instruction groups. Raw uses compile-time operand matching, while Grid uses dynamic associative operand matching queues, and ILDP's dynamic scheme uses full-empty bits on distributed register files. Accordingly, using the *AsTrO* categorization (Assignment, Transport, Ordering) from [413], Raw, Grid, and

ILDP can be classified as SSS, SDD, and DDS architectures, respectively, where S stands for static and D for dynamic. Both the Grid and ILDP designs project lower network 5-tuples than Raw, but the final numbers should be forthcoming as their implementations mature. Taken together, Grid, ILDP and Raw represent three distinct points in the scalar operand network design space, ranging from the more compile-time oriented approach as in Raw, to the dynamic approach as in ILDP.

Raw took inspiration from the Multiscalar processor [385], which uses a separate one-dimensional network to forward register values between ALUs. Raw generalizes the basic idea, and supports a two-dimensional programmable mesh network both to forward operands and for other forms of communication.

Both Raw and SmartMemories [273] share the philosophy of an exposed communication architecture, and represent two design points in the space of tiled architectures that can support multiple forms of parallelism. Raw uses homogeneous, programmable static and dynamic mesh networks, while SmartMemories uses programmable static communication within a local collection of nodes, and a dynamic network between these collections of nodes. The node granularities are also different in the two machines. Perhaps the most significant architectural difference, however, is that Raw (like Scale [242]) is *modeless*, while SmartMemories and Grid have modes for different application domains. Another architecture that represents a natural extreme point in modes is Tarantula [121], which implements two distinct types of processing units for ILP and vectors. Raw's research focus is on discovering and implementing a minimal set of primitive mechanisms (e.g., scalar operand network) useful for all forms of parallelism, while the modes approach implements special mechanisms for each form of parallelism. We believe the modeless approach is more area efficient and significantly less complex. We believe the issue of modes versus modeless for versatile processors is likely to be a controversial topic of debate in the forthcoming years.

Finally, like VIRAM and Imagine, Raw supports vector and stream computations, but does so very differently. Both VIRAM and Imagine sport large memories or stream register files on one side of the chip connected via a crossbar interconnect to multiple, deep compute pipelines on the other. The computational model is one that extracts data streams from memory, pipes them through the compute pipelines, and then deposits them back in memory. In contrast, Raw implements many co-located smaller memories and compute elements, interconnected by a mesh network. The Raw computational model is more ASIC-like in that it streams data through the pins and on-chip network to the ALUs, continues through the network to more

ALUs, and finally through the network to the pins. Raw's ALUs also can store data temporarily in the local memories if necessary. We believe the lower latencies of the memories in Raw, together with the tight integration of the on-chip network with the compute pipelines, make Raw more suitable for ILP.

Raw chip implementation

The Raw chip is a 16-tile prototype implemented in IBM's 180 nm 1.8 V 6-layer CMOS 7SF SA-27E copper process. Although the Raw array is only 16 mm × 16 mm, an 18.2 mm × 18.2 mm die is used to allow for the high pin-count package. The 1657-pin ceramic column grid array (CCGA) package provides 1080 high speed transceiver logic (HSTL) I/O pins. Measurements indicate that the chip core averages 18.2 W at 425 MHz (unused functional units, memories, and tri-state unused data I/O pins are quiesced). The target clock frequency was 225 MHz under worst-case conditions, which is competitive with other 180 nm lithography ASIC processors, such as VIRAM, Imagine, and Tensilica's Xtensa series. The nominal running frequency is typically higher – the Raw chip core, running at room temperature, reaches 425 MHz at 1.8 V, and 500 MHz at 2.2 V. This compares favorably to IBM-implemented microprocessors in the same process: the PowerPC 405GP runs at 266–400 MHz, while the follow-on PowerPC 440GP reaches 400–500 MHz.

The processor is aggressively pipelined, with conservative treatment of the control paths in order to ensure that only reasonable efforts would be required to close timing in the backend. Despite these efforts, wire delay inside a tile was still large enough to warrant a special infrastructure to place the cells in the timing and congestion-critical data paths. More details on the Raw implementation are available in [411].

As one can infer from the empirical results that follow, moving from a single issue compute processor to a two-issue compute processor would have likely improved performance on low-ILP applications. Estimates indicate that such a compute processor would have easily fit in the remaining empty space within a tile. The frequency impact of transitioning from 1-issue to 2-issue is generally held to be small.

A prototype motherboard (shown in Figure 14.4) using the Raw chip was designed in collaboration with the Information Sciences Institute (ISI) East. A larger system, consisting of 64 Raw chips, connected to form a virtual 1024-tile Raw processor, is also being fabricated in conjunction with ISI East.

Fig. 14.4. Photos of the Raw chip (top) and Raw prototype motherboard (bottom).

Methodology for performance analysis

The evaluation presented here makes use of a validated cycle-accurate simulator of the Raw chip. The evaluation is focused on three application domains that are predominant in embedded systems: stream, ILP, and bit-level computing.

Using the validated simulator, as opposed to the actual hardware, facilitates the normalization of differences with a reference system, e.g., DRAM memory latency, and instruction cache configuration. It also allows for exploration of alternative motherboard configurations. The simulator was meticulously verified against the gate-level RTL netlist to have *exactly* the same timing and data values for all 200,000 lines of the handwritten assembly-code test suite, as well as for a number of C applications and randomly generated tests. Every stall signal, register-file write, SRAM write, on-chip network wire, cache state-machine transition, interrupt signal, and chip signal pin matches in value on every cycle between the two. This gate-level RTL netlist was then shipped to IBM for manufacturing. Upon receipt of the chip, a subset of the tests was compared to the actual hardware to verify that the chip was manufactured according to specification.

Reference processor

It is important for the evaluation to ground the empirical data to an existing commercial system. For fairness, this comparison system must be implemented in a process that uses the same lithography generation, 180 nm. Furthermore, the reference processor needs to be measured at a similar point in its lifecycle, i.e., as close to first silicon as possible. This is because most commercial systems are speedpath or process tuned after first silicon is created [79]. For instance, the 180 nm Intel Pentium 3 (P3) initial production silicon was released at 500–733 MHz and gradually was tuned until it reached a final production frequency of 1 GHz. The first-silicon value for the P3 is not publicly known. However, the frequencies of first-silicon and initial-production silicon have been known to differ by as much as 2x.

The P3 is especially well suited for comparison with Raw because it is in common use, because its fabrication process is well documented, and because the common-case functional unit latencies are almost identical. The back ends of the processors share a similar level of pipelining, which means that relative cycle-counts carry some significance. Conventional VLSI wisdom suggests that, when normalized for process, Raw's single-ported L1 data cache should have approximately the same area and delay as the P3's dual-ported L1 data cache of half the size. For sequential codes with working sets that fit in the L1 caches, the cycle counts should be quite similar. And given that the fortunes of Intel have rested (and continue to rest, with the Pentium-M reincarnation) upon this architecture for almost ten years, there is reason to believe that the implementation is a good one. In fact, the P3, upon release in 4Q'99, had the highest SpecInt95 value of any processor [171].

Normalization details

The selection of a reference CPU implementation is followed by a selection of an enclosing computer. A pair of 600 MHz Dell Precision 410 workstations were used to run our reference benchmarks. These machines were outfitted with identical 100 MHz 2-2-2 PC100 256 MB DRAMs, and several microbenchmarks were used to verify that the memory system timings matched.

To compare the Raw and Dell systems more equally, the Raw simulator extension language was used to implement a cycle-matched PC100 DRAM model and a chipset. This model has the same wall-clock latency and bandwidth as the Dell 410. However, since Raw runs at a slower frequency than the P3, the latency, measured in cycles, is less. The term *RawPC* is used to describe a simulation which uses eight PC100 DRAMs, occupying four ports on the left-hand side of the chip, and four on the right-hand side.

Because Raw is designed for streaming applications, it is necessary to measure applications that use the full pin bandwidth of the chip. In this case, a simulation of CL2 PC 3500 DDR DRAM was used as it provides enough bandwidth to saturate both directions of a Raw port. There are 14 physical ports on the Raw chip and 16 logical ports: 1 logical port for each tile side that borders the chip periphery. A few of the logical ports share the same physical port. The simulation is configured to use 16 PC 3500 DRAM modules, 1 for each of the 16 logical ports on the chip, in conjunction with 16 memory controllers, implemented in the chipset, that support a number of stream requests. This configuration is called *RawStreams*. A Raw tile can send a message over the general dynamic network to the chipset to initiate large bulk transfers from the DRAMs into and out of the static network. Simple interleaving and striding is supported, subject to the underlying access and timing constraints of the DRAM.

The placement of a DRAM on a Raw port does not exclude the use of other devices on that port – the chipsets have a simple demultiplexing mechanism that allows multiple devices to connect to a single port.

Except where otherwise noted, gcc 3.3 -O3 was used to compile C and Fortran code for both Raw[1] and the P3.[2] For programs that do C or Fortran stdio calls, newlib 1.9.0 was used for both Raw and the P3. Finally, to eliminate the impact of disparate file and operating systems, the results of I/O system calls for the Spec benchmarks were captured and embedded into the binaries as static data using [410].

One final normalization was performed to enable comparisons with the P3. The cycle-accurate simulator was augmented to employ conventional 2-way associative hardware instruction caching. These instruction caches are modeled cycle-by-cycle in the same manner as the rest of the hardware. Like the data caches, they service misses over the memory dynamic network. Resource contention between the caches is modeled accordingly.

Tables 14.2 and 14.3 show functional unit timings and memory system characteristics for both systems, respectively. Table 14.4 shows Raw's measured power consumption [231]. Table 14.5 lists a breakdown of the end-to-end message latency on Raw's scalar operand network. The low 3-cycle inter-tile ALU-to-ALU latency and zero cycle send and receive occupancies are critical for obtaining good ILP performance.

Table 14.2. Functional unit timings. Commonly executed instructions appear first. FP operations are single precision.

Operation	Latency		Throughput	
	1 Raw Tile	P3	Raw	P3
ALU	1	1	1	1
Load (hit)	3	3	1	1
Store (hit)	–	–	1	1
FP Add	4	3	1	1
FP Mul	4	5	1	1/2
Mul	2	4	1	1
Div	42	26	1	1
FP Div	10	18	1/10	1/18
SSE FP 4-Add	–	4	–	1/2
SSE FP 4-Mul	–	5	–	1/2
SSE FP 4-Div	–	36	–	1/36

[1] The Raw gcc backend, based on the MIPS backend, targets a single tile's compute and network resources.

[2] For P3, the -march=pentium3 -mfpmath=sse flags were added.

Table 14.3. Memory system data.

	1 Raw Tile	P3
CPU frequency	425 MHz	600 MHz
Sustained issue width	1 in-order	3 out-of-order
Mispredict penalty	3	10–15
DRAM freq (RawPC)	100 MHz	100 MHz
DRAM freq (RawStreams)	2 × 213 MHz	–
DRAM access width	8 bytes	8 bytes
L1 D cache size	32K	16K
L1 D cache ports	1	2
L1 I cache size	32K	16K
L1 miss latency	54 cycles	7 cycles
L1 fill width	4 bytes	32 bytes
L1/L2 line sizes	32 bytes	32 bytes
L1 associativities	2-way	4-way
L2 size	–	256K
L2 associativity	–	8-way
L2 miss latency	–	79 cycles
L2 fill width	–	8 bytes

Table 14.4. Raw power consumption at 425 MHz, 25°C.

	Core	Pins
Idle – full chip	9.6 W	0.02 W
Average – per active tile	0.54 W	–
Average – per active port	–	0.2 W
Average – full chip	18.2 W	2.8 W

Table 14.5. Breakdown of the end-to-end latency (in cycles) for a one-word message on Raw's static network.

	Latency
Sending processor occupancy	0
Latency from ALU output to network	0
Latency per hop	1
Latency from network to ALU input	2
Receiving processor occupancy	0

Stream computation

Stream computations arise naturally out of real-time I/O applications as well as from embedded applications. The data sets for these applications are often large and may even be a continuous stream in real-time, which makes them unsuitable for traditional cache based memory systems. Raw provides more natural support for stream-based computation by allowing data to be fetched efficiently through a register-mapped, software-orchestrated network.

We present two sets of results for stream computation on Raw. First we show the performance of programs written in a high-level stream language called StreamIt. The applications are automatically compiled to Raw. Then, we show the performance of some handwritten streaming applications.

StreamIt

StreamIt is a high-level, architecture-independent language for high-performance streaming applications. StreamIt contains language constructs that improve programmer productivity for streaming, including hierarchical structured streams, graph parameterization, and circular buffer management. These constructs also expose information to the compiler and enable novel optimizations [421]. The StreamIt programming model allows the programmer to build an application by connecting components together into a stream graph, where the nodes represent filters that transform the data communicated along the edges.

The StreamIt compiler includes a Raw backend that performs automatic load balancing, graph layout, communication scheduling and routing

[160,161]. The partitioning adjusts the granularity of a stream graph to match the number of tiles in the target architecture. A layout phase then maps the partitioned stream graph to a given network topology. Lastly, the scheduling and routing generates a fine-grained static communication pattern for each computational element. The results presented here leverage the on-chip static networks for routing data between cores: filters are fused together to match the number of Raw cores, and then layed out in space with data streamed between the cores. There are alternative ways of performing stream computation on Raw that also take advantage of the dynamic network and its massive bandwidth off chip. The interested reader can review [160] for details.

StreamIt experiments

We evaluate the performance of RawPC on several StreamIt benchmarks, which represent large and pervasive DSP applications. Table 14.6 summarizes the performance of 16 Raw tiles vs. a P3. For both architectures, we use StreamIt versions of the benchmarks; we do not compare to hand-coded C on the P3 because StreamIt performs at least 1-2X better for four of the six applications (this is due to aggressive unrolling and constant propagation in the StreamIt compiler). The comparison reflects two distinct influences: (1) the scaling of Raw performance as the number of tiles increases, and (2) the performance of a Raw tile vs. a P3 for the same StreamIt code. To distinguish between these influences, Table 14.7 shows detailed speedups relative to StreamIt code running on a 1-tile Raw configuration.

Table 14.6. StreamIt performance results.

Benchmark	Cycles Per Outputon Raw	Speedup vs. P3	
		Cycles	Time
Beamformer	2074.5	7.3	5.2
Bitonic sort	11.6	4.9	3.5
FFT	16.4	6.7	4.8
Filterbank	305.6	15.4	10.9
FIR	51.0	11.6	8.2
FMRadio	2614.0	9.0	6.4

Table 14.7. Speedup (in cycles) of StreamIt benchmarks relative to a 1-tile Raw configuration. From left, the columns indicate the StreamIt version on a P3, and on Raw configurations with one to 16 tiles.

Benchmark	StreamIt on *n* Raw tiles					StreamIt on P3
	1	2	4	8	16	
Beamformer	1.0	4.1	4.5	5.2	21.8	3.0
Bitonic sort	1.0	1.9	3.4	4.7	6.3	1.3
FFT	1.0	1.6	3.5	4.8	7.3	1.1
Filterbank	1.0	3.3	3.3	11.0	23.4	1.5
FIR	1.0	2.3	5.5	12.9	30.1	2.6
FMRadio	1.0	1.0	1.2	4.0	10.9	1.2

The primary result illustrated by Table 14.7 is that StreamIt applications scale effectively for increasing sizes of the Raw configuration. For FIR, FFT, and Bitonic, the scaling is approximately linear across all tile sizes (FIR is actually super-linear due to decreasing register pressure in larger configurations). For Beamformer, Filterbank, and FMRadio, the scaling is slightly inhibited for small configurations. This is because (1) these applications are larger, and IMEM constraints prevent an unrolling optimization for small tile sizes, and (2) they have more data parallelism, yielding speedups for large configurations but inhibiting small configurations due to a constant control overhead.

The second influence is the performance of a P3 vs. a single Raw tile on the same StreamIt code, as illustrated by the second column in Table 14.7. In most cases, performance is comparable. The P3 performs better in two cases because it can exploit ILP: Beamformer has independent real/imaginary updates in the inner loop, and FIR is a fully unrolled multiply-accumulate operation. In other cases, ILP is obscured by circular buffer accesses and control dependences.

In all, StreamIt applications benefit from Raw's exploitation of parallel resources and management of wires. The abundant parallelism and regular communication patterns in stream programs are an ideal match for the parallelism and tightly orchestrated communication on Raw. As stream programs often require high bandwidth, register-mapped communication serves to avoid costly memory accesses. Also, autonomous streaming components can manage their local state in Raw's distributed data caches and register banks, thereby improving locality. These aspects are key to the scalability demonstrated in the StreamIt benchmarks.

Hand-optimized stream applications

ISI East, the MIT Oxygen Team, and MIT CAG have coded and manually tuned a wide range of streaming applications to take advantage of Raw as an embedded processor. These include a set of linear algebra routines implemented as Stream Algorithms, the STREAM benchmark, and several other embedded applications including a real-time 1020-node acoustic beamformer. The benchmarks are typically written in C and compiled with gcc, with inline assembly for a subset of inner loops. Some of the simpler benchmarks like the STREAM and FIR benchmarks were small enough that coding entirely in assembly was the most expedient approach. This section presents the results.

Streaming algorithms

Table 14.8 presents the performance of a set of linear algebra algorithms on RawPC vs. the P3.

The Raw implementations are coded as Stream Algorithms [198], which emphasize computational efficiency in space and time and are designed specifically to take advantage of tiled microarchitectures like Raw. They have three key features. First, stream algorithms operate directly on data from the interconnect and achieve an asymptotically optimal 100% compute efficiency for large numbers of tiles. Second, stream algorithms use no more than a small, bounded amount of storage on each processing element. Third, data are streamed through the compute fabric from and to peripheral memories.

With the exception of Convolution, we compare against the P3 running single precision Lapack (Linear Algebra Package). We use clapack version 3.0 [16] and a tuned BLAS implementation, ATLAS [453], version 3.4.2. We disassembled the ATLAS library to verify that it uses P3 SSE extensions appropriately to achieve high performance. Since Lapack does not

Table 14.8. Performance of linear algebra routines.

Benchmark	Problem Size	MFlops on Raw	Speedup vs. P3	
			Cycles	Time
Matrix multiplication	256 × 256	6310	8.6	6.3
LU factorization	256 × 256	4300	12.9	9.2
Triangular solver	256 × 256	4910	12.2	8.6
QR factorization	256 × 256	5170	18.0	12.8
Convolution	256 × 16	4610	9.1	6.5

Table 14.9. Performance (by time) of STREAM benchmark.

Problem Size	Bandwidth (GB/s)			Raw/P3
	P3	Raw	NEC SX-7	
Copy	0.567	47.6	35.1	84
Scale	0.514	47.3	34.8	92
Add	0.645	35.6	35.3	55
Scale & Add	0.616	35.5	35.3	59

provide a convolution, we compare against the Intel Integrated Performance Primitives (IPP).

As can be seen in Table 14.8, Raw performs significantly better than the P3 on these applications even with optimized P3 SSE code. Raw's better performance is due to load/store elimination, and the use of parallel resources. Stream Algorithms operate directly on values from the network and avoid loads and stores, thereby achieving higher utilization of parallel resources than the blocked code on the P3.

STREAM benchmarks

The STREAM benchmark was created by John McCalpin to measure sustainable memory bandwidth and the corresponding computation rate for vector kernels [277]. Its performance has been documented on thousands of machines, ranging from PCs and desktops to MPPs and other supercomputers.

We hand-coded an implementation of STREAM on RawStreams. We also tweaked the P3 version to use single precision SSE floating point, improving its performance. The Raw implementation employs 14 tiles and streams data between 14 processors and 14 memory ports through the static network. Table 14.9 displays the results. As shown in the right-most column, Raw is 55x–92x better than the P3. The table also includes the performance of STREAM on NEC SX-7 Supercomputer, which has the highest reported STREAM performance of any single-chip processor. Note that Raw surpasses that performance. This extreme single-chip performance is achieved by taking advantage of three Raw architectural features: its ample pin bandwidth, the ability to precisely route data values in and out of DRAMs with minimal overhead, and a careful match between floating point and DRAM bandwidth.

Other stream-based applications

Table 14.10 presents the performance of some hand-optimized stream applications on Raw. We are developing a real time 1020-microphone

Acoustic Beamformer which will use the Raw system for processing. On this application, Raw runs 16 instantiations of the same set of instructions (code) and the microphones are striped in a data-parallel manner across the array. Raw's software-exposed I/O allows for much more efficient transfer of stream data than the DRAM-based I/O on the P3. The assumption that the stream data for the P3 is coming from DRAM represents a best-case situation. The P3 results would be much worse in an actual system where the data is transmitted over a PCI bus. For FIR, we compared to the Intel IPP. Results for Corner Turn, Beam Steering, and CSLC are discussed in the previously published [392].

ILP computation

This section examines how well Raw is able to support conventional sequential applications. Typically, the only form of parallelism available in these applications is ILP-level parallelism. For this evaluation, we select a range of benchmarks that encompass a wide spectrum of program types and degrees of ILP.

Much like a VLIW architecture, Raw is designed to rely on the compiler to find and exploit ILP. We have developed Rawcc [39,257,258] to explore these compilation issues. Rawcc takes sequential C or Fortran programs and orchestrates them across the Raw tiles in two steps. First, Rawcc distributes the data and code across the tiles in a way that attempts to balance the tradeoff between locality and parallelism. Then, it schedules the computation and communication to maximize parallelism and minimize communication stalls.

Table 14.10. Performance of handwritten stream applications.

Benchmark	Machine Configuration	Cycles on Raw	Speedup vs. P3	
			Cycles	Time
Acoustic beam-forming	RawStreams	7.83M	9.7	6.9
512-pt Radix-2 FFT	RawPC	331K	4.6	3.3
16-tap FIR	RawStreams	548K	10.9	7.7
CSLC	RawPC	4.11M	17.0	12.0
Beam steering	RawStreams	943K	65	46
Corner turn	RawStreams	147K	245	174

Table 14.11. Performance of sequential programs on Raw and on a P3.

Benchmark	Source	# Raw Tiles	Cycles on Raw	Speedup vs. P3	
				Cycles	Time
Dense-Matrix Scientific Applications					
Swim	Spec95	16	14.5M	4.0	2.9
Tomcatv	Nasa7:Spec92	16	2.05M	1.9	1.3
Btrix	Nasa7:Spec92	16	516K	6.1	4.3
Cholesky	Nasa7:Spec92	16	3.09M	2.4	1.7
Mxm	Nasa7:Spec92	16	247K	2.0	1.4
Vpenta	Nasa7:Spec92	16	272K	9.1	6.4
Jacobi	Raw bench. suite	16	40.6K	6.9	4.9
Life	Raw bench. suite	16	332K	4.1	2.9
Sparse-Matrix/Integer/Irregular Applications					
SHA	Perl Oasis	16	768K	1.8	1.3
AES Decode	FIPS-197	16	292K	1.3	0.96
Fpppp-kernel	Nasa7:Spec92	16	169K	4.8	3.4
Unstructured	CHAOS	16	5.81M	1.4	1.0

Rawcc is a prototype research compiler and is, therefore, not robust enough to compile every application in standard benchmark suites. Below are the results for a selection of benchmarks that it can compile. The speedups attained in Table 14.11 shows the potential of automatic parallelization and ILP exploitation on Raw. Of the benchmarks compiled by Rawcc, Raw is able to outperform the P3 for all the scientific benchmarks and several irregular applications.

Table 14.12 shows the speedups achieved by Rawcc as the number of tiles varies from 2 to 16. The speedups are compared to performance of a single Raw tile. Overall, the improvements are primarily due to increased parallelism, but several of the dense-matrix benchmarks benefit from increased cache capacity as well (which explains the super-linear speedups). In addition, Fpppp-kernel benefits from increased register capacity, which leads to fewer spills.

Table 14.12. Speedup of the ILP benchmarks relative to single-tile Raw.

Benchmark	Number of Tiles				
	1	2	4	8	16
Dense-Matrix Scientific Applications					
Swim	1.0	1.1	2.4	4.7	9.0
Tomcatv	1.0	1.3	3.0	5.3	8.2
Btrix	1.0	1.7	5.5	15.1	33.4
Cholesky	1.0	1.8	4.8	9.0	10.3
Mxm	1.0	1.4	4.6	6.6	8.3
Vpenta	1.0	2.1	7.6	20.8	41.8
Jacobi	1.0	2.6	6.1	13.2	22.6
Life	1.0	1.0	2.4	5.9	12.6
Sparse-Matrix/Integer/Irregular Applications					
SHA	1.0	1.5	1.2	1.6	2.1
AES Decode	1.0	1.5	2.5	3.2	3.4
Fpppp-kernel	1.0	0.9	1.8	3.7	6.9
Unstructured	1.0	1.8	3.2	3.5	3.1

For completeness, we also compiled a selection of the Spec2000 benchmarks with gcc for a single tile, and ran them using MinneSPEC's [233] *LgRed* data sets to reduce the length of simulations. The results, shown in Table 14.13, represent a lower bound for the performance of those codes on Raw, as they only use 1/16 of the resources on the Raw chip. The numbers are quite surprising: on average, the simple in-order Raw tile with no L2 cache is only 1.4x slower by cycles and 2x slower by time than the full P3. This suggests that in the event that the parallelism in these applications is too small to be exploited across Raw tiles, a simple 2-way Raw compute processor might be sufficient to make the performance difference negligible.

Bit-level computation

We measure the performance of RawStreams on two bit-level computations [52]. Table 14.14 presents the results for the P3, Raw, FPGA, and ASIC implementations. The FPGA implementations use a Xilinx Virtex-II 3000-5 FPGA, which is built using the same process generation as the Raw chip. The ASIC implementations were synthesized to the IBM SA-27E process that the Raw chip is implemented in. For each benchmark, we present three problem sizes: 1024, 16384, and 65536 samples. These problem

Table 14.13. Performance of SPEC2000 programs on one tile on Raw.

Benchmark	Source	# Raw Tiles	Cycles on Raw	Speedup vs. P3	
				Cycles	Time
172.mgrid	SPECfp	1	0.240B	0.97	0.69
173.applu	SPECfp	1	0.324B	0.92	0.65
177.mesa	SPECfp	1	2.40B	0.74	0.53
183.equake	SPECfp	1	0.866B	0.97	0.69
188.ammp	SPECfp	1	7.16B	0.65	0.46
301.apsi	SPECfp	1	1.05B	0.55	0.39
175.vpr	SPECint	1	2.52B	0.69	0.49
181.mcf	SPECint	1	4.31B	0.46	0.33
197.parser	SPECint	1	6.23B	0.68	0.48
256.bzip2	SPECint	1	3.10B	0.66	0.47
300.twolf	SPECint	1	1.96B	0.57	0.41

Table 14.14. Performance of two bit-level applications: 802.11a Convolution Encoder and 8b/10b Encoder. The hand-coded Raw implementations are compared to reference sequential implementations on the P3.

Benchmark	Problem Size	Cycles on Raw	Speedup vs. P3			
			Raw		FPGA	ASIC
			Cycles	Time	Time	Time
802.11a ConvEnc	1024 bits	1048	11.0	7.8	6.8	24
	16408 bits	16408	18.0	12.7	11	38
	65536 bits	65560	32.8	23.2	20	68
8b/10b Encoder	1024 bytes	1054	8.2	5.8	3.9	12
	16408 bytes	16444	11.8	8.3	5.4	17
	65536 bytes	65695	19.9	14.1	9.1	29

sizes are selected to fit in the L1, L2, and miss in the cache on the P3, respectively. We use a randomized input sequence in all cases.

On these two applications, Raw is able to excel by exploiting fine-grain pipeline parallelism. To do this, the computations were spatially mapped across multiple tiles. Both applications benefited by more than 2x from Raw's specialized bit-level manipulation instructions, which reduce the latency of critical feedback loops. Another factor in Raw's high performance

Table 14.15. Performance of two bit-level applications for 16 streams: 802.11a Convolution Encoder and 8b/10b Encoder. This test simulates a possible workload for a base-station that processes multiple communication streams.

Benchmark	Problem Size	Cycles on Raw	Speedup vs P3	
			Cycles	Time
802.11a ConvEnc	16*64 bits	259	45	32
	16*1024 bits	4138	71	51
	16*4096 bits	16549	130	92
8b/10b Encoder	16*64 bytes	257	34	24
	16*1024 bytes	4097	47	33
	16*4096 bytes	16385	80	56

on these applications is Raw's exposed streaming I/O. This I/O model is in sharp contrast to having to move data though the cache hierarchy on a P3.

Table 14.15 presents the results for the operation on 16 parallel input streams. This is to simulate a potential workload for a base-station communications chip that needs to encode 16 simultaneous connections. For this throughput test, a more area-efficient implementation was used on Raw. This implementation has lower performance on a single stream but utilizes fewer tiles, achieving a higher per-area throughput. Instantiating 16 copies of this implementation results in the maximum total throughput.

Conclusion

This chapter describes the architecture and implementation of the Raw microprocessor. Raw's exposed ISA allows parallel applications to exploit all of the chip resources, including gates, wires and pins. Raw supports ILP by scheduling operands over a scalar operand network that offers very low latency for scalar data transport. Raw's compiler manages the effect of wire delays by orchestrating both scalar and stream data transport. The Raw processor demonstrates that existing architectural abstractions like interrupts, caches, and context-switching can continue to be supported in this environment, even as applications take advantage of the low-latency scalar operand network and the large number of ALUs.

Our results demonstrate that the Raw processor performs at or close to the level of the best specialized machine for each application class. When compared to a Pentium III, Raw displays one to two orders of magnitude more performance for stream applications, while performing within a factor of 2 for low-ILP applications. It is our hope that the Raw research will provide insight for architects who are looking for ways to build versatile processors that leverage the vast silicon resources while mitigating the considerable wire delays that loom on the horizon.

Acknowledgments

We thank our StreamIt collaborators, specifically M. Gordon, J. Lin, and B. Thies for the StreamIt backend and the corresponding section of this paper. We are grateful to our collaborators from ISI East including C. Chen, S. Crago, M. French, L. Wang, and J. Suh for developing the Raw motherboard, firmware components, and several applications. T. Konstantakopoulos, L. Jakab, F. Ghodrat, M. Seneski, A. Saraswat, R. Barua, A. Ma, J. Babb, M. Stephenson, S. Larsen, V. Sarkar, and several others too numerous to list also contributed to the success of Raw. The Raw chip was fabricated in cooperation with IBM. Raw is funded by Darpa, NSF, ITRI, and the Oxygen Alliance.

15 Processor Clock Generation and Distribution

Stefan Rusu

Intel Corporation

Introduction

The clock is a periodic synchronization signal used as a time reference for data transfers in synchronous digital systems. Since the clock plays a central role in the operation of a synchronous system, significant effort is invested in the design, optimization and verification of high-performance clock distribution schemes. Because the clock edges determine the state updates in a synchronous system, higher clock frequencies are generally (but not always) associated with a higher system performance. Clock frequency for mainstream microprocessors has increased significantly over time, driven by the process technology scaling, aggressive circuit design techniques and deeper pipelines.

The clock distribution is particularly affected by process scaling. Smaller process geometries allow designers to pack more functionality on a single die. The number of sequential elements that need the clock is constantly increasing, thus making the clock distribution a more difficult task. Transistors are getting smaller and faster, so clock deskew or compensation circuits are cheaper to design. However, the metal interconnect does not scale well and that requires careful extraction and modeling of the clock tree lines resistance and capacitance. This trend is aggravated by the increase in die size since the clock needs to be distributed to all the sequential circuits on the die, therefore the clock lines are getting longer and require more buffering levels. For high frequency clocks (generally above 1 GHz) the inductive effects in the clock distribution lines must also be modeled.

J. Nurmi (ed.), Processor Design: System-on-Chip Computing for ASICs and FPGAs, 339–366.

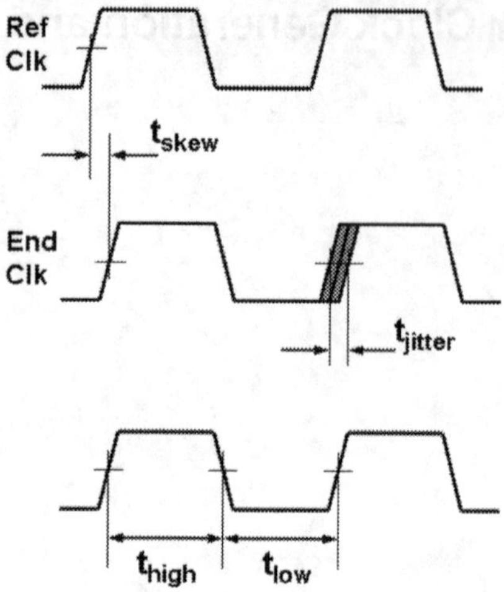

Fig. 15.1. Characteristic parameters for a clock system.

This chapter is organized as follows: A section on clock parameters and trends defines the main characteristics of a clock distribution network and examines their evolution over time. A section on clock distribution networks reviews several clock distribution schemes, with specific examples from high-performance microprocessor designs. A section on deskew circuits presents the evolution of deskew circuits and their benefits, while section jitter reduction techniques describes jitter reduction circuit techniques. Since power is a limiting factor in most digital designs today, we review several low-power clock distribution ideas in section low power clock distribution. Finally, section further directions in clock distribution discusses several future directions in clock distribution, including distributed VCOs and PLLs, as well as rotary and standing wave clock distribution.

Clock parameters and trends

Figure 15.1 shows the main parameters of a clock distribution network. The *clock skew* is the spatial variation of the clock signal as distributed through the chip. We distinguish between global (chip-level) and local (block level) skew. A proper clock distribution design typically attempts to reduce the clock skew to zero. However, as we will show in Section 4, intentional skew insertion is sometimes used to relieve timing critical paths, which allows a chip to run at a higher frequency. An ideal clock distribution network would minimize the skew across the entire die. However, it is important to recognize that clock skew is relevant only when there is a data transfer from one sequential element to another. Since the speed of data

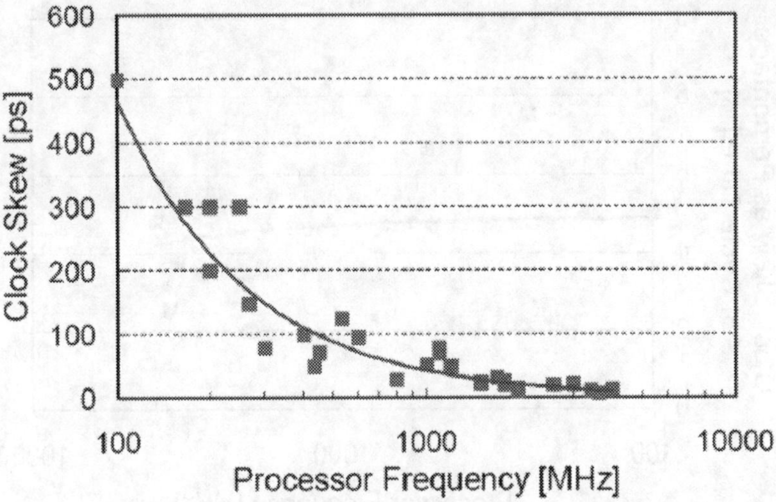

Fig. 15.2. Historical clock skew trends.

traveling across the chip is also limited by the RC delay of the metal wires, we cannot have data signals traveling across the entire die in a single cycle. Therefore, the clock skew between two corners of the chip is not relevant and will not affect the operating frequency of the design. Detailed modeling of the clock skew must take into account the actual cycle-level data transfers across the chip. Figure 15.2 shows the clock skew as a function of the cycle time. The points in the chart represent the skew vs. operating frequency, as reported in International Solid-State Circuits Conference (ISSCC) or Journal of Solid-State Circuit (JSSC) published papers on large microprocessor designs by major industry players. As expected, the skew decreases as the frequencies are increased.

A better way to measure the clock skew is to express it as a percentage of the cycle time, as shown in Figure 15.3. Notice that the skew has been averaging about 5% of the cycle time, although a wide variability exists between different designs (2–8%). As the frequencies continue to increase in the multi-GHz space, designers have to pay more attention to containing skew and that could explain the slightly downward trend seen in the figure. One skew reduction technique that is gaining popularity in large microprocessor designs is the use of deskew circuits, which will be described in a later section.

The main sources of clock skew are shown in Figure 15.4 [143]. Notice that more than half of the skew is caused by device mismatches. The difference in the local supply levels for the intermediate clock buffers accounts for about a quarter of the skew.

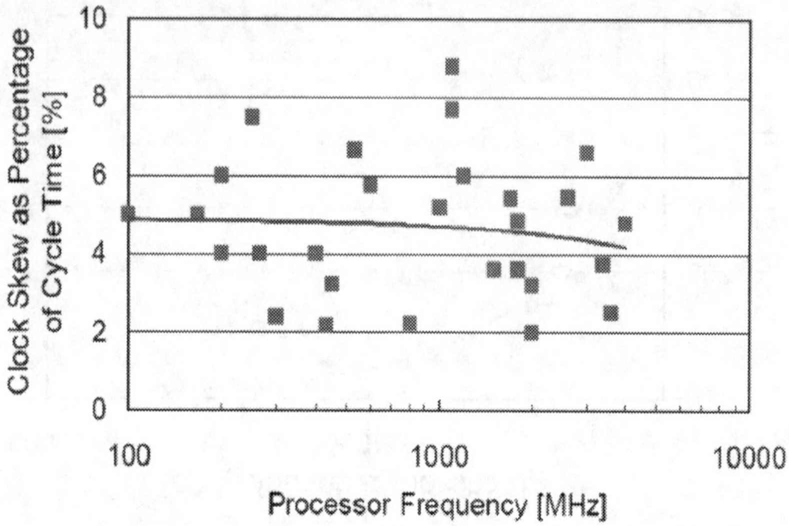

Fig. 15.3. Clock skew as percentage of the cycle time.

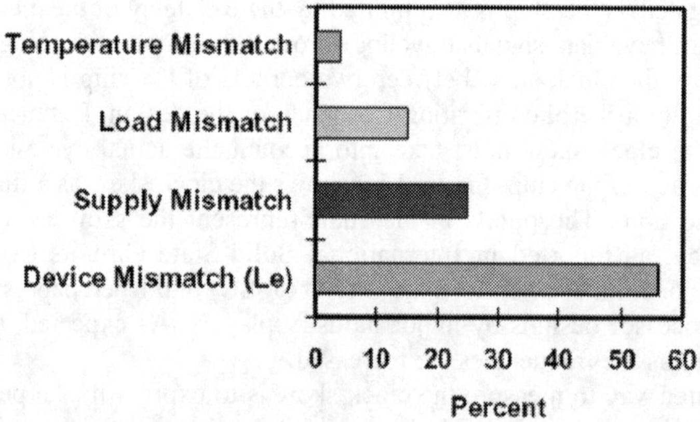

Fig. 15.4. Clock skew sources [143].

The load mismatch is due to imbalances in the clock distribution trees and can be corrected by investing additional design effort. However, most of this extra effort is manual, so typically design teams accept the residual skew that remains after balancing out the clock trees using automatic tools. Another approach is to use a clock grid that shorts all the clock end-points. As we will see in Section 3, clock grids have lower skews but consume significantly more power. Finally, the temperature mismatch has a small impact on the clock skew. This is good news for the clock designers, since large temperature gradients exist in modern high-performance designs.

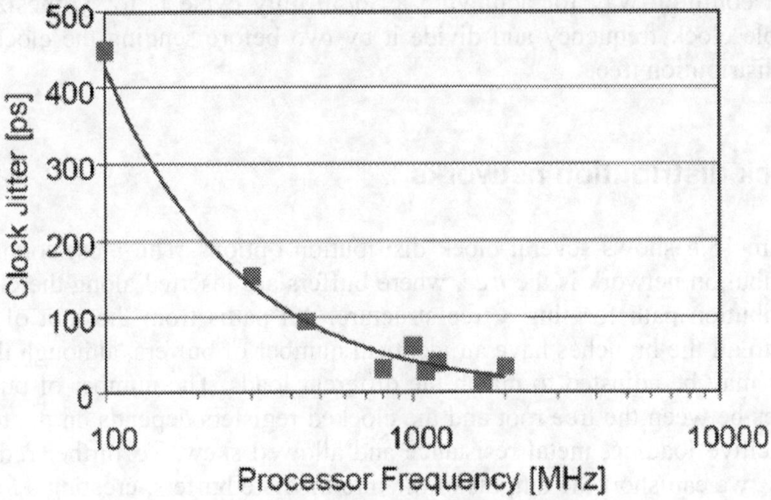

Fig. 15.5. Historical clock jitter trend.

The *clock jitter* is the temporal variation of the clock signal with respect to a reference edge. We differentiate between long-term jitter (that accumulates over a long period of time) and cycle-to-cycle jitter (which is measured between adjacent clock cycles). Long-term jitter accumulates as a phase error relative to the reference clock and degrades setup and hold timings for the entire circuit. The cycle-to-cycle short-term jitter is seen by the logic circuits as a frequency shift, similar to the clock skew.

Figure 15.5 shows the clock jitter trend as a function of the processor frequency as reported in major processor papers at ISSCC or JSSC. Notice that this graph has a lot fewer data points, since not all papers report on the clock jitter results. This could be explained by the fact that clock jitter is very difficult to measure in a VLSI circuit. Most probing techniques like e-beam or laser probing rely on time averaging, thus removing any jitter information. The most common technique is to drive a clock node to an external pin and measure the jitter with an oscilloscope. However, the additional clock buffers needed to bring the clock out introduce additional jitter and the scopes introduce their own jitter too.

The main sources of clock jitter are the power supply noise coupling into the voltage-controlled oscillator (long-term jitter) and the supply noise modulation of the clock network buffer delay (short-term cycle-to-cycle or multi-cycle jitter). Jitter reduction techniques will be discussed later.

The *clock duty cycle* is the ratio of the clock high and low times. Ideally we would like the duty cycle to be 50/50, although small intentional deviations may enable higher operating frequencies in phase-based designs. The most common way for achieving a 50/50 duty cycle is to synthesize a double clock frequency and divide it by two before sending the clock to the distribution tree.

Clock distribution networks

Figure 15.6 shows several clock distribution options. The most common distribution network is the *tree*, where buffers are inserted along the clock distribution path forming a tree structure. All paths from the root of the tree to all the branches have an identical number of buffers, although their sizes may be adjusted to match the different loads. The number of buffer stages between the tree root and the clocked registers depends on the total capacitive loading, metal resistance and allowed skew. To further reduce skew, we can short the outputs of the intermediate buffers, creating a *mesh* clock structure. The clock *grid* can be driven from two sides (as shown in the figure) or from all sides.

Another approach to ensure zero clock skew uses hierarchical *H* or *X-tree* structures. In this approach, the clock is driven from the center of an H structure to its four corners. Each corner drives another, smaller H structure. This distribution process is repeated through progressively smaller, hierarchical H structures. The end-points of the smallest H shape drive the

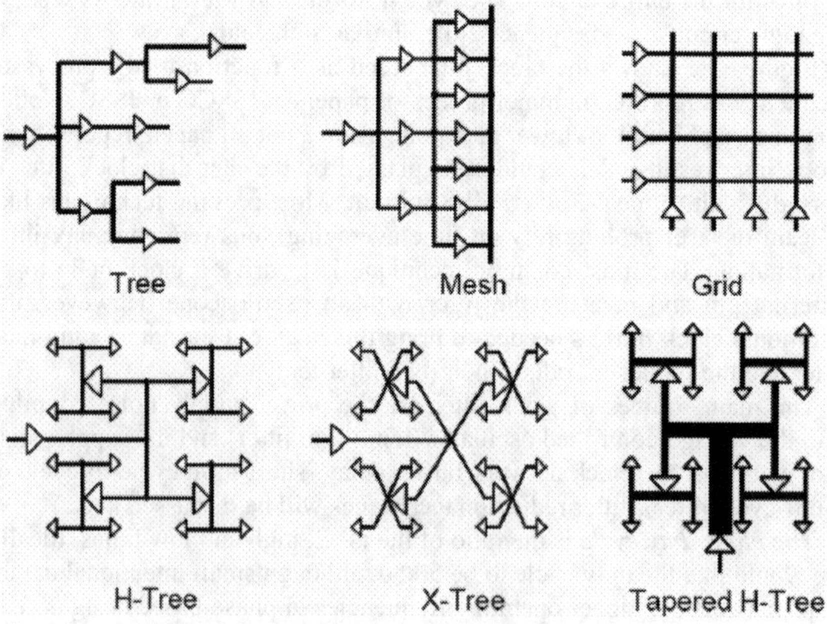

Fig. 15.6. Common clock distribution structures.

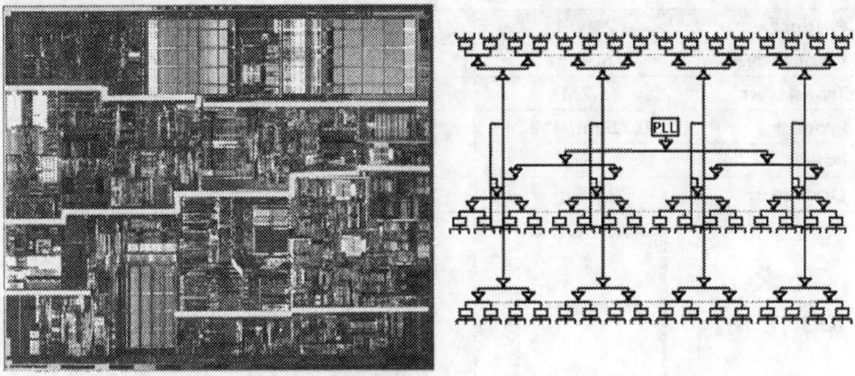

Fig. 15.7. Clock distribution for the Pentium® 4 Processor [247].

clocked registers. Due to the symmetry of the H or X-trees, the path from the clock source to each end-point has the same delay. As the interconnect resistance increases with the technology scaling, adequate intermediate buffers are required along the H-tree distribution.

The *tapered H-tree* matches the clock line impedance to minimize reflections at branching points. The width of the clock trunk decreases as the signal propagates through the tree. The impedance of the line exiting each branch point is designed to be twice the impedance of the line feeding the branch point. Notice that perfectly symmetrical H-trees are difficult to implement in actual designs due to floorplan constraints. Actual clock trees require careful extraction and characterization to achieve a balanced design.

Figure 15.7 shows the clock distribution of the Pentium® 4 Processor [247]. The clock is distributed using a triple spine approach to cover the large die. Each spine contains a binary distribution tree, with each of the 47 leaf nodes providing an independent domain clock. Local clock drivers are used to buffer the clock load as well as produce the proper frequency and clock type for each particular block. The drivers are connected to the appropriate domain clocks through delay-matched taps. The maximum RC delay from the output of the local drivers to the input of a latch is restricted in order to minimize the local clock skew.

Figure 15.8 shows the evolution of the clock distribution network for the Alpha microprocessor [163] through three consecutive generations of the design.

The 21064 had a two-phase single-wire clocking scheme. A single driver was located in a stripe distributed across the center of the die. The clock accumulated skew as it propagated towards the edges of the die. To handle the large transient currents in the power grid when the clock driver

Product	21064	21164	21264
Frequency	166MHz	300MHz	600MHz
Transistors	1.7M	9.3M	9.3M
Process	0.75um, 4ML	0.5um, 4ML	0.35um, 6ML
Power	25W	50W	72W
Clock load	2.75nF	3.75nF	2.8nF

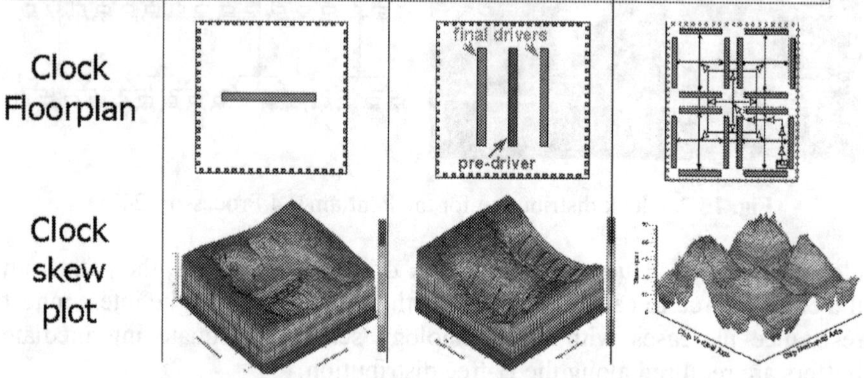

Fig. 15.8. Evolution of the clock distribution network for the Alpha processors [163].

switched, on-chip decoupling structures were placed around the clock driver. Roughly 10% of the chip area was allocated to decoupling capacitance.

In the 21164, the main clock driver was split into two banks and placed midway between the center of the die and the edges. The pre-driver was located in the center of the die to distribute the clock to the two main drivers. The clock skew was reduced by a factor of two using this approach. In addition, by distributing the main clock driver over a larger area, the localized heating seen on the 21064 was reduced.

In the 21264, the power consumption became a major concern in designing the clocking system. To reduce it, a single wire global clock (GCLK) was routed over the entire chip as a global timing reference. The GCLK drivers were distributed around the four quadrants to reduce clock grid delay and distribute clock power. The GCLK drives a hierarchy of thousands of buffered and conditioned local clocks used across the chip. There are several advantages to this clocking scheme. First, conditioning the local clocks saves power. Second, circuit designers can take advantage of multiple clocks to add local skew that benefits timing critical paths. Finally, using local buffering significantly lowers the GCLK load, which reduces clock skew to less than 75 ps.

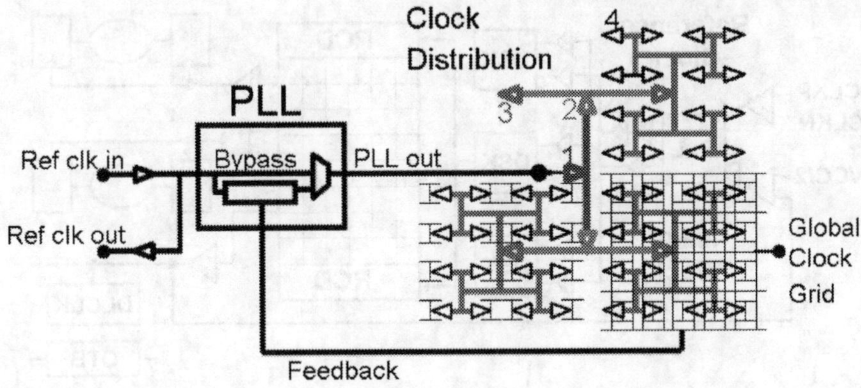

Fig. 15.9. Power4 microprocessor clock distribution [349].

The Power4 microprocessor clock distribution is shown in Figure 15.9 [349]. Considering the complexity of this 174 million transistor dual-processor chip, the clock distribution is relatively simple. A single chip-wide clock domain is used, with no active or programmable skew-reduction circuitry. A single PLL is used near the center of the chip to minimize the global clock distribution delay. The clock is distributed by 64 tuned trees driving a single full-chip clock grid at 1024 points. The grid smoothes out clock skew caused by across-chip process variations, but it does consume more power than a balanced tree structure. Experimental measurements using pico-probes at 19 locations across the die showed a maximum skew of 25 ps. Optical probing on 9 of the 64 sector buffers confirmed less than 18 ps skew at the leading edge of the photon pulses.

The clock distribution of the first Itanium® Processor [408] is shown in Figure 15.10. The clock topology is partitioned into three segments. The global distribution consists of the clock synthesis using an on-die phase-locked loop (PLL) and the distribution of the core clock and the reference clock from the PLL clock generator to the deskew buffers (DSK). The regional distribution includes the clock distribution from the DSKs to the 30 regional clock grids. Finally, the local distribution consists of the local clock buffers (LCBs) taking the input from the regional clock grid and the local interconnect to support the clocked elements. The clock deskew function will be discussed in detail in Section 4. The regional clock grid is implemented using metal 4 horizontal and metal 5 in the vertical direction. As with the global clock network, the regional clock grid contains full lateral shielding to ensure low capacitance coupling and good inductive return paths. The regional clock grid utilizes up to 3.5% of the available metal 5 and up to 4.1% of the available metal 4 routing over a region.

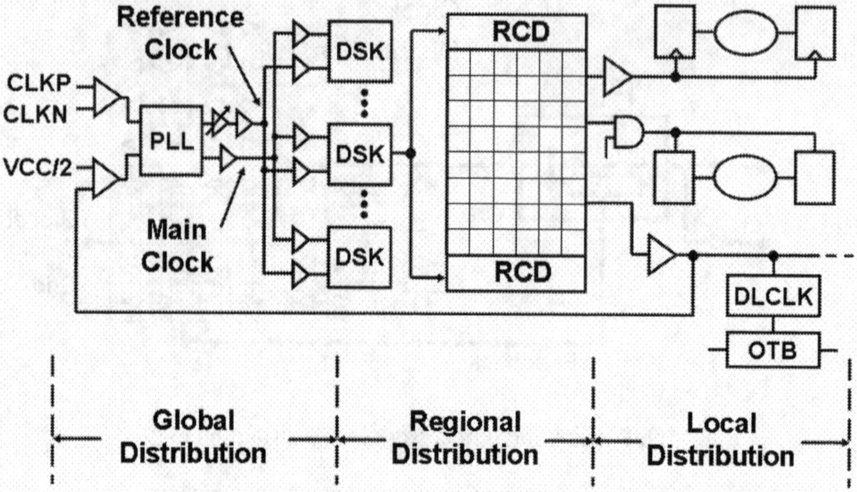

Fig. 15.10. Clock distribution for the Itanium® Processor [408].

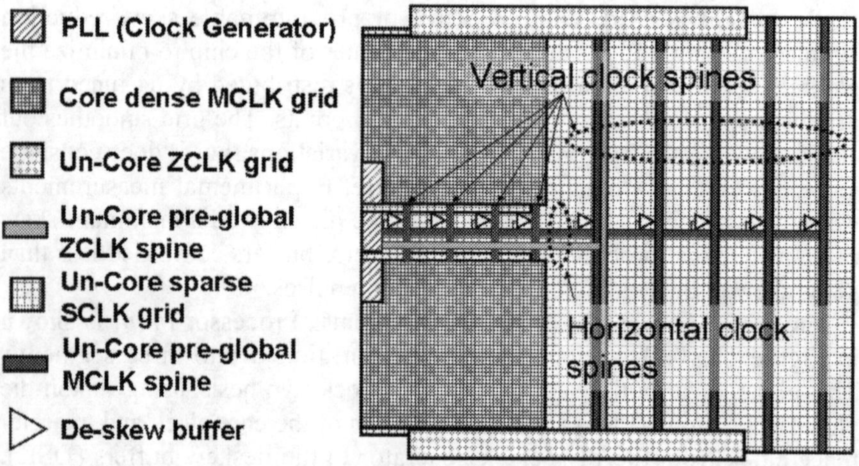

Fig. 15.11. Clock distribution for the 65nm dual-core Xeon® Processor [358].

The clock distribution of the 65 nm dual-core Xeon® Processor [358] is described in Figure 15.11. Each core has its own independent high frequency core grid. Within the uncore area, there are two horizontal clock spines and nine vertical spines responsible for distributing the various clocks. The ZCLK grid supports the front side bus (FSB) areas with the quad-pumped FSB clock. A binary tree embedded inside a horizontal and a vertical clock spine delivers the pre-global FSB clock from the output of the IOPLL to the FSB areas and the isolated FSB clock islands. The un-core clock SCLK grid covers the entire un-core area. Since minimizing power is a primary objective, two sparse grid structures are used. To minimize the within-die clock skew, fuse-programmable deskew buffers are inserted at the base of the vertical spines to provide a simulated de-skew range of 60 ps with four selectable steps. The fuse-based deskew is an efficient and robust method to address skews caused by structural design mismatches or to introduce intentional skew to improve the operating frequency. A regression flow incorporating a complete set of test vectors and system validation suite is used to derive and validate the optimal fuse settings.

The last topic in this section is the inductive effect in clock distribution networks and the need to carefully model these effects when designing high frequency clock trees.

Low resistance copper interconnects together with fast clock edge rates result in inductive wire impedance comparable to the pure resistive impedance. As a result, high frequency clock grids are modeled as two-dimensional distributed RLC transmission line structures, as shown in [464]. Detailed 3D field simulations are used to extract accurate per unit length RLC clock grid values in the presence of finite coplanar current return paths.

Transmission line effects, like signal reflections near the clock drivers as well as overshoot and undershoot at the far end of the grid, are clearly observed in Figure 15.12. Distributed wire inductance increases the delay between the early clock (near the driver) and the late clock (at the end of the grid).

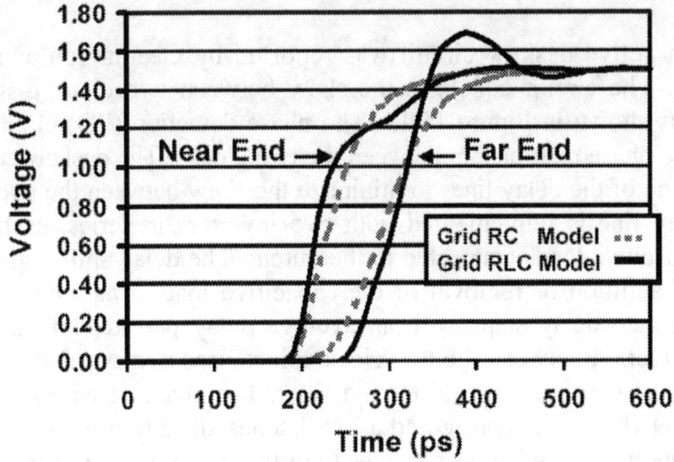

Fig. 15.12. Inductive effects on clock distribution grids [464].

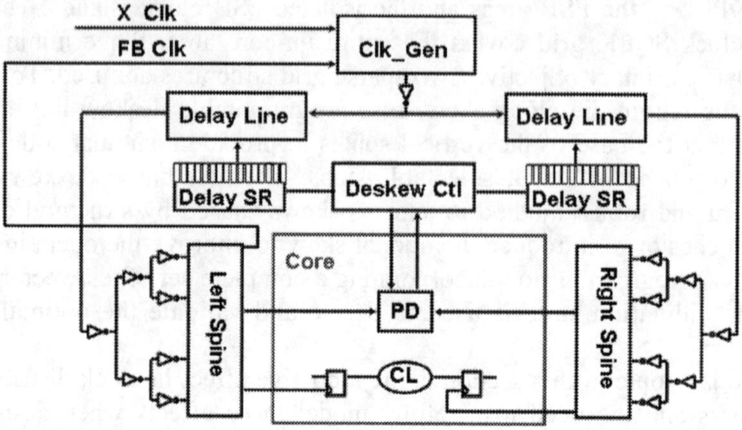

Fig. 15.13. Block diagram of a two-spine clock deskewing circuit [143].

Clock loads near the driver are inductively shielded from remote loads during the clock transition and observe a faster clock. On the other hand, remote loads receive the clock later due to additional signal phase imposed by wire inductance. Optimal clock driver sizing requires that the equivalent output impedance be smaller than the grid characteristic impedance to guarantee a high amplitude incident wave, and yet that the clock rise and fall times at the drivers are sufficiently slow to ensure that voltage overshoot and undershoot are not excessive at the far ends.

Deskew circuits

The first active deskew circuit was reported by Geannopoulos and Dai in [143]. The circuit equalizes the skew between two clock distribution spines, as shown in Figure 15.13. The phase detection (PD) circuit determines the phase relationship between the two clocks. The deskew controller adjusts one of the delay lines to minimize the skew between the two spines. The delay line is implemented with two inverters in series, each having eight capacitive loads connected to the output. The delay shift register controls the addition or removal of the capacitive loads, enabling 17 monotonic discrete delay steps with an average delay per step of 12 ps. The phase detection between the two clock spines uses two symmetrical phase detectors and an adaptive noise band filter. The phase detectors are cross-coupled NAND gates configured as RS latches sized to minimize metastability. The noise band filter removes high frequency power supply voltage noise to avoid false corrections that would add to the phase error.

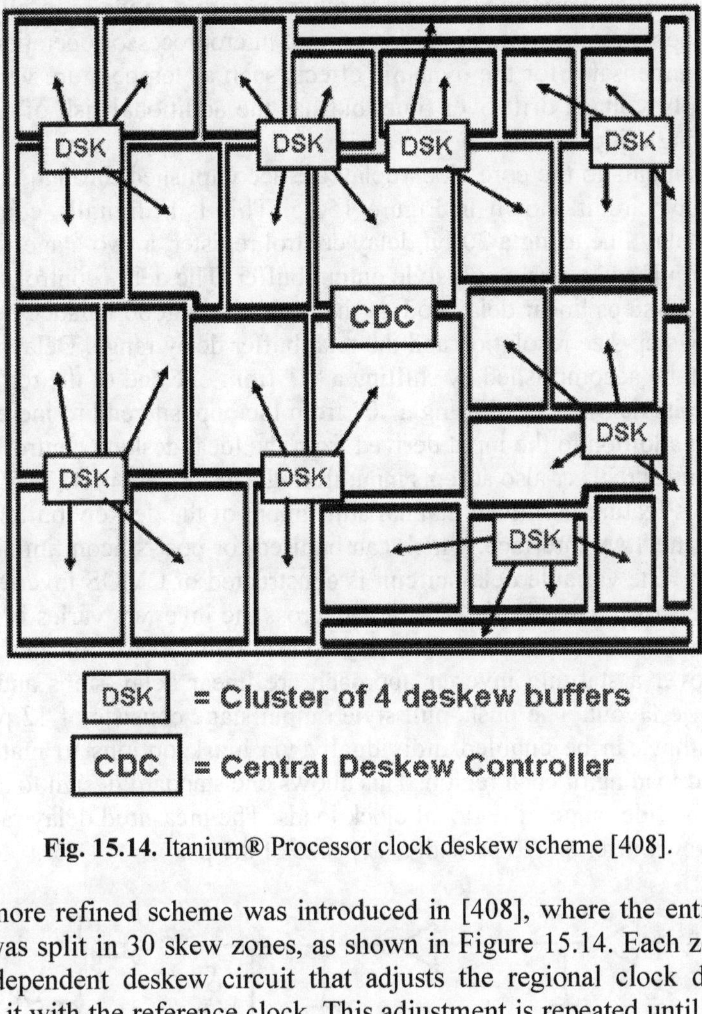

Fig. 15.14. Itanium® Processor clock deskew scheme [408].

A more refined scheme was introduced in [408], where the entire chip area was split in 30 skew zones, as shown in Figure 15.14. Each zone has an independent deskew circuit that adjusts the regional clock delay to match it with the reference clock. This adjustment is repeated until a minimum phase error is achieved. Therefore, any load mismatches and within-die variations in the core clock distribution are automatically compensated. Since all the clock regions use the same reference clock, the residual skew of the reference clock, the uncertainty of the phase detector and the mismatches of the feedback clocks determine the overall skew across these regions.

The deskew operation is executed in parallel for the 30 clock regions during the initial microprocessor reset. The global deskew controller monitors the progress and signals the deskew completion. Once this occurs, the DSK delay register settings are fixed until the next power up sequence. This mode compensates for the process variations and most of the voltage and temperature variations. An alternative operating mode is to allow the deskew operation to continue during normal microprocessor operation. This mode compensates for the dynamic effects such as temperature variations and supply voltage drift over time, but has the additional risk of creating new timing critical paths.

Adjustments to the core clock delay are accomplished through the variable delay circuit shown in Figure 15.15. This is a digitally controlled analog delay line using a 20-bit delay control register, a two-stage variable delay circuit and a push–pull style output buffer. The delay control register forms a 20-steps linear delay coding that provides a good balance between the delay step-size resolution and the total buffer delay range. Delay adjustment can be accomplished by shifting a "1" from one end of the register to decrease its delay or by shifting a "0" from the opposite end to increase its delay. In addition to the input derived from the local deskew controller, the delay control register also accepts input from the test access port (TAP) interface. This feature permits a manual adjustment of the deskew buffer delay through the TAP interface, which can be used for post-silicon timing optimization. The variable delay circuit is constructed of CMOS inverters and two arrays of passive loads. The delay across the inverters varies in accordance to the setting stored in the delay control register. Advantages of this design over a starving inverter approach are linear delay steps and more symmetric layout. The push–pull style output stage consists of 12 parallel drivers that can be enabled individually via mask options to match the extracted loading of each region. This allows one standard design to accommodate a wide range of regional clock loads. The measured delay range of the deskew buffer is 170 ps with a step size of 8.5 ps.

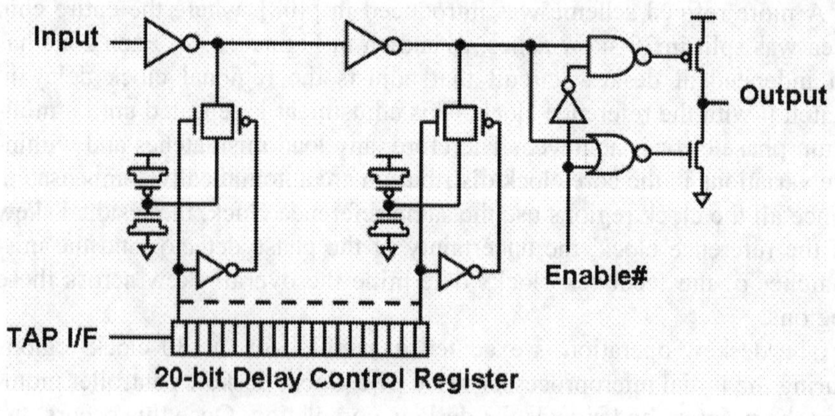

Fig. 15.15. Deskew buffer schematic [408].

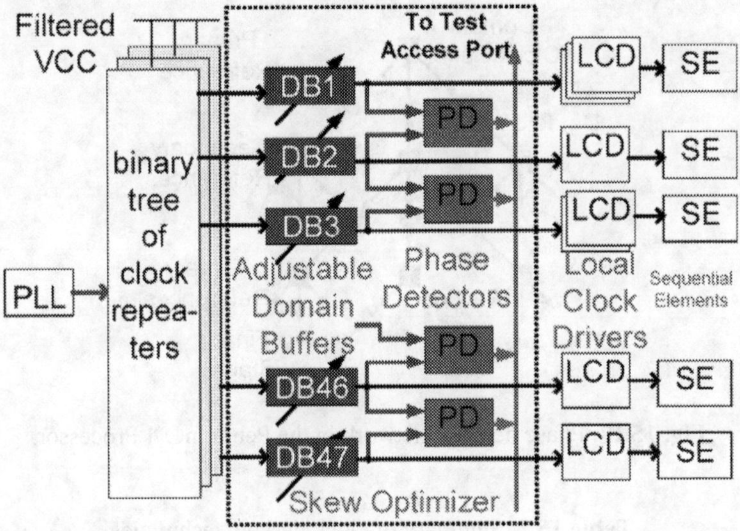

Fig. 15.16. Logical diagram of the skew optimization circuit in the Pentium® 4 Processor.

Figure 15.16 presents the schematic of the deskew circuit implemented in the Pentium® 4 Processor [247]. The main components of the skew optimization circuit are 47 adjustable delay domain buffers and a phase-detector network of 46 phase detectors. The delay adjustment control for the domain buffers and the output of the phase detectors are accessible from the TAP.

One domain buffer at the center of the die is chosen as the primary reference. The remaining buffers are categorized as secondary, tertiary, and final buffers. Figure 15.17 shows the phase-detector network, which coupled with the TAP, aligns the domain buffers to the primary reference. To limit the phase detector accumulation errors, the domain buffers go through at most three levels of phase detection. First, phase detectors adjust the delay of the secondary references to the primary reference. The phase detector outputs a high or low based on the leading or lagging inputs. The output is read out into a scan chain controlled by the TAP. Based on the outcome, the clock domain buffers are adjusted. This is repeated until all the secondary reference clocks are deskewed. Then, after the secondary reference delays have been adjusted, a second set of phase detectors adjust the delay of tertiary references. Similarly, the final stage buffers are adjusted to the tertiary references. With this scheme, the skew is adjusted to within an error of about 8 ps, limited mainly by the resolution of the adjustable delay elements.

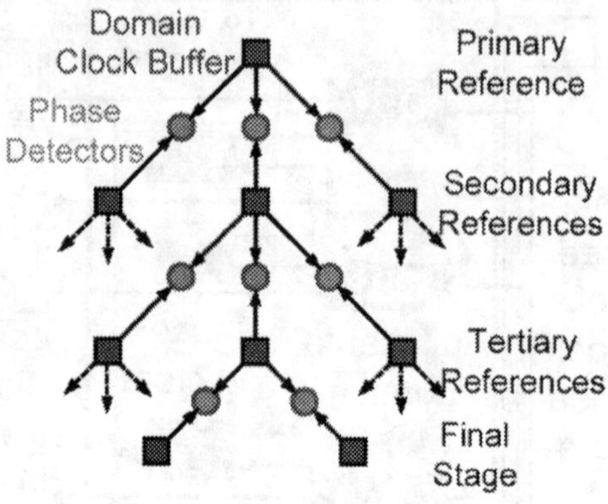

Fig. 15.17. Phase detector network in the Pentium® 4 Processor.

Table 15.1. Summary of clock deskew techniques.

Author	Source	Zones	Skew Before	Skew After	Step Size
Geannopoulos	ISSCC 1998	2	60 ps	15 ps	12 ps
Rusu	ISSCC 2000	30	110 ps	28 ps	8 ps
Kurd	ISSCC 2001	47	64 ps	16 ps	8 ps
Stinson	ISSCC 2003	23	60 ps	7 ps	7 ps

Table 15.1 summarizes all published clock deskew designs. Notice that all designs manage to reduce the clock skew to less than a quarter of the value measured without the deskew mechanism. As the process technology shrinks, the step size is reduced, without requiring any additional control bits.

Jitter reduction techniques

Clock jitter can originate in the PLL and in the clock distribution tree. To minimize the jitter form the PLL, all modern processor designs include a special filter for the PLL supply that can be a simple LC filter all the way to a sophisticated on-die regulator. The clock distribution jitter is due to the fact that intermediate clock buffering stages are connected to the noisy core supply and are distributed all over the chip. The supply noise causes these buffering stages to have slightly different delays depending on the core switching pattern. To reduce this jitter, we want to keep the delay of the global clock distribution to a minimum. Another technique is to filter the supply voltage for the clock buffering stages. Figure 15.18 shows a

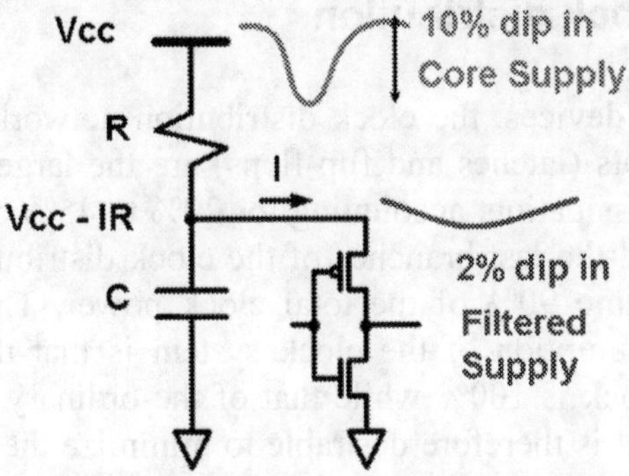

Fig. 15.18. Low jitter, RC-filtered power supply for clock drivers [247].

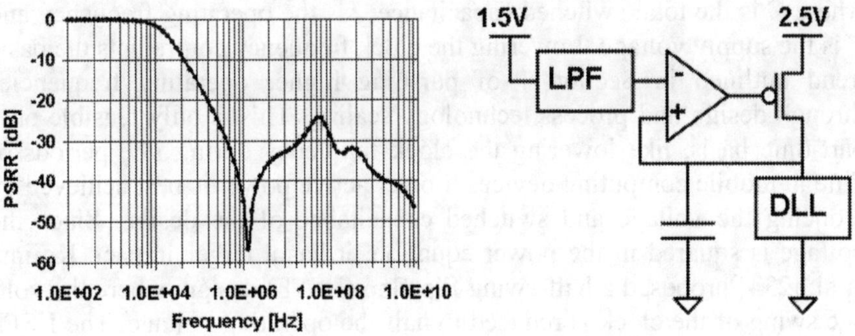

Fig. 15.19. Active clock supply regulator [358].

simple low-pass RC filter designed to reduce the core supply noise for the clock buffers [247]. The resistance is implemented using PMOS devices. The optimal design has an IR drop of 70mV with an RC constant of 2.5 ns, while minimizing the layout area required by the capacitor. The actual component values for the RC filter were adjusted for the different clock buffer types to a fixed IR drop and RC time constant. The filter circuit model simulations with typical supply noise waveforms show up to 5X noise amplitude reduction on the filtered supply.

Another technique is to use an active voltage regulator as described in [358] and shown in Figure 15.19. The core voltage of 1.5 V is used as a reference after passing it through a low-pass filter (LPF). The I/O voltage of 2.5 V is used to generate a clean, local supply for the delay-lock loop (DLL) circuits. The regulator attenuates supply noise frequencies in excess of 1 MHz by more than 15 dB, while lower supply noise frequencies are easily tracked by the DLLs themselves.

Low power clock distribution

In modern VLSI devices, the clock distribution network and the clocked sequential elements (latches and flip-flops) are the largest components of the total power dissipation, accounting for 20% to 45% of the total power. The flip-flops and the last branches of the clock distribution network that drive them consume 90% of the total clock power. The reason for this large power consumption of the clock system is that the transition probability of the clock is 100% while that of the ordinary logic is less than 10% in average. It is therefore desirable to minimize the power consumed by the clock distribution and sequential elements.

To accomplish this, we start from the dynamic power equation

$$P = CfV^2 \tag{15.1}$$

where C is the total switched capacitance, f is the operating frequency and V is the supply voltage. Lowering the clock frequency contradicts the basic trend outlined in Section 1 of pursuing higher operating frequencies through design and process technology scaling. This is only feasible on a part-time basis, like lowering the clock frequency during idle periods of time in mobile computing devices. Lower active power is best achieved by reducing the voltage and switched capacitance of the design. Since the voltage is squared in the power equation, it has a larger impact. Kojima et al. [234] proposed a half-swing flip-flop (HSFF) design, where the voltage swing of the clock is reduced to half the operating voltage. The HSFF requires four clock signals as shown in Figure 15.20. Two clock phases with a swing between Vdd and Vdd/2 drive the PMOS devices, while the other two phases with a swing between Gnd and Vdd/2 drive the NMOS transistors. A theoretical analysis of this scheme shows that the clocking power is reduced by 75% compared to the full clock swing distribution.

Experimental savings of 67% were demonstrated on a 0.5 μm CMOS test chip with only 0.5 ns degradation in speed. However, this scheme requires additional area for the special clock drivers and suffers from skew problems between the four clock phases.

Kawaguchi and Sakurai [222] proposed a reduced clock swing flip-flop (RCSFF) that needs only one clock signal that swings between Gnd and Vck, where Vck is lower than Vdd, as shown in Figure 15.21a. To control the leakage of the pull-up P-transistors driven by this low swing clock, a well-biasing technique is used to increase the threshold voltage of these devices. Several reduced swing clock drivers can be used, as shown in Figure 15.21b. Type A drivers use the same Vdd supply as the rest of the

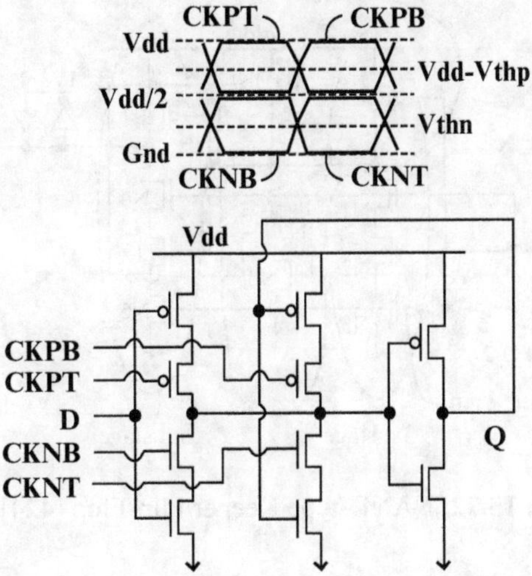

Fig. 15.20. Flip-flop design using half-swing clock [234].

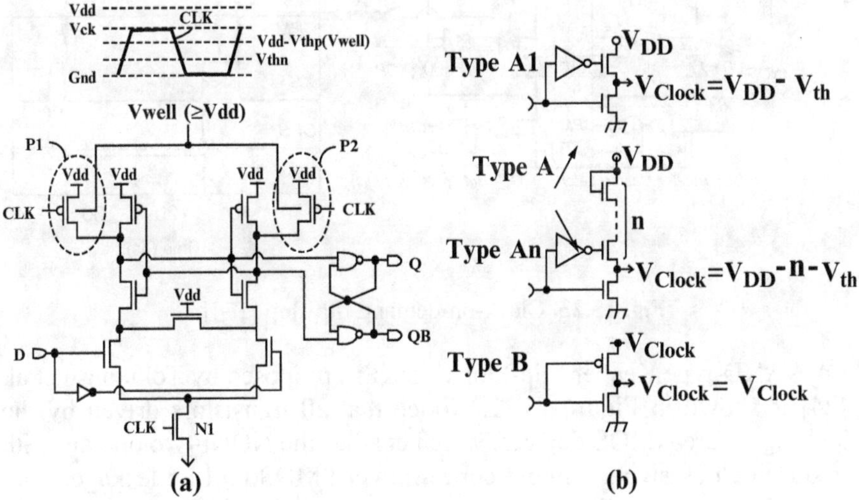

Fig. 15.21. Reduced clock swing flip-flop [222].

core, while Type B use an external Vclock supply. While this scheme can achieve up to 63% clock power reduction, it requires additional layout area for the well biasing scheme.

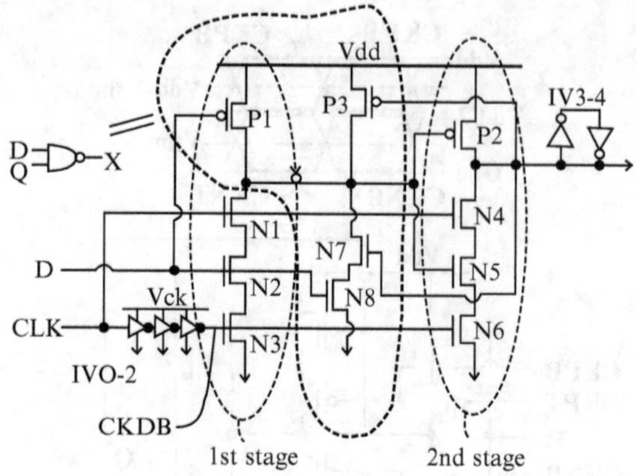

Fig. 15.22. NAND-type Keeper Flip-Flop [424].

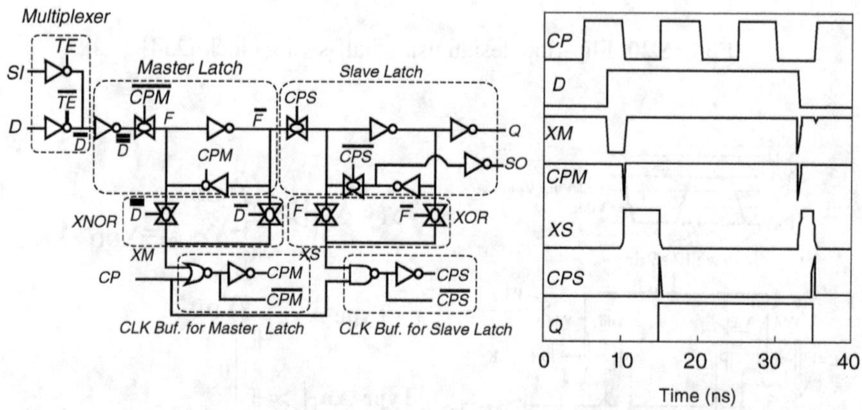

Fig. 15.23. Clock-on-demand flip-flop [174].

A NAND-type Keeper Flip-Flop (NDKFF) proposed by Tokumasu et al. [424] is shown in Figure 15.22. Notice that all transistors driven by the clock signal are NMOS devices, which enables the NDKFF to operate with a reduced clock swing without concern over PMOS pull-up leakage. The-NAND-type keeper eliminates unnecessary transitions at the internal node X, further reducing the power consumption.

Hamada et al. [174] propose a conditional clocking flip-flop shown in Figure 15.23. In this design, the internal clock is activated only when the incoming data will change the state of the flip-flop. This amounts to the finest granularity (single bit level) of clock gating. The conditional clocking flip-flop generates a self-aligned pulsed clock internally, that enables the latch circuit to operate like an edge-triggered flip-flop. The design has an exclusive-NOR whose output gates the clock buffer for the master latch and an exclusive-OR whose output gates the clock of the slave latch. The proposed design consumes less power than a conventional flip-flop when the data transition probability is less than 55%. The setup time is more than double that of the conventional flip-flop because the conditional clocking flip-flop has to decide whether the clock pulse is required before the external clock rises. On the other hand, the hold time is a larger negative value. The cell area increases by 33%. Experimental measurements on a 90 nm MPEG4 codec chip using the conditional clocking flip-flop for 24% of the total storage elements showed that there is no degradation of the maximum operating frequency due to the worse setup time of the conditional clocking flip-flops. Power is reduced by 8–31% depending on the input vectors.

Another approach to reduce the clock power is to use dual-edge triggered storage elements that can achieve the same throughput as single-edge clocked flip-flops at half the clock frequency. Nedovic et al. [308] proposed the design shown in Figure 15.24. The first stage consists of two symmetric pulse-generating latches that create data conditioned clock pulses on each edge of the clock. The second stage is a 2-input NAND gate, effectively used as a multiplexer.

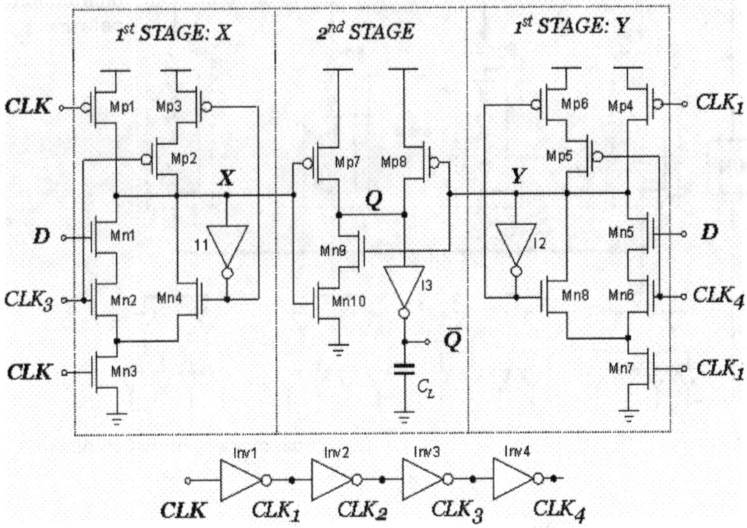

Fig. 15.24. Dual edge flip-flop circuit [308].

Simulation results in 0.11 μm technology show an energy-delay product and delay comparable to the best single-edge designs. The clock load of this design is similar to the clock load of single-edge flip-flops used in high-performance processor designs, allowing a power savings of about 50%. The main drawback of dual-edge clocking is that it requires tight control of the clock duty cycle, as any variation in the clock duty cycle becomes clock skew.

Future directions in clock distribution

Mizuno and Ishibashi [292] proposed using distributed voltage-controlled oscillators (VCO) to generate local clocks as shown in Figure 15.25. Metal lines of equal length l short the outputs of the VCOs to minimize the skew between the multiple oscillators. The voltage Vc is the frequency control signal for the VCOs that is distributed across the chip instead of the global clock signal. Careful shielding and filtering are required to insure the noise immunity for this analog voltage level.

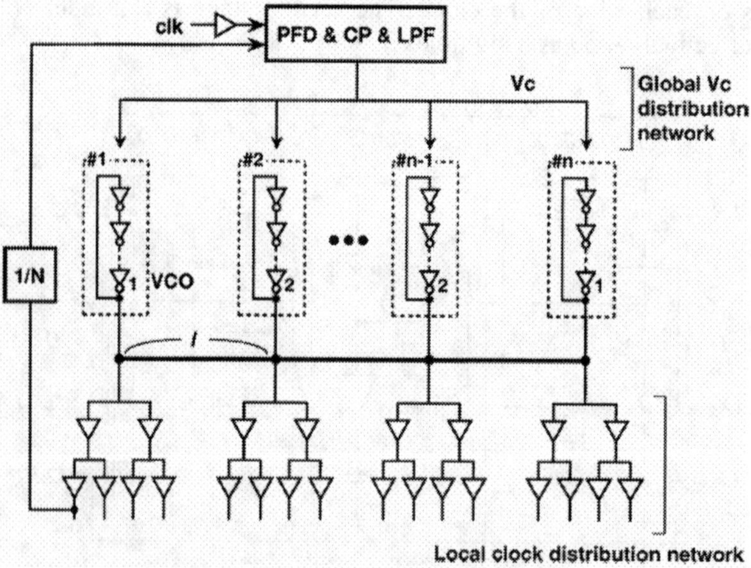

Fig. 15.25. Clock distribution using synchronous distributed oscillators [292].

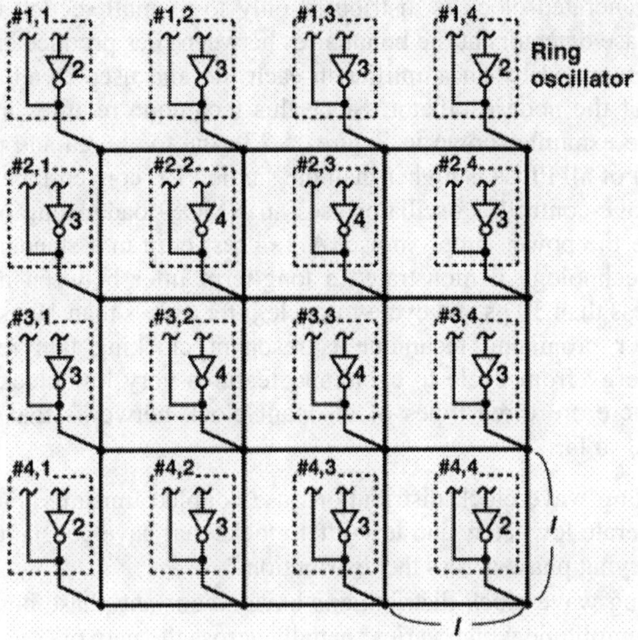

Fig. 15.26. Matrix configuration of synchronous distributed oscillators [292].

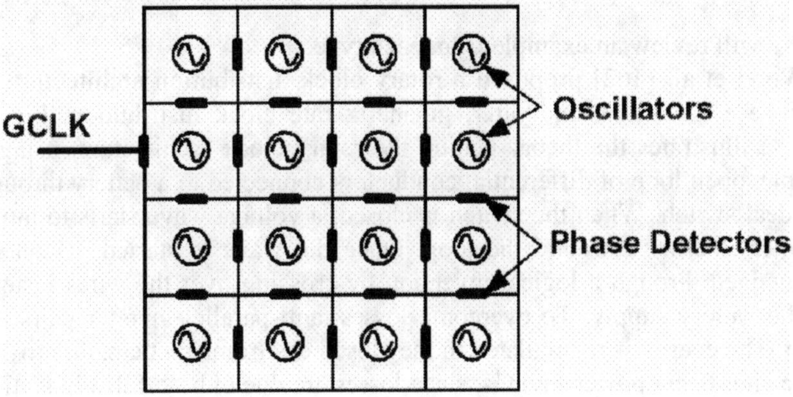

Fig. 15.27. Distributed phase locked loops [170].

A two-dimensional matrix configuration distributes the VCOs over the entire chip area, as shown in Figure 15.26. Using this scheme, each VCO is placed close to the local clock distribution network. A test chip fabricated in 0.25 μm CMOS technology achieved a mean skew of 17 ps.

Gutnik and Chandrakasan [170] propose distributing phase-locked loops (PLLs) at multiple points across the chip, as shown in Figure 15.27. Each locally generated clock is distributed only to a small section of the chip (tile). Phase detectors at the boundaries between tiles produce error signals that are summed by an amplifier in each tile and used to adjust the frequency of the node oscillator. Since this technique requires many nodes (16 in the example shown in Figure 15.27), the total area and power consumption of all PLLs is higher than the single PLL conventional approach. The voltage-controlled oscillator uses an NMOS-loaded ring oscillator to minimize the power supply noise. A 4×4 test chip in a standard 0.35 μm CMOS technology demonstrated a long-term jitter between neighboring tiles of less than 30 ps and cycle-to-cycle jitter of less than 10 ps.

Another promising technique is resonant clocking that recycles the clock energy from cycle to cycle and features very low clock skew and jitter. There are three types of resonant clock networks that have been published so far:

- Traveling wave clock distributions use coupled transmission line rings to generate low-skew and low-jitter clocks that have constant amplitude but varying phase across the distribution
- Standing wave clock distributions have a constant phase but produce a clock amplitude which varies spatially across the network
- Coupled LC oscillators that have uniform phase and uniform amplitude across the clock grid

We will review an example from each type.

Wood et al. [462] proposed a rotary clock distribution architecture to achieve a low skew and jitter, gigahertz rate clock distribution. Figure 15.28a illustrates the theory behind the rotary clock architecture, using a simple, open loop of differential conductors connected to a battery through an ideal switch. When the switch is closed, a voltage wave starts to move counter-clockwise around the loop. Once the wave is started, it can be maintained through a logical inversion by crossing over the wires instead of the battery supply. To overcome losses, anti-parallel inverter pairs are used. The energy is recirculated in the closed electromagnetic path, providing a significant power savings, since losses are due only to I2R dissipation in the wires and not CfV^2 dissipation as in the conventional clock distribution. Figure 15.28b shows the layout of a rotary clock with 25 interconnected rings. Each ring consists of a differential line driven by anti-parallel inverters distributed around the ring. The clock wave frequency depends on the electrical length of the ring and the inductance and capacitance of the lines. A prototype circuit was built in a 0.25 μm 2.5 V CMOS process and has a 12000 μm long ring with 60 μm conductors on a 120 μm pitch. Simulations predicted a clock frequency of 925 MHz, while measured waveforms clocked at 965 MHz. Jitter was measured to be 5.5 ps rms.

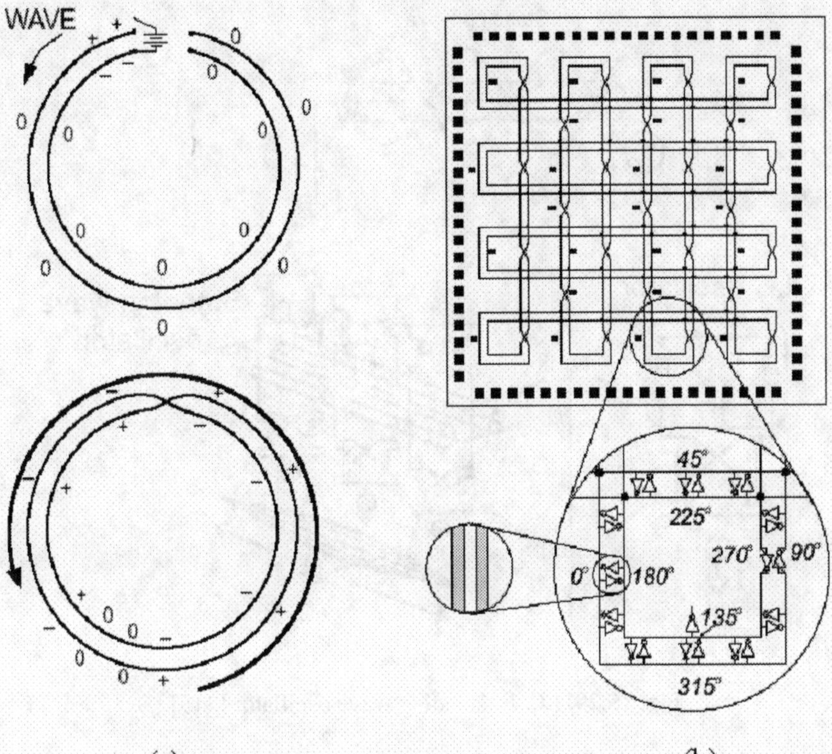

(a) (b)

Fig. 15.28. Rotary clock distribution architecture [462].

O'Mahony et al. [315] describe a 10 GHz standing-wave clock distribution system that achieves sub-picosecond skew and jitter using on-chip interconnects and distributed buffers to create a network of coupled oscillators. Standing waves have the same phase at all points, as opposed to the rotary clock scheme discussed earlier that generates traveling waves. A standing-wave oscillator is similar to a differential LC oscillator (both shown in Figure 15.29) where the gain and tank are distributed. The NMOS cross-coupled pairs provide enough gain to compensate for wire losses. The PMOS diode-connected loads set the common mode voltage and allow injection of an external clock reference.

Figure 15.30 shows a prototype standing wave oscillator clock network implemented in a 0.18 μm, 1.8 V CMOS process with six AlCu metal layers. The differential λ/2 lines are 3 mm long, 14 μm wide and are 4 μm apart in metal 6. The clock jitter added by the clock grid is below 0.5 ps. The measured skew is 0.6 ps (0.6%) when the grid is tuned to 10GHz with

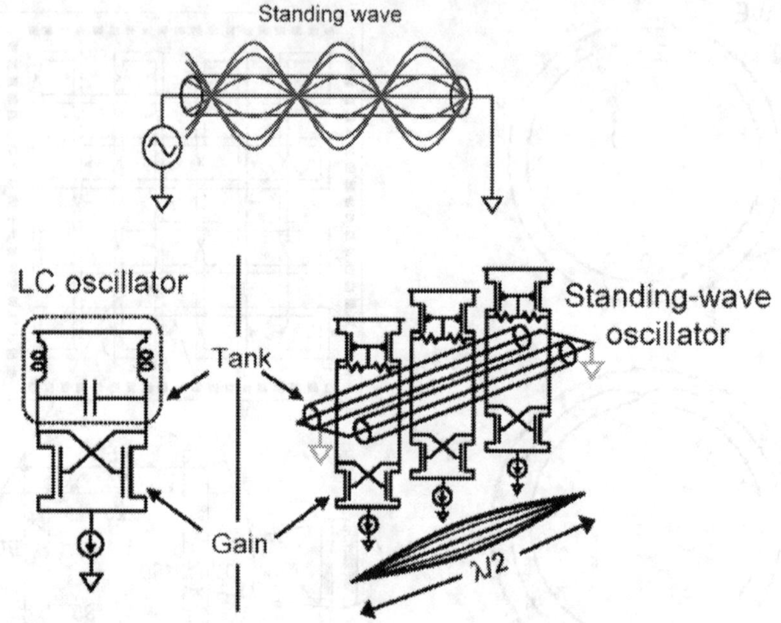

Fig. 15.29. LC vs. standing wave oscillator [315].

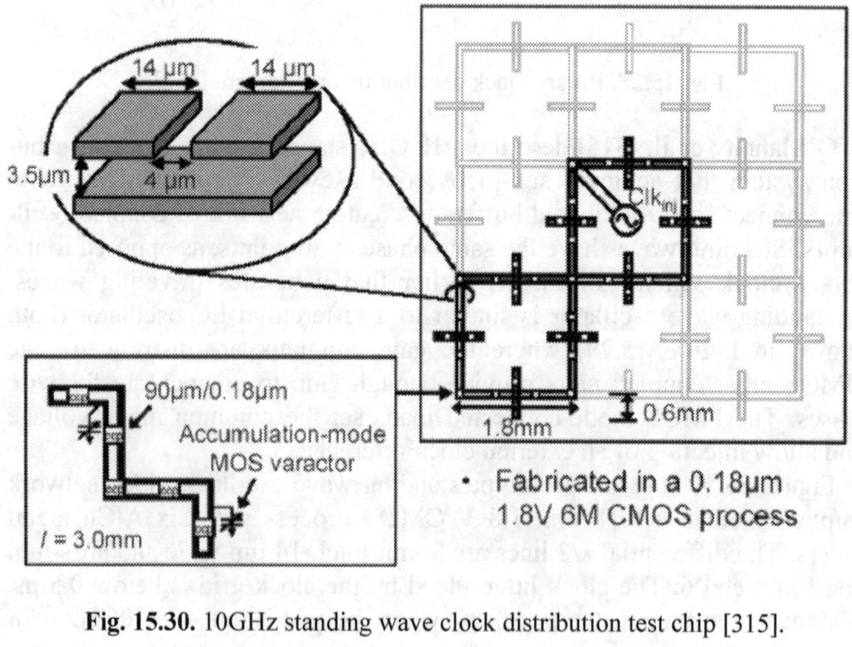

Fig. 15.30. 10GHz standing wave clock distribution test chip [315].

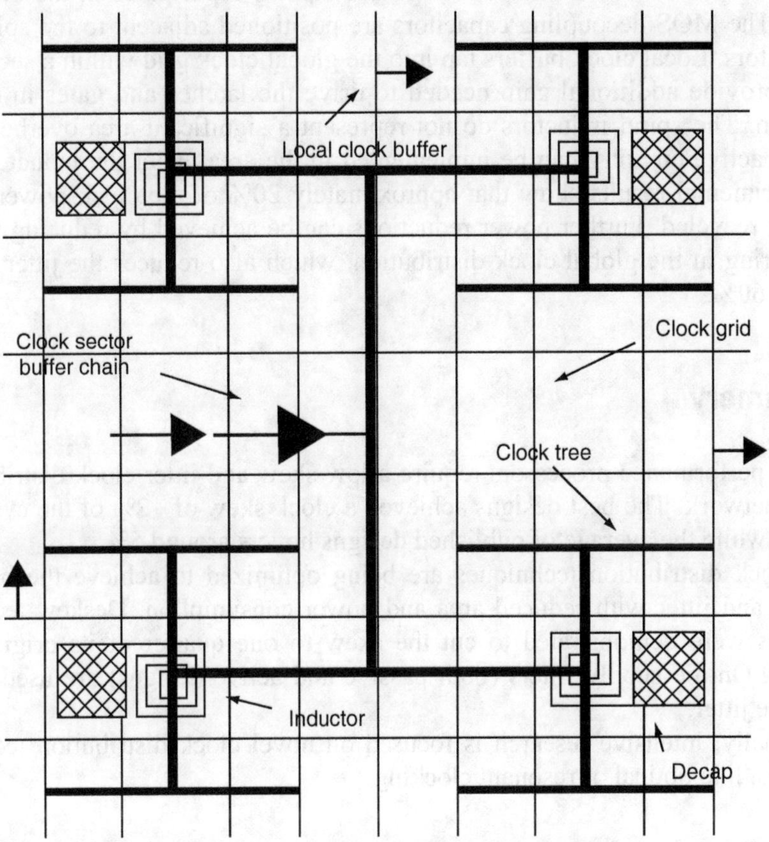

Fig. 15.31. Components and topology of a resonant clock sector [71].

a single control voltage for all varactors and 3.2 ps when half of the grid is de-tuned by 1%. The worst-case skew between any two adjacent points is 1.4 ps for the de-tuned grid.

Researchers from IBM and University of Columbia designed a uniform-phase, uniform-amplitude resonant-load global clock distribution that preserves the frequency and area scalability of a traditional design [71]. In this implementation, a tree-driven grid is rendered resonant with a set of discrete on-chip spiral inductors distributed throughout the clock network. The large clock capacitance resonates with this inductance, reducing the gain required in the distribution, which in turn saves power, reduces skew-through a reduction in clock latency, and reduces power-supply noise-induced jitter. Figure 15.31 shows the topology of a resonant clock sector using a single-ended topology. Four spiral inductors are attached to the clock tree on one end and to a large decoupling capacitance at the other end. The MOS decoupling capacitors are positioned adjacent to the spiral inductors. Local clock buffers tap into the global clock grid within a sector and provide additional gain needed to drive the latches and gates in the design. The spiral inductors do not represent a significant area overhead, since active circuitry can be implemented in the area under the inductors. Experimental results show that approximately 20% of the clock power is being recycled. Further power reductions can be achieved by reducing the buffering in the global clock distribution, which also reduces the jitter by up to 60%.

Summary

High-performance processors require a low skew and jitter clock distribution network. The best designs achieved a clock skew of ~2% of the cycle time, while the average of published designs hovers around 5%.

Clock distribution techniques are being optimized to achieve the best skew and jitter with reduced area and power consumption. Deskew techniques were demonstrated to cut the skew to one quarter of its original value. On-die supply filters (both passive and active designs) are used to reduce jitter.

Finally, intensive research is focused on novel clock distribution techniques, like optical or resonant clocking.

16 Asynchronous and Self-Timed Processor Design

Jim Garside and Steve Furber

The University of Manchester, UK

The great majority of microprocessors are clocked; that is, their internal operations are controlled and timed by an external periodic timing reference known as a clock signal. Clocks are convenient, simplify design, and are the foundation upon which most design automation tools are built. However, clocks are not all good news. They cause a circuit to generate excessive electromagnetic interference, to dissipate excessive power, and they force all circuit functions to operate at the same rate however unnatural that may be for any particular function. The challenge of designing a microprocessor that operates without a central clock has appealed to some designers for many years, and the results of their research have now made fully clockless designs not only feasible but also commercially available, albeit still as a minority interest. In this chapter we survey some of the developments in asynchronous processors and speculate as to where these might lead in the future.

Motivation for asynchronous design

Asynchronous circuits are a Bad Idea; all electronics students are taught this early in their courses. This is because the synchronous model removes one of the more unpleasant 'unknowns' from the design process, simplifying debugging and allowing the designer to concentrate on the logical correctness of the system. Asynchronous inputs, such as interrupts, should be synchronised as soon as possible to confine timing problems to tiny areas of the device.

J. Nurmi (ed.), Processor Design: System-on-Chip Computing for ASICs and FPGAs, 367–389.

So why would anyone want to make something as complex as a microprocessor without a clock? There are several possible answers to this.

Power consumption

The majority of the power in CMOS circuits is dissipated during switching. A fundamental assumption of the synchronous model is that all state holding elements are clocked regularly and simultaneously, yet many change relatively infrequently. The necessity of distributing a fast clock to arrive at every part of a system with the same phase can itself consume significant power. Up to 40% of the overall power budget can be expended on the clock signal. In power sensitive applications – an increasing proportion of all applications – clock gating may be applied to reduce the consumption in parts of the device which are temporarily idle. However this, in itself, is a complication, especially to the clock distribution network as it can introduce undesirable skew if not managed correctly.

An asynchronous circuit is inherently *event driven*: instead of being clocked it processes on demand and stops when not needed. This applies to every part of the device and is inherent and automatic, thus any subcircuit – or, indeed, the whole processor – will shut down and restart instantaneously, using only the power required to do the job.

Clock distribution and modularity

As integration levels continue to rise an increasing number of different subsystems can be integrated onto a single device. Often these take the form of commercial IP (Intellectual Property) blocks making up an SoC (System-on-Chip) where processors and peripherals may be imported from a number of sources. Achieving *timing closure*, i.e. getting all these devices to work together at the desired clock rate, is an increasingly difficult task, especially as the clock frequency tends to rise as well as the number of devices increasing. Of course it may not be necessary to run all the subsystems at the speed of the fastest, but synchronous design makes this inherently attractive.

Even in terms of processor design this is a problem. A modern processor is a complex system in itself and there may be several processing cores in a single 'processor'. Getting these to work together at high frequencies is an expensive business.

In an asynchronous system each subsystem processes at its own rate; a slow module will therefore slow down the system – at least when that module is in use – but will not cause a functional failure. Asynchronous systems can therefore integrate different subsystems more conveniently than a globally synchronous system. Of course, if certain timing constraints are to be met then faster modules may be needed, but these can be inserted relatively painlessly because there is no global effect. In software terms, the clock is a global variable and global variables always cause problems!

The degree of asynchronicity varies from system to system. Currently an attractive compromise appears to be GALS (Globally Asynchronous, Locally Synchronous) systems where timing closure is addressed using an asynchronous bus or network between clocked components. However this chapter concentrates on the extreme example where there are no clocks within the processor either.

EMC

In some applications, especially those involving radio, electromagnetic compatibility is a big issue. In a synchronous circuit the peak current demands on the power supplies are correlated by the clock. This produces large, regular current spikes which are ideal for radiating radio-frequency energy concentrated in particular clock harmonics. In an asynchronous circuit switching occurs at different times in different parts of a device and, often, will be at irregular intervals. The supply current demands are therefore much more constant and, with lesser *changes* in current, the radiation is much lower in intensity as well as being broad spectrum. In particular, 'spikes' at harmonics of the clock are not apparent (Figure 16.1).

Transistor variation

The tremendous impact of digital electronics has been achieved by the huge progress made in integration by shrinking components. However transistors have now shrunk to the size where their 'strength' is becoming harder to predict, because variations in the doping and gate length are, proportionately, large. Coupled with the enormous number of transistors on a chip it is hard to determine the maximum clock frequency at which each chip can function reliably.

Self-timed logic offers a possible way to bypass this problem; if the chip is correct by design it will function when manufactured. Operations which turn out faster than typical will be exploited automatically; those which are slower will impact performance, but not significantly if they are rare.

Of course this does not help in marketing where there is no 'GHz' number to quote and two 'identical' devices will show different speed behaviour in a side-by-side comparison!

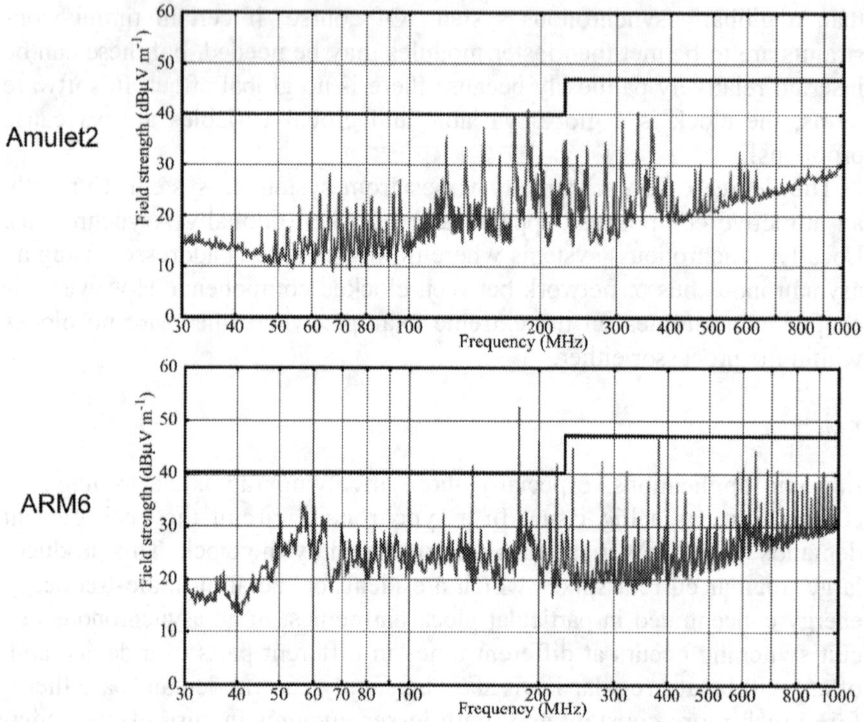

Fig. 16.1. Electromagnetic emissions of comparable synchronous and asynchronous ARM devices.

The development of asynchronous processors

Before looking at some detailed aspects of asynchronous processor design it is interesting to look at the history of the subject, especially developments through the 1990s.

Many of the early vacuum tube computers were asynchronous in their design principles, but they employed a number of ad-hoc techniques and manual adjustments that made their design approaches difficult to scale to the complexity of modern design.

Probably the first systematic approach to asynchronous design was that employed on the Macromodules project at Washington State University, completed in 1974 [82]. Here a number of basic functional modules were developed with asynchronous interfaces that allowed the modules to be reconfigured to implement a wide range of computational functions. The system was robust, scalable, and flexible in a way that would have been very hard to deliver using the synchronous technology of the time.

The first asynchronous microprocessor (that is, processor on a single chip) was designed at Caltech by a group led by Alain Martin and fabricated in 1989 [275]. This milestone design incorporated a 16-bit datapath and performed all of the functions expected of a microprocessor apart from handling interrupts and exceptions.

The first asynchronous microprocessor that implemented a commercial instruction set architecture was Amulet1 [463], which was developed between 1991 and 1993 at the University of Manchester in the UK. Amulet1 was an asynchronous re-implementation of the ARM 32-bit RISC architecture [133]. It executed the ARM6 instruction set, and had full support for interrupts and exceptions (including support for memory management faults). Amulet1 was followed in 1996 by Amulet2, which was integrated into the Amulet2e embedded controller [134], and in 2000 by Amulet3 [142], an ARM9-class processor developed in collaboration with a German telecommunications company who wished to use it in an ISDN-DECT base-station on account of its superior electromagnetic emissions properties compared with a clocked equivalent. Unfortunately the industrial collaborator ran into financial difficulties just as the part became available on silicon, and Amulet3 never went into a commercial product. The complete chip incorporating Amulet3 was called DRACO. Die plots of the Amulet processor family are shown in Figure 16.2.

During the 1990s, in parallel with the Amulet developments at Manchester, a group first at Tokyo Institute of Technology and later the University of Tokyo developed the TITAC series of asynchronous microprocessors. TITAC1 was an 8-bit microprocessor with a bespoke instruction set [306], whereas TITAC-2 was an asynchronous re-implementation of the MIPS instruction set architecture [406].

All of the microprocessors mentioned so far were implemented using full custom manual layout techniques, but by the late 1990s the mainstream ASIC and SoC businesses were leaving full custom 'hard core' processors behind. They were steadily being replaced by synthesisable 'soft' IP cores. The problem for asynchronous processors was (and to some extent still is) that synthesis tools for asynchronous circuits are some way behind their equivalents for clocked circuits. However, the future clearly lay with synthesis, so this problem had to be addressed.

At the lower performance levels asynchronous synthesis tools emerged during the 1990s: the most advanced was Tangram, developed by Philips in the Netherlands. Tangram was used to synthesise an asynchronous 80C51 [224] and this processor has enjoyed considerable product success since its introduction. It was first used on pager chips, but Philips pulled

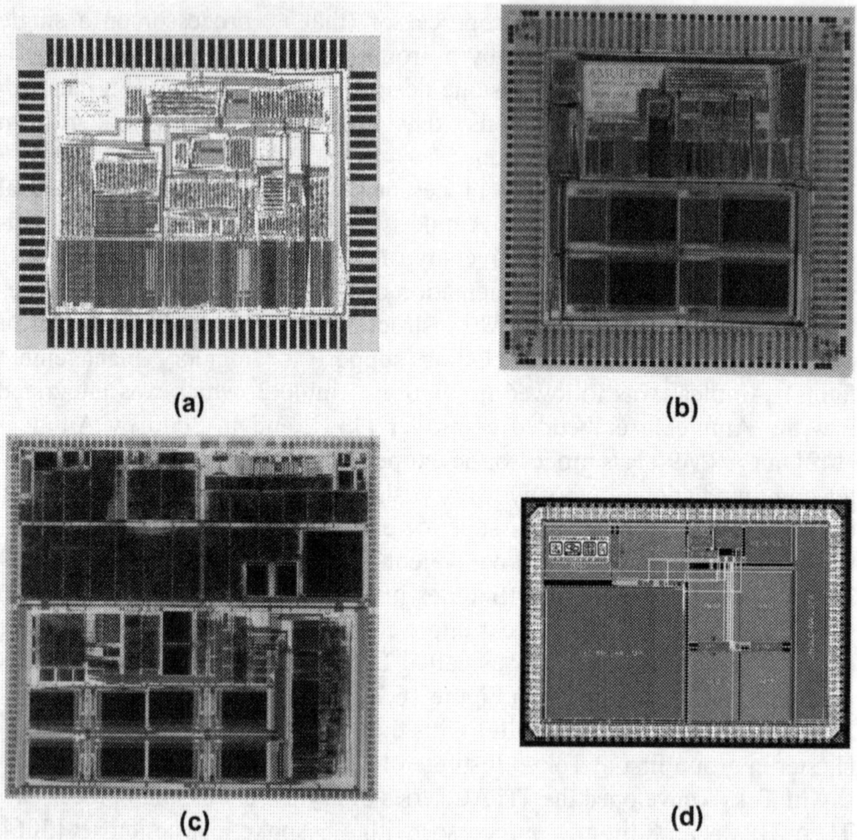

Fig. 16.2. The Amulet processor family: (a) Amulet1; (b) Amulet2e; (c) DRACO; (d) SPA.

out of the pager business soon thereafter. It was then used to great effect on a range of contactless smart cards, where its excellent power-efficiency enabled the on-card system to make the most of the meagre power delivered through the RF link. Since then it has found use on a number of small microcontrollers and related products, and has led to Philips becoming by far the biggest commercial producer of asynchronous semiconductor products world-wide.

During the late 1990s, the Manchester group were also developing a synthesis system for asynchronous circuits along the same lines as Tangram, in this case called Balsa [116]. Balsa was used to develop the fourth Amulet processor, SPA [337], as part of a European-funded project investigating the contribution asynchronous logic could make to smart-card security, in particular through the intrinsic resistance of well-designed asynchronous logic to differential power analysis and EMC attacks [470].

A die plot of the SPA chip is shown in Figure 16.2. SPA was very slow compared with the earlier, custom Amulet processors. Some of this lack of performance could be put down to the primary research objective of the work, which was security, not speed. Quite a lot of performance was lost due to this being the first use of Balsa to develop a 32-bit processor, and a fair number of (in retrospect) avoidable mistakes were made in the course of the design work. But after all of these factors have been removed, it is still the case that it is difficult to synthesise an asynchronous processor that can compete with a clocked processor in performance.

Finally, however, in 2006 the ARM966 was announced. ARM966 is the result of a collaboration between ARM Ltd. and Handshake Solutions, a Philips company that was formed to commercialise the Tangram (now renamed 'Haste') tools. Although it is still not competitive with a clocked ARM9 for performance, the ARM966 has respectable performance combined with excellent power-efficiency, and therefore meets a market need. Following on from the established success of the Philips asynchronous 80C51 at the low end, there is now a commercially available and supported 32-bit asynchronous processor.

The asynchronous microprocessors mentioned above all implement conventional instruction set architectures, but these were designed with clocked processors in mind. Might it not be the case that asynchrony offers an opportunity for a radical departure in instruction set architecture? Many have felt this to be the case, and there have been a number of highly innovative attempts to exploit this opportunity. Of particular note are the Sun Labs counterflow pipeline architecture [387] and the FLEET communications-oriented architecture [84]. None of these more radical designs have yet been proven to offer advantages on silicon, but the world of silicon is forever moving on, and asynchronous design may be just what is required to cope with the variability and unreliability that is forecast to be a feature of near-future deep submicron process technologies.

Asynchronous design styles

The defining characteristic of asynchronous logic is, of course, the absence of a clock. Bar that there are numerous and varied approaches to constructing the circuits themselves. Perhaps fortunately, these need not be examined in detail here, but a brief overview is useful.

There are two 'main' philosophies to asynchronous logic: the first is usually called 'bundled data' and is closely analogous to conventional logic design. In a bundled-data circuit a datapath is produced and its timing is modelled by a delay circuit which substitutes for the clock in a synchronous circuit. This timing is local, so different parts of a device will have their own timing models and run at their own speeds. The delay can also be varied on a 'per cycle' basis according to the function being performed: for example an ALU might perform an addition with one delay but a logical operation with a different, shorter delay.

The production of a reliable delay model is the most obvious problem with this approach. However it is not always as difficult as it might appear; in many circuits it is possible to add an extra 'bit' to the datapath which has the same delays as the data evaluation and, because it is in close proximity on the silicon, tracks manufacturing, temperature and supply voltage variations appropriately. The silicon overhead for timing is therefore small.

An alternative philosophy is to embed the timing information with the data. The 'purest' realisation of such codes are known as 'delay insensitive' because the circuit is guaranteed to function even with arbitrary delays added to any gate or wire. Unfortunately, constructing real circuits in such a manner is very difficult, but a close approximation, known as 'quasi delay insensitive' (qDI) – where it is assumed that an isochronic fork[1] is possible – has been widely explored.

The simplest illustration of a qDI code is a dual-rail encoding, as shown in Figure 16.3. Here a single bit is encoded on two wires. The wires remain 'at rest' until a bit needs to be signalled, then one makes a transition. When the transition arrives – after any arbitrary delay – the bit has arrived, and which wire was used indicates the state of the bit. The handshake is completed with an acknowledgement, only after which can the data change again. A single acknowledgement can be used for more than one data bit if the bits are resynchronised by the receiver first.

Such a scheme carries a significant overhead both in silicon area (which is roughly doubled over a synchronous or bundled-data implementation) and power consumption. However it is immune from any variations in delays which makes it attractive for synthesis where wire delays may be unknown before layout. It could also be useful in future processes where manufacturing variations in transistors make the exact prediction of circuit speeds difficult.

The above styles are not mutually exclusive. A nice example is a ripple-carry adder: in a synchronous system this is a very slow circuit because the clock period must accommodate the worst-case delay, even though that occurs very rarely because, typically, carries are 'generated' and 'killed' at

[1] An isochronic fork is where it is assumed that a signal reaches all its destination gates/modules at essentially the same time, i.e. the wire delays are the same for all branches of a fan-out.

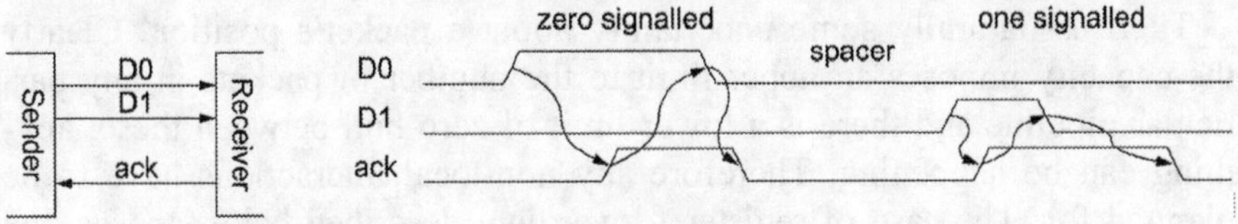

Fig. 16.3. A dual-rail encoded handshake interface.

intervals throughout the word. In an asynchronous ('self-timed') adder it is possible to detect that the carry signal has arrived at every bit position using a dual-rail signal. Modelling the time for a single-bit addition is straightforward; therefore an adder can be built without expensive carry look-ahead (or similar) which can perform, *on average*, competitively with a more expensive synchronous one [140].

The exact style of the implementation is not particularly relevant to the architecture of an asynchronous processor. Whilst bundled data may not be 'pure' at the circuit level, the handshakes connecting modules are still delay-insensitive. Implementation detail is therefore largely neglected in subsequent sections of this chapter.

Asynchronous pipelines and handshakes (locality)

Before describing some implementations specific to asynchronous processors a brief summary of the way asynchronous systems have been constructed to date may be useful. Foremost amongst these is the extensive use of pipelining, particularly as this is very easy in this paradigm.

An asynchronous pipeline is like a queue of traffic: a number of packets (vehicles) travel serially along a pipeline (road); when there is a gap ahead a packet can move forwards into it, reproducing the gap behind it. There is no compulsion that a packet has to move at any particular time but in order to do so it has to satisfy two conditions: it has to be there and there has to be a gap ahead. It can choose to move any time after this local synchronisation has been performed.

Synchronisation is performed by handshake signals. In a bundled-data implementation these will be explicit 'request' and 'acknowledge' signals; with a dual-rail protocol the request is implicitly coded with the data. The principles are the same in each case:

- the sender asserts request;
- when the receiver is ready it accepts the data;
- the receiver asserts acknowledge.

(It is often convenient to de-assert the signals too, although it is not essential to the signalling process and single (either edge) transitions have been used in some implementations [398].)

There is naturally some uncertainty about a packet's position. Clearly the capacity imposes an upper limit to the number of packets in any particular pipeline and there is a lower limit of zero but, between these, anything can be happening. Therefore any non-local interactions have to be planned for. The case of register forwarding, described below, poses one such problem because it is not known at issue time if, when and to where an instruction's results should go (other than to the stated destination) without predicting future instructions.

Features of asynchronous design

The following section describes some of the more novel possibilities which are on offer to the asynchronous processor designer. In most cases examples of their employment are given. However, before beginning this, it is worth looking at some of the 'tricks' available to a synchronous designer which must be abandoned. Some of the opportunities available in an asynchronous architecture are then outlined with, in many cases, specific examples of their exploitation.

Assumptions in synchronous design

Whatever detailed design style is used, asynchronous circuits generally communicate by handshaking with their neighbours. At a macroscopic level, this is the significant difference from synchronous circuits. A synchronous designer can make assumptions as to where data will be at a given time and *know* that a copy can be taken from a particular place at a particular time.

A good – and relevant – example is register forwarding in a microprocessor. If an instruction needs to read a particular register in a given cycle it can often recover the relevant data from an intermediate pipeline stage because it is known that the data will be there at that time. In an asynchronous processor this is not the case: the data *may* be there, but they may not have reached that point, may have already moved on or even may be in transit at the critical time. Thus a different mechanism must be sought.

Other mechanisms have been exploited in synchronous designs, not all of which have turned out to be good ideas. Related to the forwarding example, the original MIPS design relied on the ability to resolve dependencies in software; this seemed a good idea until a re-implementation changed the pipeline structure. A similar example is visible in the ARM architecture, where reading the PC (which appears in the register file as R15) yields a value two words advanced from the address of the current instruction; this was a consequence of the original pipeline structure but is more awkward for other implementations, both synchronous and asynchronous.

Elasticity

In a synchronous environment a designer must be aware of the effect of pipeline stalls at the system level. Stalls are undesirable not only because of the performance penalty but because all parts of the system need to adapt their behaviour. This can influence the macroscopic design; for example the desire for 'single-cycle execution' may exclude certain desirable instructions from an ISA because the implementation cost is unreasonably high.

The ARM architecture has always contained multiple register loads and stores (LDM/STM) which can transfer up to the whole 16 registers from or to memory. Clearly these are cases for multi-cycle operations. However in an asynchronous pipeline only the instruction decoder has to recognise this. When the instruction is decoded the decoder can iterate, producing a sequence of sub-instructions for later pipeline stages. The 'upstream' process will indeed stall, but this is simply a consequence of the LDM/STM appearing to be slow at clearing the decoder and is handled in the normal fashion by the pipeline 'backing up'. 'Downstream' the additional cycles are not distinguishable from any other form of instruction.

Another example of this process, exploited in Amulet3 [142], was to split certain (rare) multiplication instructions into two at the decoder. Due to the microarchitecture of the multiplier – which, internally, operated in several cycles although appearing, externally, to take one elongated cycle – this imposed no performance penalty, required no special measures in other parts of the system but saved both a read and a write port on the main register bank.

Just as it is possible to insert cycles within the pipeline it is equally possible to remove them. Perhaps the commonest example is a compare (CMP) instruction which is similar to a subtraction but does not produce a (full width) result. This instruction can 'disappear' after the ALU, saving the need for a 'dummy' writeback cycle, reducing power consumption and, possibly, increasing speed.

Halt

It has been suggested that asynchronous processors may not be the fastest devices for doing 'something' but are very good at doing nothing! Because the system consists of handshaking components, if one part stalls then interacting parts will have to wait and inactivity spreads very rapidly across the whole system.

In low power CMOS technologies a halted system dissipates very little power. However in the synchronous world this involves not only disabling registers but disabling the clock and, possibly an oscillator and PLL. In this last case the latency involved in resuming operation can be very significant and is often a deterrent to using 'deep sleep' modes, even when they are available; however if it is not done the power consumption can be too high for many embedded applications.

In contrast halting an asynchronous processor is extremely easy. The halt can be instigated anywhere in the system (i.e. wherever it is most convenient) and no other parts will need to be modified. Halt spreads 'naturally' as units stall waiting for input or output and, when the halt is rescinded, the system frees up in a similar manner. Full speed operation recommences instantly from the deepest sleep. This makes such systems attractive in applications where there are frequent short bursts of activity, perhaps in some response to a real-time input [433].

Data-dependent timing

Some operations take longer than others to perform. For example an addition will take longer than a bit-wise AND because of the need for carry propagation. Although it is not necessary to do so, it is possible to exploit this in an asynchronous system. This is particularly attractive when common operations can be performed quickly and the 'worst-case' is rare. As has been mentioned already, it may also be possible to exploit the data patterns within the adder to allow early completion or, alternatively, reduce hardware complexity at the cost of the occasional longer delay.

An incrementer is an excellent example of a structure which can exploit data-dependent timing. An incrementer always changes the least significant bit of a word: then, successively either it changes the next most significant bit (if the current bit went to '0') or it has finished. Although, in the worst-case, this process can be very slow the *average* number of bit changes is (just under) two, regardless of word length and every alternate incrementation in a series changes only one bit. Given a small amount of elasticity in the pipeline, implementing (e.g.) a program counter with a ripple carry adder is cheap yet effective.

Other opportunities to exploit data-dependent timing can be found at all levels of the system. Some examples, in brief, can include:

- Multiplier: this may be a sequential machine (internally) with early termination determining the number of cycles performed; externally it is just a block with a variable delay.

- Memory access: it is possible to exploit (for example) sequential access in a cache line where a hit may be on the line last read. This can then be faster than a 'random' hit which, in turn, is faster than a cache miss. Such behaviour will influence the performance of a synchronous system (in terms of stalling for a number of clock periods) but is already implicit in an asynchronous system and is useable at much finer 'grain'.

Non-determinism

With the possible exception of inputs such as interrupts, a synchronous microprocessor is a fully deterministic system; given a particular system state its future behaviour can be predicted exactly. The same can be true of an asynchronous microprocessor, but the latter can also be made to include some non-deterministic behaviour whilst still executing a program correctly.

The advantage of allowing non-determinism comes in the scheduling of operations; as timing is not predictable it is sometimes expedient to allow the order in which events happen to vary. A reasonable example is instruction prefetch. In a processor pipeline the PC generates a sequence of addresses, each of which flows down the pipeline being translated successively into instructions, operations and, finally, results. In an asynchronous processor each of these can proceed at its own rate so the pipeline occupancy may vary; this is irrelevant whilst the pipeline flows unimpeded.

If an instruction turns out to be a branch the pipeline flow must be altered at source. One approach to this is to control the pipeline occupancy by circulating tokens so that each retired instruction allows another to be fetched; this leads to a deterministic prefetch depth. An alternative approach is for the pipeline to run freely but allow branches to interrupt the prefetch process to redirect the flow; as the time of the interruption is imprecise this leads to a non-deterministic prefetch depth.

The second approach imposes fewer constraints on the system as the pipeline normally flows freely from source to sink. However it leaves the designer with two problems: how to 'interrupt' the flow reliably with an asynchronous input and how to determine what the prefetch depth was so that speculatively fetched instructions can be discarded.

The first problem can be solved with an arbiter. In the synchronous domain arbitration is a difficult problem: making a decision before the next clock edge is always subject to a chance of failure. In the asynchronous domain it is possible to defer the decision indefinitely so that a reliable decision can always be made. (The disadvantage is that, theoretically, this resolution could take unbounded time.)

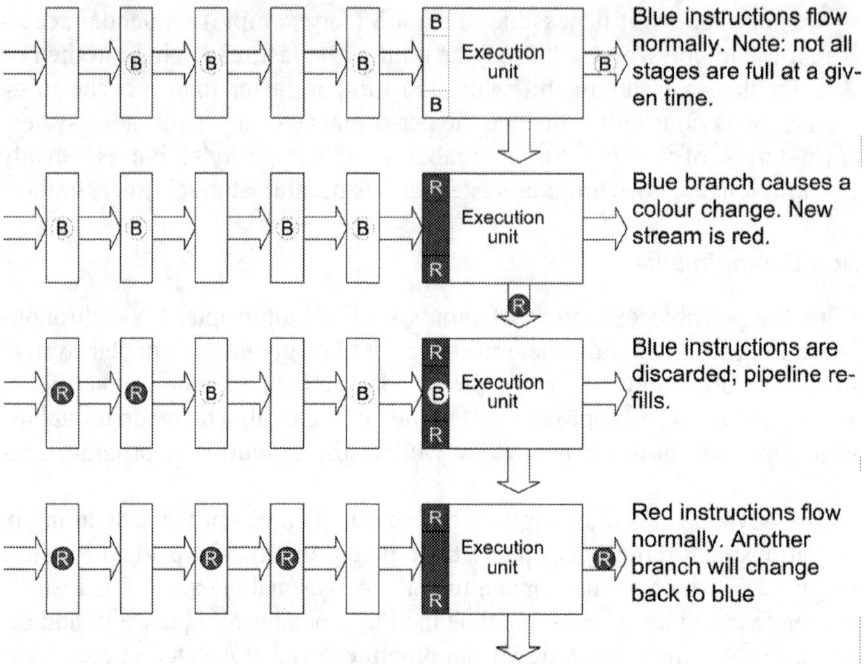

Fig. 16.4. Colour-change flush mechanism. R = red, B = blue.

The problem of an unknown prefetch depth means that the execution unit cannot simply discard the next N instructions following its committal to a branch. Instead an 'asynchronous' solution has to be found. One such – used in the Amulet devices – involves 'colouring' the addresses at the prefetch unit and changing colour when the flow changes. It operates as follows:

- prefetch a sequence of 'blue' instructions;
- execute 'blue' instructions until a branch is taken;
- interrupt the prefetch unit with the branch target address; subsequent prefetches are 'red';
- discard all 'blue' instructions; only commit to instructions which are 'red'.

Because nothing is completed following a branch until the new stream arrives it is possible to implement this scheme with a single 'colour' bit, switching back to 'blue' when a 'red' branch is taken. This sequence is illustrated in Figure 16.4.

In Amulet3 this mechanism was extended to employ two colour bits to allow late occurring data aborts. Imagine a data load is issued followed by a branch; the branch changes one colour bit as described above and the

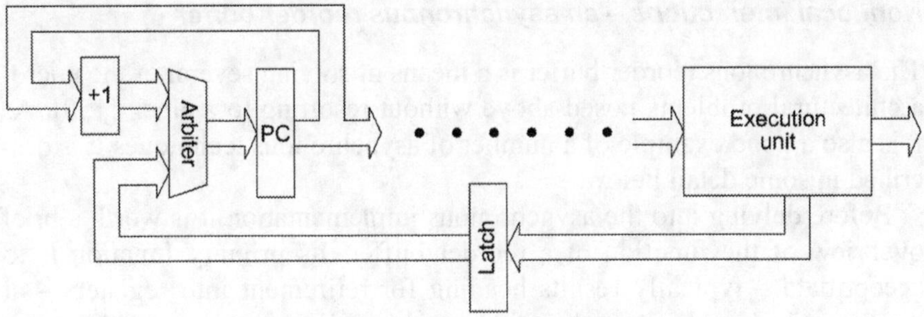

Fig. 16.5. Arbitrating branch mechanism.

prefetch colour switches. Subsequently the load aborts which means the branch should not have been taken (or at least not yet). Toggling a second colour bit allows instructions from both the original stream and the branch target stream to be discarded so that execution resumes cleanly after the trap to the abort handler.

Although non-determinism and arbitration can be convenient, they should be exploited with caution. As with any asynchronous process it is possible to introduce cyclical dependencies which result in deadlock. To continue with the branch example, suppose that the arbiter decides to ignore the interruption for a while – it can, after all, make arbitrary decisions! The branch will wait patiently whilst the prefetch unit continues to feed addresses into the pipeline. If the branch is still blocking the pipeline it will 'back up' and can reach the point where there is no free space for anything to move; the branch cannot now break the cycle because there is no free space and the system halts. In this instance, the problem is easily resolved by moving the taken branch off the main pipeline; this can then continue to flow (in theory, indefinitely) discarding prefetched operations until the branch finally manages to change the direction, as illustrated in Figure 16.5. This cannot deadlock because a second branch cannot be taken until the first has been accepted. Comfortingly, the simplest arbiters select the first arriving request so will not always choose the 'wrong' alternative.

Another 'problem' with the use of arbitration occurs during system verification. If the designer wishes to explore all states reachable by the system – a reasonable approach to a safe design – each non-deterministic element multiplies up the number of possible legal states. In general not all these states will be explored by simulation, which has its own set of timing rules, therefore other tools and models are desirable to guarantee that errant behaviour is forbidden.

Non-local Interactions – an asynchronous reorder buffer

The asynchronous reorder buffer is a means of solving several of the micro-architectural problems posed above without resorting to a clock [150]. As it is also a good example of a number of asynchronous techniques it is described in some detail below.

Before delving into the asynchronous implementation it is worth a brief overview of the function of a reorder buffer. Its primary function is to accept data – typically results heading for retirement into registers – at arbitrary times and output them in a predetermined sequence. This means that its outputs are sequenced correctly. In a processor's register bank this prevents write-after-write (WAW) hazards; it also enables a failure, such as an abort, to be processed in sequence and thus aids fault recovery. Because it may delay the retirement of data the reorder buffer's secondary function is to forward values on demand to avoid stalls whenever possible.

In the following description it is assumed that the reorder buffer is sorting results being returned to a microprocessor's registers and is able to forward the appropriate values on demand. A unit of this form was included in the Amulet3 processor, shown in Figure 16.6.

In its asynchronous implementation, four different timing domains connected with the reorder buffer can be identified, as depicted in Figure 16.7. The first domain can be subdivided into two alternative phases: in the first phase requests for forwarding are processed; this phase will be described later. Following this is an allocation phase (during instruction decode) where the result register(s) of an instruction are identified. These are then associated with numbered 'slots' in the reorder buffer in the order that they are to be read out. The slot associated with the calculation is carried forwards with the instruction packet.

An arbitrary time later the packet arrives at the reorder buffer. As each slot in the buffer has a unique number, and only one outstanding packet can have this number at any time, the value can be stored immediately. This is irrespective of other packets which may be arriving before, after, at the same or at overlapping times. Once written the reorder buffer notes the arrival in *two* places: these are signals to the other timing domains.

The third domain contains the straightforward task of moving around the reorder buffer, waiting for the subsequent data arrival and writing it back to the register bank. This process relies on an important observation: although it is not possible to predict when something will arrive or depart in an asynchronous pipeline stage it is possible to *wait* for a packet's arrival without need for arbitration. Having copied the data out the specific arrival flag is reset to await reallocation and refilling.

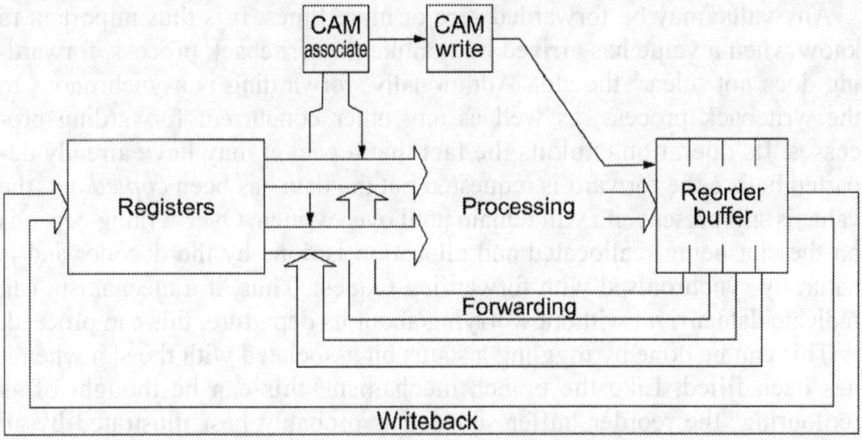

Fig. 16.6. Amulet3 reorder buffer structure.

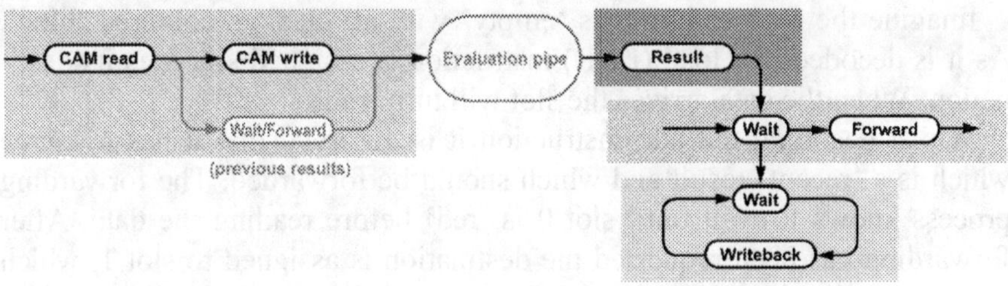

Fig. 16.7. Data flows in the decode/execution pipeline.

A simple analogy illustrates this: if someone is to open a door for you it is safe to proceed once you see that the door is open. If the door is opened before you arrive this causes no problems. The risk of injury only comes if someone tries to close the door after you have committed to going through. This last case is prevented if closure is *your* task.

The final timing domain deals with forwarding requests. These are issued from the decoder as a reverse chronological list of the register slots which have been allocated to the required register. (The same destination register may have been specified in several instructions. In an ARM implementation, such as Amulet3, this is a list because any instruction can be conditional, thus the first expected source for forwarding may not have been used.) The process then waits until a candidate slot has been filled and forwards the value – or if it is invalid moves back to the previous candidate. Again the process merely waits for an expected (or past) event. If all forwarding attempts fail an older value is available from the register bank.

Any value may be forwarded zero or more times. It is thus important to know when a value has arrived but, unlike the writeback process, forwarding does not 'clear' the slot. Additionally, forwarding is asynchronous to the writeback process, as well as any other concurrent forwarding processes. Its operation exploits the fact that a packet may have already departed before the forward is requested but the data has been *copied* out; the value is still present and will remain until over-written. Over-writing depends on the slot being reallocated and allocation is done by the decoder and is naturally synchronised with forwarding request. Thus, if a mechanism can indicate data arrival without worrying about its departure, this can proceed.

This can be done by toggling a status bit associated with the slot when it has been filled. Like the branch mechanism, this can be thought of as 'colouring' the reorder buffer slot and is probably best illustrated by an example:

```
LDR    R0, [R1]       ; Instigate a load
ADD    R1, R0, #1     ; Increment loaded value
```

Imagine the reorder buffer is 'empty' with all its slots coloured 'blue'. As it is decoded, the load (LDR) instruction is assigned slot 0 as its destination. When the data arrives the slot will turn 'red'.

Whilst decoding the add instruction it is observed that it has a source which is a 'recent' result and which should be forwarded. The forwarding process knows to wait until slot 0 is 'red' before reading the data. After forwarding has been requested the destination is assigned to slot 1, which will turn 'red' when the increment is complete. Forwarding will take place sometime after slot 0 has turned 'red', even if the result is being copied out at the time, because the processes are entirely independent.

As the reorder buffer is cycled, a few instructions later slot 0 will be reassigned. Imagine this instruction performs this:

```
SUB    R0, R0, #1     ;
```

The first action is to forward the existing (red) R0 value from slot 0; the slot is then reassigned for the destination, which cannot be overwritten before the forwarding is complete for obvious reasons! The result arriving will turn slot 0 'blue' again. An instruction following the subtraction and wanting to read R0 will therefore wait for *this* colour change. The colours therefore alternate on alternate cycles of the reorder buffer.

There is one asynchronous hazard remaining in the mechanism described above: it is theoretically possible to issue more instructions (strictly, more register *destinations*) than there are reorder buffer slots. Thus, in the example above, a forward request waiting for 'slot 0, blue' could be issued *before* it has turned 'red'. This could cause extreme embarrassment! To prevent this, allocation of a slot is throttled.

The throttle is a simple mechanism. It can be thought of as a FIFO through which the slot numbers are recycled although, in practice, there is no need to pass data because allocation is always cyclic. At reset the FIFO holds a number of tokens not exceeding the number of reorder buffer slots. Each time a register is to be written to, the decoder collects a token from this queue. If no token is available this process waits until one arrives. Tokens are replenished by the writeback process after it has retired the reorder buffer entry back to the register bank (in order). This 'dataless' FIFO is a very cheap asynchronous structure and the FIFO throttle is a 'natural' asynchronous mechanism.

Cacheing and memory systems

Although a slight diversion from the microprocessor core there are some interesting features in implementing an asynchronous cache which help to illustrate aspects of general asynchronous hardware design [200].

When a cache miss occurs it is necessary to fetch a new line from the memory. (For the moment it will be assumed that the rejected line can simply be abandoned.) For optimum performance the required word should be fetched, after which two processes proceed: one is the processor continuing execution, the other is the continuing fetch until the line is complete. Unfortunately these processes may not be independent, in that the processor may demand a word which is being fetched concurrently, thus requiring a (n unpredicted) synchronisation.

One solution to this is to assemble the fetched words in a specialised 'line fetch latch' (LFL) rather than the cache RAM, see Figure 16.8 [141].

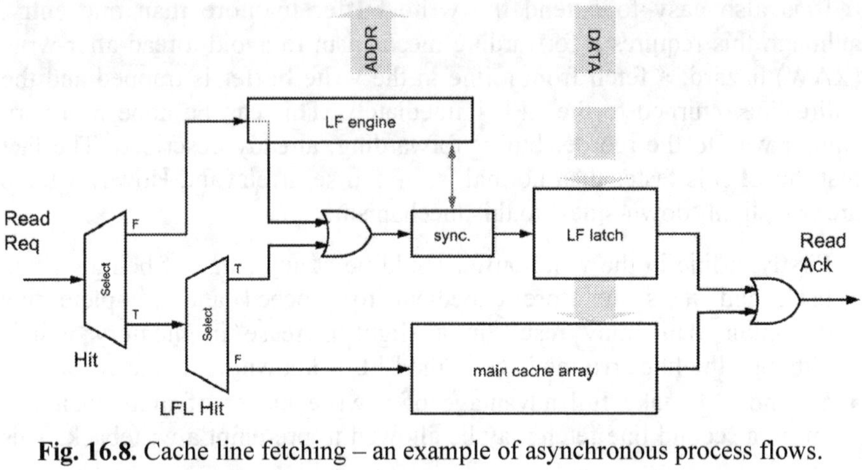

Fig. 16.8. Cache line fetching – an example of asynchronous process flows.

The line fetch process then runs like this:

- wait for any previous line fetches to be complete;
- instigate line fetch;
- concurrently, copy any LFL contents into the cache RAM before allowing any words to be received;
- as the LFL is now 'empty', cause the processor to wait for its desired word;
- as the words arrive, assemble them in the LFL;
- when the desired word arrives the processor is free to proceed;
- when the fetch is complete, stop, leaving the data in the LFL.

A subsequent cache hit will either be in the cache RAM – in which case it proceeds independently from the line fetch process – or in the LFL – in which instance it waits until the desired word arrives, as above. In all cases any desynchronised processes are merely caused to *wait* for another event (which may have already happened) and this can be done without recourse to arbitration. It is interesting to note that this mechanism automatically provides features such as non-blocking access, streaming and hit-under-miss.

An assumption made above was that a rejected cache line could simply be over-written. This is valid for an instruction cache, a write-through cache or an unmodified data line. In a write-back cache the mechanism is slightly more complicated but still quite tractable. The rejected line is first written back before the LFL is copied over it. Of course this 'writeback' may be very quick if the line is 'clean' (unmodified) where the process can be 'short-circuited' internally. It is also sensible to sideline the data in a write buffer to avoid obstructing the more urgent line fetch. In practice this will all occur whilst the memory is still contemplating the first data fetch.

It is also easy to extend the write buffer to more than one entry, although this requires a forwarding mechanism to avoid a read-after-write (RAW) hazard; a fetch from a line in the write buffer is trapped and the entire line returned to the LFL immediately. This can be done in a very similar way to the reorder buffer forwarding, already described. The fact that the LFL is faster than normal is, of course, irrelevant. However there are two slight 'down-sides' to this mechanism:

- Firstly, a line in the write buffer could be at any stage of being written back and it is therefore expedient to proceed and complete this operation. This may result in a slight increase in memory writes, although the line arriving back in the LFL is known to be 'clean'.
- Secondly, to take full advantage of a write buffer of more than one entry, a second line fetch may be allowed to pre-empt a writeback. This

decision requires arbitration as the requests are desynchronised. A throttle is then needed to prevent the write buffer filling up with rejected lines, stalling the read process and deadlocking the system.

A bonus is that the forwarding write buffer, which retains valid data during and after the writeback, automatically provides a victim cache.

An interesting consequence of the asynchronous implementation of both processor and local memory is the ease of exploitation of timing differences. Amulet3 has separate instruction and data buses to a local RAM which are subsequently unified onto an on-chip bus, as shown in Figure 16.9. The local memory is split into interleaved blocks so that access conflicts are unlikely and the two buses can operate largely independently. When a (rare) conflict does occur arbitration is necessary and a cycle may be stretched; this is easily accommodated by the processor without need for modification or explicit 'wait' cycles. The result is a local memory that provides performance close to that of a dual-port RAM at a cost similar to a single-port RAM.

Another consequence is that the processor's two buses can run at slightly different cycle rates: the instruction bus – which is the busier of the two – is unidirectional and simplified for speed; the data bus is somewhat slower as it supports not only processor writes but external accesses from the system bus. However the ratio of the bus speeds does not have to be a rational number, allowing the instruction bus to run faster than it could if everything were synchronised.

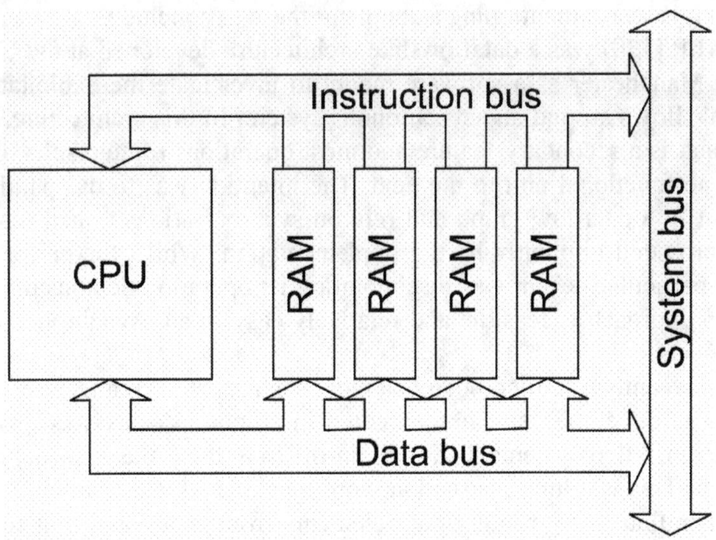

Fig. 16.9. Amulet3 local memory interface.

Summary and conclusions

It has been demonstrated that, at least in some domains, it is perfectly feasible to produce asynchronous processors which are competitive with their synchronous counterparts. No 'show stopping' features have been encountered which suggest that the synchronous paradigm is the only successful model for processor construction. In certain areas, such as EMC, it seems that asynchronous processors provide advantages, although these are not great enough to discard accepted practice yet. The promise of extremely high speed through fine-grained asynchronous pipelining has also yet to materialise. That said there are asynchronous processors both commercially available and embedded in products right now.

At the same time it needs to be noted that asynchronous processor design is difficult. Partly this is its unfamiliarity to designers, partly the lack of direct support from CAD tool vendors. It may be that tools such as *Haste* (formerly known as *Tangram*) will change this by providing easier design flows. Developing technology may also reduce the desirability – or even feasibility – of distributing a global clock at the desired rates.

This chapter has concentrated on 'conventional' processors, largely looking at asynchronous implementations of features common in synchronous processors. However asynchronous design also encourages thoughts about 'different' ways to realise an architecture. After a while a designer may think more in terms of message passing than finite state machines. This has led to other architectures into which such dataflows may naturally fit, and it is worth mentioning a couple of these to conclude.

SCALP [120] was a dataflow-like architecture developed at the University of Manchester as a software model to investigate the exploitation of 'natural' flows around an asynchronous system. In this architecture, a register bank is a secondary feature and most operations route results directly from one functional unit to the next. The intention was to use parallelism inherent in a dataflow graph and rely on sychronisations (when required) being provided implicitly by the implementation. Whilst the performance of this particular design was disappointing its operation demonstrated that such an approach is feasible and relatively easy in an asynchronous environment.

More recently a similar, but much more sophisticated architecture, Vortex [268] was developed and implemented in silicon. Vortex was produced by Fulcrum Microsystems, Inc. – a spin-off from the Caltech asynchronous research. This is a long instruction word machine which, like SCALP, exploits parallelism by routing data elements from functional unit to functional unit (or registers, or memory if storage is needed). Whilst the first, experimental device did not meet all its designers' expectations it still provided respectable speed due both to its parallel nature and implementation methodology.

In conclusion, although the great majority of the world's processors use clocks, it is entirely practical to design a processor that operates without a clock. The quality of such designs has improved steadily, and a small number are now in commercial use. So far, most progress has been made in redesigning existing processor architectures to operate without a clock, but clockless design opens up a new range of possibilities for architects that may enable asynchronous and self-timed processors to develop away from mainstream architectures into new and exciting areas that we have only just begun to explore.

17 Early-Estimation Modeling of Processors

Tero Nurmi[1], Tapani Ahonen[2], and Jari Nurmi[2]

University of Turku[1], Tampere University of Technology[2]

Introduction

We are hitting power budget and power density limitations ever harder with continuing electronics miniaturization. High power density is not a new problem: CMOS technologies were adopted in the late 70s and early 80s mainly because of their lower power density in comparison to bipolar processes. This time there is no such alternative, more energy efficient technology at sight. Thus, the solution lies in methodological and architectural innovation for more efficient energy utilization. Intel's Senior Vice President and CTO, Patrick P. Gelsinger, has noted that for every doubling of available transistors, we have gained approximately 40% in design performance. This observation suggests that performance per transistor, and hence energy efficiency, has degraded significantly.

Desktop computers deliver a high performance by means of instruction-level parallel (ILP) processing. Efficient implementation of ILP processors requires speculative processing and instruction reordering to maintain program consistency. This requires a considerable amount of supporting logic. The supporting logic itself consumes a lot of energy. The problem is compounded by discarded branches that were processed speculatively. The result is power density that exceeds physical limits of heat sinking and thus limits performance.

Portable devices incorporate multiple processors for different functions. Current CMOS processes feature a low switching energy for the transistors, but high energy utilization in the wires. Thus, even if the processors are located on the same silicon die there is a severe energy utilization penalty associated with all data transfers. This penalty manifests itself when data is transferred from memory to memory, memory to processor, processor to memory, or processor to processor. In some cases it may be

J. Nurmi (ed.), Processor Design: System-on-Chip Computing for ASICs and FPGAs, 391–404.

more energy efficient to re-compute a value rather than fetch it from the memory.

Processor throughput is typically enhanced through operation pipelining. Pipelining leads to the need to forward intermediate results from a later pipeline stage to an earlier one bypassing the register file. The forwarding mechanisms need a lot of multiplexing. Multiplexer realizations are problematic on FPGAs where they utilize the large and slow look-up table resources. The associated slowdown drastically diminishes the benefits of pipelining on FPGA. This is why FPGA optimized RISC processors usually feature from three to four pipeline stages.

This chapter presents a methodology how to model and estimate performance of processor architectures in advance and how to use this information to optimize architecture instances to meet design objectives in a best way. The focus is in architecture optimization in the presence of different constraints due to technology and cost/power limitations.

First some historical early estimation models are shortly presented, and their drawbacks to predict modern processors' performance are discussed. Some required properties are suggested to adapt those models to better meet the properties of modern architectures. After that a more detailed model for a processor based on a modified transport triggered architecture (TTA) is presented shortly and the model is used in architecture optimization in order to find the best possible architecture instance to meet the system requirements for selected case study applications. In section 5 of this chapter some issues relating to a leakage current and power are discussed and the impact of synthesis optimizations on logic structure and drive strength at 90 nm CMOS technology is presented.

Finally, there is a short overview on the physical design issues that are relevant when we are heading towards sub-100 nm technology generations, especially true with future technologies of feature sizes at 20–50 nanometers.

History of early estimation models for computer architectures

In order to predict and evaluate the performance of a microprocessor some performance estimators based on analytical equations have been introduced since the end of 1980s. H. B. Bakoglu presented the well known SUSPENS model first in a journal article [34] and later in the book [33]. Later CPU cycle-time model by Sai-Halasz [115,359,360] and also by Mii [288] were introduced. Those models are based on average wiring statistics and use different device technology, design and architecture parameters as their inputs.

Here the basic properties of those models and their possible drawbacks are examined.

SUSPENS

The SUSPENS model uses technology, design and architecture parameters as its inputs and then by using analytical equations results clock frequency, power dissipation and chip size. The calculations are based on the total number of logic gates and average length interconnects. SUSPENS lacks long interconnects, (on-chip) memory and various types of interconnect schemes. Donath's wiring statistics [107] are used to estimate the average length of on-chip interconnections. The cycle time includes the delay through a series of logic gates (logic depth) and a long interconnect that crosses the chip halfway diagonally. The model was derived to both NMOS and CMOS microprocessor circuits.

Because SUSPENS does not take various interconnect schemes into account and uses only one metal level it has to be updated to include those effects and schemes.

The Sai-Halasz performance estimator

This estimator can be considered an extension to the SUSPENS model. It uses a two-way NAND gate with a fanout of two as an average logic gate. The basic function is to calculate the delay through a given number of CMOS stages that can be regarded as a generic critical path of one clock cycle. The model relies on earlier observations of mainframe complexity uniprocessors [202]. For this reason it may not be a suitable estimator for modern on-chip circuits.

Some details have been added to the earlier models, such as taking memory size, level-to-level blockage and the power and ground distribution network into account. Anyhow, because it introduces some empirical parameters based on the actual chip designs its success in predicting some different types of architectures can be weaker.

The Mii performance estimator

The Mii performance estimator uses a hypothetical microprocessor projected from previous generations of IBM computers. This model uses a fixed interconnect structure as was the case also with the Sai-Halasz estimator. Differently from the Sai-Halasz model this model lacks level-to-level blockage which makes the model inaccurate in that sense. On-chip cache is taken into account. The model uses the same logic depth for all CPU generations assuming that the increase of CPU complexity changes only parallelism but not logic depth. Again this type of assumption may not be true for different types of processor architectures.

Adapting models to meet modern processor architectures

As noticed in the previous section that different models may have some limitations in their prediction accuracy. Clearly, one limitation in the SUSPENS model is that it does not include the internal cache memory. In modern processors there are several levels of caches on the chip which may be large in size. Two other models (the Sai-Halasz and Mii performance estimators) are based on earlier design cases of certain types of processor architectures and are not good in predicting new types of (special) processor architectures. Nowadays it is the metal pitch, not MOS gate length, that determines (logic) gate density.

The more detailed interconnect performance analysis is not covered in these models. In modern design the effects such as clock distribution network, power supply voltage variation in different parts of a circuit, signal crosstalk and even process parameter variations have to be taken into account. In addition to these technology-related features one needs to take architecture related features into account. If some modifications into the existing architectures are done or even new types of architectures are used, this becomes very important fact when evaluating the performance of different architecture instances.

Some newer estimators, such as GENESYS [113,114], BACPAC [399] and RIPE [148], are developed to better take those different interconnect (or technology) related features into account. Actually, there is a common technology extrapolation framework named GTX that combines several individual models and provides a robust, portable framework for interactive specification and comparison of modeling choices, such as system cycle time, chip size or power consumption. The GTX framework was presented in the DAC conference in 2000 [64] and the framework can also be found via Internet for creating and developing new models.

In the following section the RIPE estimator has been used as a starting point when developing estimators. After that modifications have been done to have the estimator better adapted to meet the special details of the selected architecture.

Architecture modeling

In our co-operation activities we have developed a multi-core system which uses either a synchronous or an asynchronous approach in communication between different system blocks. In our system [313], we use a RISC-based "COFFEE" processor and TTA-based "TACO" processor in handling the IPv6 data packets and decoding MPEG-coded data. Because the case study is presented in [313] we do not go into details of the case study here but instead look closer how different aspects of the architecture itself are taken into account.

Basics of TTA processors

A TTA architecture and a concept was developed in Delft university of Technology, The Netherlands, in the 1990s. Henk Corporaal's group's work there resulted in the *MOVE* TTA architecture template described in [92]. The original concept of transport triggering was earlier introduced for digital controllers by Lipovski et al. in [405] in 1980.

Traditionally processors are programmed by specifying operations which then cause data transports to occur in the processor as a "side effect". In TTA-based processors, instead of programming operations those are the data transports that are programmed. Those data transports are then followed by the operations that are determined by the specific function in a functional unit (FU) that receives the data. Here only the basics of TTA processors are told; more elaborate discussion can be found in Corporaal's book [92].

A TTA processor consists of functional units (FUs) that communicate using an interconnection network of buses, controlled by an interconnection network controller (INC) unit. The FUs connect to the buses through modules called sockets. Each FU has input and output registers, and each of these registers has a corresponding socket. If there are e.g. two input registers and one output register there are three sockets altogether. Thus, by changing the type and number of FUs and/or by changing the connectivity and capacity (bus width) of the interconnection network, a wide range of TTA processor architectures can be specified. Although there are practically unlimited range of opportunities to vary a processor configuration the experience has shown that certain configurations are better in a certain application than the others when some constraints are set for performance or cost of a processor.

TACO – a design framework for protocol processing applications

The TACO (*Tools for Application-specific hardware/software CO-design*) protocol processor design framework consist of two parts: a hardware platform optimized for protocol processing and a design methodology and a toolset for rapidly specifying, simulating, evaluating and synthesizing protocol processors based on the mentioned hardware platform. Finally, we conclude this subchapter by giving results for the case studies presented in earlier articles. In the methodology part, the main focus in this chapter is to discuss evaluation issues. However, as will be seen later, it is necessary to have some information from system simulations and synthesis of individual FUs for evaluating and predicting the performance and cost of a final processor.

TACO protocol processor architecture

Using the terminology defined in [92] TACO processors use only special functional units (SFUs) that are application-domain specific and are not very frequently needed in general purpose processing. In TACO processors there are no general purpose FUs (e.g. like ALU or multiplier).

A TACO processor in general includes SFUs, memory management units, sometimes generic registers, some internal memories (program memory, user data memory and, in the case of protocol processing domain, protocol data memory), interconnection network and of course a controller for that network (an interconnection network controller, INC).

The TACO interconnection network can be fully connected or partially connected. In the fully connected case all buses have connections to all sockets. Using the fully connected configuration ensures maximal use of bus bandwidth. There are two kinds of buses in the interconnection network: data buses and control buses. Control buses carry information on which FU acts as a data transmitter and which FU acts as a receiver.

The programming is done by using a TACO instruction word. The TACO instruction word can be divided into several subinstructions. The number of subinstructions depends on the number of buses; if there are e.g. three buses, there are three subinstructions. In each subinstructions there are a certain number of bits reserved for source and destination IDs; the source and destination IDs define the addresses from/to which data is moved. The addresses refer to registers in FUs. More information on the details of the TACO instruction word can be found in [440].

The key tasks of the interconnection network controller mentioned earlier are:

- fetching instructions from the program memory;
- maintaining the program counter (PC);
- evaluating guard signals and guard IDs for conditional execution;

- splitting long instruction words into subinstructions;
- dispatching (decoding) subinstructions onto the buses;
- extracting and dispatching immediate integers specified in program code.

As noticed it is the interconnect network controller that delivers correct source and destination IDs (accurately, source and destination identifiers) to the sockets. Additionally, the INC also fetches instructions from the program memory and divides them into subinstructions that are then sent to sockets. If there is a match between a hard-coded identifier and an identifier in the instruction bus (source ID or destination ID bus) the socket stores the result of the decoding process to be used in the next pipeline stage.

Instruction execution in TACO processors consists of four pipeline stages: *fetch*, *decode*, *move* and *execute*. First two were explained earlier, in the *move* pipeline stage FUs with enabled sockets write/store data to/from the buses and in the *execute* stage the FUs execute their operations. The operations are initialized with the data coming to the *trigger* (input) register of the corresponding FU.

More details and information of the TACO architecture and TACO processors can be found in [440].

Design methodology relating system simulations and physical design

In this chapter we concentrate on early estimation modeling of TACO processors and thus discuss only that part of TACO protocol processor design flow. Because the early estimation modeling can never predict performance of a processor with 100% accuracy, there is a need for additional information that is extracted based on the results of system simulation. In addition to system simulations we need also the information of the FUs used in the target processor architecture. That information is extracted from the synthesis results of VHDL descriptions used for FUs. If the required function is not available a new FU executing that function has to be designed (i.e. a VHDL description of that function has to be written).

In the early estimation modeling part of TACO design methodology we estimate the delays of different pipeline stages and additionally the area and power consumption of the target processor. That information is then used in the TACO protocol processor design flow together with the information on the synthesized FU blocks (i.e. the area of the block and the relative portion of combinatorial and sequential gates in the FU) and the information coming from system simulations (the clock cycle length is constrained by those results).

In the following text portions, we first discuss how to model delays of different pipeline stages, then discuss how the results from VHDL synthesis of individual FUs are utilized in early estimation modeling and what are the results of the models. We continue by discussing how the results of system simulations are affecting early estimation modeling and again what are the results of the models.

Delay modeling

As said before there are four pipeline stages in TACO processors. The cycle time of a TACO processor is defined by the delay produced by the slowest pipeline stage. So far we have assumed that the longest pipeline stage delay is either in *execute* or *move* stage. However, when heading into sub-100 nm technologies the longest delay shifts more to *move* and *fetch* stages because of the increasing dominance of interconnect delay over logic gate delay. We have assumed with TACO processors that FUs can execute their operations within one clock cycle. When clock frequencies are increasing that sets a great challenge to design interconnection network that carry signals from one FU to other FUs.

Also *fetch* stage sets challenge when clock cycle times are diminishing: memories have to be located close to the location where the data is needed because of increased interconnect delays. Program memory has to be located near the INC. If some memories for storing protocol or user data are needed they should be located near the location they are needed frequently. Also some effort has to be put into memory hierarchy in order to deal with the delay of the *fetch* stage when implementing TACO processors with future technologies.

A more detailed analysis and the equations to predict the delay of the *move* stage are found in [443].

FU modeling

Earlier we have developed some models to predict the area of individual FUs based on Rent's rule [251]. However, we noticed in our analysis that the evaluation of the area of individual is very difficult and the errors in the area can be very large even though the variation in the values of Rent's parameters (Rent's constant and Rent's exponent) is small [7].

Thus, we have omitted the FU area estimation method that is based on gate netlists of a functional unit. Instead, we get area information from VHDL synthesis results of individual FUs. In addition to area information, synthesis results reveal also very useful information about the relative amount of combinatorial and sequential logic in an FU. That information is very relevant when estimating power consumption of an FU.

The dynamic power consumption of an FU can be defined:

$$P_{tot} = f_d C_{tot} V_{dd}^2 f_{clock} \tag{17.1}$$

where f_d is the activity factor, C_{tot} represents the total nodal switching capacitance in an FU, V_{dd} is the supply voltage and f_{clock} is the clock frequency.

Thus, by knowing the relative amount of combinatorial and sequential gates in an FU we can better evaluate the power consumption in the whole FU. The key factor in this evaluation is the activity factor.

Bus traffic and power modeling

For bus traffic and power evaluations one needs always a case study. It is important to know how signals traveling in adjacent wires are switching because of crosstalk effect. The capacitance ratio of the capacitance over the ground capacitance of a single wire increases with sub-100 nm technologies. This is because the aspect ratio (height/width) of a wire increases.

So far we have investigated the power consumption in a bus using Hspice simulations. Power consumption is examined by feeding input patterns extracted from real data (*IPv6* protocol headers) into a 32-bit wide bus. Thus, power consumption varies depending on how the bit patterns change in the bus (i.e. depending on how many bits change at the clock edge and which direction, up or down, the bits are switching in adjacent wires). The results will be presented in a forthcoming article on bus traffic modeling.

The information on bus traffic is extracted from system simulations. System simulations are conducted by using SystemC. Those results give a constraint for a maximum allowed clock cycle time. The results tell how many clock cycles a required application task takes. By changing processor configuration the number of clock cycles varies. Processor configuration can be changed by varying the number of FUs and/or the number of buses. By increasing the number of buses and using redundancy for certain FUs may decrease the total number of clock cycles considerably depending on an application.

Additionally, those system simulations reveal utilization of different buses. This information helps when one estimates the total energy consumption (in picoJoules) of an application task. Finally, system simulations reveal also register transfer statistics that can also be used for estimating the power consumption in the bus. Thus, if two registers are inputting/outputting information very frequently, they (or the FUs they are located in) should be placed near to each other. As noticed, the information received from system simulations helps to design a floorplan for a processor.

TACO case study results

We have used the TACO processor for three case study applications in protocol processing domain. Later TACO has been used also for digital signal processing applications but the case studies were not so extensively covered. In protocol processing application, TACO has been used in *ATM AIS cell processing*, *IPv6 router* and *IPv6 client* applications.

For the ATM AIS cell processing case study [443] we used 0.35 μm technology. In this case study we explored few processor configurations where we had either 1, 2 or 3 buses and additionally we used redundancy for some FUs by duplicating their amount in the processor. In that case study the shortest time to execute a given task was achieved by using *double-2* (certain FUs were duplicated and two buses were used) configuration. We also noticed that bus utilization was nearly 100% when double FU configuration was used (compared to the case when there were one FU of each type). This was the result of using instruction-level parallelism (ILP). We made also some area and energy consumption estimates but due to the lack of extensive synthesis results those estimates could not be verified.

In the second case study, IPv6 client as a part of a larger system [313], we noticed that the use of Rent's rule to estimate the FU areas gives sometimes quite accurate results, sometimes not. This is because our FUs do not follow normal random logic idea, they are customized so that they execute their function in one clock cycle. Also power estimates were conducted for each FU.

In the third case study, IPv6 router, we examined some routing functions used in an IPv6 router and compared the results to a commercial network processor [442]. We noticed that especially checksum calculation was considerably with our TACO processor than with a commercial processor. This was due to the fact that our TACO processor had a special unit for checksum calculation. Some area estimates were also done and they predicted rather well the size of FUs. From the estimates of individual FUs area estimates for the whole processor were done but not verified.

In both IPv6 case studies 0.18 μm technology was used.

Processor logic optimization at 90 nm technology

Leakage issues

CMOS transistor is considered to be in a non-conductive high resistance state called cut-off when the gate to source voltage (V_{gs}) is less than the threshold voltage (V_{th}, also called pinch-off voltage and denoted with V_p). When V_{gs} is greater than V_{th} the transistor is in a conductive state, which can be divided into two regions. If the drain to source voltage (V_{ds}) is greater than $V_{gs} - V_{th}$, the transistor is in a highly conductive low resistance state called saturation. With V_{ds} below $V_{gs} - V_{th}$, the transistor is in a semi-conductive variable resistance state described with three different names: triode, linear or ohmic region.

When voltage is supplied to a CMOS circuit, it consumes power even in an idle state. There are two substantial sources of this static power consumption:

- Subthreshold leakage current results from the channel resistance not being infinite at cut-off.
- Gate leakage occurs when there is voltage over the insulation layer causing some current flow (electron tunneling) through it.

Out of these two sources of static power consumption, subthreshold leakage is the more problematic one. It increases with shorter effective channel length L_{eff} resulting from downscaling, and gets more prominent at high temperatures. Gate leakage can be effectively lowered through high-κ dielectrics. The International Technology Roadmap for Semiconductors (ITRS) [208] projects that high-κ dielectrics are in mainstream production after 2010.

Subthreshold leakage can be reduced through higher V_{th} to provide higher channel resistance at cut-off. A higher V_{th} can be achieved through a thicker gate insulation layer, which effectively lowers the gate leakage at the same time. With a high V_{th} also the dynamic power consumption decreases thanks to shorter short circuit times during logic switching. On the other hand, a high V_{th} lengthens the time required for a transistor to enter conductive state and hence limits the achievable operating speed.

Reference [322] discusses this power-delay tradeoff with 90 nm technologies. With 90 nm technologies currently at the market, the static power consumption of extremely high speed technologies is ten thousand times higher in comparison to very low power technologies. The high speed technologies are approximately three times faster than the slowest low power technologies.

The leakage power of course causes the device to heat up and limits the power budget left for the dynamic operation. With increasing gate densities, power density approaches the physical limits of heat sinking. For the very high end products the best modern heat sinking solutions have approached the capability of dissipating a hundred watts per square centimeter. In mobile devices however, the practical limits of heat dissipation are much lower. For this and battery lifetime reasons the laptop computers now feature an increasing amount of custom designed low-power devices.

Impact of synthesis optimizations on logic structure and drive strength

A processor's implementation characteristics are affected most by decisions made at the higher levels of abstraction. However, these decisions must be backed by thorough knowledge of the lower level issues discussed

Table 17.1. Impact of Synthesis Optimizations on Processor Logic Structures.

Design	Coffee RISC	Milk FP	MOVE TTA
Equiv. gates	+45.18%	+81.75%	+82.34%
Std. cells	+40.90%	+105.68%	+26.54%
Leakage power	+83.32%	+152.90%	+159.64%
Net count	+35.31%	+84.59%	+24.26%
Avg. fanout	–10.05%	–8.29%	–11.73%

above. Here we take a look at the effects of optimizations during register-transfer level (RTL) to gate-level synthesis.

Three different processor architectures were chosen for this study: the Coffee RISC processor core, the Milk floating-point (FP) co-processor, and a *MOVE* architecture instance designed for radix-2 FFT execution. *MOVE* is a template and design tool framework for transport triggered architecture (TTA) processor instantiation.

First step is to fix the target technology and the targeted operating environment, or the PVT (acceptable Process variability, supply Voltage, and junction Temperature) corner. In this study the processor architectures were synthesized to a slow/slow process corner (worst case process variability and operating environment) of a 90 nm technology. Results from optimization for the highest achievable operating speed were compared against the results from optimization for the smallest achievable area. The relative differences between the two resulting logic structures are summarized in Table 17.1.

It is obvious that the impact of optimizations in conjunction with logic synthesis is considerable. Optimization for low delay results in utilization of gates with low input degree. Reason for this is that fastest implementations are usually achieved with two-input logic.

Higher increase in leakage power than in logic area is the other obvious result from delay minimization. This happens because higher drive strength is used to achieve shorter rise and fall times. High drive strength translates into wide channel output transistors for high capacitive loads, which in turn increases leakage per area. Hence high speed is always associated with exponentially increasing power consumption whether it is achieved through faster logic structure or higher supply voltage.

Speed of the *MOVE* TTA processor is dominated by the crossbar bus that connects the FUs. Therefore the delay optimization process results in only minor changes in logic structure. This can be seen in Table 17.1. as low increase in standard cell instance and net (interconnect) count. However, high drive strength cells and buffers are instantiated to achieve lower interconnect delay. This shows up in Table 17.1. as considerable increase in leakage power and area (equivalent gate count).

The complexity of the longest combinational paths determines the degree of freedom allowed to the optimizer. Out of the three designs, most change in logic structure takes place with the most complex design, the Milk floating-point co-processor. Standard cell instance and net (interconnect) count both roughly double as a result of delay optimization. Fabrication cost (required silicon area, and amount of metal) increases accordingly, while leakage power penalty is also high.

The impact of optimizations is the most uniform with the Coffee RISC processor. Complexity of the logic structure (standard cells and interconnects) as well as drive strength of critical cells is increased. As a result, area and leakage power penalties are the lowest among the three architectures.

Physical design issues in the era of sub-100 nm technologies

When technologies enter onto a sub-100 nm and even sub-50 nm era there will be certain physical and electrical design issues which have to be taken into account when designing circuits with high dependability. In this subsection some of those issues are shortly presented and their effect on processor design at architectural level is investigated.

- Front-end-of-line challenges

Transistor performance does not anymore follow the earlier rule, i.e. that performance would be proportional to the reciprocal of gate length. Additionally, lower supply and threshold voltages result to increasing leakage currents. That means that in a processor more power is consumed in standby mode outside clock transitions.

- Back-end-of-line (BEOL) challenges

For line widths below 100 nm there is a great increase in resistivity of metal wires. Because the proportion of the conductor cross-sectional area of barrier material with higher resistivity is larger the effective resistivity increases. At the same time, the contribution of surfaces in the properties of a wire increases; the effect of surface and edge scattering has to be taken into account. Additionally, the variation in wire dimensions due to chemical–mechanical planarization (CMP) processes used leads into the variation in wire performance that has to be taken into account when designing e.g. wiring strategy and tolerance against variation in wire performance. Signal crosstalk is a natural result from technology scaling when wire separation diminishes. Delay dependence on crosstalk and signal transitions in adjacent wires has to be evaluated. Use of low-κ dielectrics together with problems in contact and especially via (contact) scaling conclude BEOL issues to be addressed in processor design at sub-100 nm technology nodes.

- Process control and reliability

  As mentioned already process control is essential for assuring reliable operation of future processors. With decreasing dimensions features as line-edge roughness (LER) is becoming an increasing concern. LER affects many transistor parameters and increases also an overlap capacitance (C_{gd}) and reduces channel length [460].

  V_{th} variation is influenced by random dopant fluctuations and gate critical dimension (CD) variation. Thinner gate oxide causes many effects such as negative bias temperature instability (NBTI).

- Lithographic issues

  After 180 nm technology generation, the CD has been smaller than the ultraviolet wavelength used in lithography. This is called as a subwavelength regime and a gap between the CD and UV wavelength as a subwavelength gap. This means that a new lithographic generation has to be developed. Below the 90-nm technology node aggressive optimal proximity correction (OPC) is necessary. Other possible resolution enhancement techniques are e.g. specialized illumination patterns, subresolution assist features and alternating phase-shift masks [460].

  As a result of the widening subwavelength gap, mask and lithographic costs will increase exponentially in subsequent generations. This will affect and decrease the number of manufacturing companies that can afford to expensive lithographic equipment.

- Modeling challenges

  Finally, new models have to be created to take different scaling-related parameters into account. For example, BSIM4 models include new features that were not included in BSIM3 models, such as a halo or pocket implant, gate-induced drain leakage (GIDL), gate direct tunneling, and trench isolation stress effects.

  Wong et al.'s book [460] presents many solutions and design procedures how to get along with the aforementioned physical features in mixed-signal circuit design, electrostatic discharge protection design, input/output design and memory (DRAM) design. It also addresses challenges in interconnect design that were shortly discussed in this subsection ("Back-end-of-line challenges").

18 System Level Simulations

Seppo Virtanen[1], Dragos Truscan[2], Sanna Määttä[3], Tomi Westerlund[1], Jouni Isoaho[1], and Jari Nurmi[3]

University of Turku[1]

Åbo Akademi University[2]

Tampere University of Technology[3]

Introduction

Due to the increasing complexity of system specifications, new methods are required for detecting design errors at the early stages of the development process, as well as for ensuring the performance characteristics of the final product. Among these are estimation of physical characteristics (discussed in the previous chapter, *Early Estimation Modeling of Processors*) and system simulation at a high abstraction level, that is, the system level as defined in [344]. High abstraction level estimation and simulation have both become necessities for the design activity. System level simulation enables the evaluation of system specifications against requirements at early stages of the development, before proceeding to hardware implementation. The approach eliminates costs and shortens the design cycle of new products. According to [294], most of the integrated circuits developed today require at least one return to early phases of the development, due to errors. Furthermore, simulation contributes to reducing the testing effort that is performed at different stages of the design process.

Simulation allows one to execute the system specification at different levels of abstraction. There are two conflicting trends in simulating embedded systems. On the one hand, the simulation process should be performed as early in the development process as possible, in order to avoid the propagation of design errors to later phases. On the other hand,

J. Nurmi (ed.), Processor Design: System-on-Chip Computing for ASICs and FPGAs, 405–426.

performing the simulation at later phases of development, when the system specification is more complete, allows one to check a wider range of the system's properties. System level simulation enables the verification of the correct functionality of the system's specification with respect to its functional requirements. Due to the low level of detail in the executable specification, the simulation process provides less accurate estimations of the specification, yet these results are obtained in a relatively short time frame. In addition, eventual modifications of the system specification at this level are fast and inexpensive. In fact, system level simulation is typically used in taking strategic decisions for providing a specific implementation solution.

Modeling and simulation

A *system* does not exist in the real world before being implemented. During the development of that system, designers create more or less abstract specifications (that is, *models*) of that system, which enable them to focus on the relevant aspects of the system at a given stage of the development process. The *simulation* is the process of executing a given system specification in a computer based environment. As such, the system specification has to be *executable*. Even though specifications are not always executable [183], in the rest of this chapter we consider models to be executable specifications.

Processor-based embedded systems consist of a coherent combination of hardware and software. As such, simulation has to be applied not only to each of the hardware and software partitions and components in part, but also to the entire system as a whole (*co-simulation*).

The idea behind executable specifications is that instead of reading through a large quantity of documents describing the desired functionality of a system, the system designer could simply run the executable specification and see how the system is supposed to work. Such an executable specification is gradually refined to contain more and more implementation details during system development. This refinement of the original executable specification can be targeted towards reaching a high abstraction level simulator of the target system. So, ideally an executable specification serves as documentation throughout the design project, and after it contains enough detail, it becomes an accurate system simulator while all the coding has been done in a single specification and simulation language. Executable specifications are expected to provide unambiguity, completeness and correctness to system specification [146].

As the system design size and complexity increase, high abstraction level design methods are needed to rapidly explore the design space and verify system functionality. At a high level of abstraction, we can implement models that describe the system functionality without specific details. System models can then be used for simulating the system and evaluating its functionality and performance. Based on the results obtained from simulating the system models, the initial models of the system may be refined and improved until satisfactory (with respect to the requirements of the system) ones are obtained. Moreover, system verification can be done before the real life implementation using the high-level models, which alleviates the burden of verification, and makes it less costly and time consuming.

System models that are created during the development process should satisfy the requirements of the real designs. Models need to fulfill their behavioral and Quality-of-Service (QoS) requirements, such as performance, timing, and power consumption. Therefore, accurate behavioral models of the system are needed to assist in verifying system functionality. Moreover, we need efficient and flexible models to rapidly explore the design space in order to make optimal design choices for the target architecture early at the design stage. For example, Virtual System Prototypes (VSP) have been suggested for creating executable system specifications. A VSP is an accurate, high performance, and complete system model defining the hardware architecture, and it fulfills the product's business and functional requirements. Therefore, it serves as a golden reference model of the design. VSPs remarkably advance the high-level design and modeling [184]. Consequently, several companies, such as VaST, Virtio, Virtutech, CoWare, and Carbon Design Systems, have developed VSPs during the past 10 years.

Several tools and languages for system level modeling have been suggested in the recent past. As examples, we briefly discuss Polis, Esterel, Ptolemy, VCC, COSY, Matlab, and Simulink here. Polis [177], developed at the University of California in Berkeley, is a framework for hardware/software co-design of embedded systems. Designers can use a high-level language, such as ESTEREL [48,207], to write a specification, which can be translated directly into a co-design finite state machine (CFSM). The Ptolemy project [419] for modeling, simulation, and design of concurrent, real-time embedded systems includes a simulator, which can perform the co-simulation of the CFSM within Polis framework.

Cadence Virtual Component CoDesign (VCC) aims at the design of system level design tools and supports platform based design. Furthermore, Philips Semiconductor has established a new system level design methodology, which focuses on HW and SW Intellectual Property (IP) reuse. Philips Semiconductor's COSY initiative employs the Cadence VCC when building a library of reusable IP blocks, whose accurate performance estimations can be used by designers at early phases of the design process and thus shorten design cycle times and the time-to-market [61].

The Mathworks's Matlab [416] is a high-level language for technical computing and an environment for algorithm development and data managing. Moreover, Matlab includes tools for design exploration. The Mathworks also has a platform for multi-domain simulation and model-based design, Simulink [417]. Simulink has full access to Matlab, for example for the purpose of data analysis and visualization.

Abstraction levels

Abstraction hides implementation details, that is, the amount of information decreases with the increasing level of abstraction. By abstracting away implementation details, a designer is able to concentrate on essential design aspects and to reduce the required design effort to evaluate a model under development. The high-level architecture exploration and design decisions steer the development towards promising architectural solutions for specific application areas. Mastering different abstraction levels is the key to an efficient design of modern multiprocessor systems and to improving productivity.

There is no unique general definition in existence for the way in which hardware abstraction levels should be categorized and used, as this depends on the characteristics of the system under design and of the methodology selected. In addition, some design methodologies may require fewer, whereas others may require more levels of abstraction. One possible way of categorizing abstraction levels is defined in [344]. Based on this definition, Figure 18.1 shows hardware abstraction levels commonly met in several design methodologies (we have combined the three lowest levels defined in [344] into a single *RT level* in the figure). Another example is Transaction Level Modeling (TLM) [149,418]. It defines four abstraction levels from which the highest and lowest levels match the system and RT levels in Figure 18.1, respectively, whereas the two middle ones together form the module level.

It is important to notice that in addition to selecting a hardware abstraction level, the designer must also make a decision on the communication abstraction to be used. At higher abstraction levels communication is often modeled using function calls. Function calls hide the implementation details of the communication between system modules, and thus allow the

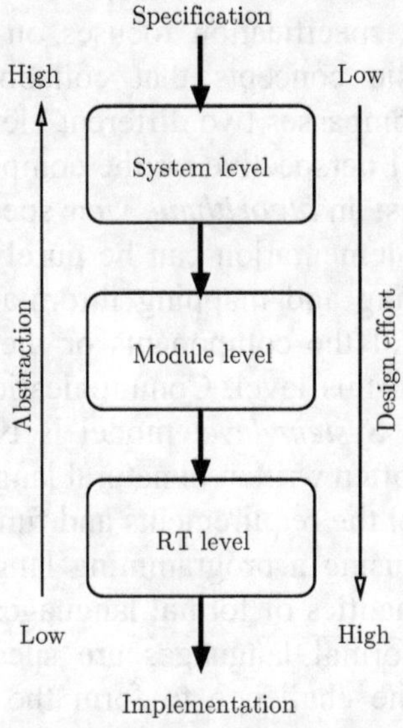

Fig. 18.1. Levels of Abstraction.

designer to concentrate on the functionality of the system under design. For example, in TLM function calls are used at the three highest levels, and only at the lowest level functions are transformed into detailed communication primitives.

The different abstraction levels at which the simulation is performed affect not only the architectural details, but also the accuracy of the simulation results and the simulation time. In other words, the simulation time that is required to estimate different architectural solutions shortens with the increasing level of abstraction. However, at the same time the accuracy of the estimation results decreases. Therefore, at high abstraction levels, one estimates architectural (system level) properties of the systems, without concern for the implementation details. The latter are in turn simulated and verified at lower abstraction levels: there the promising architectural solution is expected to have a sound basis, since alteration of high-level design decisions at lower abstraction levels is expensive.

As stated above, all the abstraction levels have their own architectural details, the higher we are, the less accurate representation of the real system we have. The amount of details grows rapidly when shifting the abstraction level towards lower ones, as both the computation and the communication details proliferate. Below we have gathered some of the main characteristics of the given abstraction levels and their effect on simulation.

- At *System level* the specification focuses on the requirements of the system and its basic concepts that collaborate in achieving these requirements. It encompasses two different views: an *architectural view* provides a high-level perspective on the components of the system and their interconnections; an *algorithmic view* specifies implementations of operations. The implementation can be purely executable model with or without partitioning and mapping information. No implementation related information of the components or their intercommunication is taken into account at this level. Communication can be modeled using functions calls. The *System level* model is typically obtained from a specification that is often written in natural language. Therefore, the first challenge is to model the requirements and functionalities of the system under development using a programming language, either informal or formal one. The semantics of formal languages are grounded in mathematics, whereas informal languages are specified using natural languages. However, the challenge to form the first model remains the same in both approaches.
- At *Module level*, or functional module level [344], the algorithms that implement the functionality of each component are selected and represented using block diagrams. In addition, an appropriate communication model is instantiated, for instance bus-based or network-on-chip (NoC), to support the communication between different components. Performing simulation at this level provides more detailed results as compared to the system level simulation. These results enable the evaluation of different implementations of algorithms and the instantiated communication medium. Depending on the needs of the design process, the estimates resulting from the module level simulation process may be *time-approximate* or *time-accurate*. The former provides general indications on the trend on which the characteristics of the system vary with architectural changes. The latter provides accurate estimations of the time the application takes to execute on a given configuration.
- At *Register Transfer level (RT level)* the module level components are refined into logical building blocks such as adders, flip flops, and latches. Furthermore, the module level communication is also refined into detailed pin-accurate communication based on the detailed description of the communication protocols and their transmission medium. At this level, real (hardware) data types are used. Moreover, often the RT level specification of a system is also synthesizable, that is, it can be input to hardware synthesis tools for hardware implementation. Simulations at this level ensure the correctness of the architectural decision and the implementation of the algorithms. Since the model is synthesizable, a

comparison between pre- and post-synthesis simulations can be performed. That is, the functional and temporal behavior of the system is verified against the system implementation. In terms of computer time this is an expensive process, due to the high effort in developing cycle-accurate models of the system components. Although simulations at this level can provide accurate results, the running times of the RT models are often too long to support an effective development process without employing high level simulation.

Simulation and languages

Modeling systems at higher abstraction levels has not been possible in the early days of hardware design. A major impact on the development and abstraction level in which systems are currently designed has been made by the introduction of *hardware description languages* (HDL's). Two popular examples of such languages are Verilog [203,438] and the VHSIC Hardware Description Language (VHDL) [27]. Like HDL's in general, VHDL enables a designer to model a system under development at RTL level (see Figure 18.1). At this level, VHDL enables one to describe the structure of the system in a hierarchical manner and to define how its subsystems are interconnected. Most of all, VHDL enables the simulation of designs before the implementation. That is, using VHDL several possible alternatives could be compared without the cost of manufacturing process, therefore reducing time-to-market and lowering the development costs.

Verilog is very similar to VHDL as a HDL, but some differences exist. Although their function is very similar, their syntax is different: Verilog resembles C, whereas VHDL resembles more Ada or Pascal. Moreover, Verilog has been unable to provide as high-level design constructs as VHDL. Verilog also lacks reusability, because it includes neither a concept of packages, nor libraries [438].

Nowadays, the traditional VHDL development method is facing the reality of being unable to answer the challenges posed by the continuously increasing system complexities and level of integration. In addition to mentioned challenges, demands of mixed hardware–software design capabilities in embedded system development must be answered. There are basically two major approaches in adopting system level design languages. One approach is to extend the existing HDLs to support higher level of abstractions (see for instance System Verilog [402,205]). Such an approach typically implies complementing the initial language constructs with constructs for hardware descriptions like module, ports, signals and support for timing and concurrency. Another approach is to adapt high-level programming languages like C and C++ for hardware description, see for instance CynLib from former CynApps, HandelC [70], CycleC from C Level Design, SpecC [137,432], or SystemC [204,265].

As mentioned, System Verilog employs the first approach, that is, an existing HDL is extended with features required in system level design. Employing the latter approach has enabled one to take benefit from object-oriented principles for answering and alleviating the modern design challenges. As such, the SystemC language has rapidly become one of the most widely adopted system level simulation environments in the industry.

System Verilog

System Verilog is built on top of Verilog 2001 by the EDA organization Accellera. System Verilog extends the Verilog 2001 to meet the system level design requirements, but it preserves all the Verilog 2001 features. Several EDA companies support System Verilog, such as Bluespec, Cadence, Mentor Graphics, and Synopsys [206].

System Verilog adds many new, mostly C-like, constructs and features to Verilog 2001, such as strings, enumerated types, new integer types, structs, unions, new logic and bit data types, dynamic processes for modeling pipelines, and assertion-based verification. Furthermore, System Verilog supports object oriented programming by providing classes, class instantiation, polymorphism, and data encapsulation. The new constructs and features improve the readability and usability of Verilog based designs. Moreover, they raise Verilog's level of abstraction, which has been lower in comparison with VHDL [1,205]. Many of the System Verilog features are originally VHDL features. Therefore, it is also possible to integrate System Verilog designs into the VHDL environment.

Despite the different approach when building System Verilog and SystemC, both languages offer the ability to high-level modeling and object oriented design. System Verilog may outperform SystemC when writing Register Transfer Level (RTL) description and Transaction Level (TL) testbenches, as well as when considering extensive tool support for the final RTL descriptions. However, SystemC is more suitable for abstract TL model writing for architectural exploration [368].

SystemC

The Electronic Design Automation (EDA) community is gradually adopting executable specifications as a potential replacement for traditional written specifications, and as a basis for building system simulators. C and

C++ have been the most popularly chosen bases for implementing executable specification, as well as high-level simulator development languages and environments already for several years, as seen in e.g. [137,146,345]. The choice to use these two languages is obvious in terms of availability and cost: existing tools and programming skills can be used, since companies already use C++ in their software development and C in their embedded system programming. Also, operating systems like Linux [425] provide C and C++ compilers and utilities free of charge. However, these languages are designed for writing computer programs, not for describing computers or other hardware devices. Therefore, they lack necessary functionality and features for describing clocks, signals, reactivity, and parallel processing. Also, most current design flows that start from a C or C++ based functional level description contain a "jump phase" in which the functional model written in C/C++ is translated into an HDL (hardware description language) [136,265]. This phase is often manual and involves refining the model down to the RTL level. To solve this problem, either a system description language should be built on top of C or C++, or a language designed specifically for system description should be created.

SystemC explores the first option by providing a C++ class library for designing executable specifications and cycle-accurate simulators of hardware. SystemC is distributed under an open license and is supported by several major EDA companies [401], such as Synopsys, Cadence, Celoxica, and Mentor Graphics. It provides support for hardware-oriented data types like modules, ports, and signals. Originally, there were two major goals in designing SystemC [265]: to provide a single language framework for co-verifying systems at varying, possibly mixed, abstraction levels, and to allow system designers to gradually refine their models towards the RT level without translating them into a HDL.

When SystemC was introduced in 1999, it seemed to have a lot of potential in system level modeling. SystemC development seemed to be on the right track, providing a modeling and simulation platform for combining the advantages of object oriented (OO) programming techniques available in C++ with the support for cycle-accurate simulations and hardware data types provided by SystemC. Surprisingly though, initially SystemC actively discouraged the use of C++'s object oriented techniques. The situation has improved in later versions (SystemC 2.0 onwards), but still certain limitations for object oriented programming are enforced (for example, ports cannot be defined inside a member function, which would in certain cases be very useful). Initially it seemed that the SystemC community widely used SystemC for constructing low level descriptions of hardware devices in a similar way as hardware is described in traditional HDLs. However, research presented in EDA conferences and discussions in online discussion groups nowadays focus on higher abstraction levels and include also the use of object oriented techniques in context with system design using SystemC.

As discussed previously, the constantly increasing system complexity in design projects has required design teams to move to higher abstraction levels in their design representations. Similarly it has become evident that the system models used for simulations necessarily become more complex, thus leading to slower simulation speeds. In addition to raising the abstraction level also in simulation model representation, the problem can be addressed by resorting to object oriented programming techniques in constructing simulators for complex hardware.

An important concept deemed problematic in terms of using object-oriented techniques in hardware simulation and modeling is the use of polymorphism: because of the late binding, one cannot tell the exact type of the object until the run-time. Some other drawbacks of object orientation have been summarized in [345] and [366]. Among these are results stating that it is difficult to implement a method call in hardware, and that protocol mechanism communication does not always fit real hardware. Another well-known problem called *inheritance anomaly* [367], reveals that synchronization mechanisms and inheritance often conflict. One possible solution for resolving inheritance anomaly has been given in [367].

However, employing mechanisms like inheritance and data encapsulation proved beneficial in terms of code reuse and error removing. Furthermore, by taking advantage of the power of abstraction provided by object orientation, one is allowed to describe the system under design at several abstraction levels, as follows:

1. *Conceptual level:* The classes represent concepts in the domain of study. This view is taken very early in the analysis phase.
2. *Specification level:* The classes specify interfaces of the system. A type type is the interface of the class. A type can have many classes that implement it, and a class can implement many types.
3. *Implementation level:* Classes represent code in a programming language.

Inheritance is an implementation technique that is used to implement subtyping in many object oriented languages. Subtyping on the other hand allows polymorphism, where a function may be called with several different argument types. The reason that polymorphism is difficult to implement in hardware is simply that in hardware all types (objects) have to be static, since they are physical objects. However, it can be claimed that in any practical hardware design these kinds of situations do not arise, or the corresponding code can be rewritten in a way that circumvents the problem.

This holds, because in an instance of a hardware system the set of objects is fixed. It does not change and thus the types of the variables in the program do not change while the system is running. Indeed this is exactly the case with the TACO SystemC simulation framework discussed in the next section.

Polymorphism and inheritance are very powerful concepts that have allowed programmers to increase their productivity substantially mainly because they enable reuse of code. Therefore, since hardware design is more and more becoming a programming activity, the importance of allowing hardware "programmers" to use these techniques should not be underestimated.

TACO configurable SystemC simulator

Tools for Application-specific hardware/software CO-design (TACO), a framework for designing and evaluating protocol processors, is an ongoing research effort at the Turku Centre for Computer Science, Finland, since 1999. In this framework we have developed tools and methods for helping the designer in specifying, simulating, evaluating and synthesizing a certain type of protocol processors, TACO processors. TACO processors are based on the Transport Triggered Architecture (TTA) [92,405]. A detailed discussion on the TACO protocol processor architecture and the way the TACO SystemC simulation model fits into the overall TACO protocol processor design flow was already provided in Chapter 12 (*Protocol Processor Design Issues*). In this chapter we focus on details of the configurable SystemC simulator of the TACO framework. This section has its foundation in our research originally presented in [440,443,444].

The TACO simulation library

To be able to rapidly simulate, evaluate and explore architectures for a given protocol processing application or algorithm, an object oriented configurable protocol processor simulation model was devised for the TACO framework. The component library based model is written in SystemC and maintained in a standard x86 PC running Linux. The model contains implementations of functional units (FUs), sockets, interconnection buses, and the interconnection network controller. Using the simulation model it is possible to construct a cycle-accurate simulator of any given TACO architecture, and to simulate both the functionality of the hardware as well as the software. The application software code is input to simulations as hexadecimal values (i.e. compiled TACO instruction words).

The level of abstraction used in implementing the TACO simulation model is heterogeneous in the following sense: the inter-module communication is handled at the RT level, whereas the internal functionality of the modules is implemented as higher level C++ using SystemC's fixed bit-string length data types (e.g. `sc_uint<32>`, 32-bit unsigned integer). The motivation for using heterogeneous abstraction in the simulation model implementation is that simulators are notably faster when the execution logic is implemented as normal C++ instead of RTL style coding; high-level C++ can be used inside the modules as long as the correct amount of cycle delays (i.e. SystemC `wait` statements) is inserted into modules requiring more than one clock cycle to complete their execution. On the other hand, since the communication is modeled at RTL level, very precise bitwise details on data transportation can be extracted from simulations.

Since the functional units of the TACO hardware platform have a very similar interface, and since the simulation model needs to be easily expandable, inheritance is used as a structuring mechanism in the TACO SystemC simulation model (see Figure 18.2). This is done by gathering the behavior and connectivity that is the same for all functional units into a parent class, and placing only the additions to the port/signal configuration required by individual modules as well as the code for the particular FU's execution logic to the child classes. The approach has obvious benefits: the code is more compact and readable (the interface code is not repeated multiple times), there are less errors to debug (only additions to the interface are coded) and adding new functional units to the SystemC component library is faster (most FUs in the hardware platform differ only in the internal implementation, not so much in the physical interface). New FUs are always verified for correctness at the time they are made available in the TACO component library. Thus, all FU descriptions in the library are known to have already been verified.

The execution of a TACO simulator for a given architecture consists of two phases. In the *setup* phase all modules are instantiated. This phase relies heavily on polymorphism to allow automatic socket instantiation and addressing, and to connect different kinds of functional units to buses through sockets. After the *setup* phase all modules in the processor have been instantiated. No more modifications to the architecture are made, and polymorphism is no longer used. The second phase, simulation, is started when the command `sc_start()` is issued in the `sc_main()` routine (SystemC's equivalent to the `main()` routine found in all C++ programs).

It is important to realize that at the moment the simulation starts, the processor architecture is completely static and no more modules are dynamically constructed. For this reason, it is possible (although it would require some effort) to mechanically remove the polymorphic code used for automatic simulator construction. Also the inheritance can be mechanically removed, thus making it possible to have static module instantiation and standalone modules without a class hierarchy. Such removal of polymorphism and inheritance would be useful, were it necessary to develop the code at RTL level for direct synthesis from SystemC; certain commercial tools nowadays support logic synthesis of system descriptions written in a predefined subset of SystemC. However, in the TACO framework this is at least currently not necessary, since we have a separate VHDL model for logic synthesis, and a design tool that sets up the synthesis model exactly according to the simulated configuration.

The classes that are used for simulating hardware are derived from the class `sc_module` provided by SystemC. This class provides among other things macros for simulating signals and ports. The TACO `SocketManager` class is not a hardware simulation module: it is used by objects from the functional unit classes during the `setup` phase for generating, connecting, and maintaining sockets, socket ID's and signals dynamically. During the *simulation* phase, `SocketManager` is used for obtaining pointers to any modules that inherit from `sc_module` and were dynamically created in the *setup* phase of simulator execution.

We recall from Chapter 12 (*Protocol Processor Design Issues*) that there are three different types of sockets in a TACO processor: Input sockets are used for writing data into operand registers in FUs, output sockets for reading data from result registers, and trigger sockets for writing data into trigger registers and simultaneously triggering FU operations. In the simulator all sockets are derived from the base class `Socket` that provides most of the socket interfacing and a state machine for each socket. The subclasses add their own internal functionality and interface requirements to the base class description.

The three level hierarchy for functional units was needed to overcome certain SystemC limitations that we have discussed in detail in [440,444]. The base classes `FuBase` and `FunctionalUnit` provide the interfacing and a state machine needed by each FU. Additions to the base FU interface (e.g. an additional register) and the actual processing task to be executed by a specific functional unit are placed into the FU leaf classes

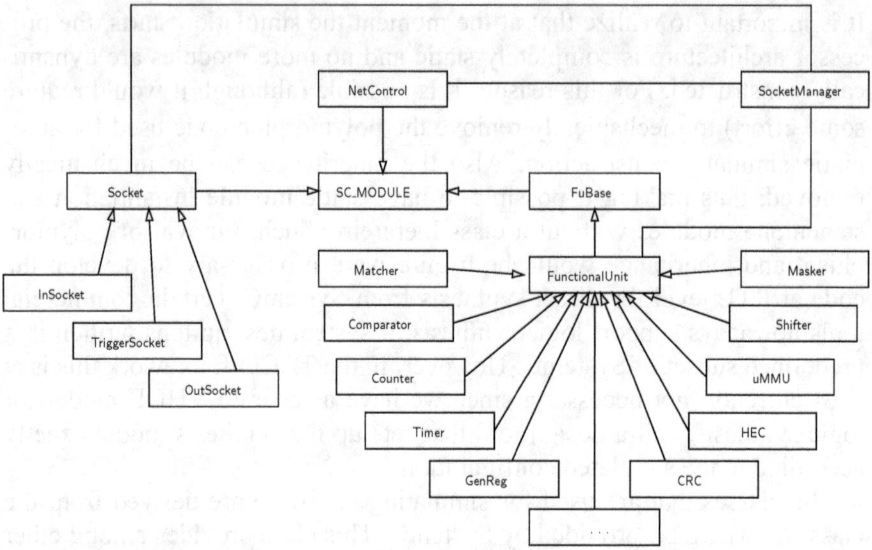

Fig. 18.2. SystemC simulator class hierarchy. Arrows indicate inheritance (arrow head points to parent class), and lines indicate association. © Inderscience Publishers, Inc., 2005 [443]

(Matcher, Comparator, Counter, etc. in Figure 18.2). A functional unit is added to the TACO SystemC simulation model component library by specifying a new leaf class under the `FunctionalUnit` unit class, specifying additional FU registers (if any), specifying the identifiers for logical trigger socket IDs, specifying whether the FU needs a result bit signal, and writing the code for the operation to be performed. However, due to limitations of the SystemC version used in implementing TACO (SystemC 1.0.1), some additional code is also needed. For example, the system clock needs to be tied to the new FU at this level of the class hierarchy.

A simple adder FU with one operand, one trigger and one result register and no result bit would be added to the library as seen in the following example. Note the use of `SocketManager` for reserving a socket ID. For clarity, bypass code for the previously mentioned SystemC issues is not included in the example.

```
class Adder: public FunctionalUnit {
  void assignTriggerIds(){
    trigger->setId(SocketManager::reserveInSocketId("TADD"));
  };
  void triggerOperation() {
    resultReg = operandReg + triggerReg;
  };
};
```

In the above code example, a new leaf class `Adder` is specified under the parent class `FunctionalUnit`. Then, a logical trigger ID called `TADD` is allocated to the new functional unit. Note that the sockets are defined in the parent class; thus, in the leaf class only the logical trigger IDs need to be defined. Next, the operation to be performed is defined. For FUs that are able to perform more than one operation on the data, additional logical trigger IDs need to be allocated. This is done by repeating the allocation line of the code example with different ID names.

The code in the function `triggerOperation()` defines that once the FU is triggered (i.e. when data is written into the trigger register), the values stored in the operand and trigger registers are added together, and the result of this addition is placed in the result register. Timing issues for register reads and writes are managed in the parent class; in the leaf class only the operation to be performed in the *execute* pipe stage needs to be defined.

The interconnection network controller (class `NetControl`) does not have any subclasses since there is always only one such module in a TACO processor. The controller is responsible for extracting bus instructions from TACO instruction words, generating immediate values and evaluating guard expressions for conditional execution. The interconnection network controller implementation in the simulation model is state machine based.

Instantiating the TACO SystemC model

The simulation model is set up for simulating a given TACO architecture by instantiating as many interconnection network buses as necessary and then instantiating as many functional units as necessary (and specifying their types). This is done either manually or using a tool like the TACO design tool [441]. The functional units are connected to the interconnection network by calling connect routines in the newly created bus objects. The creation of sockets and the signals required for connecting the sockets to buses and functional units is done automatically and dynamically by specifying which FU registers connect to which buses as seen in the code example below (line numbers have been added for commenting purposes): no code in the TACO top level file (`main.cpp`) is needed for instantiating and connecting the socket modules between buses and FUs.

```
0: NetControl* nc = new NetControl("NC");
1: Bus* bus1 = new Bus("Bus1");
2: Bus* bus2 = new Bus("Bus2");
3: Matcher* m1 = new Matcher("M1",clk);
4: bus1>insertOperand(m1);
5: bus1>insertData(m1);
6: bus2>insertData(m1);
```

Line 0: Create the interconnection network controller `nc`.
Lines 1 and 2: Create two buses and connect them to `nc`.
Line 3: Create a matcher functional unit`m1`, connect the system clock to it.
Line 4: Create an input socket for the operand register of `m1`, create signals for connecting the socket to `m1`, connect the socket to `m1` and `bus1`.
Line 5: Create an input socket for the data register of `m1`, create signals for connecting the socket with `m1`, connect the socket to `m1` and `bus1`.
Line 6: The data input socket already exists; just connect the socket to `bus2`.

As can be seen in the code example, much of the complexity in specifying the interconnections between different modules has been abstracted away from the designer by resorting to object oriented programming techniques. Still, using the TACO design tool for simulator instantiation should be preferred: the tool generates the instantiation file completely and without errors every time, which is more demanding to achieve with manual instantiation. Once the architecture has been specified in the described manner in the top level TACO SystemC file (`main.cpp`), the simulator is compiled. This is done by issuing the command `make` in the directory in which the SystemC model resides. Normally the compilation takes less than a minute. The executable produced by the compilation is the simulator for the specified architecture. The architecture can then be simulated by starting the executable. The SystemC simulations of TACO processor architecture provide the following results for design quality evaluation:

- *Functional verification.* The key result of any simulation is of course verification of correct system functionality. Such is also the case in TACO: the most important goal in a processor simulation is to find out whether the simulated architecture functions correctly when executing the target protocol processing application.
- *Clock cycle count.* TACO simulators count the number of clock cycles used in each simulation run. This information along with the network speed requirement of the target application (100 Mbps Ethernet, 622 Mbps ATM, etc.) and the estimated achievable clock speed determines whether the architecture being simulated is able to execute the application fast enough.
- *Bus utilization.* During each simulation, TACO simulators calculate the number of possible data transfers and the number of actual data transfers for each bus in the interconnection network. Thus, when a simulation ends, the simulator is able to report relative bus utilization values for each bus (e.g. 150 data transfers actually made out of 200 possible ones: relative bus utilization = 75%).
- *Register transfer statistics.* During simulations TACO simulators also record the source and destination addresses for each data move into a

database. At the end of each simulation, the values in the database are used for calculating the frequency of data moves between individual registers. This list of data move frequencies is then sorted into descending order and reported as register transfer statistics to the designer.

The results extracted from the simulation process are returned either in a human-readable format or are written into a computer-readable format (i.e. XML file) that can be used for passing the simulation results on to a dedicated analysis tool.

By studying the source code of the SystemC class library it can be observed that by using the internal calls in the SystemC implementation a cleaner design of the TACO simulation model could probably have been obtained. However, since the SystemC internals were not really documented and standardized at the time, the decision to use the official SystemC API was made for the TACO SystemC simulation model implementation. If the TACO simulation framework development would start today with the most recent SystemC version, we would probably still decide on using the official SystemC API, among other reasons to ensure compatibility with SystemC-enabled EDA tools nowadays in existence.

We have performed several simulation case studies, for example those presented in [5,442,443]. In these studies we have been able to determine that TACO simulators execute very fast: one run of a packet processing loop can be performed in less than a second or at most a few seconds on a standard PC. At least two reasons for this can be identified. First, the implemented class hierarchy of the system level simulation framework seems to be well suited for simulating this type of protocol processors. Second, since it has been possible to implement certain parts of the SystemC simulation model in high-level C++, fast simulations can be expected. Naturally, the simulation speed in terms of clock cycles per second in the simulated processor depends on the complexity of the architecture being simulated, and on the performance of the PC used for simulation. As an example of simulation speed from one of our previous case studies, we have determined that a simple TACO processor (an IPv6 client processor) can be simulated at 2950 clock cycles per second in a 1000 MHz PC with 256 MB RAM [5]. In this particular case study this translates to about 1.5 IPv6 datagrams per second. Obviously the total simulation time depends also on the number of protocol data units (PDUs) processed in the simulation; in the IPv6 client processor simulator the simulated processing of 1000 IPv6 datagrams takes 10–11 minutes. We would expect to see similar performance figures in any simulations involving the use of high-level C++ with SystemC.

The simulation process allows the designer to detect bottlenecks in the performance of the system by analyzing simulation results for different processing tasks. Consequently, optimizations of the simulated TACO configurations can be deemed necessary and incorporated into the simulated configuration. Optimizations can be made either by increasing the parallelism level or by designing even further optimized FUs. It is important to remark, though, that each newly designed FU provides its own functional primitives, which may be used to express the application specification during the process of mapping the application to hardware and software partitions. Therefore, the mapping process has to be performed again for the processing tasks that benefit from the newly designed FUs.

High-level instruction set simulator generator for COFFEE Risc Core

An *instruction set simulator (ISS)* is a tool that mimics target processor's behavior when running an application program. ISSs can be either *compiled* or *interpretive*. Compiled simulators are fast, but not flexible, whereas interpretive simulators can dynamically adapt to program code changes.

ISSs are useful tools for exploring new processor architectures and for verifying embedded systems. ISS generation allow designer to easily customize and configure them for different target architectures. Commercial approaches of ISS generators are for instance Tensilica's Xtensa [414], ARC Cycle-Accurate simulator (CAS) [20], Target's Chess/Checkers tool suite, which includes a retargetable ISS generator [147], and CoWare Processor Designer [93]. Also academic approaches exist, such as Aachen Lisa Tools [431] and PEAS III ASIP Design Environment [210].

At the Institute of Digital and Computer Systems at Tampere University of Technology (TUT), we have a recently started a research project for implementation of an instruction set simulator generator to create instruction set simulators for different Reduced Instruction Set Computer (RISC) architectures. The purpose of our work is to alleviate the burden of application verification and to evaluate the consequences of changes in instruction set. Moreover, considering that the simulator generation is nearly fully automated, we can easily generate various simulators in order to direct application software to appropriate target processor. At first, we aim at generating ISSs for COFFEE Risc Core.

COFFEE RISC Core [248] is an open source processor core developed at the Institute of Digital and Computer Systems at TUT. The objective of the COFFEE project is to design a core as an Intellectual Property (IP) block and to define a "perfect" instruction set. For the purpose of instruction set exploration, our simulator generator is able to generate flexible and reusable ISSs. That is, instructions, number of registers, memories, their sizes, and different operating modes among other processor-specific information are generated according to the processor description. Let us next present the generator flow and the structure of generated simulators more in detail.

Simulator generation

Figure 18.3 depicts the generator and simulator flow. We describe the *instruction set* of the application target processor in XML format. The instruction set should contain at least the definition of the instructions, but it

Fig. 18.3. Simulator generator and application simulation flow.

may also contain information of the required parameters, their types, and possible restrictions. A Parser *reads* the instruction set file and extracts information of each instruction, which the *Simulator generator* uses when generating the body of the *Simulator* (that is, instruction header files, which function as the interface to other classes). Generating the functionality of the instructions is most likely nearly impossible due to the diversity of their possible functionality. Therefore, we have an *Instruction library* from which the already verified implementations of different instructions can be selected. In case the instruction implementations cannot be found or they are very different to the example implementations, they must be written manually and connected to the simulator.

As seen in Figure 18.4, generated simulators are class based and each instruction is a class. Hence, the number and implementation of instructions can be changed without regenerating the simulator, only recompilation is required.

Simulator structure and application simulation

We have used object oriented C++ to implement the generated simulators. A simulator consists of the base class *Instruction* from which all the instructions (from *Inst_0* to *Inst_n*) are inherited. The benefits of object oriented design and inheritance were discussed earlier in this chapter.

Instructions are connected to *Arithmetic library*, *Simulator*, *Register*, and *Memory* classes. The *Arithmetic library* contains implementations for arithmetic, Boolean, and bitfield manipulation operations. The *Simulator* class controls the simulation of instructions and for instance the register and memory read and write operations, whereas *Register* and *Memory* classes only implement them.

The right-hand side of Figure 18.3 illustrates the application simulation flow. At first, *Application software* traverses through a processor *Compiler flow*, which produces the *Application binary* file. The *Instruction set simulator* reads the binary and executes the application instructions. As a result, the simulator produces separate text files for *register* and *memory* use for register and memory operation analysis. In the future, we will implement a hardware interface, which allows the running of application code in the simulator and on hardware in parallel, as well as a model of a pipeline to make the high-level model more consistent with real hardware.

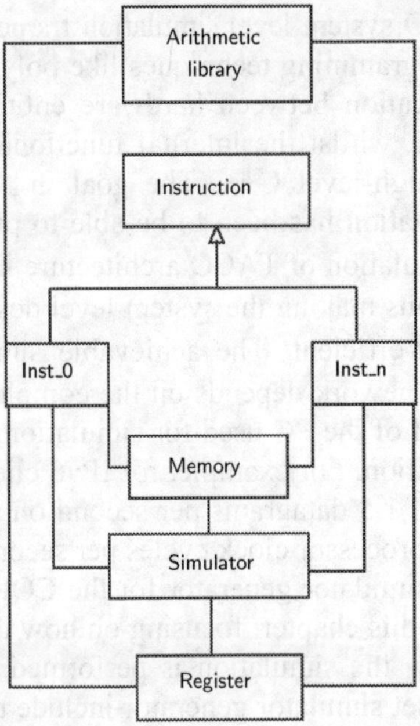

Fig. 18.4. Class hierarchy of a generated simulator.

Conclusion

In this chapter we have discussed the general principles behind system level simulation and positioned the system level simulation abstraction in the design flow relative to other levels of abstraction that may be used in modeling and simulating processors. System level modeling and simulation languages are nowadays often based on C or C++. With C++ based languages, the powerful concepts and benefits provided by object oriented programming techniques become available for the system designer modeling hardware. The design community has regarded some object oriented techniques as problematic, but as discussed in this chapter, this can also be seen as a misinterpretation of the way the system is represented by objects and classes. We discussed these issues, and system level design in general, especially in terms of the SystemC modeling language.

As examples of system level simulation frameworks we discussed the modeling and simulation of TACO protocol processors using SystemC, and instruction set simulator generation for the COFFEE RISC processor using C++. The TACO system level simulation framework takes advantage of object-oriented programming techniques like polymorphism and inheritance. The communication between hardware entities is modeled at the register transfer level, whilst the internal functionality of the entities is implemented using high-level C++. The goal in the TACO simulation framework implementation has been to be able to provide accurate, rapid configuration and simulation of TACO architecture instances using affordable PC computers, thus making the system level design space exploration process fast and cost-efficient. The achievable simulation speed of the TACO simulation framework depends on the complexity of the simulated architecture, the speed of the PC used for simulation, and the target protocol processing application. For example, for IPv6 client operation we have measured the speed of 1.5 datagrams per second on a 1000 MHz PC with 256 MB RAM (2950 processor clock cycles per second).

An instruction set simulator generator for the COFFEE RISC processor was also discussed in this chapter, focusing on how the generation process and, respectively, how the simulation is performed. Future plans for the COFFEE instruction set simulator generator include the implementation of a hardware interface that allows running the application code in the simulator and on hardware in parallel, as well as a model of a pipeline to make the high-level model more consistent with real hardware.

As a general remark, from the two simulation experiments discussed in this chapter we can conclude that the system level simulation does not provide a magical solution for ensuring the correctness of the final system. Firstly, a system level executable specification of a system, as complete as it may be, does not represent the real system, but only an abstraction of it. This means that, in fact, we are not simulating the system as such, but only its specification. Secondly, since embedded computer systems are reactive systems (they react to stimuli from the environment and eventually provide response) their I/O behavior has to be tested. This is a nontrivial task, since generating test data to cover all possible scenarios can prove difficult.

19 Programming Tools for Reconfigurable Processors

Claudio Mucci[1], Fabio Campi[2], Claudio Brunelli[3], and Jari Nurmi[3]

ARCES, University of Bologna[1]

STMicroelectronics[2]

Tampere University of Technology[3]

The capability to tailor processor instruction set architecture (ISA) around the computational requirements of a given application is proposed today as the most appealing way to match performance with very short time-to-market, thus reduction non-recurring engineering (NRE) costs. From Mask-Time Configurable Processors (MTCPs) to Run-Time Reconfigurable Processors (RTRPs), ISA customization is performed transferring the implementation of critical kernels from software to hardware. This introduces a new design-space exploration problem that requires skills in both software and hardware design. Since adaptive processors appear as the natural extension of Digital Signal Processors (DSPs), programming tools for customizable processors need to be as similar as possible to standard software development environments, in order to enable the adaptive computing to the wide audience of DSP programmers. While fast design-space exploration can be performed using high-level description languages, programmers proficient in hardware design can further improve performance through "structural" descriptions involving, for example, the direct utilization of macro-operators or the possibility of balancing critical paths through register insertion. The widespread knowledge of ANSI C among developers suggests its usage as main entry language for both configurable and reconfigurable architectures. This in turn introduces the problem of translating C codes (or C dialects) into some kind of hardware description, be it HDL in case of MTCPs or bit-stream for RTRPs. In this context, data-flow graphs (DFGs) can be efficiently used to close the gap between

J. Nurmi (ed.), Processor Design: System-on-Chip Computing for ASICs and FPGAs, 427–446.

hardware and software design, bridging hardware and software with a computational model common to both the design environments. Standard ANSI C can also be used by the programmer for the management of the application control flow on the processor core. Custom-designed instructions are then embedded in the C code using standard, compiler-friendly mechanisms, such as intrinsics and assembly inlining.

Algorithm development on reconfigurable processors (programming issues)

Processor-based Systems-on-Chip (SoC) are becoming the most popular way to perform computation in the electronic marketplace. Today, at least one processor is present in every SoC in order to handle in a simple fashion the overall system synchronization, providing the operating system functionalities (i.e. multi-tasking management, real-time issues) and I/O communications. Usually, general-purpose embedded processors, like ARM9, PowerPC, MIPS, do not have responsibility of the computation that is demanded to high-performance co-processing engines. Depending on application constraints and on the required degree of flexibility, computation intensive parts are implemented on dedicated hardware accelerators (when non-recurring costs allow that) or on application-specific digital signal processors (DSPs). Since they are software programmable, application-specific DSPs are proposed as a way to match flexibility with high performance. Architectures like the Texas Instruments OMAP or the STMicroelectronics STW51000 (also known as GreenSIDE) are examples of state-of-the-art commercial SoCs including ARM-based system that achieve performance efficiency through one or more application-specific DSPs and one or more dedicated hardware accelerators.

One of the most interesting trends in the field of high-performance energy-aware SoC design is the introduction of dynamically (or run-time) reconfigurable hardware (e.g. embedded FPGAs, reconfigurable data-paths and reconfigurable processors) in addition or in substitution to the constellation of DSPs and dedicated hardware accelerators [38,50,97,98,274,343]. In general terms, the exploitation of such kind of architectures implies the capability to tailor the SoC functionalities around the computational requirements of the given application, extending the base instruction set of the main processor (for example, the ARM in the previously cited examples), thus making reconfigurable hardware a run-time extension of the baseline computation engine.

As in the case of DSPs and dedicated accelerators, the exploitation of any degree of parallelism at bit-, word-, instruction-, iteration- and task-level is the control lever for the effective utilization of the reconfigurable hardware. This requires from the programmer a deep knowledge of both application and system architecture to understand at best how to partition and how to map algorithms over the different available computational resources. On the other hand, this also requires to the programmer the capability to investigate a hybrid design-space including both software and hardware concepts. For application developers long used to C/assembly design environments, this requirement is not so usual given the common lack of skills in hardware design flows.

With respect to mask-time programmable hardware accelerators, reconfigurable computing offers to the programmer the capability to design after fabrication its extensions in order to fulfill application requirements. For that, the capability of providing *soft*-hardware (or hardware programmable as software) is probably the key point to enable the large market of application developers to effectively utilize reconfigurable devices [99]. In the past, reconfigurable devices borrowed tools and methodologies from the discrete FPGA world (thus utilizing design flows based on hand-coded RTL HDL), although it was clear from the beginning the severe lack of user-level programmability coupled to this approach. But very soon, the utilization of C language has been clearly established as the most promising way to approach the customers at the "reconfigurable" proposal.

A lot of C-oriented dialects have been presented including entire new object classes dedicated to hardware design, like in SystemC or Handel-C. This kind of approach moves C towards the hardware, design making hardware description friendlier for application developers. In this context, the C-based languages basically become yet another HDL, therefore requiring hardware skills to developers.

A more promising approach is to use standard ANSI C code and translate it into some kind of RTL, making use of some sort of C-to-RTL hardware compilation. Companies like Celoxica, Mentor Graphics, Impulse, Altium and CriticalBlue offer stand-alone C-to-RTL and/or C-dialect-to-RTL synthesizers that can be integrated in standard flows for FPGAs and that were used in many works on reconfigurable system implemented using commercial FPGAs.

In the case of embedded applications, the reconfigurable device is a part of a usually complex system with a rigid cost, power and area budget. This precludes the utilization of standard FPGAs, since they are too area demanding and not so appealing in terms of performance, power consumption and cost. Reconfigurability is provided either through embedded-FPGAs (small FPGA suitable for SoC integration), or reconfigurable data-paths and reconfigurable processors that offer flexibility under severe constraints in terms of area.

In fact, the area occupation of reconfigurable devices is very often considered a key issue for SoC designers. This means that the reconfigurable device needs to be as small as possible, while configuration efficiency must grow up to the peak performance offered by the device.

On the architecture side, area limitation can be accomplished by accurate trade-off between logic and interconnect, reducing for example the impact of programmable interconnect with respect to the area required for computational logic. In island-style programmable architectures it is possible to achieve better area figures increasing the grain of the basic logic element with respect to the interconnect structure, or decreasing the interconnect capabilities, limiting the connection at level of rows or supporting only communication among the neighbors logic elements [380,445]. This implies an undeniable reduction in term of flexibility, paid to the need of guarantee small area budget.

On the programming side, the increase of design constraints and the reduction of degree of freedom in the mapping of algorithms imply that any inefficiency of the automated high-level synthesizer lead to a dramatic performance loss. To avoid this, many reconfigurable devices provide "structural" languages in which operators are directly mapped into the device without synthesis. Application designers can tune, refine or re-write from scratch the implementation in order to maximize the performance benefit in the same way that the DSP programmer can use the assembly language.

All these preliminary considerations can be summarized in few points that we can see as requirements for an application development environment in the field of reconfigurable computing:

- to be appealing for the wide "world" of software/DSP programmers, such environment needs to be as similar as possible to traditional software-only environments;
- to be effective and compliant to the huge investment in term of area and costs required by reconfigurable hardware, such environment needs to provide capability to exploit as much as possible architectural features.

Instruction set extension implementation on a standard compilation tool-chain

The extension of a standard software tool-chain in order to support instruction set metamorphosis implies to analyze the role played by each tool and the efficiency required by each step, with the final goal of proposing to application developers a tool-chain in which hardware and software can be handled together. The introduction of instruction set extensions implies modification in each step of the compilation process, and the addition of bitsream generation tools dedicated to the mapping of instruction set extensions in the reconfigurable hardware. In this section we focus on the software support necessary to handle instruction set reconfiguration from a C compiler. Aspects concerning the extension definition and its mapping on the hardware will be dealt with in the next sections.

In general terms, modifications in non-optimizing assembler and in the linker are kept as minimal as possible, since the assembler can be reduced to a simple mnemonic translator and the linker needs to include the eventual bit-stream for the hardware customization. On the contrary, high-level compilers or optimizing back-end tools (for example, instruction schedulers) needs to be aware of the reconfigurable parts in order to help the user in the optimization process. We can require to programming tools for reconfigurable processors many tasks:

- provide to the user the capabilities to define and instance an extended instruction;
- schedule extension instruction accurately;
- automatically recognize user-defined extended instructions in a general-purpose code;
- detect critical kernels and automatically generate a set of extension instructions.

The definition of extension instructions is usually accomplished by dedicated tools, while the capability of instancing extension instructions in a software code can be obtained by using the same functionalities provided for assembly inlining. Although apparently similar to software tools required for configurable processors, the last three points are very specific of reconfigurable computing. The main difference is in that cost metrics and constraints management change significantly between an application-specific custom design and a soft-programmable hardware design. Accurate scheduling and identification of custom instructions, be it performed in compiler front-end or optimizing back-end tools, can be handled by modifying the machine description and the intermediate representation (a sort of virtual machine-independent assembler) of the compiler.

In the case of traditional C tool-chains, like, GNU GCC [157], this implies the complete recompilation of the compiler front-end since the machine description, as well as the pattern matching automata detecting the assembly instruction, is implemented statically. It is thus possible to deal with extended instruction in the same way that a compiler handles floating-point extensions, describing the required functional units in the intermediate representation [56]. Of course, this proves to be a hard obstacle for most application developers, also in terms of time required during the design-space exploration when the instruction set extension is under definition.

Alternative approaches have been proposed in research projects on advanced high-level compilers like Impact [72] and SUIF [393]. In these cases, machine descriptions and intermediate representations can be dynamically extended without rebuilding the tools, since the target description is read before each compilation. Consequently, pattern recognition and instruction scheduling are accomplished by mean of generic algorithms or implemented by dynamically built optimizing structures. State-of-the-art compiler capable to handle optimized scheduling of long latency custom instructions can be found, for example, in the MOLEN project [436] and in the DRESC framework [282], respectively based on SUIF and Impact. The Trimaran framework [426] proposes a scheduling mechanism based on simulation/profiling back-annotations to reduce stalls in a computation-aware environment, although this has a significant impact on compilation time.

This point introduces the last issue that reconfigurable computing imposes on programming tools: the reconfiguration of the simulator. Similarly to the case of the compiler, the simulator needs to be adapted to the change/extension of the instruction set. Language for Instruction Set Architecture (LISA), commercially available from CoWare and Axys, as well as open-source architecture description languages like Arch-C, are examples of frameworks where cycle-accurate instruction set simulators are built with the support of native structures implementing typical processor objects, like the pipeline or the register file. This approach requires a rebuilding of the instruction set simulator every time the instruction set is changed. In [300] an alternative approach is proposed. A dynamically linked library is used to model the instruction set extension, while the main processor is modeled by standard simulator support. The mechanism is applied on both functional and cycle-accurate simulation, integrating the mechanism on a LISA/System-C environment and on a pure-functional debugging environment based on the GNU GDB simulator.

Figure 19.1 shows a simplified and very general block diagram for a programming environment supporting reconfigurable computing. It includes

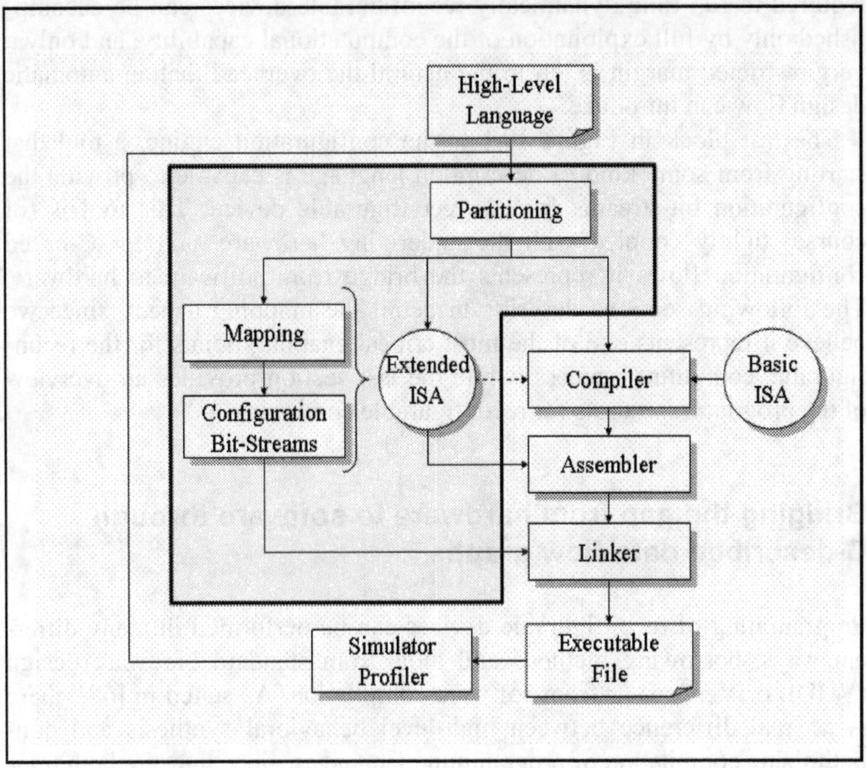

Fig. 19.1. Basic software toolchain extension to support reconfigurability issues.

the basic software support (compiler + assembler + linker + simulator) previously described, and the partitioning and configuration parts. The partitioning is the process, automatic or not, of design-space exploration in which critical tasks (kernels) are moved from software implementation to hardware (and/or vice versa) depending on the required performance constraints (speed, energy, …). Today, this process is usually under the complete control of the programmer, although it can be helped by the usage of tools. Research is going in the direction of full automation of the partitioning, since this may represent the key enabling step towards true soft-programmable hardware (e.g. [83]). Despite this, very few works are at the moment leaving the academic/research environment to challenge the market, and these few works are focused in the field of mask-time programmable devices (e.g. [35 6]). As explained in the introduction, the configuration efficiency required to run-time dynamically reconfigurable devices can be accomplished only by full exploitation of the computational capability, and only a very restricted margin is left to the natural the overhead that an automatic design flow can introduce.

The last block in Figure 19.1 is the configuration engine, a tool that starting from some kind of description language is capable to provide the configuration bit-streams for the reconfigurable device. This tool is (of course) tightly coupled with the underlying hardware, and for C-based configuration flows it represents the bridge from software to hardware. The following sections describe in detail the mapping aspect, since we believe it represents one of the most critical enabling points for the reconfigurable computing success, while the last section provides an overview of the programming tools for reconfigurable computing.

Bridging the gap from hardware to software through C-described data-flow graphs

Programming of reconfigurable devices can be performed in many different ways, borrowing methods and tools from standard hardware design (VHDL or Verilog) or from software compilation. As stated in [68], there is no real difference between high-level behavioral synthesis and non-optimizing compilation of programming languages, since they are both translations of an initial language to an intermediate representation. On the contrary, the optimization is a very different step in hardware synthesis from the software synthesis, with different metrics and cost-functions. Another common point between compilation and synthesis is that graphs are most often used as a mean for internal representation. In software programs, we can distinguish between two kinds of graphs: control- and data-flow graphs (respectively CFG and DFG). The CFG is the representation of the paths that might be traversed in a program during its execution. Each node of the CFG is known as basic block and its graph representation is a DFG. The DFG describes the dependencies among the set of operations required in the data processing. As shown in the example in Figure 19.2, branches of a conditional statement (if…then…else…) are represented as nodes of the CFG, while the operations performed in each branch are described by a DFG "attached" to CFG nodes.

In hardware description languages we have co-existence of both sequential and concurrent definitions of operations. As an example, the behavior

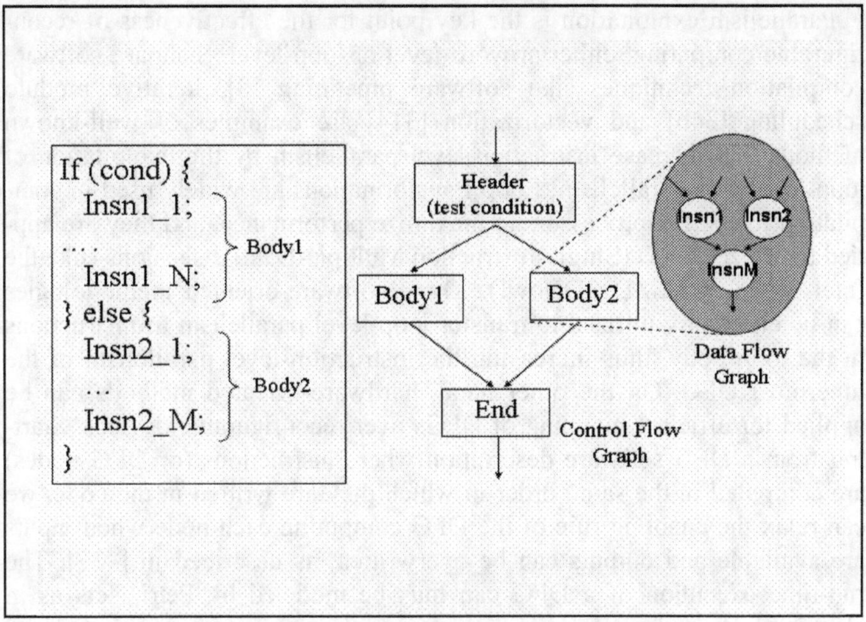

Fig. 19.2. Example of control- and data-flow graphs.

of a process or the expression assigned to a signal follow a sequential paradigm, although this not means that the same semantic of the software languages is used. Operations can be viewed as nodes in the DFG, as well as sub-graphs of a DFG, depending on the granularity we choose to assign to the node. Hardware description languages use an event-driven activation mechanism in which more than one DFG and more than one DFG node can be active per time natively. This represents the most significant difference with respect to control flow resolution in software languages. Of course, during the compilation for processors featuring some degree of parallelism (e.g. VLIWs, Superscalars, TTAs, ...) this constraint is heavily relaxed, bringing software implementation near to the hardware even though with different optimization metrics.

For the definition of a suitable bridge between hardware and software in the field of reconfigurable computing, the DFG represents the most natural choice. In the case of reconfigurable processors, control is typically managed by the processor core, while DFGs are implemented through hardware acceleration. Hence, the DFG suitable for the mapping on the reconfigurable device can be described by a sequential language, like C, while it can be viewed as an abstract circuit representation.

Parallelism exploitation is the key point for the effectiveness of reconfigurable computing, either at word-level or loop-level. Standard software compilation techniques like software pipelining [8], iterative modulo scheduling [346] and vectorization [314] are examples of well-known methods that increase instruction-level parallelism by the exploitation of loop-level data parallelism. Loop transformations are widely used in compilation for VLIW processors to maximize performance, and they are applied to utilize SIMD (Single-Instruction Multiple-Data) extensions (like the Intel MMX or AMD 3DNow!). These software-oriented methodologies can be efficiently utilized to transfer loop-level parallelism to instructions in the loop body, thus increasing the instruction-level parallelism of the innermost DFG. On the other hand, hardware-oriented methods can be applied for efficient mapping of DFGs over reconfigurable devices. Starting from a DFG software description where instructions (or DFG nodes) are computed in the same order in which they are written in the code, we can relax the enabling rule of the DFG computing each node when inputs are available and outputs can be overwritten, as described in [437]. The run-time execution of a DFG can thus be modeled by Petri Nets as in [139,302]. By nodes scheduling and register insertion it is possible to build the DFG in a pipelined form, without affecting the functionality. In this case, it is possible to overlap the execution of successive DFG activations (if data-dependencies allow that) hence improving performance by the exploitation of parallelism at level of iteration, as in [384,449].

So far, we have discussed about the role played by DFG as bridge between software and hardware. One more point is of course represented by the way in which the DFG can be described in order to meet the requirements of effectiveness and friendliness posed as basis of a programming tool-chain for reconfigurable processor. We said that the entry language must be appealing to software programmer and must be effective in term of hardware utilization. An interesting option is to use the C language for that goal. C of course allows describing DFGs since DFGs are representations of the basic blocks. It also allows handling the DFG topology under simple restrictions. For example, the utilization of a single-assignment form, in which each variable is assigned exactly once, can help the user in the DFG modeling, providing a simple way of handling efficiently all the data-dependencies, as proposed in [305]. Single assignment is today introduced in many compilation frameworks as an important intermediate representation in order to simplify and/or optimize internal steps. Starting from the version 4, also the GNU GCC toolchain makes extensive use of single-assignment representations, although the conversion to single assignment is performed (in our knowledge) only for scalar register values (everything except memory) at level of basic block. For this reason, conditional statements (if…then…else) are converted computing concurrently each branch of the statement and introducing *merge*-nodes that select the correct outputs among the branch-replication, similarly to multiplexers in the hardware design. In general, the translation of standard C and C-dialects into some kind of hardware description is a complex problem addressed by many research programs [100,132,167,221,394], especially if the memory access issues are considered (i.e. pointers) [371]. In the case of reconfigurable processors, the processor core can normally handle memory access, eventually with the help of DMAs to speed up the memory access, thus simplifying synthesis requirements [380].

Summarizing, single-assignment forms are restrictions of the C semantic, and they are useful to accurately handle and optimize DFG performance (parallelism and pipeline structure). Their significant advantage is that they can be extracted from high-level C compilers. The application developer can thus start the implementation over a reconfigurable processor from the application description written in C, selecting the critical kernels suitable for the reconfigurable hardware mapping. Depending on the efficiency required, the application developer can choose to use an automatic translation mechanism or to hand-code the kernel with a low-level description language such as single-assignment C stripped of non-hardware-friendly constructs. This evaluation introduces a third trade-off point represented by the time spent to the development.

Overview of programming tools for reconfigurable processors

Programming frameworks for reconfigurable architectures are highly dependent on the structure, the hardware granularity and the language used. Although far from being an ideal hardware description language, C was selected as an appealing entry-point for the configuration of reconfigurable processors since the first architectures (e.g. PRISM [29]).

Milestones of the research on reconfigurable processors, like the Garp [65] processor, and commercial state-of-the-art reconfigurable processors [25,32,119,364] propose C-based design environments envisioning the possibility to offer the end-user the capability of automatically partition the source code, and then to co-compile the same code over both the processor core and the reconfigurable logic. The Nimble compiler [264], targeting the Garp processor, is one of the first tools that automatically transfers critical kernels from a processor core to reconfigurable hardware accelerator, selecting them from the basic blocks found in the innermost loops.

PipeRench [59, 159], one of most popular coarse-grained reconfigurable data-paths, is configured using a single-assignment language with C operators (called DIL, Dataflow Intermediate Language). RaPiD [95] features a C-based proprietary language, RaPiD-C, that consists of nested loops describing pipelines. Language extensions allow the programmer to explicitly handle synchronization, and specify parallelism and data movement (that is stream-based). Another example of popular coarse-grained architecture is represented by the RAW architecture [411] developed from the MIT: in this case a SUIF-based compiler partitions the application over a sort of RISC-based multiprocessor, rather than performing technology mapping. Another programming approach based on C language was provided for the NAPA architecture [158], including a C-programmed reconfigurable device as I/O co-processor.

Many reconfigurable devices are programmable at assembly-level and/ or by graphical tools for manual mapping, in a way that seems to trade part of the programmability offered by high-level languages the programming efficiency (MOPS/mm^2), as reported in the Hartenstein's retrospective [180]. In general, the underlying architecture has a strong impact on the technology mapping, on the placement and to a lesser term on the routing algorithm.

Direct mapping is probably the most used methods for coarse-grained architectures: operators are mapped to the programmable elements that compound the device without a real logic synthesis step. PACT XPP [445] and MorphoSys [380] are examples of this kind of approach. In both cases, a tentative to virtualize the underlying mapping layer using C-based high-level compiler flows is provided [282,449]. This notwithstanding, for full exploitation of the architecture capabilities, low level, architecture specific approaches are utilized. PACT XPP is programmed through the Native Machine Language (NML), a structural event-based netlist description language. Specific tools are proposed for the place-and-route phase [390] where the strongly pipelined structure requires to pipeline also interconnections across rows by dedicated registers. For the MorphoSys architecture, a SUIF-based compiler is provided for the host processor, while partitioning between hardware and software is performed manually by the programmer. MorphoASM, a structural assembly-like language, is used to configure each programmable element to the functionality required. Usually, the programmer need to take into account also the interconnect capabilities of each programmable element in order to distribute processing elements in the device pipeline in a way compliant to timing requirements.

In general, application mapping on coarse-grained architectures different from the island style of FPGAs, requires specific management constructs. As an example, in the Garp processor, the GAMA tool [65] maps the DFG using a tree covering algorithm that splits the original graph into sub-trees with single fanout nodes, introducing significant overhead in resource utilization. Furthermore, only acyclic graphs are supported. Modules detected by the tree covering are placed in Garp array rows (only one module per row) using bit-slice methods proposed for data-paths synthesis in regular architectures. The DRESC compiler [282] is an example of high-level compiler targeting a MorphoSys-like coarse-grain architecture. It focuses on the exploitation of loop-level parallelism and perform place-and-route for the reconfigurable hardware using an extended iterative modulo scheduling algorithm. A simulated annealing strategy is used to determine when a legal configuration can be accepted or not, helping to escape from local minimum.

In some cases, where the reconfigurable processor is integrating an embedded FPGA or it is implemented on a stand-alone FPGA (like in MOLEN [436]), VHDL and Verilog are used for the hardware customization. The optimization process is typical of hardware design flows: behavioral RTL HDL descriptions are substituted by FPGA-specific macros, when expected performance are not achieved directly from synthesis. For what concerns programmability issues, since HDL is the entry-point all C-based languages and tools generating VHDL can be applied to provide a more software-oriented approach, but the optimization process is performed under the hardware design paradigm, for example analyzing timing constraints and critical paths.

An example of algorithm development environment for reconfigurable processors: the Griffy-C approach

This section describes, as an example of algorithm development environment, the Griffy-C approach proposed for the XiRisc reconfigurable processor and developed at the Arces/ST joint lab of the University of Bologna. In this context [270] reconfiguration is performed at level of assembly instruction, while the functionality associated to each extension instruction is modeled as a DFG. The reconfigurable device is fit in the processor pipeline as an additional functional unit, triggered by a specific assembly directive (*pgaop*).

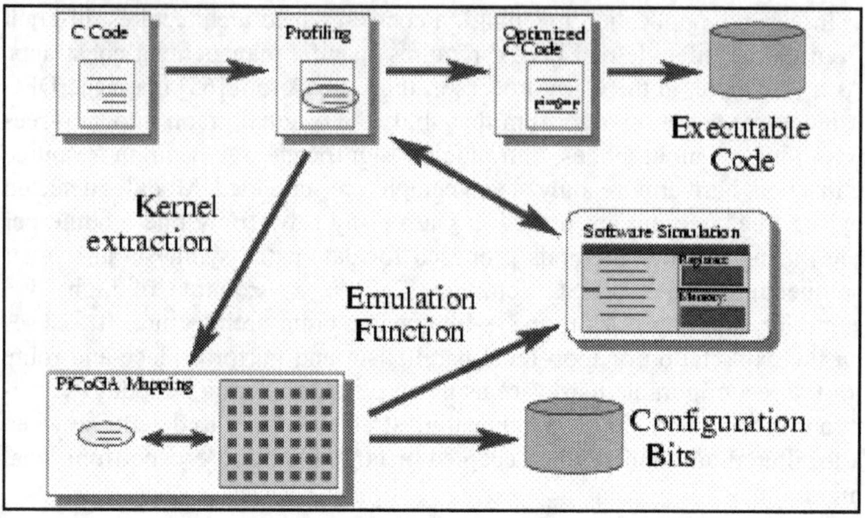

Fig. 19.3. XiRisc algorithm development environment.

Figure 19.3 overviews the programming environment proposed to the application developer. In particular, through profiling analysis the programmer manually identifies critical kernels suitable for the mapping on reconfigurable hardware (pgaops). The compiler tool-chain is based on a retargeted version of the GNU GCC and instruction set extensions are handled through assembler inlining. No specific scheduling support is provided for the extended instruction set. Software simulation is provided for both functional debugging on the GNU GDB and cycle-accurate instruction set simulation in the LISA/SystemC environment. With respect to the compiler, the simulation environment supports instruction set metamorphosis through the utilization of dynamically linked shared library (so library under Linux environment). The emulation library of the instruction set extension is automatically generated from DFG compilation, and plugged in both GDB and LISA/SystemC environment [300].

The functionality of each instruction set extension is described starting from a single-assignment manually-dismantled C syntax called Griffy-C [299]. Griffy-C is a structural description in which basic C operators (e.g. sum, subtraction, bitwise logical operation and comparisons) are directly instanced on hardware resources, without logic synthesis. Figure 19.4 shows an example of the Griffy-C code used to implement a sum of absolute differences (SAD) from video encoding and the corresponding (non-optimized)

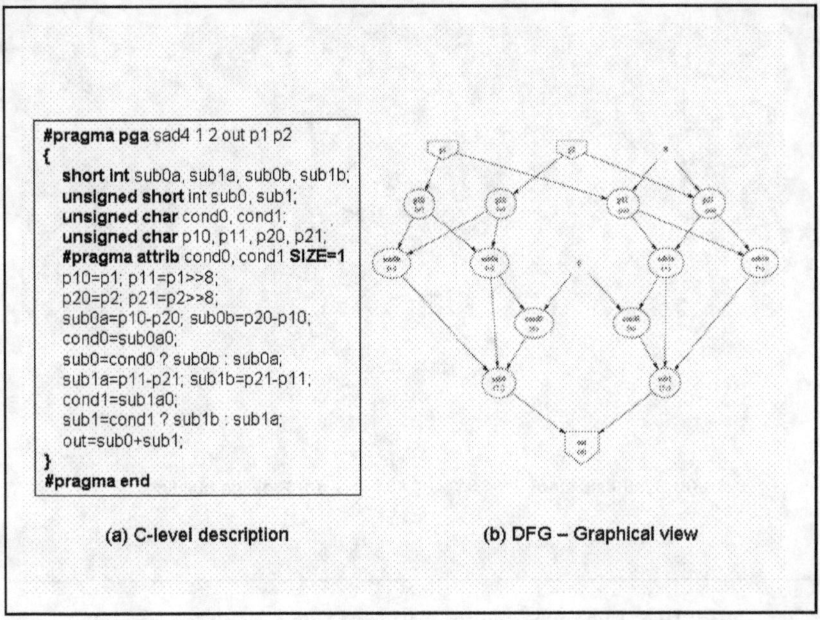

Fig. 19.4. Example of Griffy-C code and the corresponding DFG.

DFG. Griffy-C does not support control flow statements, with the only exception of the conditional assignment ("?:") used to implement multiplexers, in hardware terms, that is a *merge*-node under the data-flow paradigm.

The C-oriented description implies that some operations with constant operands may be resolved by constant folding and collapsing on following nodes. This kind of operators does not need explicit instantiation of processing elements. This can be regarded as a very basic synthesis step. An example of such approach is the utilization of the routing resources to implement constant amount shifts in fine-grained routing architectures. Figure 19.4(a) shows the collapsing of shifts used in the previous SAD example for unpacking the input variables, thus providing the optimized pipelined DFG depicted in Figure 19.4(b). Horizontally aligned nodes represent a single pipeline stage, and dotted nodes represent collapsed operators.

The single-assignment syntax used in Griffy-C allows the user to handle accurately the pipeline structure and at the same time can be automatically generated from a high-level compiler tool-chain as proposed in [254]. Specific extensions for the bit-level definition of the variable size are provided

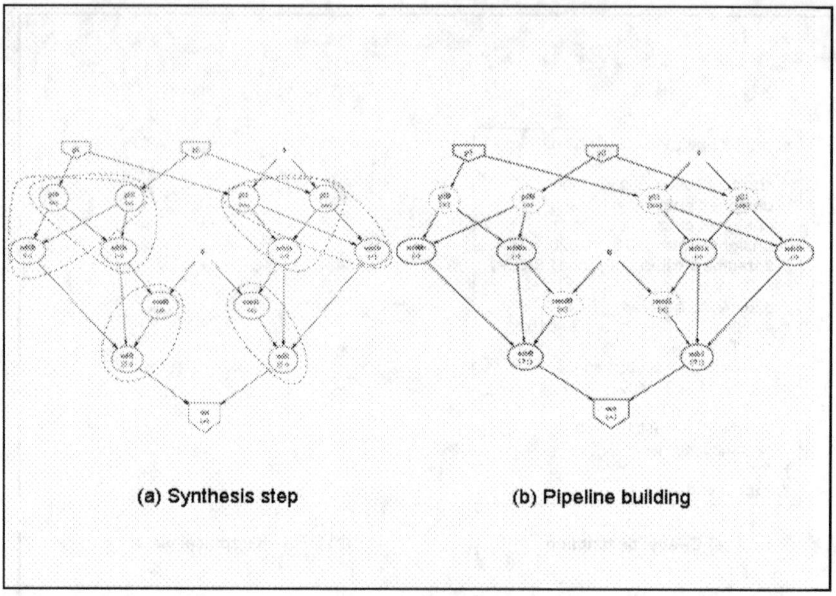

Fig. 19.5. Example of optimization of routing-only operators.

through *#pragma* directives in order to reduce area occupation. The mapping process can be divided in five main steps:

- *Instruction-Level Parallelism extraction*. Starting from the data-dependencies of the DFG an optimized scheduling algorithm builds the pipeline structure. Griffy-C code is analyzed and scheduled in pipeline stages applying an earliest enabling rule, in which instruction are executed as soon as possible. Detection of routing-only instructions allows to build optimized pipeline stages, although the presence of internal feedbacks (e.g. described by static variable in the C syntax) requires special management (see Figure 19.5).
- *Physical Mapping*. The arithmetic/logic operations that require computational resources on the array are generated with a proper configuration. The result is a netlist, annotated with configuration bits, where elements are hierarchically organized for pipeline stages and macro-elements (i.e. set of basic computational blocks implementing a Griffy-C operation).
- *Placement, routing, and pipeline synchronization*. In this phase the netlist is arranged on available hardware resources, and synchronization mechanisms required for the pipeline evolution are programmed (Figure 19.6).

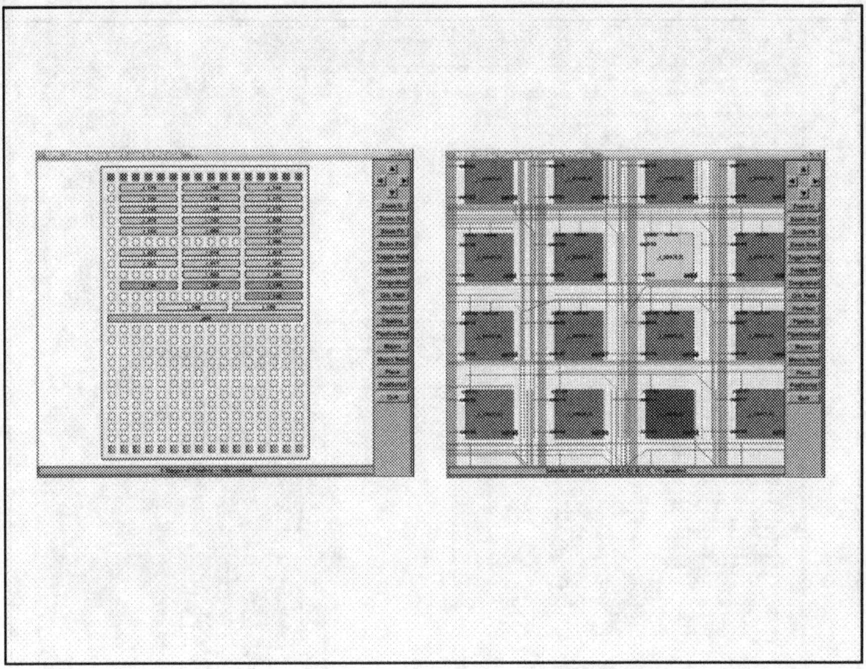

Fig. 19.6. Placement and routing visualization.

- *Bit-stream generation* is the last step in the configuration process, and provides the set of bits necessary for hardware configuration in a C vector form that can be included in any standard processor tool-chains.

The validation process or debugging is an focal capability required to any algorithm development environment. In reconfigurable architectures controller by a standard processor core, the overall simulation can be managed by a software debugger such as the one provided in the GNU environment (GDB and DDD) and/or by the cycle-accurate simulation frameworks such as those based on LISA/System-C. In both cases, in the Griffy-C environment the validation of the reconfigurable part is handled by a separate viewer that shows the same Griffy-C code written by the user annotated with intermediate results. Breakpoints and, control flow management in general are handled by the processor debugger, and the application developer can inspect in every moment the status of the reconfigurable unit. As an example, Figure 19.7 provides a screen-shot of the GDB-based debugging environment.

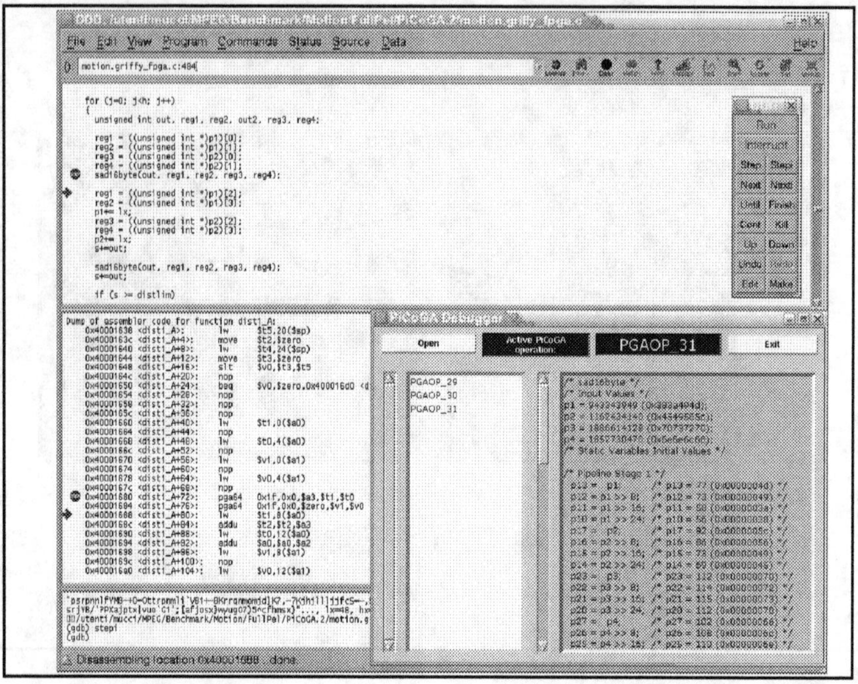

Fig. 19.7. Griffy-C debugging and validation environment.

Application development under the Griffy-C approach is a process in which the programmer can iteratively move application kernels from software to hardware in a sort of continuous space. In fact, the user, starting from the original C code, is required to rewrite manually his kernels in Griffy-C usually working with C-based operators and then start performance analysis. The partitioning between C-code on the processor and Griffy-code on the reconfigurable device is an iterative process of refinement where experience and knowledge play a crucial role. But, differently from methodologies borrowed from FPGA design, this kind of approach is mainly software-oriented. The user can change the partitioning and/or optimize the kernel moved to the reconfigurable hardware in the same way that DSP programmers use assembly for speeding-up their applications. Optimization of Griffy-C code can leverage on two main factors: pipeline re-arrangement and intrinsic optimization. In the first case, the pipeline structure is modified playing with data-dependency in order to retime the graph or to adjust the write-back point in pipeline, for example using software pipelining methods [384,449]. In the second case (intrinsic optimization), the programmer can substitute part of the code with optimized operations like the direct instance of a look-up table. For software programmers this seems the assembly-level optimization in which high-level code is substituted by built-in functions, linear assembly or assembly, since the syntax remains strongly sequential and imperative, without any kind of direct parallelism exposition. Only tools are responsible of that.

A set of applications has been developed measuring the relationship between performance achieved and time spent for the development. The experimental analysis has been obtained monitoring application developers of different skills and back-ground, including students and researchers. Results are provided in Figure 19.8 where it is possible to distinguish between two main regions.

The first region shows a trend, typical of software programming, in less than 10 days it is possible to achieve an average speed-up of ~3x with respect to a processor-only implementation, while spending additional time (more than one month) it is possible to obtain up to an order of magnitude performance improvement. The reference, in this case, is the same RISC processor used to handle the reconfigurable device.

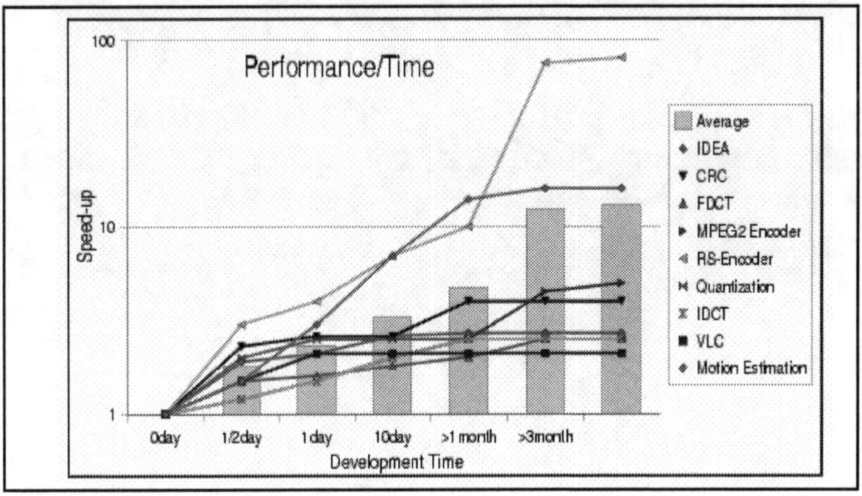

Fig. 19.8. Performance vs. Development Time trade-offs.

This methodology, originally developed in the context of the XiRisc project [270], was also applied to an evolution of the XiRisc processor described in [271], including an additional embedded FPGA as co-processing engine. As a proof of concept, a first prototype of tool generating behavioral VHDL from Griffy-C descriptions was implemented as described in [301].

With respect to other approaches, Griffy-C allows also user unexperienced of hardware design to achieve interesting performance improvements (2–3x speed-ups) in a relatively short time, since it is based on the same optimization principles of digital signal processors. For experienced users, on the other hand, the possibility to explore hardware/software codesign methodologies allows to spend most of the development time on few critical kernels implemented with an hardware perspective, whereas other kernels can be optimized according to faster implementation strategies. This allows developers to consider and engineer the time spent on the implementation as a cost, and to consequently exploit a further trade-off in addition to performance and energy consumption.

20 Software-Based Self-Testing of Embedded Processors

Nektarios Kranitis,[1] Antonis Paschalis,[1] Dimitris Gizopoulos,[2] and George Xenoulis[2]

University of Athens,[1] University of Piraeus[2]

No silicon integrated circuit (IC) manufacturing process is perfect. Therefore, IC testing is used to screen imperfect devices before shipping them to customers. Chips containing manufacturing defects are potentially malfunctioning chips that may cause system crashes and lead to financial deficit, environmental disaster, and/or jeopardize human life. Moreover, if manufacturing defects are not detected early, the cost of repair is increased by an order of magnitude at each step after the chip fabrication line. It comes naturally that chip testing is an important factor of the business in computer and communications industries, since customers demand reliable products at a reasonable cost and manufacturers, in order to stay competitive in business, must find the means to provide the best products at the lowest cost.

IC testing is done in several phases of chip realization process. When a new chip is first designed and fabricated, *first silicon debug and validation* of early prototypes of the chip in the design laboratory should verify that the design is correct and meets all specifications. During this phase, *functional tests* are applied, comprehensive AC and DC measurements are made, design errors are corrected, final specifications are set (e.g. the limits of chip operating values), as well as the manufacturing process and corresponding yield are improved. Besides, during this phase, test development for *manufacturing testing* is verified and improved, as well. Successful first silicon debug and validation marks the beginning of large scale manufacturing for the new chip. Every fabricated chip is subjected to manufacturing testing in the factory which can only be less comprehensive than first silicon debug and validation, but it verifies that the chip meets all

J. Nurmi (ed.), Processor Design: System-on-Chip Computing for ASICs and FPGAs, 447–481.

relevant specifications. Automatic Test Equipment (ATE) is the usual term for instruments used to apply test vectors (i.e. binary patterns) to the input of the chip, analyze its responses and mark the chip as *good* when its responses match with the expected ones, or *faulty* otherwise. The time interval where the chip remains on the ATE is termed *test application time* and should be as short as possible to reduce the chip test cost. The test vectors may not cover all possible functions and input data patterns, but it is mandatory to provide a very high detection coverage of modeled faults where a modeled fault is a formal, convenient representation of the effect of the physical faults on the system operation. For example the most popular fault model is the single stuck-at (SSA) fault model at the logic gates level of representation, where the circuit is modeled as an interconnection of logic gates and only one input or output of a gate is permanently set to either 0 or 1. The single stuck-at fault model is still widely used today as a solid basis for manufacturing testing.

The shrinking device dimensions that very deep submicron (VDSM) technologies offer, have revised the cost models of modern System-on-Chip (SoC) manufacturing, since very complex SoCs are designed and manufactured at reasonable costs, reusing existing cores based on mature electronic design automation (EDA) tools. The SoC design process supported by highly sophisticated EDA tools that integrate pre-designed, pre-verified, intellectual property (IP) cores around industry standard on-chip buses, dramatically improves design productivity and reduces time-to-market and design cost. However, as the cost for designing complex SoCs has been reduced, the percentage of the total cost attributed to testing has been increased significantly. Many system companies consider testing to be 50% to 60% of their total manufacturing cost and according to Intel the combined cost of first silicon debug and validation and manufacturing testing is its major capital cost, and not the multi-billion silicon fabrication lines. In order to control testing cost, the design and test engineers must consider the design and test complexity and adopt testability solutions that reduce the testing cost without imposing excessive hardware, performance, and/or power consumption overheads.

Embedded processors integrated with other IP cores constitute the heart of today's complex SoCs. Manufacturing testing of a SoC built around one or more processor cores is a new challenging task. Test data volume (test patterns and test responses) required for external ATE-based testing of embedded processors and SoCs is becoming excessively large [208] resulting in very long test application times, while the test application cost using high-performance ATE (also known as functional testers) is very high. Furthermore, new types of defects appear in deep-submicron technologies and affect the functionality as well as the performance of the processor and the SoC built around. These new types of defects require *at-speed testing*[1] in order to achieve high test quality. The Murphy experiments [279] corroborate that at-speed tests identify more defective chips. However, the increasing gap between ATE frequencies and SoC operating frequencies makes external at-speed testing almost infeasible. Moreover, ATE measurement accuracy problems can lead to serious yield loss [208].

Traditional *hardware-based self-test* (or built-in self-test – BIST) moves the testing task from external resources (ATE) to internal hardware, synthesized to generate test patterns and to evaluate test responses of the circuit under test. Hardware-based self-test achieves at-speed testing reducing the overall test costs of the chip [208] and is a reusable approach since self-test hardware can be used in different stages of the chip's life cycle. Recent applications of commercial hardware-based Logic BIST techniques in industrial designs [191] and microprocessors [75] reveal that extensive and manual design modifications have to be performed in order to make the design hardware BIST-ready. In particular, design adjustments should prevent the circuit from reaching an unknown state that will corrupt the compacted test response.[2] Logic BIST applies pseudorandom patterns, and test points are inserted to enhance the testability of random pattern resistant circuits. These design modifications, increase the circuit area and degrade its performance. Therefore, the use of Logic BIST on high-performance and power-optimized embedded processors imposes several limitations.

Software-based self-test (*SBST*), also called instruction-based self-test, is the process of detecting physical defects (or faults that model them) in a processor or processor-based system by executing processor instructions in its normal mode of operation. SBST has recently emerged as an effective methodology for the manufacturing testing of microprocessors and embedded processors along with other components in Systems-on-Chip (SoCs). SBST is a non-intrusive approach that embeds a "*software tester*" with the form of a self-test program in the processor's on-chip memory. It leverages the use of low-speed, reduced pin-count external ATE providing high quality, at-speed testing virtually without introducing any hardware or performance overheads. An outline of the software-based self-test concept is shown in Figure 20.1. Numerous test technology research groups as well as key microprocessor companies such as Intel [326] and Sun [41] have recently recognized the potential of SBST adopting it in their test flows. We provide a comprehensive elaboration on all recent approaches in the next section of the chapter.

---

[1] Test application at the actual operating frequency of the device.

[2] Test responses of a circuit under test may be compacted in *test signatures* to reduce the total size of test response data.

Initially, the self-test program is loaded into the processor's on-chip memory using a low-speed, low-cost *structural tester* (Figure 20.1a). Secondly, during test application (Figure 20.1b), the processor executes the self-test program from its on-chip memory at its normal clock frequency, thus achieving full at-speed testing. During this phase, the processor collects the test responses (possibly compressed in a test signature), and stores them in its on-chip data memory (Figure 20.1c). Finally, the low-speed, low-cost tester is used again to unload the test responses from the on-chip memory for further external analysis (Figure 20.1d). Since modern microprocessors integrate large caches on the same die, execution from on-chip cache is considered a further advantage provided that a cache-loader mechanism exists to load the test program and unload the test response(s).

SBST changes the role of the external ATE from actual test application to a simple interface with the on-chip memory before and after the test

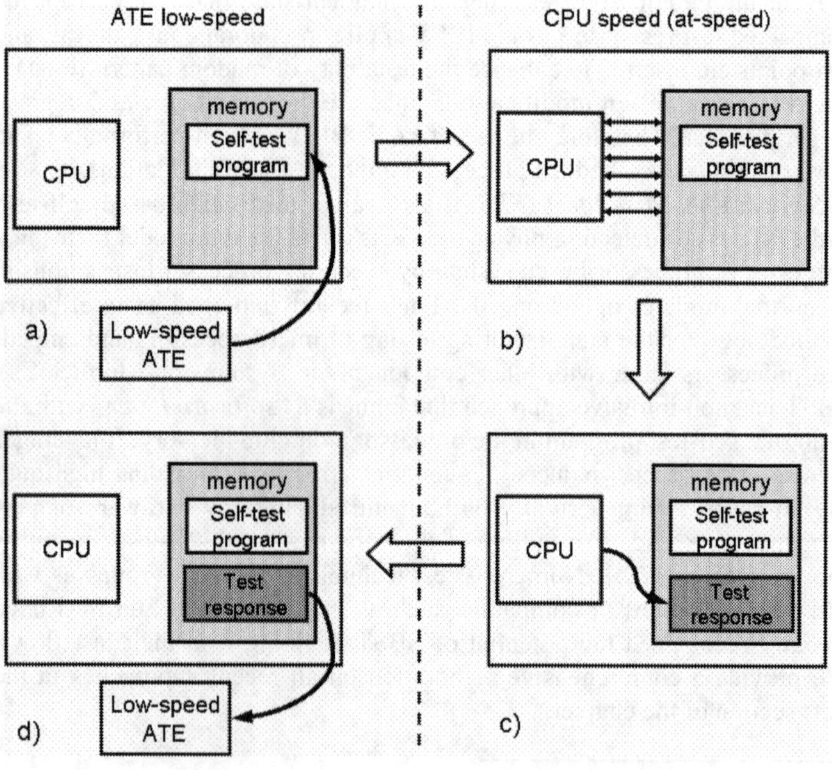

Fig. 20.1. Software-based self-testing concept outline.

execution. Therefore, SBST achieves the goal of at-speed testing using low-speed ATE. In addition, since the means for applying SBST programs are existing processor instructions, at-speed testing is feasible without the risk of thermal damage due to excessive signal activity in special test modes of circuit operation. Furthermore, by utilizing the processor's Instruction set architecture (ISA) and complying with all the restrictions enforced by both the ISA and the designers' decisions, SBST avoids *over-testing* (for faults that do not appear during normal circuit operation) and saves valuable yield.

SBST is a scalable, portable, and reusable methodology for high quality testing at virtually zero performance, power or circuit area overhead. SBST can be reused at different stages of the microprocessor or microprocessor-based system life cycle. SBST routines can be used during both *first silicon debug and validation* of early prototypes of a chip and *manufacturing testing* when a chip moves to full production. Besides, SBST can be used during the operation of the chip in the application field via *periodic on-line testing* for the detection of failures that did not exist or did not manifest themselves during manufacturing. In this case, SBST routines may be stored in on-chip ROM or Flash memory. On-line periodic SBST can be applied to improve reliability of low-cost systems based on embedded processors where hardware, software or time redundancy can not be applied due to their excessive cost in terms of silicon area and/or execution time. Table 20.1 summarizes the different application stages of SBST and the different requirements of each stage in terms of self-test code and data size, application time and power consumption.

This chapter gives a review of the state-of-the-art on the emerging area of SBST of embedded processors that recently captured the interest of processor design and test engineers. The several advantages of SBST over traditional structured design-for-testability (DFT) and hardware-based self-testing techniques made SBST a very attractive testing approach, and

Table 20.1. Application stages of SBST and corresponding requirements.

Stage	Self-Test Code/Data Stored in	Test Program Size	Test Application Time	Test Power Consumption
First silicon debug and validation	Low-speed ATE On-chip cache	Large to very large	Long to very long	High
Manufacturing testing	Low-speed ATE On-chip cache	Small to medium	Short to medium	Average to high
On-line periodic testing	ROM or flash memory	Small	Short	Low

numerous SBST methodologies have been proposed by research groups in universities, research centers and industry. In this chapter, several different SBST strategies proposed in the research literature are briefly discussed showing the evolution of SBST and experimental data sourcing from successful applications of the SBST approach are provided wherever available. Subsequently, a high-level component-based SBST methodology for embedded processors that aims to high structural fault coverage of the processor at a minimum test cost is presented. The high-level SBST methodology is demonstrated through its complete application to several processor benchmarks with escalating complexity.

Evolution of software-based self-test

At-speed functional testing

Traditionally, processor testing resorted in *functional testing* approaches. Functional test program development is based on either functional fault models or just the reuse of test sequences developed originally for design verification.

The latter approach has been extensively used in industry over the last two decades. Test programs generated by verification suites to verify the functionality of the processor design, are *reused* for at-speed functional manufacturing testing in an ATE-based setup. The drawback of verification-based functional testing is that it does not take account of the actual structural testability requirements of the processor, which are related to the physical defects and are formally described by fault models. Since the development of verification-based test sequences does not target structural faults (for instance single stuck-at faults) but rather processor functionality and compliance with the processor's ISA, when fault graded with respect to a structural fault model, the resulting fault coverage does not usually meet the required test quality goals. To increase the structural fault coverage, the functional-based test programs are usually augmented with manually written code by engineers with substantial knowledge of the processor architecture. Despite this additional test development effort, functional test programs cannot achieve acceptable levels of fault coverage by themselves.

In functional testing, external ATE (also called *functional testers*) is used to supply test patterns to the processor, mimicking the test program execution and the interaction between the processor and the main memory, i.e. the processor's functionality. First, simulations with a processor model are performed to capture the trace at processor I/O during the execution of test program (see Figure 20.2). Afterwards, the simulation trace is translated into ATE test language and stored in the ATE memory. Finally, during test application, the ATE applies the test patterns to the processor input pins to mimic the execution of instructions while at the same time it captures the test responses at the processor output pins.

It has already been mentioned that at-speed testing is mandatory for achieving high test quality in today's deep-submicron manufacturing technologies. Thus, the ATE used for at-speed functional testing of a processor must have the following characteristics:

- ability to supply test patterns at-speed (i.e. ATE technology needs to follow high-end microprocessors technology);
- high pin count to drive all processor I/O pins;
- large memory to store the test patterns and test responses.

However, the increasing gap between ATE frequencies and processor or SoC operating frequencies, the large test data volume, the difference between external and internal bandwidth along with the limited access to deeply embedded processor cores in complex SoCs, make external at-speed functional testing extremely costly and in many cases almost infeasible. All these drawbacks including not acceptable fault coverage as well,

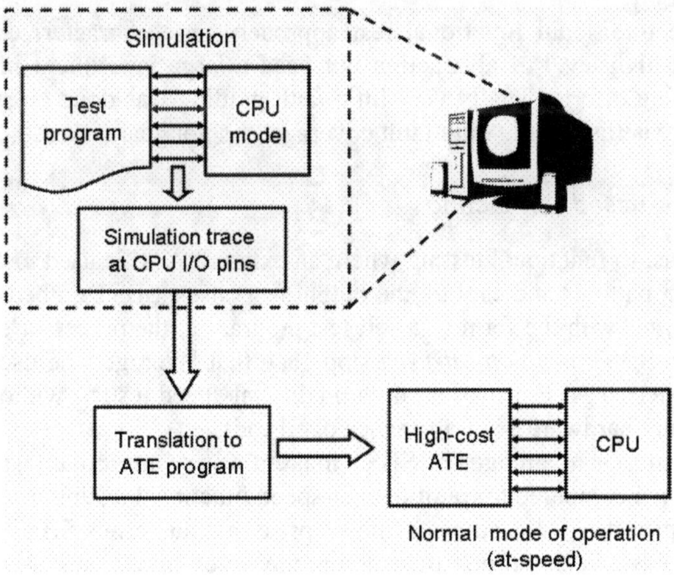

Fig. 20.2. Traditional at-speed functional testing.

lead microprocessor industry to move slowly (when compared e.g. with ASIC industry) towards more intrusive, structural DFT test approaches, such as scan-based testing and BIST. However, such techniques usually have a non-trivial impact on a circuit's performance, size and power consumption and are applied with serious consideration and careful incorporation into the processor design.

The pioneer work of Thatte and Abraham [415], is considered a landmark paper in processor functional testing. Based on the register transfer (RT) level description of the processor the authors introduced a functional fault model and considered the processor as a graph model. Since then, many processor functional testing methodologies were proposed. Those approaches were either based on a functional fault model (the model of [415] or other similar ones), or based on verification principles without assuming any functional fault models at all. The functional testing work of [415] was complemented by the work of Brahme and Abraham [53] which reduces the complexity of the generated tests for the processor's instruction sequencing and execution logic. A functional model based on a reduced graph is used for the microprocessor and a classification of all faults into three functional categories is given. Tests are first developed for the registers read operations and then for all remaining processor instructions. The developed tests are proposed for execution in a self-test mode by the processor itself.

These traditional functional test approaches are characterized by the required high level of abstraction but need a large investment in manual test writing effort. Usually very little fault grading was done on structural processor netlists while high fault coverage was not guaranteed.

Software-based self-testing

In contrast to functional testing where an external ATE is used to drive the input test patterns and capture the output responses, SBST embeds a "*software tester*" with the form of a self-test program in the processor's on-chip memory. SBST is a *non-intrusive* approach that leverages the use of low-speed external ATE providing high quality, at-speed testing without introducing any hardware or performance overheads.

The various advantages of SBST make it a very attractive testing approach when compared to traditional at-speed functional testing or structural DFT approaches, so it comes as no surprise that numerous SBST methodologies have been proposed; a comprehensive survey can be found in [152]. Experimental results provided in [75] demonstrate several advantages of SBST for processors over traditional structured DFT approaches such as full scan design and hardware Logic BIST.

Although the single stuck-at fault model dominates among the SBST approaches presented so far, SBST has been proved to be very effective on delay fault testing [244,250,381], speed binning[3] [450], interconnect cross talk faults testing [74], fault diagnosis [47,76], and validation [91, 373] as well.

The SBST approaches presented so far in the literature can be classified into three main categories. The first category includes the SBST approaches [40,41,326,353,373] that have a high level of abstraction and are functional in nature. A common characteristic of such SBST approaches is the almost exclusive use of randomized instructions and/or operands. The second category includes the SBST approaches [75,77,89,90,216,239,240, 241,329,341,362] which are structural in nature and require structural fault-driven test development. The third category includes the SBST approaches [168,169,451] which combines the previous two categories such that randomized instruction test programs are followed by test programs that apply ATPG deterministic tests targeting hard-to-detect structural faults, thus constituting a "hybrid" SBST approach that provides improved fault coverage. A comprehensive list of SBST approaches from all three categories is briefly discussed in the following subsections.

SBST approaches using randomized instructions and/or operands

The development of functional SBST programs, based on *randomized* instruction sequences and random operands has a major advantage. Due to its high level of abstraction, SBST development requires only basic knowledge of the processor architecture, and therefore requires limited test development effort and cost. However, in most cases, manual intervention is required to determine an efficient mix of instruction sequences (possibly by defining and fine-tuning instruction frequency biases) along with architecture expertise to increase fault coverage. Instruction sequences characterized as corner cases are usually targeted by specific handwritten code. Functional self-test code development does not consider any fault model and the test programs are randomly generated; thus a long test program is typically required to achieve an acceptable level of fault coverage. Despite the large number of instruction sequences, saturating behavior in fault coverage is usually observed due to pseudorandom operands used. Further increase of the random test program size is usually proved ineffective in targeting the remaining hard-to-detect faults and manual test development is a necessary supplement.

---

[3] Speed binning is the process that classifies processors according to their actual silicon speed so that they are marketed accordingly.

In [373], Shen and Abraham proposed a functional self-test methodology, which generates a random sequence of instructions that enumerate all the combinations of the processor operations and systematically selected operands. Test development is performed at a high level of abstraction based on ISA. However, since test development is not based on an a priori fault model, the generated tests – applicable for design validation as well – cannot achieve high fault coverage without the use of large code sequences and a considerable manual effort. When applied on the GL85 processor (model of Intel's 8085) consisting of 6,300 gates and 244 FFs, a test program consisting of 360,000 instructions was derived and the attained single stuck-at fault coverage was 90.2%. The fault coverage was reduced to 86.7% when the responses were compressed in a signature.

In [40], Batcher and Papachristou proposed *instruction randomization self-test* (IRST) for processor cores, a pseudorandom self-testing technique. Self-test is performed with processor instructions that are randomized by a special circuit designed outside the processor core. Randomization occurs at the operand level as well. IRST does not add any performance overhead to the processor and the extra hardware is relatively small compared to the processor size (3.1% hardware overhead is reported for a 27,860 gates DLX-like RISC processor core). The obtained fault coverage after an iterative process considering different parameters for the processor core following the execution of a random instructions sequence running for 50,000 instruction cycles is 92.5%, and after the execution of 220,000 instruction cycles it is 94.8%.

In [326], Parvathala et al. proposed an automated functional self-test methodology called *functional random instruction testing at speed* (FRITS) based on the generation of random instruction sequences with pseudorandom data generated by software LFSRs; on-chip cache is used for application. Instruction-based constraints are extracted and built into the generator to ensure generation of valid instruction sequences also ensuring that no cache misses and bus access cycles are produced during self-testing. The high-level functional nature of the proposed approach requires a large amount of cycles to be applied that makes fault grading a non-trivial task. The methodology achieved 70% fault coverage when applied on the Intel Pentium® 4 processor in an industrial environment and helped to detect the defects that escaped the normal test flow. Also, application of the approach to the integer and floating point units of Intel Itanium™ processor led to 85% single stuck-at fault coverage.

In [353], Rizk et al. proposed a self-test program design technique for embedded DSP cores. The method requires minimal knowledge of the core's internals and minimal insertion of external LFSR hardware. The test program consists of a small set of instructions which operate iteratively on pseudorandom data generated by the LFSRs to test the DSP core components. The method introduces instruction-based testability analysis metrics, namely the controllability and observability metrics. The instruction metrics are computed through simulation with random data with respect to internal modules of the processor. Whenever a module is considered for test generation, the methodology selects the most appropriate instructions for test vector application and fault effect propagation. If, however, instructions with good metrics cannot be found through simulation, the high-level description of the processor is used to identify instruction sequences with good testability for the module under test (MUT). Experimental results are provided and the proposed methodology is evaluated on a pipelined DSP core achieving high test coverage of 98.33% with a small test program of 34 instructions which is looped for 6,000 times. The test program applies a total of 204,000 test vectors.

Recently, Bayraktaroglu et al. [41] proposed the conversion of existing legacy tests, either handwritten or randomly-generated to *cache resident* tests aiming to eliminate cache misses. The basic objective of this work was to apply SBST fully avoiding the non-determinism of memory accesses in high-end microprocessors with several cache memory levels based on the use of low-cost ATE. They demonstrated their method, called *Load&Go*, to an 8-core, 32-thread Sun UltraSPARC T1 processor model.

Structural SBST approaches

The development of SBST programs targeting structural faults using a deterministic approach clearly results a higher fault coverage when compared to randomized instructions/operands SBST approaches where no fault model is considered at test development phase. Although SBST approaches targeting sequential fault models (such as the path delay fault model) have been presented [244,250,381], the single stuck-at fault model dominates among the SBST approaches since it reduces significantly the complexity; it is implementation technology independent while test patterns for stuck-at faults are proved effective to target most manufacturing defects. The structural SBST approaches that will be discussed in the following paragraphs, target the SSA fault model.

The contribution of the work presented by Chen and Dey in [75] is twofold. First, it demonstrates the superiority of SBST for embedded processors over traditional DFT approaches such as Full Scan design and hardware Logic BIST. This is shown by applying Logic BIST to a very simple 8-bit accumulator-based processor (Parwan) and a stack-based 32-bit soft processor core that implements the Java Virtual Machine (picoJava). In both cases, Logic BIST adds more hardware overhead compared to full scan, but is not able to obtain satisfactory structural fault coverage even when a very high number of test patterns are applied. Secondly, an SBST methodology is proposed which is structural in nature, targeting specific components and fine-tuning test development to the low, gate-level details of the processor core. Initially, pseudorandom pattern sequences are developed for each processor component in an iterative method taking into consideration manually extracted constraints imposed by its instruction set. Then, test sequences are encapsulated into self-test signatures that characterize each component. Alternatively, component tests can be extracted by structural automatic test pattern generation (ATPG) and downloaded directly in embedded memory by the tester. The component self-test signatures are then expanded on-chip by a software-emulated LFSR (test generation program) into pseudorandom test patterns, stored in embedded memory and finally applied to the component by software test application programs. Application to an accumulator-based CPU core, Parwan, consisting of 888 gates and 53 FFs, resulted in 91.4% fault coverage in 137,649 cycles using a test program of 1,129 bytes.

In [77], Chen et al. proposed a methodology that extends previous work [75] by automating the complex constraint extraction phase, while emphasizing in ATPG-based test development instead of pseudorandom. Statistical regression analysis is applied on the RT-level simulation results using manually coded instruction templates, to derive a model of the surrounding logic of the MUT. The learned model is converted into virtual constrained circuit (VCC) followed by ATPG on the VCC-MUT in an iterative way. Application of the methodology on the combinational logic in the execution stage of a processor from Tensilica (Xtensa™) with 24,962 faults resulted in 288 ATPG test patterns and 90.1% fault coverage after constrained ATPG. When the tests are applied using processor instructions in a test program of 20,373 bytes, the fault coverage for the targeted component is increased (due to collateral coverage) to 95.2% in 27,248 cycles.

In [89], Corno et al. proposed a partially automated test development approach. First, a library of macros is generated manually by experienced assembly programmers from the ISA, consisting of instruction sequences using operands as parameters. Then, a greedy search and a genetic algorithm are used to optimize the process of random macro selection among the macros set, along with selecting the most suitable macros parameters to build a test program that maximizes the attained fault coverage when the test program is applied and fine-tuned on the gate-level netlist of the processor. The approach attained 85,2% fault coverage when applied on a 8051 8-bit microcontroller design of 6,000 gates using 624 instructions.

In [90], Corno et al. proposed an automated test development approach based on evolutionary theory techniques (MicroGP), that maximizes the attained fault coverage when the evolved test program is applied on the gate-level netlist of the processor. It utilizes a directed acyclic graph for representing the syntactical flow of an assembly program and an instruction library for describing the assembly syntax of the processor ISA. Manual effort is required for the enumeration of all available instructions and their possible operands. Experiments on a 8051 8-bit microcontroller design of 12,000 gates, resulted in 90% fault coverage.

In [216], Kambe et al. proposed a template generation methodology for hierarchical test generation targeting structural faults. According to the methodology, gate-level test generation is performed for each MUT, and a test program is generated to justify test patterns from primary input to the MUT and propagates test responses at instruction level. The proposed methodology enumerates possible templates considering dependence of instructions each of which involves one or more data transfers between registers. In order to justify value of MUT inputs, a concept of adjacent registers of the MUT is introduced that makes it possible to consider input spaces of the MUT determined by signals from other modules as well as signals directly from registers. Templates are generated considering dependence of instructions each of which invokes one or more data transfers between registers. The approach is demonstrated on an accumulator-based 8-bit CPU core, Parwan. Out of 276 templates generated for testing the ALU of Parwan, 12 templates contributed to the fault coverage, and the fault coverage achieved for the ALU was 99.44%.

In [239], Kranitis et al. introduced a high-level structural SBST methodology, showing for the first time that small deterministic test sets, applied by compact test routines provide significant improvement when applied to the same simple accumulator-based processor design, Parwan, which was used in [75]. Compared to [75], the methodology described in [239] requires 20% smaller test program using 923 bytes, 75% smaller test data and almost 90% smaller test application time using 16,667 cycles. Both methodologies achieve single stuck-at fault coverage slightly higher than 91% for the simple accumulator-based Parwan processor.

Despite the successful first application of the approach of [239], scaling from simple accumulator-based processor architectures to more realistic ones in terms of complexity like contemporary complex processors implementing commercially successful ISAs (i.e. RISC), brings out several test challenges that remained unsolved. These challenges arise when high-level test development is applied to complex processor architectures that contain large functional components (i.e. fast parallel multipliers, barrel shifters, etc.) and large register banks, while trying to keep the test-cost as low as possible. In [240], Kranitis et al. addressed low-cost SBST challenges by defining different test priorities for processor components, showing that high-level self-test code development based on ISA and RT-level description of a processor can lead to low test cost without sacrificing fault coverage independently of the gate-level implementation. The methodology was applied on two processors: Plasma/MIPS with simple 3-stage pipeline and MIPS R3000 application specific instruction set processor (ASIP) with 5-stage pipeline designed using the ASIP/Meister design environment.

In [329], Paschalis and Gizopoulos identified the stringent characteristics of an SBST test program to be suitable for on-line periodic testing of embedded processors. SBST for on-line periodic testing can be applied to improve reliability of low-cost embedded systems based on embedded processors where hardware, software or time redundancy cannot be applied due to their excessive cost in terms of silicon area and execution time. A new classification and test priority scheme more fine-grained than in [240] was proposed. Both types of permanent and intermittent faults are detected by a small embedded test program with test execution time much less than a quantum time cycle.

In [362], Sanchez et al. proposed an automatic methodology to transform a test set originally developed for manufacturing test in a test set suitable for on-line testing. The generated programs are suitable for non-concurrent periodic on-line testing as well as for shutdown or startup testing. While the new test set is likely to contain a larger number of programs, these programs are shorter and completely independent (i.e. they can be executed at different times and do not rely each on the results of the previous ones), and thus perfectly fit a non-concurrent on-line test scheme. The transformation of the test set is performed in two phases: first the original programs are simulated with a special instruction-set simulator that for each instruction generates a spore, i.e. a small program able to fully replicate the processor behavior. Second, an evolutionary algorithm is used to collapse the set of spores into a test set. The proposed approach is able to guarantee the same fault coverage on all functional units. Experimental results were provided targeting the ALU and Control Units of an 8-bit 8051 processor core. The initial test set is compact in size but requires a long time to be executed and is usually designed to be run without regarding sharing constraints. The final on-line test set is larger in size, but composed of small and extremely fast programs that can be freely scheduled. Both test sets guarantee the same fault coverage on the target units.

In [341], Psarakis et al. identified testability hotspots in processor pipeline logic and proposed a generic SBST methodology that enhances existing SBST programs [240], to target more effectively the pipeline logic. The methodology was applied on the miniMIPS and OpenRISC 1200 processor cores. Results show fault coverage improvements of up to 12% on average for the entire processor, and fault coverage improvements of 22% for the pipeline logic.

The contribution of the work presented by Kranitis et al. in [241] is two-fold. First, a reliability analysis and a cost function was introduced in order to minimize the test cost incurred when selecting a periodic SBST strategy, and achieve high detection probability. Reliability analysis was based on a two-state Markov model for the probabilistic modelling of intermittent faults for optimal periodic testing is introduced. Then, an SBST strategy for on-line SBST of pipelined embedded processors was proposed that enhances SBST programs for manufacturing [240] and on-line testing [329]. The proposed strategy was demonstrated by applying it to a 5-stage fully pipelined RISC embedded processor, Athena. Experimental results provided showed 8.2% fault coverage improvement for the entire processor and fault coverage improvements of 26% for the pipeline logic.

"Hybrid" SBST approaches

Recent work in SBST includes [168,169,451]. A common characteristic among these recent SBST approaches is that randomized instruction test programs are followed by test programs that apply ATPG deterministic tests targeting hard-to-detect structural faults, thus constituting a "hybrid" SBST approach.

In [451], Wen et al. introduced an SBST methodology that employs random test program generation (RTPG) as a baseline with deterministic target test program generation (TTPG) as a supplement, in order to provide tests specifically targeting faults that are hard-to-test for RTPG. The proposed TTPG method utilizes simulation results to develop learned models for the surrounding modules of the block under test. Simulation-based TTPG is performed similar to previous works; however, arithmetic and Boolean learning techniques are used instead of statistical regression to develop learned models for the surrounding logic of the MUT. These techniques offer the advantage of being deterministic in nature, in contrast to regression that is a statistical method. Additionally, Boolean learning can also handle logic-intensive modules in which regression is not effective. Then, the learned models replace the surrounding modules around the block in the actual test generation process. Because the learned models are much simpler to handle, this method minimizes the cost of functional TPG. The methodology is applied on the controller and ALU of the OpenRISC 1200 processor. When RTPG is applied in the "controller" module fault coverage saturates around 62.14%, while on the other side, TTPG generates 134 valid test patterns and detects 4967 faults including all faults that RTPG can detect, for an overall fault coverage of 69.39%. For the ALU module, after application of 100K RTPG test patterns, TTPG is applied and the combined fault coverage is 94.94%.

In [168], Gurumurthy et al. introduced a novel technique to map pre-computed test patterns, generated by commercial ATPG tools, into sequences of instructions, based on the ISA of the processor under test. The technique applies at the RT-level source code of the processor, at the module level. It uses bounded model checking in order to produce automatically a counter-example which will contain an instruction sequence that generates the pre-computed test pattern. First, a bound is defined for the bounded model checker (BMC) for each step of the process, taking into consideration the pipeline depth, the stall/reset mechanism of the processor, the forwarding mechanism and the number of cycles of the instructions. Then, every test pattern for each module should be manually transformed into linear temporal logic (LTL) property. LTL property is negated and passed to the BMC. Additionally, the instruction set of the processor is passed to the BMC in order to constrain its input space. BMC checks partial correctness of the property and generates a counter-example in case the property fails within the bound. If a counter-example is not generated, pre-computed test patterns are characterized as functionally infeasible, otherwise processor instruction sequence containing test pattern is included in counter-example. The technique is applied to both controllability and observability stages and results to a combination LTL property of both stages. Although controllability is fully controlled in this technique, in observability stage spurious counter-examples can be generated that do not ensure propagation of outputs to observable points, thus they have to be refined entirely manually. Experiments were performed on OpenRISC 1200. Initially, a random test program of 36,750 instructions was generated in order to fault grade the processor. The fault coverage saturated around 68% and the remaining hard-to-detect fault list was split based on modules and passed through a commercial ATPG tool in order to obtain the pre-computed patterns. Those pre-computed test patterns were applied to the presented technique. In a total of 22,633 test sequences, 6,765 were identified to be functionally infeasible uncontrollable sequences. On the remaining, sequences of instructions were generated for some of the patterns in ALU, Control and Operandmuxes modules of OpenRISC 1200 and example instruction sequences were given.

In [169], Gurumurthy et al. proposed a new technique that fully automates the process of functional test generation targeting specific faults. The technique supplements the observability part of the automated mapping technique of pre-computed test patterns, generated by commercial ATPG tools, into sequences of instructions proposed in [168] that required manual effort for the propagation of test responses. The proposed technique applies at the RT-level source code of the processor in module-level.

It uses Boolean difference, LTL and bounded model checking (BMC) in order to map module-level test responses into instruction sequences. Experiments were performed on OpenRISC 1200 processor as in [168]. Again, in order to focus on hard-to-detect faults, a random test program of 36,750 instructions was generated and the processor was fault-graded. The fault coverage saturated around 68% and the hard-to-detect fault list formed the base list of the proposed technique. The base list was sorted based on module-level and the overall technique was used for every module. Even though the mapping efficiency of most of the modules is above 90%, the overall mapping efficiency was 71% due to low efficiency of ALU and LSU modules. In a total of 17,319 test sequences, 9,708 were found to be not mappable within the bound, thus no counter-example was produced and were rejected. The remaining test sequences were successfully mapped and increased the fault coverage of the processor to 82%.

Table 20.2 provides a summary of representative SBST methods for manufacturing testing of processor cores. A brief description of each methodology and the year of publication is accompanied by information such as the test experiment performed (whole processor benchmark or specific processor component) and test statistics such as test program size, test execution time and fault coverage achieved. The methods listed in Table 20.2 have been proposed for manufacturing testing and used single stuck-at fault coverage measurements for the processor benchmarks.

High-level SBST methodology

In this section, we will discuss in detail our perspective of SBST as a high-level, structural methodology for high-quality and low-cost processor self-testing. The key properties of a Register-Transfer (RT) level, component-based, SBST methodology for embedded processors are the following:

- it should follow a divide-and-conquer approach using *component-based* test development;
- test development should be based only on the ISA of the processor and its RT-level description, which is in almost all cases available, without the need of low gate-level fine-tuning.

Although the main target always remains the high structural fault coverage, test development and application cost should be considered as a very important aspect. Therefore, a high-level SBST methodology must have two main objectives both aiming to low test cost:

Table 20.2. Summary of key SBST approaches for manufacturing testing.

Ref	(Year)	Methodology	Test Program Size	Test Execution Time	Benchmark Processor	FC (%)
[373]	(1998)	Random instructions and systematic operands	Large	Long	GL85	90.2
[40]	(1999)	Hardware-based, instruction randomization and random operands	Large	Long	DLX	94.8
[326]	(2002)	Random instructions and software LFSR operands with constraints	Large	Long	Pentium4 Itanium	70 85
[90]	(2003)	Semi-automated based on evolutionary theory and netlist	Medium	Average	8051	90
[75]	(2001)	Component based, random/ ATPG patterns based on netlist	Medium	Long	Parwan	91.4
[77]	(2003)	Component based, ATPG patterns based on netlist, automatic constrained extraction based on regression analysis	Medium	Average	Xtensa (only ALU)	95.2
[216]	(2004)	Netlist and module based template generation	Medium	Average	Parwan (only ALU)	99.4
[239]	(2002)	High-RTL, component based, deterministic operands for functional components	Small	Short	Parwan	91.1
[240]	(2005)	High-RTL, components test prioritization, deterministic operands for functional components, verification tests for control	Small	Short	Plasma ASIP MIPS	95.3 92.6
[341]	(2006)	Generic solutions for address-related and pipeline-related logic	Small	Short	miniMIPS OpenRISC	95.1 90.0
[451]	(2005)	Component based, random tests and ATPG patterns based on netlist, automatic constrained extraction based on learning techniques	Medium	Long	OpenRISC (only ALU)	94.9
[169]	(2006)	Random tests and automated mapping of precomputed ATPG tests based on Boolean difference, LTL and BMC	Medium	Long	OpenRISC	82

- generation of as small and as fast as possible self-test code routines (reduced *test application* time and cost);
- as small as possible engineering effort and test development time (reduced *test development* time and cost).

The first objective leads to smaller download times at the low frequency of the external tester as well as to smaller test execution times of the routines, thus reducing the total processor test time. The second objective reduces test development cost and time-to-market, leading to significant improvements in product cost-effectiveness and market success.

An RT-level self-test development approach is well suited to the high RT-level flow of the design cycle. Since design, simulation, and synthesis are usually carried out at the high RT-level, test development can also be carried out at the same level providing high convenience and flexibility. In this case, processor cores can be easily integrated into a SoC environment, configured and re-targeted in a variety of silicon technologies without any specific need for fine-tuning the test development to specific synthesis optimization parameters using a specific technology library. Experimental results show that the gate-level independent test strategy is very efficient and achieves similar high fault coverage results for different gate-level implementations of the RT-level processor core.

High-level SBST development consists of three phases (Figure 20.3):

- Information extraction (Phase A)
- Component classification and test priority (Phase B)
- Self-test code development (Phase C)

Component classification and test priority phase (Phase B) uses a generic pipelined processor model. The model consists of a sequence of unique pipeline stage models like the one given in Figure 20.4. For a specific ISA, a specified number of pipeline stages are stacked to compose the processor core. Each pipeline stage consists of *datapath* and *control* logic. Datapath combinational data/address logic implements the data/address operations and transfers defined by the ISA micro-operations. Data/address operations and transfers are implemented by data/address functional components. Control logic in each pipeline stage controls the data/address operations and transfers implemented by the functional components, by evaluating the execution conditions. Pipeline registers transfer operation results of functional components and control logic to the next pipeline stage.

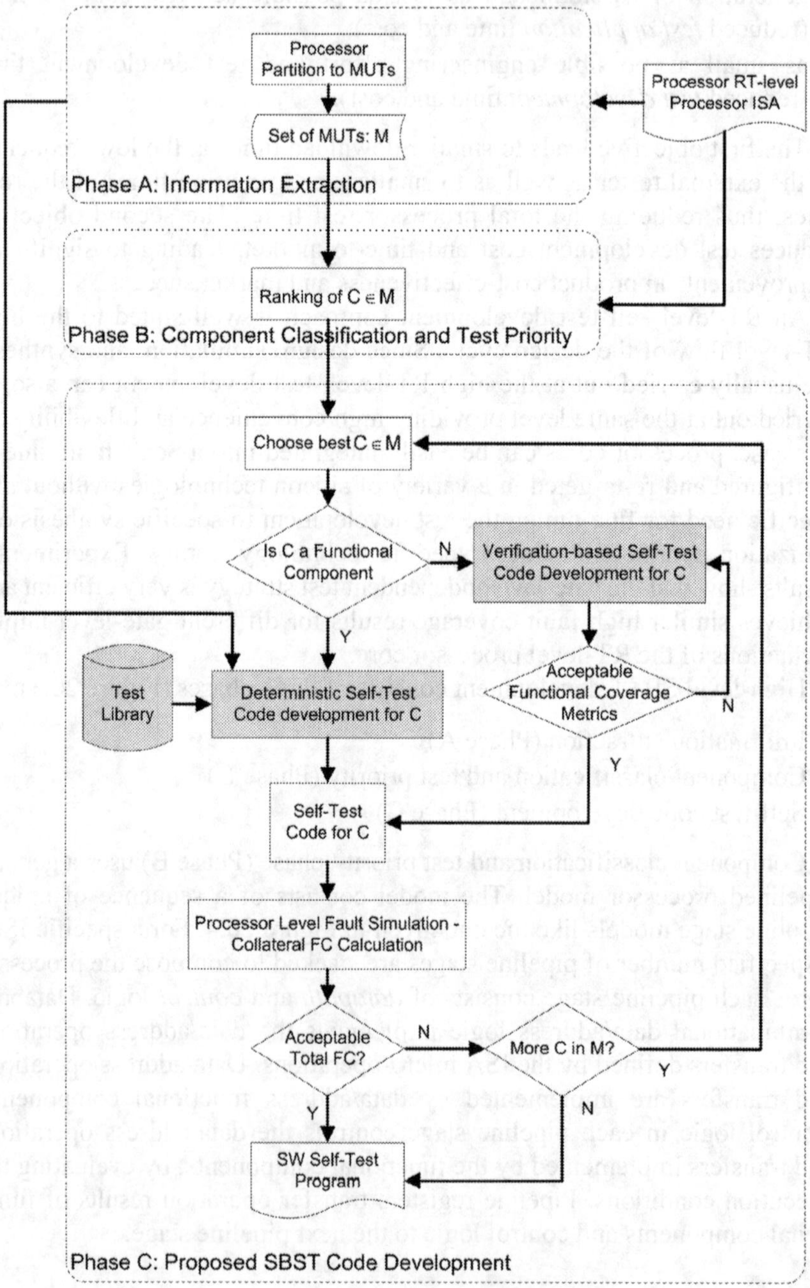

Fig. 20.3. High-level SBST methodology.

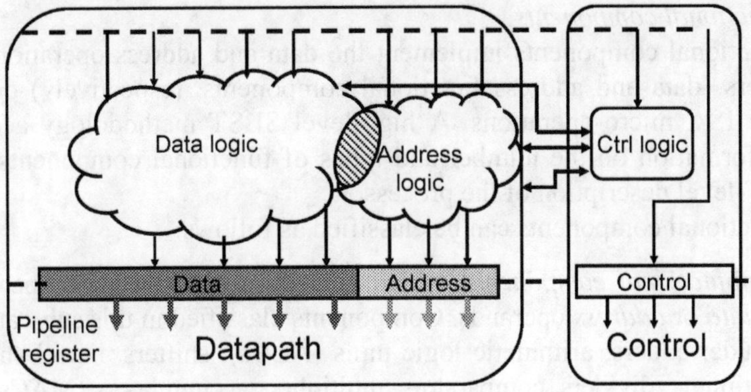

Fig. 20.4. Processor components per pipeline stage.

Phase A: information extraction

In Phase A (*information extraction,* Figure 20.3) the processor is partitioned to a set M of processor components C using the processor RT-level description.

During Phase A the following information is extracted using the ISA and the RT-level description of the processor under test:

- the sets of component operations O_C for each component C of set M (for some components this set may consist of several different operations, i.e. a multi-functional ALU, while for other components the set may consist of only one operation);
- the sets of *basic test instructions* that excite these component operations (there may be several instructions exciting different operations in different processor components);
- the sets of *peripheral test instructions* (or instruction sequences) for controlling or observing processor registers (these instructions supplement the previous ones in order to perform test application and component response observation).

Phase B: component classification and test priority

In Phase B (*component classification and test priority*, Figure 20.3), the processor components are categorized in classes and ranking is performed among the processor components to determine the component *test priority*.

Component classification

The processor components are classified in the three main classes: *Functional, Control and Hidden.*

Functional components

Functional components implement the data and address operations and transfers (data and address functional components, respectively) defined by the ISA micro-operations. A high-level SBST methodology acquires the information on the number and types of functional components from the RT-level description of the processor.

Functional components can be classified as follows:

- *Computational components*, which perform arithmetic/logic operations on *data* or *address* operands. Components classified in this sub-category include: adders, arithmetic logic units (ALUs), shifters, barrel shifters, multipliers, dividers, comparators, multipliers/accumulators (MACs), etc. An example of *data computational component* is the multiplier while the adders that perform the next address calculation or the branch address calculation are examples of *address computational components*.
- *Interconnect components*, which serve the flow of *data* or *address* operands in a processor's datapath. Components classified in this sub-category include multiplexers. An example of *data interconnect components* are the forwarding multiplexers at the inputs of an ALU at the execution stage that select data operands from following to the execution stage pipeline stages. An example of *address interconnect component* is the write address multiplexer at the write address input of a register file.
- *Storage components*, which are processor *data* and *address* holding elements that feed the inputs of the data or address computational components and capture their output. Components classified in this sub-category include: general processor registers (register files), special purpose registers, pipeline registers, etc. Examples of *data storage components* are the special purpose registers for storing the 64-bit multiplication result, the general purpose registers of the register file, etc. Such data storage components are also known as *architectural registers* and are visible to the assembly language programmer. Pipeline registers that hold data information between pipeline stages are data storage components; however, they are not visible to the assembly language programmer. Examples of *address storage components* are the program counter (PC) and pipeline registers that hold address information between pipeline stages.

Control components

Control components control either the flow of instructions or data inside the processor core or from or to the external environment (memory, peripherals). These components include the processor's main control unit that implements the instruction decoding and produces the control signals for subsequent pipeline stages, the local pipeline stage controllers and control registers, the instruction and data memory controllers that implement instruction fetching, control logic for stalling the pipeline, control logic for data hazard detection and enabling forwarding mechanisms, etc.

Hidden components

Hidden components are usually employed in a processor's implementation to improve its performance. The term *hidden* denotes the fact that such components *are not visible to the assembly language programmer*. Actually, hidden components are functional or control components. For example, pipeline logic is not visible to the assembly language programmer and is considered as hidden logic. The pipeline hidden logic includes the following components: pipeline registers (storage functional components), pipeline multiplexers implementing forwarding/bypassing mechanisms (interconnect functional components) and pipeline stage controllers (control components). Other logic considered as hidden include those components related to other performance increasing mechanisms like instruction level parallelism (ILP) and speculative mechanisms to improve processor performance such as branch prediction schemes.

Component test priority

Components are ranked in descending order of *test priority* to determine the order in which test routines will be developed for each component. Since the basic aim of the high-level SBST approach is to reach high fault coverage at an as small as possible test development effort and cost, prioritization of processor components is particularly useful so that the test development process first deals with the most important components that are likely to have the largest contribution to the overall fault coverage.

High priority components will be considered first while low priority components will be considered afterwards only if the achieved overall fault coverage result is not adequate. In many cases, test development for top priority components leads to sufficient fault coverage for not targeted components as well due to collateral coverage. This is particularly true in processors because the execution of a computation in a functional unit also excites many of the control subsystem components. The criteria that are used for component test prioritization for low-cost software-based self-testing are discussed and analyzed in the subsequent paragraphs.

Test priority criterion 1: component size

Component size is an intuitive criterion that should be first considered for low-cost test development. It gives the following simple but very important advice: *component self-test code development should give higher priority to processor components that have the largest contribution to the overall processor fault coverage* (*or equivalently gate count*).

The following observations are valid for the majority of processor implementations:

- The *register file* of the processor is one of the largest components. This is particularly true in RISC processors with a classic load/store architecture and a large number of general-purpose registers. Large register files offer many advantages enabling compilers to increase performance by reducing memory transactions.
- The *parallel multiplier* or *multiplier-accumulator* is usually one of the largest components (particularly true in RISC processors and DSPs).
- The *functional components* of the processor that perform all arithmetic and logic operations of the processor are much larger than the corresponding control logic which controls their operation. This size difference increases when processor word sizes move from 8-bits and 16-bits towards the 32-bits or 64-bits domains. Additionally, in DSPs with multiple instances of functional components of the same type, the processor area is dominated by the size of the functional components.
- The *hidden components* of the processor (mainly these related to the data and address information flow in the pipeline logic) may occupy a significant portion of the processor's area. Since these components are not visible to the assembly language programmer, they are more difficult to test than the visible components. Specific and generic solutions for the effective testing of hidden components should complement the high-level SBST methodology.

Test priority criterion 2: component accessibility

The second criterion to be considered for prioritization of processor components for low-cost test development, is the component's *accessibility* (controllability and observability) by processor instructions. The *controllability* of a processor component is related to how easily an instruction sequence can apply a test pattern to the component inputs while the *observability* is related to how easy an instruction sequence propagates component output values to the primary outputs of the processor. Usually, the controllability and observability of processor components is directly mapped to the controllability and observability of the registers that drive or are driven by the components inputs and outputs.

The data functional components usually provide easy and full accessibility since their adjacent registers are fully accessible registers (i.e. general purpose registers of register file or special purpose registers like accumulator).

Testability of address functional components (i.e. program counter (PC), address part of pipeline registers, branch target adders, etc.) is very poor [341]. The controllability issues with address functional components are as follows: fault excitation of address storage functional components (i.e. program counter) requires the application of test vectors that set all address bits to both 0 and 1. However, for example, setting the two least significant bits of PC to 1 during normal mode of operation is not feasible due to alignment restrictions. On the other hand, poor observability of the address storage functional components like the address part of pipeline registers is due to the fact that their faults cannot be directly propagated to data memory. The testability problems of the address-related logic are not only a concern for pipelined processors but also for non-pipelined processors as well. The problem is accentuated in pipelined implementations, because multiple instances of address information flow through the pipeline stages. Careful and cumbersome testability analysis is required to extract faults that cannot be detected during normal mode of operation on address-related logic due to memory mapping constraints.

The sequential nature of control components imposes several accessibility difficulties while control components that implement interface functions with memory (instruction fetch, memory handshaking, etc.) impose serious additional controllability problems.

Hidden components are functional or control components which are not directly visible to the assembly language programmer. Thus, hidden component accessibility depends on whether the hidden component is functional (data or address) or control. Testability of data functional hidden components can be very high. Consider for example the case of forwarding/bypassing logic implemented by forwarding multiplexers. Despite the fact that such logic is not directly visible to the assembly language programmer, forwarding multiplexers are data functional components with full accessibility (controllability and observability) that can be targeted effectively by deterministic test routines that apply regular test patterns to the forwarding multiplexers and thus guarantee complete fault coverage.

As a concluding remark, for a low-cost high-level SBST methodology, the data functional components (including the hidden ones) of the processor should have the highest test priority for test development since their size dominates the processor area and they are easily accessible (the data functional components are more accessible than their address counterparts). When total processor fault coverage is not sufficient, test development should proceed to the other processor components with lower test priority (address functional components and control components).

Phase C: Self-test code development

In Phase C (*self-test code development*, Figure 20.3), self-test routines are developed for each processor component C and the most critical-to-test components (components C that ranked with higher test priority) are targeted first. A suitable test strategy is followed for each component C. After each iteration, processor-level fault simulation is performed only to evaluate the total fault coverage and estimate the collateral coverage. The iterative process is repeated on remaining modules until desirable processor-level fault coverage is achieved. The self-test program consists of several component-based self-test routines.

In the following paragraphs, we show that self-test code development for functional components is fundamentally different from self-test code development for control components. It should be reminded that hidden components are functional or control with the attribute that are not directly visible to the assembly language programmer. Self-test code development for hidden components follows a functional or control component test strategy accordingly.

Self-test code development for functional components

If C is a functional component, we follow a *deterministic self-test code development* approach. The methodology applies all possible component operations O_C with deterministic operands. Application of component operations O_C is performed by selecting the *basic test instruction* which requires the shortest instruction sequences (*peripheral test instructions*) to apply specific operands to component inputs and propagate the outputs to processor primary outputs.

The key for selecting the most appropriate deterministic operands lies beneath the architecture of most critical-to-test processor components. Such components have an inherent regularity, which can lead to very efficient test algorithms for any gate-level implementation. This inherent regularity is not exploited either by pseudorandom test development or by ATPG-based test development approaches. Many processor components, in particular the vast majority of functional components like computational (arithmetic and logic operation modules), interconnect (multiplexers) and storage components (registers, register files) have a very regular or semi-regular structure. Regular structure appears in several forms like in the form of arrays of identical cells (linear or rectangular), tree-like structures of multiplexers, memory element arrays, etc. Such components can be efficiently tested with small and regular test sets that are gate-level independent, i.e. provide high fault coverage for any different gate-level implementation.

A component *test library* has been developed, including test algorithms that generate small deterministic tests and provide very high fault coverage for most types and architectures of functional processor components. The nature of these deterministic tests is fundamentally different from ATPG-generated test patterns since a gate-level ATPG tool is not capable to identify regular structures and generate patterns optimized for compact-code software-based test application. Based on these small deterministic tests, test routines are developed that, additionally to the small number of tests, take advantage of test vectors' regularity and algorithmic nature resulting in efficient compact loops [328]. These loop-based compact test routines require a very small number of bytes while the small number of tests results in low test routine execution time. The *test library* routines that describe a test algorithm in assembly pseudocode, for almost all generic functional components, are tailored each time to the instruction set and assembly language of the processor's under test.

In the remainder of this section we briefly describe how regular structures in datapath functional components that implement the most usual arithmetic, logic, interconnect or storage operations, can be efficiently tested with short deterministic tests without the need of gate-level details.

Arithmetic components testing

For components that implement arithmetic operations, we have developed deterministic tests for every type of arithmetic operations like addition, subtraction, multiplication, and division for various word lengths. Deterministic tests are also available for several architecture and algorithm alternatives. For example, for the addition operation precomputed tests exist for ripple-carry adder (RCA), carry-look-ahead (CLA), etc. architectures. Likewise, for the multiplication operation precomputed tests exist for carry-save array, booth-encoded, Wallace tree summation, etc. architectures [151,327].

Logic array testing

Testing a logic array that implements multi-bit Boolean logic operations like *and, or, nor, xor, not* is performed by applying simple well known necessary patterns while propagating the test response to well observable registers and primary outputs.

Interconnect/multiplexer testing

An *n*-to-1 multiplexer is usually decomposed and implemented by smaller multiplexers in a tree structure. For example, an 8-to-1 multiplexer can be classically implemented as a tree of seven 2-to-1 multiplexers. The implementation of an *n*-to-1 multiplexer tree is not unique and in an RT-level design, logic synthesis tools can generate different implementations depending on the technology mapping algorithms and technology libraries.

We have developed deterministic tests for the case of n-to-1, m-bit wide multiplexers [240]. This implementation independent deterministic test set is linear, the minimum number of $2n$ test vectors is required, and provides 100% single stuck-at fault coverage. The deterministic tests provide 100% single stuck-at fault coverage for multiplexer-based logic structures (i.e. shifter units).

Register testing

We have developed deterministic tests for the flip-flops and write enable multiplexers that implement general purpose registers, special purpose registers and pipeline registers. This implementation independent deterministic test set is constant and provides 100% single stuck-at fault coverage for the flip-flops and write enable multiplexers.

Register file testing

The register file component can be considered as a hierarchical design with sub-components. These sub-components are the write address decoder, the register array, and the two large read ports. We have developed deterministic tests [240] for the two large multiplexer trees (usually 32 input, 32-bit wide multiplexers for processor architectures with 32 registers with each register 32-bit wide) that implement the two register file read ports. The register array is tested by deterministic tests developed for registers discussed above. This implementation independent deterministic test set results in near complete (>99.5%) single stuck-at fault coverage for the entire register file.

Address components testing

The excitation of the faults of address computational and storage functional components (i.e. PC+4 adder, branch target adders, PC, etc.) requires the application of a rich set of address values that potentially must cover the entire memory space of the processor. The only way to achieve this, is to execute instructions and to access (read or write) data that are stored in several different regions of the memory space of the processor.

In [341] a generic solution was proposed for the excitation (controllability) of address related faults by partitioning of the SBST code to segments that are virtually stored and executed from different memory regions along with implementation of specific mapping logic (at the behavioral RT-level for fault simulation and/or FPGA-based loadboard attached to the memory interface of the processor). Propagating address functional component faults towards the data bus can be accomplished by the use of *link instructions* in the processor ISA. When a *jump-and-link* or *branch on condition and link* instruction is executed, the value of the PC is stored in a return address (RA) register. Using a store instruction, the value of the RA register can be propagated to the data bus and therefore to the data memory.

Self-test code development for control components

If C is a control component, a *verification-based self-test code development* approach is adopted with self-test development still performed at high-level.

Often, verification-based functional tests cannot guarantee acceptable fault coverage while the manual effort required to derive simple instruction sequences that verify control subsystem functionality is substantially higher when compared to the effort required for processor functional components, where reusing the *test library* minimizes costly manual effort. Satisfaction of high-level RT-level verification metrics supported by industry standard simulation tools like RT-level *statement, branch, condition*, and *expression coverage* helps to improve verification manual effort. Besides the overhead in test development cost, testing control components imposes a substantial overhead in the test program size and test application time as well. Such overheads characterize functional testing approaches, since it is very difficult for such components to be tackled by small and fast deterministic routines. Thus, verification-based functional testing techniques are adopted for the control components with the test cost/fault coverage trade-off in mind.

Let's briefly discuss an example of verification-based self-test code development for pipeline control components. The basic concept is to apply test instruction sequences that exercise the functionality and increase the activity of the pipeline components: cause hazards of different types, and activate all forwarding paths multiple times with different data. In [341], a systematic method is proposed that processes existing SBST routines (routines for functional components guarantee the diversity of operands) and creates multiple instantiations of a given routine (code variants) so as to activate the pipeline control logic.

In many cases, verification-based functional self-test programs based on templates and instruction biases are generated automatically by *Architectural Verification Suites (AVS)* developed for processor verification. The test engineer can *reuse* appropriate parts of the verification routines, for the testing of control components with no additional manual effort involved or in the worst case substantially alleviating any manual self-test routine development effort. However, corner cases are usually targeted by proper handwritten instruction sequences. In case that verification-based test programs are reused, it should be taken care that test responses are propagated to memory since as opposed to verification, in SBST the internal state of the processor is not directly observable, thus making propagation imperative.

Case studies – experimental results

Several experiments have been performed on a diverse set of different processors benchmarks with varying complexity. Some of these processor core models are publicly available benchmarks while one of them was developed in-house to overcome architectural and functional limitations of the publicly available benchmarks. Table 20.3 lists the processor benchmarks used, the usefulness of each processor benchmark and some remarks on the architecture and implementation of each benchmark.

In order to evaluate the fault coverage, a test evaluation framework was developed and used for every processor core benchmark, based on industrial tools for synthesis, functional and fault simulation. It should be noted that the fault coverage reports that follow are smaller than the actual fault coverage, since many of the undetected faults are *structurally testable*

Table 20.3. Processor core benchmarks used.

Processor Name	Architecture Implementation	Complexity	Usefulness of Experiment
Plasma	Simple RISC processor, Princeton architecture, 3-stage simple pipeline, no data forwarding, public available.	Simple	First application of the methodology to a RISC processor. Many different synthesized versions used.
ASIP	RISC processor, Harvard architecture, 5-stage pipeline. Public available limited version.	Average	Application of the methodology on a RISC processor generated automatically by an ASIP suite.
Athena	RISC processor, Harvard architecture, 5-stage full pipeline, component hierarchy maintained thoughout synthesis, in-house.	Average	Application of the methodology on a fully-pipelined processor with optimized coding style.
miniMIPS	RISC processor, Princeton architecture, system co-processor, 5-stage full pipeline, public available.	Average	Application of the methodology on a fully-pipelined public processor.
OR1200	"Real world" RISC processor (used in industrial SoCs), configurable Harvard architecture, 5-stage full pipeline, public available.	High	Application of the methodology to a successful, industrial embedded processor.

and functionally untestable.[4] There are various reasons for the existence of functionally untestable faults in processor architectures. Some of them are due to microarchitectural and ISA constraints and other due to design techniques used. Testing of such functionally untestable fault is considered overtesting that results to yield loss. However, there is no commercial tool that can identify such faults.

Test strategy

For all the processor core benchmarks used we applied the high-level SBST methodology discussed in previous paragraphs. All highest priority functional components were targeted using Deterministic self-test code development by simply tailoring the *test library* algorithms that apply pre-computed gate-level independent deterministic tests, to the processor benchmark assembly language. The self-test program was composed targeting the computational functional components (parallel multipliers, ALUs, shifters, etc.), the interconnecting functional components (pipeline forwarding multiplexers) and the storage functional components (register files, and pipeline registers). The self-test program was enhanced by self-test routines targeting control components including the pipelined control unit, the local pipeline stage controllers and the hazard detection unit. All control components were targeted using a coverage-driven verification-based self-test code development approach aiming at maximizing functional coverage metrics by taking advantage of existing software verification platforms where available. For example, *Athena Processor Verification Suite (APVS)* which was developed at the design phase of Athena CPU, automatically generated verification-based functional test code. Through reusability of verification-generated self-test code, any manual verification-based self-test code development effort was substantially alleviated. Likewise, *OR1200 Processor Verification Suite (OPVS)* which we developed for OpenRISC 1200 design verification, generated automatically verification-based self-test programs based on templates and instruction biases. Furthermore, a small number of corner cases are targeted by proper handwritten instruction sequences. In the following paragraphs, we will briefly discuss each processor benchmark used. Test statistics for the five processor benchmarks including different gate-level implementations are summarized in Table 20.4.

[4] Faults that cannot be detected at normal mode of operation while are identified as undetected by the fault simulator.

Table 20.4. Test statistics for the processor benchmarks.

Processor Core	Gate Count	FC (%)	Size (Words)	Time (Clock Cycles)
Plasma (Synthesis A)	26,080	95.3	853	5,797
Plasma (Synthesis B)	27,824	95.3	853	5,797
Plasma (Synthesis C)	30,896	94.5	853	5,797
ASIP	37,400	92.6	1,728	10,061
Athena	26,122	96.3	3,014	11,637
miniMIPS	32,817	95.1	1,565	7,162
OpenRISC 1200	44,476	86.4	2,947	108,429
OpenRISC 1200 (Hybrid)	44,476	91.6	15,693	210,167

Plasma

Plasma was the first publicly available 32-bit RISC processor core benchmark we used to evaluate the high RT-level methodology on processor cores. Plasma supports interrupts and all MIPS I user mode instructions except unaligned load and store operations (which are patented) and exceptions. The synthesizable CPU core is implemented in VHDL with a simple 3-stage pipeline [338]. The CPU has been enhanced by adding a parallel multiplier. The fast multiplier module was generated using a public available module generator and has the following characteristics: Booth recoding, Wallace trees for partial product summation and fast carry look-ahead addition at the final stage. The Plasma processor VHDL model was synthesized in two different technology libraries with different synthesis parameters. Synthesis A (26,080 gates) was optimized for area, in a 0.35 μm technology library and the design runs at a clock frequency of 57 MHz. Synthesis B (27,824 gates) was also optimized for area, in a 0.50 μm technology library and the design runs at a clock frequency of 42 MHz. Finally, Synthesis C (30,896 gates) was optimized for delay, in a 0.35 μm technology library and the design runs at a clock frequency of 74 MHz.

The developed self-test program consists of 853 words and is executed for 5,797 clock cycles. Table 20.4 shows that very similar fault coverage results (≈95%) were obtained when the processor core was synthesized in different technology libraries with different optimization scripts, optimizing either for area or performance. This set of experiments on Plasma CPU was performed to show that test development for the targeted components which is performed at high RT-level, results in high structural fault coverage that does not depend in the gate-level synthesis implementation. Therefore, the self-test program is gate-level independent, i.e. independent of the logic synthesis parameters and technology library used.

ASIP

A 5-stage pipeline processor was designed using the *application specific instruction set processor (ASIP)* design environment of [28] for evaluating the proposed methodology in a totally different processor implementation of the MIPS-I ISA, i.e. as generated automatically by an ASIP design environment. A 52 instruction subset was implemented while co-processor and interrupt instructions were not implemented in this experiment. It should be noted that in the current educational release version of the ASIP design environment of [28], data hazard detection and register bypassing are not implemented. The automatically generated RT-level VHDL model of the processor has been synthesized, optimized for area, targeting a 0.35 μm technology library and the design runs at a clock frequency of 44 MHz with a total gate count of 37,400 gates.

The self-test program developed consists of 1,728 words that execute in 10,061 clock cycles. Almost complete coverage was achieved for the computational functional components (ALU, Multiplier, and Shifter), the storage functional components (register file, HI-LO registers) and the interconnect functional components (pipeline forwarding multiplexers). Fault coverage at processor level was 92.6%.

Athena

Athena is processor core developed at the University of Athens, to fulfill the requirements of a fully functional pipelined processor benchmark with academic and research applications. *Athena* is a 32-bit embedded RISC processor core with 32 GPRs that implements a 5-stage pipeline with hazard detection and forwarding mechanisms along with exception handling mechanisms. Athena implements the full MIPS-I ISA with the sole exception of the unaligned load and store operations that are patented. The processor core is enhanced with a high-performance Booth encoded, Wallace-tree parallel multiplier (same multiplier implementation as in the Plasma CPU).

The RT-level coding style partitioned the processor core into hierarchical blocks maintained throughout the synthesis process while avoiding glue-logic at the top level. Hierarchical RT-level coding style and bottom-up compile allowed for critical components to be constrained and compiled separately. For example, finite state machine (FSM) logic and datapath time critical components were isolated and optimized for synthesis to meet specific area and timing constraint requirements. Furthermore, partitioning the processor into modules maintained throughout synthesis process, allowed development and evaluation of SBST routines at the component level targeting pipeline logic. The RT-level processor model was synthesized targeting a 0.18 μm technology library. Synthesis was optimized for area and the netlist gate-count was 26,122 gates with 1,515 FFs. The design runs at a clock frequency of 82 MHz.

The self-test program developed consists of 3,014 words that execute in 11,637 clock cycles. Almost complete coverage was achieved for the computational functional components (ALU, Multiplier, and Shifter), the storage functional components (register file, HI-LO registers) and the interconnect functional components (pipeline forwarding multiplexers). Fault coverage at processor level was 96.3%.

miniMIPS

miniMIPS is a public available 32-bit RISC processor model, implementing the MIPS I ISA with a classic 5-stage pipeline that supports hazard detection and incorporates a system co-processor which handles exceptions and interrupts [289].

The RT-level processor model was synthesized targeting a 0.18 μm technology library. The gate-count was 32,817 gates and the design runs at a clock frequency of 100 MHz. The self-test program developed consists of 1,565 words that execute in 7,162 clock cycles. Fault coverage of 95.1% was achieved at processor level.

At this point it should be noted that all three processor core benchmarks considered so far (Plasma, miniMIPS, and Athena) implement almost the same MIPS I ISA; however, their microarchitecture differs considerably. The microarchitecture complexity of miniMIPS is comparable to the Athena CPU, which is much more complex than Plasma CPU.

All these four sets of experiments on four different implementations of the same ISA, were performed to show that test development which is performed at high RT-level based on the ISA of the processor, can result in high structural fault coverage that does not depend in the specific implementation of the ISA.

OpenRISC 1200

OpenRISC 1200 is a public available processor core [317] which has been employed in numerous industrial applications. Its complexity is the higher in the complexity scale among the processor benchmarks used for the experimental results. The currently available version is a 5-stage pipelined 32-bit RISC processor core with Harvard architecture. It is configurable and a configuration supporting 56 instructions and a multiply–accumulate (MAC) unit was synthesized targeting a 0.18 μm technology library. The netlist gate-count was 44,476 gates with 2,021 state-holding elements.

The self-test program developed according to the high RT-level methodology consists of 2,947 words that execute in 108,429 clock cycles. Fault coverage of 86.4% was achieved at processor level.

Although the test program size is comparable to previous processor benchmarks, the number of cycles is increased. This is due to the fact that OpenRISC 1200 ISA includes multi-cycle instructions (i.e. multiply–accumulate instruction). Furthermore, OR1200's much higher complexity requires an increased number of verification-based test sequences for targeting processor's control logic.

If the derived fault coverage attained is not considered adequate, a hybrid SBST approach that includes: first high RT-level test program generation, then constrained-ATPG test program generation similar to [75] for the hard-to-detect faults and finally random test program generation (RTPG), it can lead to fault coverage of 91.6%, at an increased test program size (15,693 words) and test execution time (210,167 cycles).

Conclusions and perspective

Software-based self-testing has been recently proposed as a very promising alternative to classic hardware-based self-testing, for actual at-speed testing of microprocessors, embedded processors and processor-based systems. Many academic research groups and, recently, major processor companies, have proposed interesting approaches in the framework of software-based, instruction-based or cache-resident manufacturing self-testing of processors. In this chapter, we have described the key concepts of this emerging testing philosophy, reviewed several approaches recently presented in the literature in this topic and have elaborated on our high-level software-based self-testing methodology. The methodology has been successfully applied to several embedded processors and a detailed elaboration on the experimental setup and results has been given.

Software-based self-testing has a strong potential to eventually evolve into a mature and widely adopted self-test approach, because of its non-intrusive nature and its applicability at different stages of the chip life cycle (prototyping, production/manufacturing, in the field).

Intense research efforts and investments are necessary for the near future so that software-based self-testing is adopted in the emerging processor architectures (including chip multiprocessor architectures) and automation of existing or new methodologies comes true.

21 Future Directions in Processor Design

Jari Nurmi

Tampere University of Technology

An outlook of the future directions in processor design is provided in this chapter. Shortly, there will be

- More processors
- More application-tailored processors
- More parallelism in different forms

If we think of the technology development as predicted in the ITRS roadmap [208], it is clear that more and more processors will be crammed onto a single chip. The prediction of the roadmap is that we will see hundreds or even thousands of processors integrated within the next ten … fifteen years, e.g., 424 processing elements per chip in 2017. The trend can be confirmed by looking at some contemporary ambitious high-end projects in multi-core and multi-processor development. For example, Rapport, Inc. is shipping a chip with 256 processing elements on board and is developing a 1024-core processor, however these are only 8-bit elements [281]. Even in workstation processor complexity, Sun has announced an 8-core processor [109], and Intel and AMD have both announced quad-core ones [218]. Each of these processors also uses multi-threading on its cores. The CELL architecture with eight data processing cores and one control processor is a well-known development for high-end embedded systems, already in mass production [165].

So, there will be more processors in an integrated embedded system. But what kind of processors? My bet is that we will see more specialized processors for different specific tasks. The "one size fits all" approach simply cannot provide cost and power efficient enough solutions for the embedded sector. Thus, we will see various special-purpose off-the-self cores emerging. At the leading edge of applications, customizable cores and processor generators will be used to unleash the processing power

J. Nurmi (ed.), Processor Design: System-on-Chip Computing for ASICs and FPGAs, 483–485.

provided by new technologies. As time will pass, some of the application-specific cores will move to the standard part depository, some will die, and new emerging applications will be accelerated by tailored solutions. This does not mean that there would not be space for general-purpose processors in future SoCs, definitely their services will still be needed in parallel with the emerging application-specific cores.

In future System-on-Chip we will see unparalleled parallelism. Data-Level Parallelism (DLP) such as in split-data multimedia instructions, and Instruction-Level Parallelism (ILP) as in superscalar and VLIW processing, will be complemented with Thread-Level Parallelism (TLP) and Process-Level Parallelism (PLP). The former was introduced already in Chapter 2. By the latter I mean Chip-Multi-Processing (CMP) type parallel processing of loosely (if at all) interrelated processes in a massively parallel computation system.

In the search for higher efficiency in the use of processor resources, different forms of multi-threading will find foothold in the embedded processing era. Within single processors simultaneous multi-threading (SMT) provides the highest efficiency. It is noticeable that SMT is merely a small extension from a multi-issue (superscalar or VLIW/EPIC) processor architecture. In true multi-processor environments, it is natural to use coarser granularity and thus to run different threads on different processors and only switch to a new thread at a cache miss or other discontinuity point. As pointed out in Chapter 3, running several parallel threads speculatively in parallel is waste of energy and thus intolerable in many of the embedded applications. Instead, basing the multi-threading on independent parallel threads (thus approaching PLP in the case of a multi-processor system) provides the energy efficiency needed.

A debate going on in the SoC designer community is whether homogeneous or heterogeneous multi-processor systems will prevail. In fact, there is room for both. I foresee that general-purpose computing is more amenable to homogeneous processors, while highly optimized embedded SoCs will be mostly constructed out of heterogeneous processors – some of them even tailored to the specific application of the SoC.

In conclusion, the future of processor design includes multiple "multi-" aspects – multi-issue, multi-threading, multi-core, multi-processor, and even multi-million transistor circuits, multi-discipline multi-site multi-national design teams, and fabrication in multi-billion euro/dollar fabs.

One interesting and very different type of direction for embedded processor design will be the low-end market. As trends such as electronics printed on paper, plastic or textile will be realized in increasing complexities, simple and extremely inexpensive processors in such bulk applications will be needed. The cost requirements will be orders of magnitude lower than in consumer electronics which is already considered a cost-sensitive segment.

Further in future there will be gigantic challenges in implementing processor-like structures in nanoelectronics such as carbon nanotubes or biologically reproduced nanomaterials. Such challenges will require a major paradigm shift that is already far beyond the scope of this book.

References

1. Accellera. (2004) SystemVerilog 3.1a Language Reference Manual – Accellera's Extension to Verilog. Available at http://www.eda-stds.org/sv/SystemVerilog_3.1a.pdf
2. Agarwal V, Hrishikesh MS, Keckler SW, Burger D (2000) Clock Rate versus IPC: The End of the Road for Conventional Microarchitectures. In Proc. ISCA, pp 248–259
3. Aho AV, Ulman JD (1999) Principles of Compiler Design. Addison-Wesley, Narosa
4. Ahonen T (2006) Designing Network-Based Single-Chip System Architectures, DrTech Thesis, Tampere University of Technology. TUT Publication 625
5. Ahonen T, Virtanen S, Kylliäinen J, Truscan D, Kasanko T, Sigüenza-Tortosa DA, Ristimäki T, Paakkulainen J, Nurmi T, Saastamoinen I, Isännäinen H, Lilius J, Nurmi J, Isoaho J (2004) A Brunch from the Coffee Table – Case Study In Noc Platform Design. In Nurmi J, Tenhunen H, Isoaho J, Jantsch A (eds) Interconnect-Centric Design for Advanced SoC and NoC, Kluwer Academic Publishers, Chapter 16, pp 425–453
6. Ahonen T, Nurmi J (2005) Integration of a NoC-Based Multimedia Processing Platform. In Proc. International Conference on Field Programmable Logic and Applications, pp 606–611
7. Ahonen T, Nurmi T, Nurmi J, Isoaho J (2003) Block-wise Extraction of Rent's Exponents for an Extensible Processor. In Proc. IEEE Computer Society Annual Symposium on VLSI, pp 193–199
8. Allan V, Jones R, Lee R, Allan S (1995) Software Pipelining. ACM Computing Surveys, 27:3
9. Allen E, Chase D, Hallett J, Luchangco V, Maessen JW, Ryu S, Sttele GL Jr, Tobin-Hochstadt S (2007) The Fortress Language Specification v. 1.0 β. Technical Memo, Sun Microsystems, March. Available at http://research.sun.com/projects/plrg/fortress.pdf
10. Allen R, Kennedy K (2001) Optimizing Compilers for Modern Architectures. Morgan Kaufmann Publishers
11. Altera (2007) company web page http://www.altera.com
12. Altera (2006) Stratix II vs. Virtex-4 Density Comparison. Consulted on 16 February 2007. Altera white paper at http://www.altera.com
13. Altera (2007) Nios Embedded Processor (visited on January 2007). Available at http://www.altera.com/products/ip/processors/nios/nio-index.html
14. Altera (2007) Nios II. http://www.altera.com/nios2

15. Altera (2007) FPU DFPAU (visited on January 2007). Available at http://www.altera.com/products/ip/dsp/arithmetic/m-dcd_dfpau.html
16. Anderson E, Bai Z, Bischof C, Demmel J, Dongarra, Du Croz J, Greenbaum A, Hammarling S, McKenney A, Sorensen D (1990) LAPACK: A Portable Linear Algebra Library for High-Performance Computers. In Proc. Supecomputing, pp 2–11
17. Andrews J, Baker N (2006) Xbox 360 System Architecture. IEEE Micro, 26(2):25–37
18. Annaratone M, Arnould E, Gross T, Kung HT, Lam M, Menzilicioglu O, Webb JA (1987) The Warp Computer: Architecture, Implementation and Performance. IEEE Transactions on Computers, 36(12):1523–1538
19. Appel AW (2000) Modern Compiler Implementation in C. Cambridge University Press
20. ARC (2007) Cycle-Accurate and Instruction Set Simulator (CAS & ISS). Available at http://www.arc.com/software/simulation/casandiss.html. Verified 2007-01-25
21. ARM (2000) ARM9TDMI, Technical Reference Manual. Advanced RISC Machines Ltd., Revision 3
22. ARM (1995) An Introduction to Thumb. Advanced RISC Machines Ltd., Version 2.0
23. ARM (2007) Jazelle Technology, consulted 18 January 2007. Available at http://www.arm.com/products/esd/jazelle_home.html
24. ARM (2007) VFP9-S coprocessor. Available at http://www.arm.com/products/CPUs/VFP9-S.html (visited on January 2007)
25. Arnold JM (2005) S5: The Architecture and Development Flow of a Software Configurable Processor. In Proc. IEEE International Conference on Field-Programmable Technology, pp 121–128
26. Arnold M, Corporaal H (2001) Designing Domain Specific Processors. In Proc. 9th International Symposium on Hardware/Software Codesign (CODES'01), pp 61–66
27. Ashenden PJ (1996) The Designer's Guide to VHDL. Morgan Kaufmann Publishers, San Francisco
28. ASIP Meister (2007) http://www.eda-meister.org
29. Athanas R, Silvermann H (1993) Processor Configuration Through Instruction Set Metamorphosis. IEEE Computer, 26(3):11–18
30. Atkins M (1991) Performance and the i860 Microprocessor. IEEE Micro, October: 11(5):24–27, 72–78
31. Babb J, Frank M, Lee V, Waingold E, Barua R, Taylor M, Kim J, Devabhaktuni S, Agarwal A (1997) The RAW Benchmark Suite: Computation Structures for General Purpose Computing. In Proc. FCCM, pp 134–143
32. Baines R, Pulley D (2003) A Total Cost Approach to Evaluating Different Reconfigurable Architectures for Baseband Processing in Wireless Receivers. IEEE Communication Magazine, 41(1):105–113
33. Bakoglu HB (1990) Circuits, Interconnections and Packaging for VLSI. Addison-Wesley

34. Bakoglu HB, Meindl JD (1987). A System Level Circuit Model for Multi- and Single-Chip CPUs. In ISSCC Digest of Technical Papers, pp 308–309
35. Ball J (2005) The Nios II Family of Configurable Soft-core Processors. In Proc. Hot Chips 17, Stanford, CA, 2005
36. Ball S (2002) Embedded Microprocessor Systems: Real World Design, 3rd edn. Newnes (Elsevier Science), Burlington
37. Ball T, Larus JR (1993) Branch Prediction for Free. In Proc. SIGPLAN Conference on Programming Language Design and Implementation (PLDI), pp 300–313
38. Barat F, Lauwereins R, Deconinck G (2002) Reconfigurable Instruction Set Processors from a Hardware/Software Perspective. IEEE Transactions on Software Engineering, 28(9):847–862
39. Barua R, Lee W, Amarasinghe S, Agarwal A (1999) Maps: A Compiler-Managed Memory System for Raw Machines. In Proc. ISCA, pp 4–15
40. Batcher K, Papachristou C (1999) Instruction Randomization Self Test for Processor Cores. In Proc. VLSI Test Symposium, pp 34–40
41. Bayraktaroglu I, Hunt J, Watkins D (2006) Cache Resident Functional Micro-processor Testing: Avoiding High Speed IO Issues. In Proc. IEEE International Test Conference, paper 27.2
42. BDTI (2000) Choosing a DSP Processor. Technical Report, Berkeley Design Technologies, Inc.
43. BDTI (2000) Evaluating DSP Processor Performance. Technical Report, Berkeley Design Technologies, Inc.
44. BDTI (2004) Buyer's Guide to DSP Processors, 2004 Edition. Berkeley Design Technology, Inc.
45. Becker J, Thomas A, Vorbach M, Baumgarte V (2003) An Industrial/Academic Configurable System-on-Chip Project (CsoC): Coarse-Grain XPP-/Leon-Based Architecture Integration. In Proc. DATE, pp 11–12
46. Bell G, Strecker WD (1976) What Have We Learned from The PDP-11? In Proc. the 3rd Annual Symposium on Computer Architecture. Pittsburgh, pp 1–14
47. Bernardi P, Sanchez E, Schillaci M, Squillero G, Sonza Reorda M (2006) An Effective Technique for Minimizing the Cost of Processor Software-Based Diagnosis in SoCs. In Proc. DATE, pp 412–417
48. Berry G, Gonthier G (1998) The Esterel Synchronous Programming Language: Design, Semantics, Implementation. Technical Report 842, Institut National de Recherche en Informatique et en Automatique, France, May 1988. Available at http://www.inria.fr/rrrt/rr-0842.html. Verified 2007-01-24
49. Bolsen I (2002) Challenges and Opportunities of FPGA Platforms. In Proc. FPL, pp 391–392
50. Bondalapati K, Prasanna VK (2002) Reconfigurable Computing Systems. Proceedings of the IEEE, 90(7):1201–1217
51. Borgatti M, Lertora F, Foret B, Cali L (2003) A Reconfigurable System Featuring Dynamically Extensible Embedded Microprocessor, FPGA, and Customizable I/O. IEEE Journal of Solid-State Circuits (JSSC), 38(3):521–529

52. Both A et al. (1994) Hardware-Software-Codesign of Application Specific Microcontrollers with the ASM Environment. In Proc. Conference on European Design Automation, pp 72–76
53. Brahme D, Abraham JA (1984) Functional Testing of Microprocessors. IEEE Transactions on Computers, 33(6):475–485
54. Brown S and Rose J (1996) Architecture of FPGAs and CPLDs: A Tutorial. IEEE Design Test of Computers, 13(2):42–57
55. Brunelli C (2003) Design of a Floating-Point Unit for a RISC Microprocessor, MSc Thesis, Tampere University of Technology
56. Brunelli C, Garzia F, Campi F, Mucci C, Nurmi J (2005) A FPGA Implementation of an Open-Source Floating-Point Computation System. In Proc. International Symposium on SoC, pp 29–32
57. Brunelli C, Cinelli F, Rossi D, Nurmi J (2006) A VHDL Model and Implementation of a Coarse-Grain Reconfigurable Coprocessor for a RISC Core. In Proc. PRIME, pp 229–232
58. Brunelli C, Garzia F, Nurmi J (2006) A Coarse-Grain Reconfigurable Machine with Floating-Point Arithmetic Capabilities. Invited paper in Proc. ReCoSoC'06, Montpellier, France, pp 1–7
59. Budiu M, Goldstein SC (1999) Fast Compilation for Pipelined Reconfigurable Fabrics. In Proc. FPGA, pp 195–205
60. Burger D, Goodman JR, Kagi A (1997) Limited Bandwidth to Affect Processor Design. IEEE Micro, 17(6):55-62
61. Cadence (2007) The Virtual Component Co-Design (VCC) Available at http://www.cadence.com/company/success_stories/success.aspx?xml=philips_ss. verified 2007-01-24
62. CaffeineMark 3.0 (2007) Benchmark Information, consulted 18 January 2007, URL: http://www.benchmarkhq.ru/cm30/info.html
63. Gagnon E (2002) A Portable Research Framework for the Execution of Java Bytecode. PhD Thesis, McGill University, Montreal
64. Caldwell AE, Cao Y, Kahng AB, Koushanfar F, Lu H, Markov IL, Oliver M, Stroobandt D, Sylvester D (2000) GTX: The MARCO GSRC Technology Extrapolation System. In Proc. DAC, pp 693–698
65. Callahan T, Hauser JR, Wawrzynek J (2000) The Garp Architecture and C Compiler. IEEE Computer, 33(4):62–69
66. Campi F, Toma M, Lodi A, Cappelli A, Canegallo R, Guerrieri R (2003) A VLIW Processor with Reconfigurable Instruction Set for Embedded Applications. In ISSCC Digest of Technical Papers, pp 250–251
67. Campi F, Mucci C, Deledda A, Vanzolini L, Ciccarelli L, Pizzotti M, Vitkovski A, Lodi A, Rolandi P (2007) A Dynamically Adaptive DSP for Heterogenous Reconfigurable Platforms. In Proc. DATE.
68. Camposano R (1990) From Behavior to Structure: High-Level Synthesis, IEEE Design & Test of Computers, 7(5):8–19
69. Carlson CR (1975) A Survey of High-Level Language Computer Architecture. In Chu Y (ed) High-Level Language Computer Architecture. Academic Press, New York, Chapter 3

70. Celoxica (2007) Technical Library. Available at http://www.celoxica.com/techlib/default.asp?Action=1&CatID=9&CatType=2&OrderBy=1. Verified 2007-01-24
71. Chan SC, Shepard KL, Restle PJ (2005) Uniform-Phase Uniform-Amplitude Resonant-Load Global Clock Distributions. IEEE Journal of Solid-State Circuits, 40(1):102–109
72. Chang PP, Mahlke SA, Chen WY, Water NJ, Hwu WW (1991) IMPACT: An Architectural Framework for Multiple-Instruction-Issue Processors. In Proc. 18th Annual Int'l Symposium on Computer Architecture, pp 266–275
73. Charles P, Donawa C, Ebcioglu K, Grothoff C, Kielstra A, Saraswat V, Sarkar V, von Praun C (2005) X10: An Object-Oriented Approach to Non-Uniform Cluster Computing. In Proc. OOPSLA, pp 519–538
74. Chen L, Bai X, Dey S (2002), Testing for Interconnect Crosstalk Defects Using On-Chip Embedded Processor Cores. Journal of Electronic Testing: Theory and Applications, 18(4):529–538
75. Chen L, Dey S (2001) Software-Based Self-Testing Methodology for Processor Cores. IEEE Transactions on CAD of Integrated Circuits and Systems, 20(3):369–380
76. Chen L, Dey S (2002) Software-Based Diagnosis for Processors. In Proc. DAC, pp 259–262
77. Chen L, Ravi S, Raghunathan A, Dey S (2003) A Scalable Software-Based Self-Testing Methodology for Programmable Processors. In Proc. DAC, pp 548–553
78. Cherry S (2004) Edholm's Law of Bandwidth. In IEEE Spectrum, July 2004, pp 58–60
79. Chinnery DG, Keutzer K (2002) Closing the Gap Between ASIC and Custom. Kluwer Academic Publishers
80. Choi H et al. (1998) Synthesis of Application Specific Instructions for Embedded DSP Software. In Proc. ICCAD, pp 665–671
81. Ciricescu S, Essick R, Lucas B, May P, Moat K, Norris J, Schuette M, Saidi A (2003) The Reconfigurable Streaming Vector Processor. In Proc. Intl Symposium on Microarchitectures (MICRO-36), pp 141–150
82. Clark WA, Molnar CE (1974) Macromodular Computer Systems. In Waxman BD, Stacey R (eds) Computers in Biomedical Research, Vol IV, pp 45–85, Academic Press, New York
83. Clark NT, Zhong H, Mahlke SA (2005) Automated Custom Instruction Generation for Domain-Specific Processor Acceleration. IEEE Transactions on Computers, 54(10):1258–1270
84. Coates B, Lexau J, Jones I, Fairbanks S, Sutherland IE (2001) FLEETzero: An Asynchronous Switching Experiment. In Proc. Async'01, pp 173–182
85. Cocke J, Markstein V (2000) The Evolution of RISC Technology at IBM. IBM Journal of Research and Development 44:48–55
86. Comer DE (2004) Network Systems Design Using Network Processors. Prentice Hall, Upper Saddle River, NJ
87. Cooley J, Tukey J (1965) An Algorithm for the Machine Calculation of Complex Fourier Series. Mathematics of Computation, Vol 19, pp 297–301

88. Cooper KD, Harvey TJ (1998) Computer-Controlled Memory. In Proc. ASPLOS-VIII, October, pp 2–11
89. Corno F, Sonza Reorda M, Squillero G, Violante M (2001) On the Test of Microprocessor IP Cores. In Proc. DATE, pp 209–213
90. Corno F, Cumani G, Sonza Reorda M, Squillero G (2003) Fully Automatic Test Program Generation for Microprocessor Cores. In Proc. DATE, pp 1006–1011
91. Corno F, Sanchez E, Reorda MS, Squillero G (2004) Automatic Test Program Generation: A Case Study. IEEE Design & Test of Computers, 21(2):102–109
92. Corporaal H (1998) Microprocessor Architectures: From VLIW to TTA. John Wiley & Sons, Chichester
93. CoWare (2007) Processor Designer. Available at http://www.coware.com/products/processordesigner.php. Verified 2007-01-25
94. Cramer T, Friedman R, Miller T, Seberger D, Wilson R, Wolczko M (1997) Compiling Java Just in Time. IEEE Micro, 17(2):36–43, May–June
95. Cronquist DC, Franklin P, Berg SG, Ebeling C (1998) Specifying and Compiling Applications for RaPiD. In Proc. IEEE Workshop on FPGAs for Custom Computing Machines (FCCM), pp 116–125
96. Day JD, Zimmermann H (1983) The OSI reference model. In Proceedings of the IEEE, 71:1334–134
97. DeHon A (2000) The Density Advantage of Reconfigurable Computing. IEEE Computer, 33(4):pp 41–49
98. DeHon A, Wawrzynek J (1999) Reconfigurable Computing: What, Why and Implications for Design Automation. In Proc. DAC, 1999, pp 610–615
99. DeHon A, Adams J, DeLorimier M, Kapre N, Matsuda Y, Naeimi H, Vanier M, Wrighton M (2004) Design Patterns for Reconfigurable Computing. In Proc. IEEE Symposium on FCCM, pp 13–23
100. De Micheli G (1999) Hardware Synthesis from C/C++ Models. In Proc. DATE
101. Dezan C, Jegot C, Pottier B, Gouyen C, Lagadec L (2006) The Case Study of Block Turbo Decoders on a Framework for Portable Synthesis on FPGA. In Proc. Hawaii International Conference on System Sciences
102. Dimitroulakos G, Galanis M, Goutis C (2005) Performance Improvements Using Coarse-grain Reconfigurable Logic in Embedded SOCs. In Proc. FPL, pp 630–635
103. Ditzel DR, Patterson DA (1980) Retrospective on High-Level Language Computer Architecture. In Proc. the 7th Annual Symposium on Computer Architecture. La Baule, France, pp 97–104
104. Ditzel DR, McLellan HR (1982) Register Allocation for Free: The C Machine Stack Cache. In Proc. the 1st International Symposium on Architectural Support for Programming Languages and Operating Systems. Palo Alto, California, pp 48–56
105. Ditzel DR, McLellan HR, Berenbaum AD (1987) Design Tradeoffs to Support the C Programming Language in the CRISP Microprocessor. In Proc. ASPLOS 1987 pp 158–163

106. Ditzel DR, McLellan HR, Berenbaum AD (1987) The Hardware Architecture of the CRISP Microprocessor. In Proc. the 14th Annual International Symposium on Computer Architecture. Pittsburgh, pp 309–319
107. Donath W (1981) Wire Length Distribution for Placements of Computer Logic. IBM Journal of Research and Development 25(3):152–155
108. Doran RW (1975) Architecture of Stack Machines. In: Chu Y (ed) (1975) High-Level Language Computer Architecture. Academic Press, New York, Chapter 4
109. Dunn D (2005) Sun Offers Servers Based on 8-Core Processors. In Information Week, December 6. Available at http://www.informationweek.com/
110. Earnest ED (1980) Twenty Years of Burroughs High-Level Language Machines. In Proc. of the International Workshop on High-Level Language Computer Architecture, pp 64–71
111. Eatherton W (2005) The Push of Network Processing to the Top of the Pyramid. Keynote Address, in Symposium on Architectures for Networking and Communications Systems (ANCS). Available at http://www.cesr.ncsu.edu/ancs/keynotes.html
112. Ebeling C, Cronquist D, Franklin P (1996) RaPiD – Reconfigurable PipeLined Datapath. In Proc. Field Programmable Logic and Applications FPL
113. Eble JC, De VK, Davis JA, Meindl JD (1996) Optimal Multilevel Interconnect Technologies for Gigascale Integration (GSI). In Proc. VMIC Conference, pp 40–45
114. Eble JC, De VK, Meindl JD (1995) A First Generation Generic System Simulator (GENESYS) and Its Relation to the NTRS. In Proc. Biennial University Government/Industry/University Microelectronics Symposium, pp 147–154
115. Edelstein DC, Sai-Halasz GA, Mii YJ (1995) VLSI On-Chip Interconnection Performance Simulations and Measurements. IBM Journal of Research and Development, 39(4):383–401
116. Edwards DA, Bardsley A (2002) Balsa: An Asynchronous Hardware Synthesis Language. The Computer Journal 45(1):12–18
117. EEMBC Benchmark Suite http://www.eembc.org/
118. Elbischger P (2001) Performance and Architectural Analysis of a Digital Signal Processor, MSc Thesis, Graz University of Technology, Austria
119. Elixent Ltd (2007) http://www.elixent.com
120. Endecott P (1996) SCALP: A Superscalar Asynchronous Low-Power Processor. PhD Thesis, University of Manchester
121. Espasa R, Ardanaz F, Emer J, Felix S, Gago J, Gramunt R, Hernandez I, Juan T, Lowney G, Mattina M, Seznec A (2002) Tarantula: A Vector Extension to the Alpha Architecture. In Proc. ISCA, pp 281–292
122. Estrin G (1960) Organization of Computer Systems: The Fixed-Plus Variable Structure Computer. In Proc. Westem Joint Computer Conference, Am. Inst. Electrical Engineers, New York, pp 33–40
123. Eysenck HJ (1991) Dimensions of personality: 16, 5, or 3? – Criteria for a Taxonomic Paradigm. Personality and Individual Differences 12:773–790

124. Fagin B, Patt Y, Sirni V, Despain A (1985) Compiling Prolog into Microcode: A Case Study Using the NCR/32-000. In Proc. the 18th IEEE Microprogramming Workshop
125. Farfeleder S, A, Steiner E, Brandner F (2006) Effective Compiler Generation by Architecture desacription. In Proc. the 2006 ACM SIGPLAN/SIGBED Conference on Languages, Compilers, and Tools for Embedded Systems (LCTES'06), ACM 2006, pp 145–152
126. Fisher JA, Faraboschi P, Young C (2005) Embedded Computing, A VLIW Approach to Architecture, Compilers, and Tools. Morgan Kaufmann Publishers for Elsevier, San Francisco
127. Flynn MJ (1980) Directions and Issues in Architecture and Language. IEEE Computer, October: 13(10):5–22
128. Flynn MJ (1995) Computer Architecture: Pipelined and Parallel Processor Design, Jones and Bartlett Publishers, Boston
129. Flynn MJ, Hung P, Rudd KW (1999) Deep-Submicron Microprocessor Design Issues. IEEE Micro, July–August: 19(4):11–22
130. Fort B, Capalija D, Vranesic ZG, Brown SD (2006) A Multithreaded Soft Processor for SoPC Area Reduction. In Proc. IEEE Symposium on Field-programmable Custom Computing Machines, pp 131–142
131. Franz G, Simar R (2004) Comparing Fixed and Floating-Point DSPs. TI white paper SPRY061, available at http://www.ti.com/etechsept04fltgptwhpaper (visited on January 2007)
132. Frigo J, Gokhale M, Lavenier D (2001) Evaluation of the Streams-C C-to-FPGA Compiler: An Applications Perspective. In Proc. FPGA, pp 134–140
133. Furber SB (2000) ARM System-on-Chip Architecture, Addison-Wesley
134. Furber SB, Garside JD, Riocreux P, Temple S, Day P, Liu J, Paver NC (1999) AMULET2e: An Asynchronous Embedded Controller. In Proc. IEEE, 87(2):243–256
135. Gaisler Research (2007) http://www.gaisler.com (visited on January 2007).
136. Gajski DD, Vahid F, Narayan S, Gong J (1994) Specification and Design of Embedded Systems. Prentice-Hall, Englewood Cliffs, NJ
137. Gajski DD, Zhu J, Dömer R, Gerstlauer A, Zhao S (2000) SpecC: Specification Language and Methodology. Kluwer Academic Publishers, Norwell, MA
138. Galanis M, Theodoridis G, Tragoudas S, Soudris D, Goutis C (2004) Mapping DSP Applications to a High-Performance Reconfigurable Coarse-Grain Data-Path. In Proc. FPL, pp 868–873
139. Gao G, Wong Y, Ning Q (1991) A Timed Petri-Net Model for Fine-Grain Loop Scheduling, Proc. ACM SIGPLAN Conference on Programming Language Design and Implementation, pp 204–218
140. Garside JD (1993) A CMOS VLSI Implementation of an Asynchronous ALU. In Proc. IFIP Working Conference on Asynchronous Design Methodologies, Manchester, England
141. Garside JD, Temple S, Mehra R (1996) The AMULET2e Cache System. In Proc. Async'96

142. Garside JD, Furber SB, Chung S-H (1999) AMULET3 Revealed. In Proc. Async'99, pp 51–59
143. Geannopoulos G, Dai X (1998) An Adaptive Digital Deskewing Circuit for Clock Distribution Networks ISSCC Digest of Technical Papers, pp 400–401
144. Gehringer EF, Colwell RP (1986) Fast object-oriented procedure calls: Lessons from the Intel 432. In Proc. the 13th Annual International Symposium on Computer Architecture. Tokyo, pp 92–101
145. Genutis M, Kazanavicius E (2001) Benchmarking in DSP. Kaunas University of Technology, ISSN 1392-2114 Ultragarsas, Nr. 2(39)
146. Gerlach J, Rosenstiel W (2000) System Level Design Using the SystemC Modeling Platform. In Proc. 3rd Workshop on System Design Automation (SDA2000), Rathen, Germany
147. Geurts W, Goossens G, Lanneer D, Van Praet J (2005) Design of Application-Specific Instruction-Set Processors for Multi-Media, Using a Retargetable Compilation Flow. In Proc. Signal Processing Conference (GSPx)
148. Geuskens B, Rose K (1998) Modeling Microprocessor Performance. Kluwer Academic Publishers
149. Ghenassia F (ed) (2005) Transaction Level Modeling with SystemC – TLM Concepts and Applications for Embedded Systems. Springer, Dordrecht, The Netherlands
150. Gilbert DA, Garside JD (1997) A Result Forwarding Mechanism for Asynchronous Pipelined Systems. In Proc. Async'97, pp 2–11
151. Gizopoulos D, Paschalis A, Zorian Y (1999) An Effective Built-in Self-test Scheme for Parallel Multipliers. IEEE Transactions on Computers, 48(9):36–950
152. Gizopoulos D, Paschalis A, Zorian Y (2004) Embedded Processor-Based Self-Test. Kluwer Academic Publishers, Frontiers in Electronic Testing series
153. Glaskowsky P (2000) Pentium 4 (Partially) Previewed. Microprocessor Report, August 28, 14(8):10–13
154. Glaskowsky P (2001) Athlon Edges Out Pentium 4. Microprocessor Report, January 8, 15(1)
155. Glossner J, Hokenek E, Mondgibl M (2002) An Overview of the Sandbridge Processor Technology. White Paper, Sandbridge Technology Inc.
156. Glossner J, Moudgill M, Iancu D, Nacer G, Jintukar S, Stanley S, Samori M, Raja T, Schulte M (2005) The Sandbridge Sandblaster Convergence Platform. Sandbridge Technologies Inc., White Plains, New York, USA
157. GNU project web pages at http://www.gnu.org/
158. Gokhale MB, Stone JM (1998) NAPA C: Compiling for a Hybrid RISC/FPGA Architecture. In Proc. IEEE Workshop on FPGAs for Custom Computing Machines (FCCM), pp 126–135
159. Goldstein SC, Schmit H, Budiu M, Cadambi S, Moe M, Taylor RR (2000) PipeRench: A Reconfigurable Architecture and Compiler. IEEE Computer, April, 33(4):70–77

160. Gordon MI, Thies W, Amarasinghe S (2006) Exploiting Coarse-Grained Task, Data, and Pipeline Parallelism in Stream Programs. In Proc. International Conference on Architectural Support for Programming Languages and Operating Systems, pp 151–162
161. Gordon MI, Thies W, Karczmarek M, Lin J, Meli AS, Lamb AA, Leger C, Wong J, Hoffmann H, Maze D, Amarasinghe S (2002) A Stream Compiler for Communication-Exposed Architectures. In Proc. ASPLOS, pp 291–303
162. Gries M, Keutzer K (eds) (2005) Building ASIPS: The MESCAL Methodology. Springer
163. Gronowski PE, Bowhill WJ, Preston RP, Gowan MK, Allmon RL (1998) High-performance Microprocessor Design, IEEE Journal of Solid-State Circuits, 33(5):676–686
164. Gross T, O'Hallaron DR (1998) iWarp, Anatomy of a Parallel Computing System. The MIT Press, Cambridge, MA
165. Gschwind M, Hofstee P, Flachs B, Hopkins M, Watanabe Y, Yamazaki T (2006) Synergistic Processing in CELL's Multicore Architecture. IEEE Micro, 26(2):10–24, March–April
166. Gupta TVK, Sharma P, Balakrishnan M, Malik S (2000) Processor Evaluation in an Embedded Systems Design Environment. In Proc. International Conference on VLSI Design, pp 98–103
167. Gupta S, Dutt N, Gupta R, Nicolau A (2003) SPARK: A High-Level Synthesis Framework for Applying Parallelizing Compiler Transformations. In Proc. International Conference on VLSI Design, pp 461–466
168. Gurumurthy S, Vasudevan S, Abraham JA (2005) Automated Mapping of Pre-Computed Module-Level Test Sequences to Processor Instructions. In Proc. IEEE International Test Conference, pp 294–303
169. Gurumurthy S, Vasudevan S, Abraham JA (2006) Automatic Generation of Instruction Sequences Targeting Hard-to-Detect Structural Faults in a Processor. In Proc. IEEE International Test Conference, paper 27.3
170. Gutnik V, Chandrakasan A (2000) Active GHz Clock Network Using Distributed PLLs. In ISSCC Digest of Technical Papers, pp 174–175
171. Gwennap L (1999) Coppermine Outruns Athlon. Microprocessor Report, October, 13(14):1–2
172. Halfhill T (1996) AMD K6 Takes On Intel P6. *BYTE*, January. URL http://www.byte.com/art/9601/sec7/art1.htm
173. Halfhill TR (2004) Tensilica Tackles Bottlenecks. Microprocessor Report, May 31, 18(5)
174. Hamada M, Hara H, Fujita T, Chen Kong The, Shimazawa T, Kawabe N, Kitahara T, Kikuchi Y, Nishikawa T, Takahashi M, Oowaki Y (2005) A Conditional Clocking Flip-Flop for Low Power H.264/MPEG-4 Audio/Visual Codec LSI. In Proc. Custom Integrated Circuits Conference, pp 527–530
175. Handshake Solutions (2007) company web page, consulted 31 August 2006 Avialable at http://www.handshakesolutions.com/
176. Handy J (1993) The Cache Memory Book. 2nd edn. Academic Press San Diego

177. Hardware/Software Codesign Group (2007) Polis: A Framework for Hardware-Software Co-Design of Embedded Systems. Available at http://embedded.eecs.berkeley.edu/research/hsc/. Verified 2007-01-24
178. Harju L (2006) Programmable Receiver Architectures for Multimode Mobile Terminals, DrTech Thesis, Tampere University of Technology. TUT Publication 604
179. Hartenstein R (1997) The Microprocessor is No More General Purpose, Invited Paper, Proc. the International Conference on Innovative Systems in Silicon, October
180. Hartenstein R (2001) A Decade of Reconfigurable Computing: A Visionary Retrospective. In Proc. DATE, pp 642–649
181. Hartenstein R, Hertz M, Hoffmann T, Nageldinger U (1998) Using the Kress Array for Reconfigurable Computing. In Proc. SPIE, pp 150–161
182. Hassitt A, Lageschulte JW, Lyon LE (1973) Implementation of a High Level Language Machine. Communications of the ACM 16:199–212
183. Hayes IJ, Jones CB (1990) Specifications Are Not (Necessarily) Executable. Technical Report 148, Key Centre for Software Technology, Department of Computer Science, University of Queensland, Australia
184. Hellestrand G (2005) The Engineering of Supersystems. IEEE Computer, 38(1):103–105
185. Hennessy JL, Jouppi N, Baskett F, Gill J (1981) MIPS: A VLSI Processor Architecture. Technical Report No. 223, Computer Systems Laboratory, Stanford University
186. Hennessy JL, Gross TR (1982) Code Generation and Reorganization in the Presence of Pipeline Constraints. In Proc. the 9th Annual Symposium on Principles of Programming Languages, pp 120–127
187. Hennessy JL, Patterson DA (1990) Computer Architecture; A Quantitative Approach, Morgan Kaufmann Publishers, San Francisco
188. Hennessy JL, Patterson DA (2003) Computer Architecture: A Quantitative Approach. 3rd edn. Elsevier Morgan Kaufmann, San Francisco
189. Henriksson T (2003) Intra-Packet Data-Flow Protocol Processor. PhD Thesis, Department of Electrical Engineering, Linköping University, Sweden
190. Henriksson T, Nordqvist U, Liu D (2000) Specification of a Configurable General-Purpose Protocol Processor. In Proc. Intl Symp. on Communication Systems, Networks and Digital Signal Processing (CSNDSP), pp 284–289
191. Hetherington G, Fryars T, Tamarapalli N, Kassab M, Hassan A, Rajski J (1992) Logic BIST for Large Industrial Designs: Real Issues and Case Studies. In Proc. IEEE International Test Conference, pp 358–367
192. Heuring VP, Jordan HF (1997) Computer Systems Design and Architecture, Addison-Wesley, California
193. Heysters P, Rauwerda GK, Smit LT (2007) A Flexible, Low Power, High Performance DSP IP Core for Programmable Systems-on-Chip. Design & Reuse Industry Articles, available at http://www.us.design-reuse.com/articles/article12159.html (visited on January 2007)

194. Hirnschrott U (2005) Software and Compiler Optimization Techniques for Reducing Energy Consumption of Embedded DSP Cores. Ph.D. Thesis, Vienna University of Technology
195. Hirnschrott U, Krall A (2003) VLIW Operation Refinement for Reducing Energy Consumption. In Proc. 2003 International Symposium on System-on-Chip (SOC'03), Tampere, Finland, pp 131–134
196. Ho R, Mai KW, Horowitz MA (2001) The Future of Wires. Proceedings of the IEEE, 89(4):490–504
197. Hoffmann A, Kogel T, Nohl A, Braun G, Schliebusch O, Wahlen O, Wieferink A, Meyr H (2001) A Novel Methodology for the Design of Application-Specific Instruction-Set Processors (ASIPs) Using a Machine Description Language. IEEE Transactions on Computer-Aided Design of Integrated Circuits and Systems, 20(11):1338–1354
198. Hoffmann H, Strumpen V, Agarwal A (2003) Stream Algorithms and Architecture. Technical Memo MIT-LCS-TM-636, LCS, MIT
199. Hofstee HP (2005) Power Efficient Processor Architecture and the Cell Processor. Invited paper in Proc. HPCA-11, pp 258–262
200. Hormdee D (2002) Copy-Back Cache Organisation for an Asynchronous Microprocessor. PhD Thesis, University of Manchester
201. Hosler MM, Hauck S, Fry TW, Kao JP (1997) The Chimaera Reconfigurable Functional Unit. In Proc. the 5th IEEE Symposium on Field-Programmable Custom Computing, pp 87–96
202. Hsi CG, Tucker SG (1990) Figures of Merit for System Path Time Estimation. In Proc. ICCD, pp 49–55
203. IEEE (2005) Std 1364-2005, Verilog Hardware Description Language
204. IEEE (2005) Std 1666-2005, SystemC Language Reference Manual
205. IEEE (2005) Std 1800-2005, System Verilog: Unified Hardware Design, Specification and Verification Language
206. IEEE P1800 (2007) System Verilog Technical Committee Page. Available at http://www.eda.org/sv/. Verified 2007-01-24
207. Inria (2007) About Esterel Language. Available at http://www-sop.inria.fr/esterel.org/Html/About/AboutEsterel.htm. Verified 2007-01-24
208. International Technology Roadmap for Semiconductors (2005) ITRS. Available at http://www.itrs.net/
209. IPFlex (2007) www.ipflex.com
210. Itoh M, Higaki S, Sato J, Shiomi A, Takeuchi Y, Kitajima A, Imai M (2000) PEAS-III: An ASIP Design Environment. In Proc. International Conference on Computer Design, pp 430–436
211. Jacome MF, de Veciana G (2000) Design Challenges for New Application-Specific Processors. IEEE Design & Test of Computers, 17(2):40–50
212. Jantsch A, Öberg J, Hemani A (1998) Is There a Niche for a General Protocol Processor Core? In Proc. Norchip Conference, pp 93–100
213. Jeong CH, Park WC, Han TD, Kim SD (1999) Cost/Performance Trade-off in Floating-Point Unit Design for 3D Geometry Processor. In Proc. AP-ASIC, pp 104–107

214. Jeong CH, Park WC, Kim SW, Han TD (2000) The Design and Implementation of CalmlRISC32 Floating-Point Unit. In Proc. AP-ASIC, pp 327–330
215. Johnson RC (2007) Intel's Teraflops Chip Uses Mesh Architecture to Emulate Mainframe. EETimes, February 12, 2007. Available at http://www.eetimes.com/news/latest/showArticle.jhtml?articleID=197004697
216. Kambe K, Inoue M, Fujiwara H (2004) Efficient Template Generation for Instruction-Based Self-Test of Processor Cores. In Proc. IEEE Asian Test Symposium, pp 152–157
217. Kane G (1989) MIPS RISC Architecture, Prentice Hall
218. Kanellos M (2006) Intel shows off its quad core, in ZDnet, February 10 Available at http://news.zdnet.com/
219. Kapasi UJ, Dally WJ, Rixner S, Owens JD, Khailany B (2002) The Imagine Stream Processor. In Proc. ICCD, pp 282–288
220. Kastrup B, van Wel A (2003) Moustique: Smaller Than an ASIC and Fully Programmable. In Proc. International Symposium on SoC, p 1
221. Kathail V, Aditya S, Schreiber R, Rau BR, Cronquist DC, Sivaraman M (2002) PICO: Automatically Designing Custom Computers. IEEE Computer, 35(9):39–47
222. Kawaguchi H, Sakurai T (1998) A Reduced Clock-Swing Flip-Flop (RCSFF) for 63% Power Reduction. IEEE Journal of Solid-State Circuits, 33(5):807–811
223. Keating M (2002) Reuse Methodology Manual for System-on-a-Chip Designs, Kluwer Academic Publishers
224. Kessels J, Peeters A, Kramer T, Timm V (2001) Descale. In Furber S, Sparsø J (eds) Principles of Asynchronous Circuit Design: A Systems Perspective Kluwer, pp 221–248
225. Keutzer K, Malik S, Newton AR, Rabaey JM, Sangiovanni-Vincentelli A (2000) System-Level Design: Orthogonalization of Concerns and Platform-Based Design. IEEE Transactions on Computer-Aided Design of Integrated Circuits, 19(12):1523–1543
226. Keutzer K, Shah N (2007) A Survey of Programmable Platforms – Network Procs. On line at http://www.eecs.berkeley.edu/~mihal/ee244/lectures/soc-prog-plat-np.pdf. Last checked 24-01-2007
227. Kiatisevi P, Azuara L, Dorsch R, Wunderlich HJ (2005) Development of an Audio Player as System-on-a-Chip Using an Open Source Platform. In Proc. International Symposium on Circuits and Systems (ISCAS), Vol 3, pp 2935–2938
228. Kienhuis B, Deprettere E, Vissers K, Van Der Wolf P (1997) An Approach for Quantitative Analysis of Application Specific Data-Flow Architectures. In Proc. Application-Specific Systems, Architectures and Processors (ASAP), pp 338–349
229. Kienhuis B, Deprettere E, Van Der Wolf P, Vissers K (2002) A Methodology to Design Programmable Embedded Systems – The Y-Chart Approach. Lecture Notes in Computer Science, January, 2268:18–37

230. Kim HS, Smith JE (2002) An ISA and Microarchitecture for Instruction Level Distributed Processing. In Proc. ISCA, pp 71–81
231. Kim JS, Taylor MB, Miller J, Wentzlatff D (2003) Energy Characterization of a Tiled Architecture Processor with On-Chip Networks. In Proc. ISLPED, pp 424–427
232. Klein M (2005) The Virtex-4 Power Play. XCELL Journal, Xilinx, Issue 52, Spring
233. KleinOsowski AJ, Lilja DJ (2006) MinneSPEC: A New SPEC Benchmark Workload for Simulation-Based Computer Architecture Research. Computer Architecture Letters, 1(1):7
234. Kojima H, Tanaka S, Sasaki K (1995) Half-Swing Clocking Scheme for 75% Power Saving in Clocking Circuitry. IEEE Journal of Solid-State Circuits, 30(4):432–435
235. Kongetira P, Aingaran K, Olukotun K (2005) Niagara: A 32-Way Multithreaded Sparc Processor. IEEE Micro, 25(2):21–29
236. Koopman PJ Jr (1989) Stack Computers: The New Wave. Ellis Horwood, Chichester, England
237. Kozyrakis CE, Patterson DA (1998) New Direction for Computer Architecture Research. IEEE Computer, 31(11):24–32
238. Krall A, Hirnschrott U, Panis C, Pryanishnikov I (2004) xDSPcore – A Compiler Based Configurable Digital Signal Processor. In IEEE Micro, Special Issue on Embedded Systems: Architecture, Design and Tools, July/August, vol 24, No.4, pp 67–78
239. Kranitis N, Paschalis A, Gizopoulos D, Zorian Y (2003) Instruction-Based Self-Testing of Processor Cores. Journal of Electronic Testing: Theory and Applications, No 19, pp 103–112, (Special Issue on 20th IEEE VLSI Test Symposium 2002)
240. Kranitis N, Paschalis A, Gizopoulos D, Xenoulis G (2005) Software-Based Self-Testing of Embedded Processors. IEEE Transactions on Computers, 54(4):461–475
241. Kranitis N, Merentitis A, Laoutaris N, Theodorou G, Paschalis A, Gizopoulos D, Halatsis C (2006) Optimal Periodic Testing of Intermittent Faults In Embedded Pipelined Processor Applications. In Proc. DATE, pp 65–70
242. Krashinsky R, Batten C, Hampton M, Gerding S, Pharris B, Casper J, Asanovic K (2004) The Vector-Thread Architecture. In Proc, ISCA, pp 52–63
243. Krewell K (2004) Intel Cancels 4GHz P4. Microprocessor Report, November 1, 18(11)
244. Krstic A, Chen L, Lai WC, Cheng KT, Dey S (2002) Embedded Software-Based Self-Test for Programmable Core-Based Designs. IEEE Design and Test of Computers, 19(4):18–26
245. Kubiatowicz J (1998) Integrated Shared-Memory and Message-Passing Communication in the Alewife Multiprocessor. PhD Thesis, MIT
246. Kuo S, Gan WS (2005) Digital Signal Processors. Pearson

247. Kurd NA, Barkarullah JS, Dizon RO, Fletcher TD, Madland PD (2001) A Multigigahertz Clocking Scheme for the Pentium® 4 Microprocessor, IEEE Journal of Solid-State Circuits, 36(11):1647–1653
248. Kylliäinen J, Nurmi J, Kuulusa M (2003) COFFEE – A Core for Free. In Proc. International Symposium on SoC, pp 17–22
249. Lafond S, Lilius J (2004) An Energy Consumption Model for Java Virtual Machine, TUCS Technical Report, No. 597
250. Lai WC, Krstic A, Cheng KT (2000) Functionally Testable Path Delay Faults on a Microprocessor. IEEE Design and Test of Computers, 17(6):6–14
251. Landman BS, Russo RL (1971) On a Pin Versus Block Relationship for Partitions of Logic Graphs. IEEE Transactions on Computers, C20(12): 1469–1479
252. Lapsley P, Bier J, Shoham A, Lee EA (1997) DSP Processor Fundamentals, Architectures and Features. IEEE Press, New York
253. La Rosa A, Passerone C, Lavagno L (2002) A Software Development Toolchain for a Reconfigurable Processor. In Proc. the Conference on Compilers, Architectures and Synthesis for Embedded Systems (CASES)
254. La Rosa A, Lavagno L, Lazarescu M, Passerone C (2006) An Optimizing C Front-End for Hardware Synthesis. In Proc. Workshop on Wireless Reconfigurable Terminals and Platforms (WiRTeP)
255. Lattice (2007) company web page http://www.latticesemi.com
256. Lattice (2007) Mico32. Available at http://www.latticesemi.com/products/intellectualproperty/ipcores/mico32
257. Lee W, Barua R, Frank M, Srikrishna D, Babb J, Sarkar V, Amarashinghe S (1998) Space-Time Scheduling of Instruction-Level Parallelism on a Raw Machine. In Proc. ASPLOS, pp 46–54
258. Lee W, Puppin D, Swenson S, Amarasinghe S (2002) Convergent Scheduling. In Proc. MICRO-35, pp 111–122
259. Leibson S (2004) Lower SoC operating frequencies to cut power dissipation. In Portable Design, February
260. Leibson S (2007) The HP 9100 Project: An Exothermic Reaction. URL http://www.hp9825.com/html/the_9100_part_2.html
261. Leibson S, Kim J (2005) Configurable Processors: A New Era in Chip Design. IEEE Computer, July, pp 51–59
262. Lenoski D, Laudon J, Gharachorloo K, Weber WD, Gupta A, Hennessy J, Horowitz M, Lam MS (1992) The Stanford DASH Multiprocessor. IEEE Computer, 25(3):63–79
263. Lewis DM et al. (2003) The Stratix™ Routing and Logic Architecture. International Symposium on Field-Programmable Gate Arrays, pp 12–20
264. Li Y, Callahan T, Darnell E, Harr R, Kurkure U, Stockwood J (2000) Hardware–Software Co-Design of Embedded Reconfigurable Architectures. In Proc. DAC, pp 507–512
265. Liao SY (2000) Towards a New Standard for System-Level Design. In Proc. 8th International Workshop on Hardware/Software Codesign

266. Liljeberg P, Plosila J, Isoaho J (2004) Self-Timed Communication Platform for Implementing High-Performance Systems-on-Chip. Integration. The VLSI Journal 38(1):43–67
267. Lindholm T, Yellin F (1997) The Java Virtual Machine Specification, 2nd edn. Addison-Wesley
268. Lines A (2007) The Vortex: An Asynchronous Superscalar Processor. In Proc. Async'07
269. Lipasti MH, Shen JP (1997) Superspeculative Microarchitecture for Beyond AD 2000. IEEE Computer, September, 30(9):59–66
270. Lodi A, Campi F, Toma M, Cappelli A, Canegallo R, Guerrieri R (2003) A VLIW Processor with Reconfigurable Instruction Set for Embedded Applications. IEEE Journal of Solid-State Circuits (JSSC), 38(11):1876–1886
271. Lodi A, Cappelli A, Bocchi M, Mucci C, Innocenti M, De Bartolomeis C, Ciccarelli L, Giansante R, Deledda A, FCampi F, Toma M, Guerrieri R (2006) XiSystem: A XiRisc-based SoC with a Reconfigurable I/O Module. IEEE Journal of Solid-State Circuits (JSSC), 41(1):85–96
272. Lodi A, Mucci C, Bocchi M, Cappelli A, De Dominicis M, Ciccarelli L (2006) A Multi-Context Pipelined Array for Embedded System. In Proc. FPL, pp 581–588
273. Mai K, Paaske T, Jayasena N, Ho R, Dally W, Horowitz M (2000) Smart Memories: A Modular Reconfigurable Architecture. In Proc. ISCA, pp 161–171
274. Mangione-Smith WH, Hutchings B, Andrews D, DeHon A, Ebeling C, Hartenstein R, Mencer O, Morris J, Palem K, Prasanna VK, Spaanenburg HAE (1997) Seeking Solutions in Configurable Computing. IEEE Computer, 30(12):38–43
275. Martin AJ, Burns SM, Lee TK, Borkovic D, Hazewindus PJ (1989) The Design of an Asynchronous Microprocessor. ARVLSI: Decennial Caltech Conference on VLSI, ed. Seitz CL, pp 351–373, MIT Press
276. Matsumoto C (2002) Net Processors Face Programming Trade-Offs. EETimes, iApplianceWeb, September. Available at http://www.iapplianceweb.com/story/OEG20020922S0003.htm. Last checked 24-01-2007
277. McCalpin JD (2007) STREAM: Sustainable Memory Bandwidth in High Performance Computers. Available at http://www.cs.virginia.edu/stream. Last checked 12-03-2007
278. McClellan J, Schafer R, Yoder M (1998) DSP First, A Multimedia Approach. Prentice Hall
279. McCluskey EJ, Tseng CW (2000) Stuck-Fault Tests vs. Actual Defects. In Proc. IEEE International Test Conference, pp 336–343
280. McFarling S (1993) Combining Branch Predictors. DEC WRL TN-36, June 1993. Available at http://citeseer.ist.psu.edu/mcfarling93combining.html
281. McLaughlin L (2006) 1,000 Cores on a Chip: Rapport's Kilocore Chip Makes Quick Work of Video Processing. In Technology Review, Massachusetts Institute of Technology, July 11. Available at http: //www.technologyreview.com/

282. Mei B, Vernalde S, Verkest D, DeMan H, Lauwereins R (2002) DRESC: A Retargetable Compiler for Coarse-Grained Reconfigurable Architecture. In Proc. IEEE International Conference on Field-Programmable Technology
283. Mei B, Vernalde S, Verkest D, DeMan H, Lauwereins R (2003) ADRES: An Architecture with Tightly Coupled VLIW Processor and Corse-Grained Reconfigurable Matrix. In Proc. FPL, pp 61–70
284. Mei B, Veredas F, Masschelein B (2005) Mapping an H.264 Decoder onto the ADRES Reconfigurable Architecture. In Proc. FPL, pp 622–625
285. Metzgen P (2004) Optimizing a High-Performance 32-bit Processor for Programmable Logic. Invited presentation in International Symposium on System-on-Chip. Available at http://www.cs.tut.fi/soc/Metzgen04.pdf
286. Meyer J, Downing T (1997) Java Virtual Machine. O'Reilly and Associates, Inc.
287. Mihal A, Keutzer K (2003) Mapping Concurrent Applications onto Architectural Platforms. In Jantsch A, Tenhunen H (eds) Networks on Chip, Kluwer Academic Publishers Chapter 3, pp 39–59
288. Mii Y (1992) Performance Considerations for the Scaling of Sub-Micron On-Chip Interconnections. In Proc. SPIE 1805, pp 332–341
289. miniMIPS CPU (2007) www.opencores.org/projects/minimips
290. Mirsky E, DeHon A (1996) MATRIX: A Reconfigurable Computing Architecture with Configurable Instruction and Deployable Resources. In Proc. IEEE Symposium on FPGAs for Custom Computing Machines (FCCM'96), pp 157–166
291. Miyamori T, Olukotun U (1998) REMARC: Reconfigurable Multimedia Array Coprocessor. In Proc. ACM/SIGDA International Symposium on FPGAs for Custom Computing Machines (FCCM), pp 2–11
292. Mizuno H, Ishibashi K (1998) A Noise-Immune GHz-Clock Distribution Scheme Using Synchronous Distributed Oscillators. In ISSCC Digest of Technical Papers, pp 404–405
293. Moore GE (1965) Cramming More Components onto Integrated Circuits. Electronics, April 19, pp 114–117
294. Moretti G (2004) The Search for the Perfect Language. EDN. Available at http://www.edn.com/article/CA376625.html. Verified 2006-08-18
295. MorphICs Technology inc. (2007) http://www.morphics.com
296. Morpho Technologies (2007) http://www.morphotech.com
297. Motorola (1995) DSP56000 – 24 Bit Digital Signal Processor Family Manual, DSP56KFAMUM/AD, Motorola Inc., Austin, Texas
298. Moulton P (1974) Microprogrammed Subprocessors for Compilation and Execution of High-Level Languages. In 7th Annual Workshop on Microprogramming. Palo Alto, California, pp 74–79
299. Mucci C, Chiesa C, Lodi A, Toma M, Campi F (2003) A C-Based Algorithm Development Flow for a Reconfigurable Processor Architecture. In Proc. International Symposium on System-on-Chip, pp 69–73
300. Mucci C, Campi F, Deledda A, Fazzi A, Ferri M, Bocchi M (2005) A Cycle-Accurate ISS for a Dynamically Reconfigurable Processor Architecture. In

Proc. 9th IEEE International Parallel and Distributed Processing Symposium (IPDPS), Vol 4

301. Mucci C, Bocchi M, Gagliardi P, Cappelli A, Innocenti M, Giansante R, Lodi A, Toma M, Campi F, Guerrieri R (2006) A C-Based Programming Environment for a Heterogeneous Reconfigurable System. In Proc. Workshop on Wireless Reconfigurable Terminals and Platforms (WiRTeP)
302. Mukund M (1992) Petri Nets and Step Transition Systems. International Journal of Foundations of Computer Science, 3(4):443–478
303. Naffziger SD, Hammond G (2002) The Implementation of the Next-Generation 64b Itanium Microprocessor. In ISSCC Digest of Technical Papers, pp 344–345, 472
304. Nagarajan R, Sankaralingam K, Burger D, Keckler SW (2001) A Design Space Evaluation of Grid Processor Architectures. In Proc. MICRO-34, pp 40–51
305. Najjar WA, Bohm W, Draper BA, Hammes J, Rinker R, Beveridge JR, Chawathe M, Ross C (2003) High-Level Language Abstraction for Reconfigurable Computing. IEEE Computer, 36(8):63–69
306. Nanya T, Ueno Y, Kagotani H, Kuwako M, Takamura A (1994) TITAC: Design of a Quasi-Delay-Insensitive Microprocessor. IEEE Design & Test of Computers, 11(2):50–63
307. NEC DRP Product Family (2007) http://www.necel.com/drp/en/index.html
308. Nedovic N, Walker WW, Oklobdzija VG, Aleksic M (2002) A Low Power Symmetrically Pulsed Dual Edge-Triggered Flip-Flop. In Proc. ESSCIRC, pp 399–402
309. Noakes M (1993) The J-Machine Multicomputer: An Architectural Evaluation. In Proc. ISCA, pp 224–235
310. Nurmi J, Eerola V, Ofner E, Gierlinger A, Jernej J, Karema T, Raitaaho T (1994) A DSP Core for Speech Coding Applications. In Proc. ICASSP, April, Vol 2, pp 429–432
311. Nurmi J, Tenhunen H, Isoaho J, Jantsch A (eds) (2004) Interconnect-Centric Design for Advanced SoC and NoC. Kluwer Academic Publishers, Amsterdam
312. Nurmi T, Virtanen S, Isoaho J, Tenhunen H (2000) Physical Modeling and System Level Performance Characterization of a Protocol Processor Architecture. In Proc. Norchip Conference, pages 294–301
313. Nurmi T, Liu J, Pamunuwa D, Ahonen T, Zheng LR, Isoaho J, Tenhunen H (2004) Global Interconnect Analysis. In Nurmi J, Tenhunen H, Isoaho J, Jantsch A (eds) Interconnect-Centric Design for Advanced SoC and NoC, Kluwer Academic Publishers, Chapter 3, pp 55–84
314. Nuzman D, Rosen I. Zaks A (2006) Auto-Vectorization of Interleaved Data for SIMD. In Proc. ACM SIGPLAN Conference on Programming Language Design and Implementation (PLDI), pp 132–143
315. O'Mahony F, Yue CP, Horowitz M, Wong SS (2003) 10GHz Clock Distribution Using Coupled Standing-Wave Oscillators. In ISSCC Digest of Technical Papers, pp 428–429

316. OMG (2001) The Unified Modeling Language Specification, Version 1.4, September. Document formal/05-04-01, Available at http://www.omg.org/docs/formal/05-04-01.pdf
317. OpenRISC CPU (2007) http://www.bsemi.com
318. Organick EI (1973) Computer Systems Organization: The B5700/B6700 Series. Academic Press, New York
319. Organick EI, Hinds JA (1977) Architecture and Programming of the B1700/B1800 Series. North-Holland, New York
320. Pact XPP Technologies (2007) http//:www.pactxpp.com
321. Palacharla S (1998) Complexity-Effective Superscalar Processors. PhD Thesis, University of Wisconsin-Madison
322. Pangrle B, Kapoor S (2004) Managing Leakage Power at 90 nm and Below. EE Times/EEdesign, November. Available at http://www.eedesign.com/article/showArticle.jhtml?articleId=51000474
323. Panis C (2004) Scalable DSP Core Architecture Addressing Compiler Requirements. PhD Thesis, Tampere University of Technology, Finland
324. Panis C, Hirnschrott U, Krall A, Laure G, Lazian W, Nurmi J (2004) FSEL – Selective Predicated Execution for a Configurable DSP Core. Proc. IEEE Annual Symposium on VLSI (ISVLSI-04), Lafayette, Louisiana, USA, pp 317–320
325. Panis C, Grünbacher H, Nurmi J (2004) A Scaleable Instruction Buffer and Align Unit for xDSPcore. IEEE Journal of Solid-State Circuits, July, 35(7): 1094–1100
326. Parvathala P, Maneparambil K, Lindsay W (2002) FRITS – A Microprocessor Functional BIST Method. In Proc. IEEE International Test Conference, pp 590–598
327. Paschalis A, Gizopoulos D, Kranitis N, Psarakis M, Zorian Y (1999) An Effective BIST Architecture for Fast Multiplier Cores. In Proc. IEEE DATE, pp 117–121
328. Paschalis A, Gizopoulos D, Kranitis N, Psarakis M, Zorian Y (2001) Deterministic Software-Based Self-Testing of Embedded Processor Cores. In Proc. DATE, pp 92–96
329. Paschalis A, Gizopoulos D (2005) Effective Software-Based Self-Test Strategies for On-Line Periodic Testing of Embedded Processors. IEEE Transactions on Computer-Aided Design of Integrated Circuits and Systems, 24(1):88–99
330. Bardeen J, Brattain WH (1948) Three-electrode circuit element utilizing semiconductive materials. US Patent 2,524,035
331. Kilby JS (1959) Miniaturized electronic circuits. US Patent 3,138,743
332. Hoff ME, Mazor S, Faggin F (1973) Memory system for a multi-chip digital computer. US Patent 3,821,715
333. Wiggins RH, Brantingham GL (1978) Speech synthesis integrated circuit device. US Patent 4,209,836
334. Patterson DA, Ditzel DR (1980) The Case for the Reduced Instruction Set Computer. Computer Architecture News, 8 October: 25–33
335. Patterson DA, Hennessy JL (1985) The Open Channel: Response to 'Computers, Complexity, and Controversy.' IEEE Computer November, 18(11):142–143

336. Patterson DA, Hennessy JL (2004) Computer Organization and Design: The Hardware/Software Interface, 3rd edn. Morgan Kaufmann Publishers, San Francisco
337. Plana LA, Riocreux PA, Bainbridge WJ, Bardsley A, Garside JD, Temple S (2002) SPA – A Synthesisable Amulet Core for Smartcard Applications. Proc. Async'02, pp 201–210
338. Plasma CPU Model (2007) http://www.opencores.org/projects/mips
339. Pnevmatikatos DN, Sohi GS (1994) Guarded Execution and Branch Prediction in Dynamic ILP Processors. Proc. the 21st Annual International Symposium on Computer Architecture (ISCA), pp 120–129
340. PowerPC FPU (2007) http://www.xilinx.com/xlnx/xebiz/designResources/ip_product_details.jsp?key=DO-DI-FPU-SP
341. Psarakis M, Gizopoulos D, Hatzimihail M, Paschalis A, Raghunathan A, Ravi S (2006) Systematic Software-Based Self-Test for Pipelined Processors. In Proc. DAC, pp 393–398
342. Rabaey JM (2000) Silicon Platforms for the Next Generation Wireless systems – What Role Does Reconfigurable Hardware Play? Proc. Field Programmable Logic and Applications Conference (FPL), pp 277–285
343. Rabaey JM (1997) Reconfigurable Computing: The solution to Low Power Programmable DSP. In Proc. ICASSP, pp 275–278
344. Rabaey JM, Chandrakasan A, Nikolic B (2003) Digital Integrated Circuits – a Design Perspective. Prentice Hall/Pearson Education International, Upper Saddle River, NJ, USA., 2nd edn.
345. Radetzki M, Nebel W (1999) Synthesizing Hardware from Object-Oriented Descriptions. In Proc. Forum on Design Languages (FDL)
346. Rau BR (1995) Iterative Modulo Scheduling. Technical Report HPL-94-115, Hewlett Packard, November
347. Raza Microelectronics, Inc. (2007) http://www.razamicroelectronics.com/products/xlr.htm
348. Razdan R, Smith M (1994) A High Performance Microarchitecture with Hardware-Programmable Functional Units. Proc. Microarchitecture (MICRO-27), pp 172–180
349. Restle PJ, Carter CA, Eckhardt JP, Krauter BL, McCredie BD, Jenkins KA, Weger AJ (2002) The Clock Distribution of the Power4 Microprocessor. In ISSCC Digest of Technical Papers, pp 144–145
350. Reuters Press Release (1989) Unisys Introduces Micro A Computer. January 19 URL: http://query.nytimes.com/gst/fullpage.html?res=950DE2DE113AF93AA25752C0A96F948260
351. Ristimäki T (2005) Reconfigurable IP Blocks: a MIMD Approach. DrTech Thesis, Tampere University of Technology. TUT Publication 573
352. Ristimäki T, Nurmi J (2003) Reprogrammable Algorithm Accelerator IP Block. In Proc. IFIP VLSI-SOC, pp 228–232
353. Rizk H, Papachristou C, Wolff F (2006) A Self Test Program Design Technique for Embedded DSP Cores. Journal of Electronic Testing: Theory and Applications, 22(1):71–87

354. Rivaton A, Quevremont J, Zhang Q, Wolkotte P, Smit G (2005) Implementing Non Power-of-Two FFTs on Coarse-Grain Reconfigurable Architectures. In Proc. Int'l Symposium on SOC, pp 74–77
355. Rixner S, Dally WJ, Kapasi UJ, Khailany B, López-Lagunas A, Mattson PR, Owens JD (1998) A Bandwidth-Efficient Architecture for Media Processing. In Proc. MICRO-31, pp 3–13
356. Rowen C, Leibson S (2004) Engineering the Complex SOC. Prentice-Hall
357. Rudd KW (1999) VLIW Processors: Efficient Exploiting Instruction Level Parallelism. PhD Thesis, Stanford University
358. Rusu S, Tam S, Muljono H, Ayers D, Chang J (2007) A 65 nm Dual-Core Multi-Threaded Xeon® Processor with 16MB L3 Cache. IEEE Journal of Solid-State Circuits, 42(1):17–25
359. Sai-Halasz G (1992) Directions in Future High-End Processors. In Proc. ICCD, pp 230–233
360. Sai-Halasz G (1995) Performance Trends in High-End Processors. Proceedings of the IEEE, 83(1):20
361. Sanchez J, Gonzales A (2000) Modulo Scheduling for a Fully-Distributed Clustered VLIW Architecture. In Proc. MICRO-33, pp 124–133
362. Sanchez E, Sonza Reorda M, Squillero G (2005) On the Transformation of Manufacturing Test Sets into On-Line Test Sets for Microprocessors. In Proc. IEEE International Symposium on Defect and Fault Tolerance in VLSI Systems, pp 494–502
363. Sangiovanni-Vincentelli A (2002) Defining Platform-Based Design. EEDesign of EETimes, February
364. Sato T, Watanabe H, Shiba K (2005) Implementation of Dynamically Reconfigurable Processor DAPDNA-2. In Proc. IEEE VLSI-TSA International Symposium VLSI Design, Automation and Test, pp 323–324
365. Scholz B, Eckstein E (2002) Register Allocation for Irregular Architectures. In Proc. Joint Conference on Languages, Compilers and Tools for Embedded Systems. ACM Press, pp 139–148
366. Schumacher G, Nebel W (1995) Inheritance Concept for Signals in Object-Oriented Extensions to VHDL. In Proc. EURO-DAC Design Automation Conference with EURO-VHDL'95, pp 428–435
367. Schumacher G, Nebel W (1998) Object-Oriented Modelling of Parallel Hardware Systems. In Proc. DATE, pp 234–241
368. Schutten R, Bergeron J (2007) Transaction-Level Modeling: SystemC on/or SystemVerilog. Synopsys, Inc. article at SOCcentral http:// www.soccentral.com/results.asp?CategoryID=488&EntryID=18241. Verified 2007-01-25
369. Schöberl M (2005) JOP: A Java Optimized Processor for Embedded Real-Time Systems. PhD Thesis, Vienna University of Technology, Austria
370. SDR forum (2007) web page at http://www.sdrforum.org/
371. Semeria L, Sato K, De Micheli G (2001) Synthesis of Hardware Models in C with Pointers and Complex Data Structures. IEEE Transactions on VLSI Systems, 9(6):743–756

372. Shah N, Plishker W, Ravindran K, Keutzer K (2004) NPClick: A Productive Software Development Approach for Network Processors. IEEE Micro, 24(5):45–54

373. Shen J, Abraham JA (1998) Native Mode Functional Test Generation for Microprocessors with Applications to Self-Test and Design Validation. In Proc. IEEE International Test Conference, pp 990–999

374. Shen JP, Lipasti MH (2005) Modern Processor Design, Fundamentals of Superscalar Processors. McGraw-Hill, New York

375. Shoemaker D, Honoré F, Metcalf C, Ward S (1996) NuMesh: An Architecture Optimized for Scheduled Communication. Journal of Supercomputing, 10(3): 285–302

376. Silc J, Robic B, Ungerer T (1999) Computer Architecture: From Dataflow to Superscalar and Beyond. Springer-Verlag, Berlin

377. Sima D, Fountain T, Karsuk P (1997) Advanced Computer Architectures: A Design Space Approach. Addison-Wesley, Boston

378. Singh A, Srinivasan S (2004), Digital Signal Processing. Thomson

379. Singh H, Lee MH, Lu G, Kurdahi FG, Bagherzadeh N, Lang T, Heaton R, Filho EMC (1998) MorphoSys: An Integrated Re-Configurable Architecture. In Proc. NATO Symposium on System Concepts and Integration, Monterey, CA, April

380. Singh H, Lee MH, Lu G, Kurdahi FJ, Bagherzadeh N, Chaves Filho EM (2000) MorphoSys: An Integrated Reconfigurable System for Data-Parallel and Computation-Intensive Applications. IEEE Transactions on Computers, 49(5):465–481

381. Singh V, Inoue M, Saluja KK, Fujiwara H (2006) Instruction-Based Self-Testing of Delay Faults in Pipelined Processors. IEEE Transactions on Very Large Scale Integration (VLSI) Systems, 14(11):1203–1215

382. Smit G, Heysters P, Rosien M, Molenkamp B (2004) Lessons Learned from Designing the Montium, a Coarse Grained Reconfigurable Processing Tile. In Proc. International Symposium on SoC, pp 29–32

383. Smotherman M (2002) Understanding EPIC Architectures and Implementations. In 40th Annual Southeast ACM Conference

384. Snider G (2002) Performance-Constrained Pipelining of Software Loops onto Reconfigurable Hardware. In Proc. FPGA, pp 177–186

385. Sohi GS, Breach SE, Vijaykumar TN (1995) Multiscalar Processors. In Proc. ISCA, pp 414–425

386. Sparsø J, Furber S (2001) Principles of Asynchronous Circuit Design – A System Perspective. Kluwer Academic Publishers

387. Sproull RF, Sutherland IE, Molnar CE (1994) The Counterflow Pipeline Processor Architecture. IEEE Design & Test of Computers 11(3):48–59

388. Stallings W (2006) Computer Organization and Architecture: Designing for Performance, 7th edn. Prentice Hall, Upper Saddle River

389. StarCore (2001) SC140 DSP Core Reference Manual. Motorola Inc., MNSC140CORE/D, Revision 3, November

390. Stock F, Koch A (2006) Architecture Exploration and Tools for Pipelined Coarse Grained Reconfigurable Arrays. In Proc. FPL, August

391. Stretch, Inc. (2007) www.stretchinc.com
392. Suh J, Kim EG, Crago SP, Srinivasan L, French MC (2003) A Performance Analysis of PIM, Stream Processing, and Tiled Processing on Memory-Intensive Signal Processing Kernels. In Proc. ISCA, pp 410–419
393. SUIF (2007) Compiler System. Available at http://suif.standford.edu
394. Sullivan C, Wilson A, Chappell S. (2004) Using C Based Logic Synthesis to Bridge the Productivity Gap. In Proc. ASP-DAC, pp 349–354
395. Sun (2007) Java Native Interface Specification, consulted 26 August 2006. Available at http://java.sun.com/j2se/1.5.0/docs/guide/jni/spec/jniTOC.html
396. Sun (2007) Java 2 Platform Specification, version 1.4.2, consulted 26 August 2006. Available at http://java.sun.com/j2se/1.4.2/docs/api/
397. Sun (2007) Java ME APIs & Docs, consulted 30 August 2006. Available at http://java.sun.com/javame/reference/apis.jsp
398. Sutherland IE (1989) Micropipelines. Communications of the ACM, 32(6): 720–738
399. Sylvester D, Keutzer K (1999) System-Level Performance Modeling with BACPAC – Berkeley Advanced Chip Performance Calculator. In Proc. SLIP, pp 109–114
400. SystemC web pages at http:// www.systemc.org
401. SystemC EDA Products (2007) http://www.systemc.org/docman2/ViewCategory.php?group_id=4&category_id=10. Verified 2007-01-25
402. SystemVerilog Homepage (2007) http://www.systemverilog.org/.Verified 2007-01-24
403. Säntti T, Plosila J (2005) Architecture for an Advanced Java Co-Processor, In Proc. ISSCS 2005, Iasi, Romania, Vol 2, pp 501–504
404. Säntti T, Plosila J (2005) Instruction Folding for an Asynchronous Java Co-Processor. In Proc. International Symposium on SoC, pp 18–21
405. Tabak D, Lipovski GJ (1980) MOVE Architecture in Digital Controllers. IEEE Transactions on Computers, 29(2):180–190
406. Takamura A, Kuwako M, Imai M, Fujii T, Ozawa M, Fukasaku I, Ueno Y, Nanya T (1997) TITAC-2: A 32-bit Asynchronous Microprocessor Based on Scalable-Delay-Insensitive Model. Proc. ICCD'97, pp 288–294
407. Talla S (2001) Adaptative Explicit Parallel Instruction Computing. PhD Thesis, Department of Computer Science, New York University, May
408. Tam S, Rusu S, Nagarji Desai U, Kim R, Ji Zhang, Young I (2000) Clock Generation and Distribution for the First IA-64 Microprocessor. IEEE Journal of Solid-State Circuits, 35(11):1545–1552
409. Tanenbaum AS (2006) Structured Computer Organization, 5th edn. Prentice Hall, Upper Saddle River
410. Taylor MB (2004) Deionizer: A Tool for Capturing and Embedding I/O Calls. Technical Memo, CSAIL/Laboratory for Computer Science, MIT. Available at http://cag.csail.mit.edu/~mtaylor/deionizer.html
411. Taylor MB, Kim J, Miller J, Wentzlaff D, Ghodrat F, Greenwald B, Hoffmann H, Johnson P, Lee JW, Lee W, Ma A, Saraf A, Seneski M, Shnidman N, Strumpen V, Frank M, Amarasinghe S, Agarwal A (2002) The Raw

Microprocessor: A Computational Fabric for Software Circuits and General Purpose Programs. IEEE Micro, 22(2):25–35

412. Taylor MB, Lee W, Amarasinghe S, Agarwal A (2003) Scalar Operand Networks: On-Chip Interconnect for ILP in Partitioned Architectures. In Proc. HPCA, pp 341–353
413. Taylor MB, Lee W, Amarasinghe S, Agarwal A (2005) Scalar Operand Networks. IEEE Transctions on Parallel and Distributed Systems, 16(2): 145–162
414. Tensilica (2007) Xtensa Instruction Set Simulator (ISS) and Xtensa Modeling Protocol (XTMP). Available at http://www.tensilica.com/pdf/iss_xtmp_v7_aug19_final.pdf. Verified 2007-01-25
415. Thatte SM, Abraham JA (1980) Test Generation for Microprocessors. IEEE Transactions on Computers, 29(6):429–441
416. The Mathworks (2007) Matlab. Available at http://www.mathworks.com/products/matlab/description1.html. Verified 2007-01-24
417. The Mathworks (2007) Simulink. Available at http://www.mathworks.com/products/simulink/description1.html. Verified 2007-01-24
418. The Open SystemC Initiative (2007) OSCI SystemC TLM 2.0 Draft 1. Available at http://www.systemc.org/web/sitedocs/TLM_2_0.html. Verified 2007-01-27
419. The Ptolemy project (2007) http://ptolemy.eecs.berkeley.edu/. Verified 2007-01-24
420. The Real-Time for Java Expert Group (2000) The Real-Time Specification for Java. Addison-Wesley
421. Thies W, Karczmarek M, Amarasinghe S (2002) StreamIt: A Language for Streaming Applications. In Proc. Compiler Construction, pp 179–196
422. TI (2000) TMS320C6000 CPU and Instruction Set Reference Guide, Texas Instruments, October
423. Todman T, Constantinides G, Wilton S, Mencer O, Luk W, Cheung P (2005) Reconfigurable Computing: Architectures and Design Methods. In IEE Proc. Computers and Digital Techniques, Vol 152, No. 1, March, pp 193–207
424. Tokumasu M, Fujii H, Ohta M, Fuse T, Kameyama A (2002) A New Reduced Clock-Swing Flip-Flop: NAND-Type Keeper Flip-Flop (NDKFF). In Proc. Custom Integrated Circuits Conference, pp 129–132
425. Torvalds L (1999) The Linux Edge. In: DiBona C, Ockman S, Stone M (eds) Open Sources: Voices from the Open Source Revolution. O'Reilly and Associates, Inc., Sebastopol, CA
426. Trimaran Consortium (2007) The Trimaran Compiler Infrastructure. Available at http://www.trimaran.org
427. Truscan D (2004) A UML Profile for the TACO Protocol Processing Platform. In Proc. Norchip Conference, pp 225–228
428. Truscan D (2007) Model-Driven Development of Programmable Architectures. PhD Thesis, Åbo Akademi University, Turku, Finland

429. Turley J (2004) A Microprocessor Report Analysis of Avispa+. Microprocessor Report, February 9. Available at Silicon Hive web site: www.silicon-hive.com
430. UPC Consortium (2005) UPC language specifications v1.2. Technical Report LBNL-59208, Lawrence Berkeley National Laboratory, May. Available at http://upc.lbl.gov/docs/user/upc_spec_1.2.pdf
431. University of Aachen (2007) Lisa Retargetable SW Development Tool Suite. Available at http://servus.ert.rwth-aachen.de/lisa/. Verified 2007-01-25
432. University of California, Irvine (2007) SpecC System. Available at http://www.ics.uci.edu/~specc/. Verified 2007-01-24
433. Van Berkel K, Burgess R, Kessels J, Roncken M, Schalij F, Peeters A (1994) Asynchronous Circuits for Low Power: A DCC Error Corrector. IEEE Design and Test of Computers 11(2):22–32
434. Van Praet J, Goossens G, Lanneer D, De Man H (1994) Instruction Set Definition and Instruction Selection for ASIP. In Proc. International Symposium on High-Level Synthesis, pp 11–16
435. Van Praet J, Lanneer D, Geurts W, Goossens G (2001) Processor Modeling and Code Selection for Retargetable Compilation. ACM Transactions on Design Automation of Electronic Systems, 6(3):277–307, July
436. Vassiliadis S, Wong S, Gaydadjiev G, Bertels K, Kuzmanov G, Panainte EM (2004) The MOLEN Polymorphic Processor. IEEE Transactions on Computers, November, 53(11):1363–1375
437. Veen AH (1986) Dataflow Machine Architecture. ACM Computing Surveys, Vol 18, No 4, December
438. Verilog Homepage (2007) http://www.verilog.com/. Verified 2007-01-24
439. Virtanen S (1999) On Communications Protocols and Their Characteristics Relevant to Designing Protocol Processing Hardware. Technical Report 305, Turku Centre for Computer Science, Turku, Finland
440. Virtanen S (2004) A Framework for Rapid Design and Evaluation of Protocol Processors. PhD Thesis, University of Turku, Finland
441. Virtanen S, Lundström T, Lilius J (2002) A Processor Design Tool for the TACO Framework. In Proc. Norchip, pp 177–182
442. Virtanen S, Paakkulainen J, Nurmi T (2005) Capturing Processor Architectures from Protocol Processing Applications: A Case Study. In Proc. 8th Euromicro Conference on Digital System Design (DSD)
443. Virtanen S, Nurmi T, Paakkulainen J, Lilius J (2006) A System-Level Framework for Designing and Evaluating Protocol Processor Architectures. International Journal of Embedded Systems, 1(1–2):78–90
444. Virtanen S, Truscan D, Lilius J (2001) SystemC Based Object Oriented System Design. In Proc. Forum on Specification and Design Languages (FDL)
445. Vorbach M, Becker J (2003) Reconfigurable Processor Architectures for Mobile Phones. In Proc. the Int'l Parallel and Distributed Processing Symposium, April
446. VSI Alliance™ (2001) Virtual Component Interface Standard Version 2 (OCB 2 2.0), On-Chip Bus Development Working Group, April

447. Waingold E, Taylor M, Srikrishna D, Sarkar V, Lee W, Lee V, Kim J, Frank M, Finch P, Barua R, Babb J, Amarasinghe S, Agarwal A (1997) Baring It All to Software: Raw Machines. IEEE Computer 30(9):86–93
448. Wanhammar L (1999), DSP Integrated Circuits, Academic Press
449. Weinhardt M, Luk W (2001) Pipeline Vectorization. IEEE Transactions on Computer-Aided Design of Integrated Circuits and Systems, February, 20(2):234–248
450. Wen CHP, Wang LC, Cheng KT, Yang K, Liu WT, Chen JJ (2005) On a Software Based Self-Test Methodology and Its Application. In Proc. IEEE VLSI Test Symposium, pp 107–113
451. Wen CHP, Wang LC, Cheng KT, Liu WT, Chen JJ (2005) Simulation-Based Target Test Generation Techniques for Improving The Robustness of a Software-Based-Self-Test Methodology. In Proc. IEEE International Test Conference, pp 936-945
452. Wentzlaff D, Agarwal A (2006) Constructing Virtual Architectures on a Tiled Processor. In Proc. International Symposium on Code Generation and Optimization
453. Whaley RC, Petitet A, Dongarra JJ (2000) Automated Empirical Optimizations of Software and the ATLAS Project. Parallel Computing 27(1–2):3–35
454. Wholey S, Fahlman SF (1984) The Design of an Instruction Set for Common Lisp. In Proc. ACM Symposium on LISP and Functional Programming, pp 150–158
455. Wikipedia (2006) Thread (Computer Science), consulted 5 August 2006. URL http://en.wikipedia.org/wiki/Thread_(computer_science), Simultaneous Multithreading, consulted 5 August. URL http://en.wikipedia.org/wiki/Simultaneous_multithreading
456. Wikipedia (2006) Very Long Instruction Word, consulted 14 September 2006. URL http://en.wikipedia.org/wiki/VLIW
457. Wilson R et al. (1994) SUIF: An Infrastructure for Research on Paralelizing and Optimizing Compiler. SIGPLAN Notices, page 31, Dec 1994
458. Wittig RD, Chow P (1996) Onechip: An FPGA Processor with Reconfigurable Logic. In Proc. IEEE Workshop on FPGAs for Custom Computing Machines
459. Wolf W (2004) FPGA-Based System Design. Prentice Hall, Upper Saddle River, NJ
460. Wong BP, Mittal A, Cao Y, Starr G (2005) Nano-CMOS Circuit and Physical Design. John Wiley and Sons
461. Wong S, Vassiliadis S, Cotofana S (2002) Future Directions of (Programmable and Reconfigurable) Embedded Processors. In Proc. the 2nd Workshop on System Architecture Modeling and Simulation (SAMOS)
462. Wood J, Lipa S, Franzon P, Steer M (2001) Multi-gigahertz Low-Power Low-Skew Rotary Clock Scheme. In ISSCC Digest of Technical Papers, pp 400–401
463. Woods JV, Day P, Furber SB, Garside JD, Paver NC, Temple S (1997) AMULET1: An Asynchronous ARM Microprocessor. IEEE Transactions on Computers 46(4):385–398

464. Xanthopoulos T, Bailey DW, Gangwar AK, Gowan MK, Jain AK, Prewitt BK (2001) The Design and Analysis of the Clock Distribution Network for a 1.2GHz Alpha Microprocessor. In ISSCC Digest of Technical Papers, pp 402–403
465. Xilinx (2007) company web page. Available at http://www.xilinx.com
466. Xilinx (2007) MicroBlaze. Available at http://www.xilinx.com/microblaze
467. Xilinx (2007) Microblaze architecture. Available at (visited on January 2007) http://www.xilinx.com/ipcenter/processor_central/microblaze/architecture.htm
468. Xilinx (2007) FPU Available at http://www.xilinx.com/xlnx/xebiz/designResources/ip_product_details.jsp?key=DO-DI-FPU-SP (visited on January 2007)
469. Yiannacouras P, Steffan JG, Rose J (2006) Application-Specific Customization of Soft Processor Microarchitecture. In Proc. International Symposium on Field-Programmable Gate Arrays, pp 201–210
470. Yu ZC, Furber SB, Plana LA (2003) An Investigation into the Security of Self-Timed Circuits, Proc. Async'03, pp 206–215
471. Zivkovic VD, Lieverse P (2002) An overview of Methodologies and Tools in the Field of System Level Design. Embedded Processor Design Challenges: Lecture Notes in Computer Science, 2268:74–88. Springer Verlag
472. Cavium Networks homepage (2007) http://www.cavium.com

Index

G

H

I

www.ingramcontent.com/pod-product-compliance
Lightning Source LLC
Chambersburg PA
CBHW071229140526
45046CB00006B/1116
* 9 7 8 1 4 0 2 0 5 5 2 9 4 *